T0329306

BRAIN LIPIDS IN SYNAPTIC FUNCTION AND NEUROLOGICAL DISEASE

ELSEVIER *science & technology books*

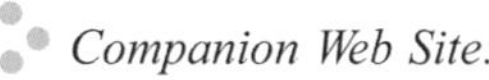

Companion Web Site:

http://booksite.elsevier.com/9780128001110

Brain Lipids in Synaptic Function and Neurological Disease
Jacques Fantini and Nouara Yahi, Authors

ACADEMIC
PRESS

BRAIN LIPIDS IN SYNAPTIC FUNCTION AND NEUROLOGICAL DISEASE

CLUES TO INNOVATIVE THERAPEUTIC STRATEGIES FOR BRAIN DISORDERS

JACQUES FANTINI
Molecular Interactions in Model and Biological Membranes Laboratory, Faculty of Science and Technology, Marseille, France

NOUARA YAHI
Molecular Interactions in Model and Biological Membranes Laboratory, Faculty of Science and Technology, Marseille, France

AMSTERDAM • BOSTON • HEIDELBERG • LONDON
NEW YORK • OXFORD • PARIS • SAN DIEGO
SAN FRANCISCO • SINGAPORE • SYDNEY • TOKYO
Academic Press is an imprint of Elsevier

Academic Press is an imprint of Elsevier
125, London Wall, EC2Y 5AS, UK
525 B Street, Suite 1800, San Diego, CA 92101-4495, USA
225 Wyman Street, Waltham, MA 02451, USA
The Boulevard, Langford Lane, Kidlington, Oxford OX5 1GB, UK

Cover image: Formation of α-helical ion channels by Aβ. At appropriate Aβ/GM1 ratios, Aβ monomers may bind to cell surface GM1 and subsequently fold into an α-helical structure. The insertion of α-helical Aβ in lipid raft domains enriched in cholesterol (chol) is followed by the oligomerization of Aβ into an oligemeric annular pore permeable to calcium. This sequence of events is summarized in the top panel; the structure of the oligomeric Aβ channels is shown in the images below (from left to right, top view, lateral view, and bottom view); the surface potential of the models is colored in red (negative), blue (positive) or white (apolar regions).

Notices

Knowledge and best practice in this field are constantly changing. As new research and experience broaden our understanding, changes in research methods, professional practices, or medical treatment may become necessary.

Practitioners and researchers must always rely on their own experience and knowledge in evaluating and using any information, methods, compounds, or experiments described herein. In using such information or methods they should be mindful of their own safety and the safety of others, including parties for whom they have a professional responsibility.

To the fullest extent of the law, neither the Publisher nor the authors, contributors, or editors, assume any liability for any injury and/or damage to persons or property as a matter of products liability, negligence or otherwise, or from any use or operation of any methods, products, instructions, or ideas contained in the material herein.

British Library Cataloguing-in-Publication Data
A catalogue record for this book is available from the British Library

Library of Congress Cataloging-in-Publication Data
A catalog record for this book is available from the Library of Congress

ISBN: 978-0-12-800111-0

For information on all Academic Press publications
visit our website at http://store.elsevier.com/

Typeset by Thomson Digital

Publisher: Mica Haley
Senior Acquisition Editor: Natalie Farra
Senior Editorial Project Manager: Kristi Anderson
Production Project Manager: Lucía Pérez
Designer: Greg Harris

Dedication

Dedicated to the memory of our wonderful friend and colleague, Nicolas Garmy, who left us too soon. How we wish, how we wish you were here…

Contents

About the Authors

Jacques Fantini was born in France in 1960. He has more than 30 years of teaching and research experience in biochemistry and neurochemistry areas. Since 1998, he has been Professor of Biochemistry and an honorary member of the Institut Universitaire de France.

Nouara Yahi was born in France in 1964. She has accumulated 25 years of fundamental research and teaching experience in virology and molecular biology areas. She is currently Professor of Biochemistry and leader of the research group Molecular Interactions in Model and Biological Membrane Systems. This group is internationally recognized for studies of lipid–lipid and lipid–protein interactions pertaining to virus fusion, amyloid aggregation, oligomerization, and pore formation.

Together, Nouara Yahi and Jacques Fantini have discovered the universal sphingolipid-binding domain (SBD) in proteins with no sequence homology but sharing common structural features mediating sphingolipid recognition. The SBD is present in a broad range of infectious and amyloid proteins, revealing common mechanisms of pathogenesis in viral and bacterial brain infections and in neurodegenerative diseases. Their current research is focused on the molecular organization of the synapse in physiological and pathological conditions. Jacques Fantini and Nouara Yahi have been working together since 1991 and have copublished 72 articles referenced in PubMed and nine patent applications. Jacques Fantini is the author/coauthor of 167 articles (PubMed), with 5800 citations and an H-index of 42. Nouara Yahi is the author/coauthor of 82 articles (PubMed), with 3500 citations and an H-index of 35.

Preface

Lipids are the most abundant organic compounds found in the brain, accounting for up to 50% of its dry weight. The brain lipidome includes several thousand distinct biochemical structures whose expression may greatly vary according to age, gender, brain region, and cell type, as well as subcellular localization. In synaptic membranes, brain lipids specifically interact with neurotransmitter receptors and control their activity. Moreover, brain lipids play a key role in the neurotoxicity of amyloidogenic proteins involved in the pathophysiology of neurological diseases.

Biochemistry provides the ultimate mechanistic explanation of most biological processes. Obviously this information may not be sufficient to understand some biological functions, which also involve integrated networks at different levels, from the cell to the body. However we rarely need to bring biology to the subatomic level. Studying biochemistry is not boring, provided that it is taught with the aim of answering clearly enunciated questions. What would be the rationale of learning an endless list of molecular structures that we would probably never meet in a scientist's life? As professors of biochemistry, we did in this book what we do in our teaching activity: provide the molecular basis required for understanding biological functions, not more, not less. It is just like learning a language: good grammar rules and just enough vocabulary to guide the traveler throughout foreign countries.

Our ambition with this book is to offer a comprehensive overview of brain lipid structures and to explain how these lipids control synaptic functions through a network of lipid–lipid and lipid–protein interactions. We also explain the role of major brain lipids (cholesterol and sphingolipids) in the pathogenesis of neurodegenerative diseases, including Alzheimer's, Creutzfeldt–Jakob, and Parkinson's. We show that these diseases involve strikingly common mechanisms of pathogenesis that are also used by pathogens to invade brain cells. This concerns especially HIV-1 and amyloid proteins, as well as bacterial toxins and amyloid oligomers.

The book has been written to provide a hands-on approach for neuroscience graduate students. Biochemical structures are dissected and explained with molecular models. Moreover, we propose a step-by-step guide to memorize and draw the biochemical structure of brain lipids, including cholesterol and complex gangliosides. To conclude the book, we present new ideas that can drive innovative therapeutic strategies based on the knowledge of the role of lipids in brain disorders.

Jacques Fantini
Nouara Yahi
Septèmes-les-Vallons, France
December 1, 2014

Acknowledgments

We would like to thank all the scientists, colleagues, and friends who supported us during the course of our scientific career, especially Francisco Barrantes, Luc Belzunces, Pierre Burtin, Henri Chahinian, Ahmed Charaï, Caroline Costedoat, Patrick Cozzone, Olivier Delézay, Coralie Di Scala, Assou El Battari, Francisco Gonzalez-Scarano, Claude Granier, Nathalie Koch, Xavier Leverve, André Jean, Jean-Marc Sabatier, Louis Sarda, and Catherine Tamalet.

We also thank the Fondation Marcel Bleustein-Blanchet pour la Vocation (Paris, France) and the scientific advisors who interviewed us, Claude Bernard and François Jacob, for supporting the young scientists that we were before our names appeared in PubMed.

It has been a pleasure to work on this book with Kristi Anderson and Natalie Farra from Elsevier/Academic Press.

Finally we thank our beloved son Driss, nephews Fahem and Lounis, sister Djamila, brother-in-law Jean-Philippe, and our late parents. Your love is our life.

CHAPTER

1

Chemical Basis of Lipid Biochemistry

Jacques Fantini, Nouara Yahi

OUTLINE

1.1 INTRODUCTION

Biochemistry (biological chemistry, chemical biology, or chemistry of living systems) is a scientific discipline that arose during the nineteenth century when progress in organic chemistry allowed the study of biological functions at the molecular level. It comprises several domains, each with its own purpose and more or less specific methods of investigation. The most important include the following:

- Enzymology: the study of biological catalysts (chiefly enzymatic proteins referred to as enzymes, but also catalytic RNAs called ribozymes)
- Molecular biology: the study of informational macromolecules (DNA, RNA, and, in the case of neurodegenerative diseases, proteins)
- Structural biology: the study of the shapes embraced by all these macromolecules (described at the atomic level) and the molecular interactions controlling the formation of functional superstructures (e.g., ribosomes).

Brain Lipids in Synaptic Function and Neurological Disease. http://dx.doi.org/10.1016/B978-0-12-800111-0.00001-1

These domains share the same disciplinary field, biochemistry, whose goal is to understand the molecular nature and functioning of living organisms. Should you want to study the properties of living matter at the subatomic level in detail, you must leave the field of biology to enter quantum chemistry. Therefore, biochemistry is the ultimate level of investigation for the study of biological functions. To study phenomena at this molecular scale requires a basic knowledge of chemistry. The study of brain lipids and their role in synaptic function and neurodegenerative disorder does not escape this rule.

1.2 CHEMISTRY BACKGROUND

Because lipids are chiefly defined on the basis of their insolubility in water, you must first understand the key features of a molecule that is soluble in water, and then try to figure out why lipid molecules are not. It is not a simple task, and we will restrict our discussion to basic rules that emerge from clear-cut chemical concepts. Studying the water molecule will be helpful to address these fundamental concepts. As encouragement to take the time to carefully read this section (in which we do not present any lipid structure yet), be aware that mastering these basic notions will give you a number of universal keys for entering the complex world of biological structures with confidence. Hopefully it will convince you of how logical this world is indeed.

Water is by far the most abundant molecule found in living organisms, accounting for approximately 70% of our own weight. Nevertheless, it is neither specific (water is also found outside living creatures) nor representative (it lacks carbon atoms that are the hallmark of organic molecules). Its chemical formula H_2O indicates that it is composed of two hydrogen atoms and one oxygen atom. Given that the valence of these atoms is 2 for oxygen and 1 for hydrogen, the chemical bonds of the water molecule can be formed in only one way: the oxygen atom establishes a bond with each hydrogen atom. Therefore, the O—H bond is the only type of chemical bond in the water molecule. This covalent bond results from the sharing of a pair of electrons, one given by oxygen, the other by hydrogen. Now, where are these electrons located? An intuitive answer could be "somewhere between the oxygen and the hydrogen atoms." If both atoms were identical (e.g., in the hydrogen molecule H—H), then the highest probability of finding the electrons of the chemical bond would be at an equal distance from each atom. However, such is not the case for the O—H bond because oxygen is more "electro-attractive" than hydrogen. As a consequence, the electrons that form the O—H bond are closer to the oxygen atom than to hydrogen. Thus, the O—H bond is polarized, whereas the H—H bond is not. This characteristic is how we distinguish heteropolar polarized bonds linking two distinct atoms X—Y versus homolar nonpolarized bonds linking two identical atoms X—X. The asymmetrical distribution of the electron pair in the O—H bond greatly benefits the oxygen atom that is surrounded by an excess of electrons, whereas the electron cloud of the hydrogen atom is in deficit. Both surplus and deficit can be energetically quantified, but we will restrict our discussion to a qualitative analysis: the excess of electrons around the oxygen atom creates a partial negative charge that is noted δ^-. Why a partial and not a real negative electrical charge? Because the O—H bond still exists. It is only when the chemical bond breaks that the oxygen atom takes full control of the electron pair (i.e., it recovers its own electron and steals the one provided by hydrogen). In this case, the oxygen atom bears a

TABLE 1.1 Linus Pauling's Electronegativity Scale for Biochemistry Students

Atom	Relative electronegativity
Hydrogen (H)	2.1
Carbon (C)	2.5
Nitrogen (N)	3.0
Oxygen (O)	3.5

bona fide negative charge and is now written as O^-. If one applies the same reasoning, the hydrogen atom of the O—H bond possesses a partial positive charge that is noted δ^+. In summary, the O—H bond is characterized by the presence of two opposite partial charges. Correspondingly, we can write this bond δ^- **O-H** δ^+, which means that the O—H bond is polarized and the oxygen atom attracts to itself the electrons pair of the chemical bond. We say that oxygen is more electronegative than hydrogen. Nobel Prize winner Linus Pauling was the first to define an electronegativity scale for all atoms.[1] In this chapter, we will simply mention the electronegativity of the four most important atoms in biochemistry: C, H, O, and N (Table 1.1). When necessary, we will extend this table to less abundant but biologically significant atoms such as P or S.

Should you be marooned on a desert island, you will need to know these values by heart if you want to recover an almost complete knowledge in biochemistry. Indeed, this table allows you to predict both the presence and distribution of partial charges in any molecule containing these atoms. For instance, we can see that oxygen has a higher electronegativity than carbon (respectively, 3.5 and 2.5). Thus, the C—O bond is polarized and is noted: δ^+ **C-O** δ^-. Similarly, the electronegativity of nitrogen is higher than that of hydrogen, so that the N—H bond is polarized: δ^- **N-H** δ^+. The C—N bond is also polarized: δ^+ **C-N** δ^-. However, when it comes to C and H, the difference of electronegativity falls below 0.5 so that the C—H bond is not significantly polarized (also due to the intrinsic low electronegativity of H and C). All you need to know in chemistry to understand the concepts developed in this book is based on the electronegativity of a handful of atoms that constitute the living matter.

1.3 MOLECULAR INTERACTIONS

Over the years biochemistry has become the science of molecular interactions between biomolecules. We will consider the water molecule as a guide to explore one of the most intriguing properties of biomolecules, that is, their capacity to assemble with one another to form complex (and generally transient) structures. Such molecular complexes are noncovalent in nature because building and breaking covalent bonds requires a high energy input. The hydrogen bond (H-bond) is probably the most famous noncovalent interaction able to stabilize a molecular complex at a minimal energy cost. H-bonds are involved in the reversible association of the two antiparallel strands of the DNA double helix. The H-bond also ensures the cohesion of water molecules and explains why so much energy (100°C) is required to separate these molecules when passing water from liquid to gaseous state. The H-bond is a bond of electrostatic type that should not be confused with an electrostatic bond between two opposite electric charges (e.g., Na^+ and Cl^- ions). It usually occurs between an electronegative atom

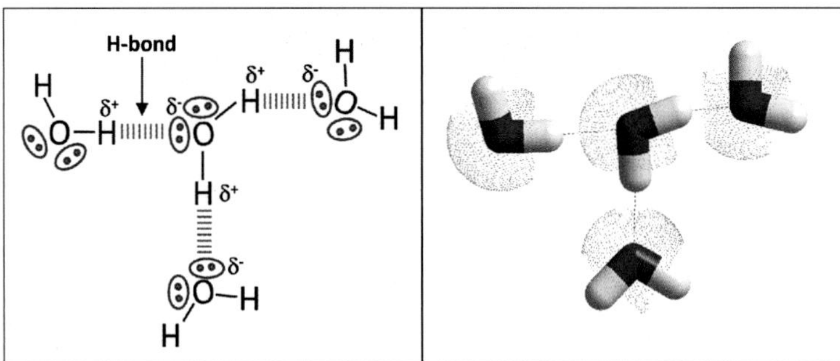

FIGURE 1.1 **Hydrogen bond (H-bond) network between water molecules.** Hydrogen bonds are indicated by dotted lines (in blue on the left panel). The right panel shows the results of molecular dynamics simulations of the hydrogen bond network involving four H_2O molecules in the same orientation as in the scheme of the left panel (obtained with the HyperChem software).

having at its periphery a free electron pair (most often N or O in biology) and a δ^+ hydrogen covalently bound to an electronegative atom (e.g., OH or NH). Both atoms involved in an H-bond move closer to each other due to the attraction of the δ^+ hydrogen atom by the electron pair of the electronegative atom. It is precisely this type of interaction that occurs between two water molecules (Fig. 1.1).

One differentiates the H-bond donor group (the one that provides the hydrogen atom) and the acceptor group (the oxygen atom that provides the electron pair). Correspondingly, the water molecule has two H-bond donor groups (two hydrogen atoms) and two acceptor groups (two pairs of peripheral electrons on oxygen). Each water molecule can thus form a maximum of four H-bonds with its neighbors. Thus, the maximum coordination number of water is equal to 4. In practice, this value of 4 is reached only in water in the form of ice. In liquid water at 25°C, it is still as high as 3.7, indicating a strong cohesion of liquid water at this temperature (vaporization requires much more energy, i.e., 100°C). Separating the water molecules to pass from the liquid state to the gaseous state implies breaking all H-bonds connecting water molecules in a given volume of water. Although each individual H-bond is of low energy, their large number compensates for this energy weakness, which explains why it is necessary to provide a high amount of energy (equivalent to 100°C) to reach the temperature of vaporization of water. In addition to the H-bonds, several other types of molecular interactions are described in subsequent chapters (in particular London dispersion forces that are involved in various lipid assemblies, including biological membranes, see Chapter 2).

1.4 SOLUBILITY IN WATER: WHAT IS IT?

A compound is soluble in water if it is able to surround itself with water so that the molecules no longer have any contact between themselves. This compound is then referred to as a "solute." Solute molecules interact with water molecules by establishing hydrogen bonds with them. Therefore, a compound is water-soluble if it possesses at its surface chemical groups capable of forming H-bonds. For instance, the urea molecule (Fig. 1.2), which can

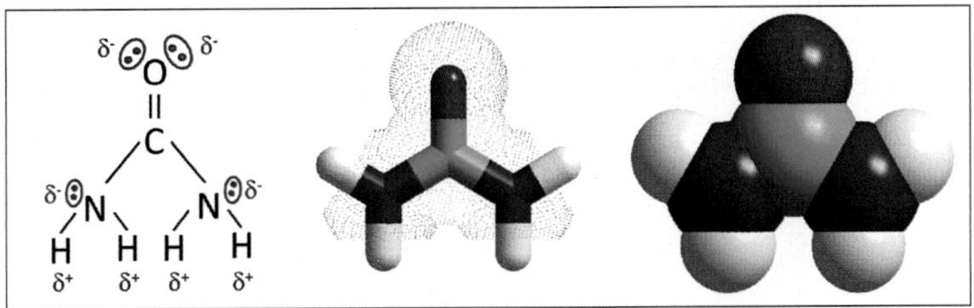

FIGURE 1.2 **Urea, a molecule that is highly soluble in water.** Three representations of the urea molecule are shown: from left to right, chemical structure with partial electric charges, tube model, and sphere model (oxygen in red, carbon in green, nitrogen in blue, and hydrogen in white/gray).

form up to eight hydrogen bonds (4 acceptor + 4 donor), is particularly soluble in water. (Urea is used at concentrations as high as 8 mol L^{-1} for denaturing proteins.)

One could wonder how it is possible for solutes to form H-bonds with water molecules that are already interacting with other water molecules through H-bonds. Indeed, as stated earlier, the coordination number of liquid water at 25°C is 3.7, which means that statistically each water molecule forms at least three H-bonds with their neighbors and that the majority of them even form the maximal number of four H-bonds. Thus, the molecular cohesion of liquid water is high. Nevertheless, the lifetime of H-bonds is short, about 1–20 ps. Broken H-bonds will reform rapidly (lifetime of 0.1 ps), most often to reform the same H-bond.[2] However, in the presence of a solute molecule, the rotation of the water molecule will allow the formation of a new H-bond with a polar group of the solute. Thus, as cleverly summarized by M. Chaplin, even though liquid water contains by far the densest hydrogen bonding of any solvent, these hydrogen bonds can rapidly rearrange in order to accommodate solute molecules.[3] This is how water-soluble molecules are dissolved in water.

In marked contrast with urea, palmitic acid CH_3–$(CH_2)_{14}$–COOH is not soluble in water because only a little part of the molecule (the carboxylic acid group –COOH) can form H-bonds with water. The long aliphatic chain (CH_3–$(CH_2)_{14}$–) containing only carbon and hydrogen atoms cannot form H-bonds because both C and H atoms have a similar low electronegativity (Table 1.1). Thus, the hydrogen atoms of such aliphatic chains do not display permanent partial δ^+ charges. Therefore, palmitic acid, as well as any apolar molecule unable to form H-bonds with water, including lipids, is not soluble in water. When such molecules are confronted with water, a phase separation often occurs. Because the density of lipids is inferior to water density, they float on the surface of water. For instance, some believe that it is a good thing to add a drop of olive oil in pasta water to prevent the noodles from sticking together. Other people add only a pinch of salt to water, arguing that the oil may coat the noodles with a thin film that will prevent the later absorption of sauce. An inattentive cook who does not remember putting salt in water has no alternative than tasting the water. However, because oil floats on the surface of pasta water, even a small drop can be detected by eye. In this case, a simple glance at the pan may allow the cook to determine whether oil has been added (Fig. 1.3). For instance, it is well known for centuries that oil has a calming effect on the sea. This calming effect was initially reported in the first century (A.D.) by Pliny the Elder in his *Natural History*, stating that "water is made smooth by oil." Later, Benjamin Franklin investigated this strange phenomenon by adding a drop of olive oil ("not more than a tea spoonful") to the water in a

FIGURE 1.3 **Drops of olive oil spread on the surface of water.** The photograph has been taken immediately after deposing a few drops of olive oil (about a half teaspoon) in a pan containing 1 L of water.

small pond in Clapham Common (London, UK). He noticed that the oil "produced an instant calm over several yards . . . making all that quarter of the pond, perhaps half an acre, smooth as a looking glass." Although Franklin's experiment passed largely unnoticed in his time, it opened the route to the famous monolayer studies of Irvin Langmuir, who, following the decisive input of Agnes Pockels (she measured the surface tension of water and was a pioneer in surface science), was awarded the Nobel Prize in 1932 "for his discoveries and investigations in surface chemistry." An account of the history of surface science has been written by Charles Tanford.[4] As we will see in Chapters 2 and 6, the floating properties of lipids on water have also been used to decipher the molecular mechanisms controlling lipid–lipid and lipid–protein interactions.

1.5 LIPID BIOCHEMISTRY

1.5.1 Definition

Although it is quite easy to define on a structural basis what are nucleic acids, proteins, and, to a lesser extent, carbohydrates, it is impossible for lipids. As a matter of fact, lipids are a category of biomolecules that are not defined on a biochemical background but on solubility basis. Lipids are insoluble in water but soluble in organic solvents such as chloroform, hexane, ethyl ether, or methanol. One can say that this definition is a poor one because some proteins (i.e., most membrane proteins) are not soluble in water, but they are not lipids either. Reciprocally, butyric acid (CH_3–CH_2–CH_2–COOH), due to its very short aliphatic chain, is highly soluble in water (more precisely it is liquid above –7.9°C and fully miscible with water), but it is still considered as a lipid based on its biochemical structure. To improve the lipid definition, we should also state that lipids generally have a relatively low molecular weight (e.g., 386 for cholesterol). If one considers membrane lipids, the situation becomes clearer because most of them belong to one of only three categories: glycerophospholipids, sphingolipids, and sterols. Moreover, none of them is soluble in water. In

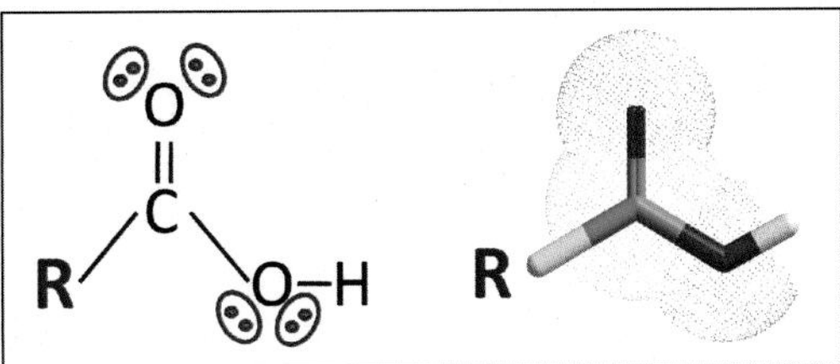

FIGURE 1.4 **Chemical formula of a fatty acid.** R (arbitrarily shown in yellow on the tube structure) represents an aliphatic chain.

this chapter we will focus our study on membrane lipids, especially those present in brain membranes. To enter this world, we will first analyze the structure of what we consider are the simplest lipids (i.e., fatty acids), which are not found in membranes but enter in the composition of membrane lipids. Understanding the biochemistry and physicochemical properties of fatty acids is mandatory to understand the structure and functions of plasma membranes.

1.5.2 Biochemistry of Fatty Acids

Fatty acids have the following general formula: R–COOH (Fig. 1.4). In this formula, the carboxylic function –COOH is constant and provides the acid properties of these molecules. The R group is an aliphatic chain of various lengths, which can be either a saturated or an unsaturated aliphatic chain. Therefore, fatty acids can be classified into two main categories: saturated and unsaturated.

1.5.3 Biochemistry of Saturated Fatty Acids

The aliphatic chain of saturated fatty acids is a simple alkyl group with the general structure $CH_3–(CH_2)_n–$. If you know the total number of carbons of a saturated fatty acid, you can deduce its complete chemical structure. For instance, palmitic acid is a saturated fatty acid with 16 carbons: its biochemical structure is thus $CH_3–(CH_2)_{14}–COOH$. The structure of palmitic acid can be drawn in several ways (Fig. 1.5), but it is helpful to memorize the fact that palmitic acid is the saturated fatty acid with 16 carbons.

The aliphatic chain is referred to as *saturated* because all carbon atoms are linked to four other atoms (sp^3 carbons in organic chemistry). When a carbon atom is linked to only three other atoms, one of the chemical bond has to be a double bond to respect the valency of 4, which is characteristic of carbon. In this case, the double bond links two carbons and it is noted C=C. The nomenclature used for saturated fatty acids refers to the number of carbon atoms, and to the lack of double bonds: correspondingly, palmitic acid, which is the saturated fatty acid with 16 carbons, is noted C16:0 (C16 for the carbon number, 0 for the number of double bond). You can train yourself to write the structure of the following fatty acids: C9:0, C14:0, and C18:0. A list of biologically saturated fatty acids can be found in Table 1.2. Note that only the limited piece of information found in the first column is useful to draw the chemical structure of all these lipids.

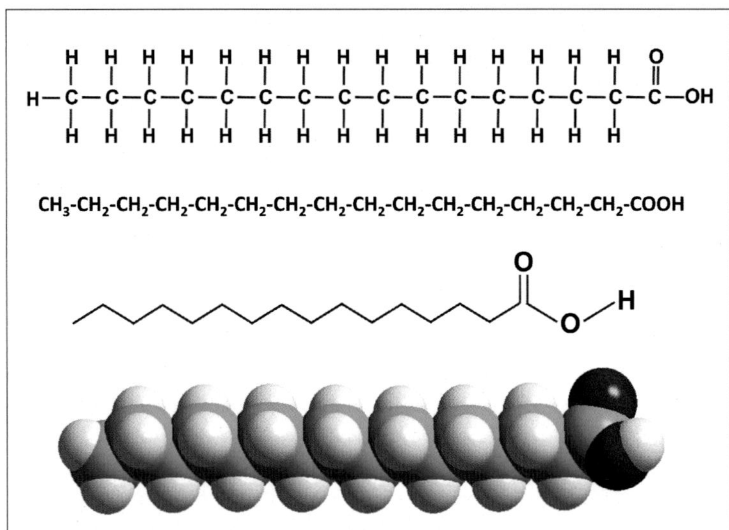

FIGURE 1.5 **The structure of palmitic acid.** Four possible representations of this fatty acid are shown. The most realistic is the sphere model in the bottom.

TABLE 1.2 The Main Saturated Fatty Acids in Biochemistry

Formula	Chemical name	Common name	Biology
C4:0	Butanoic acid	Butyric acid	Found in milk and butter triglycerides, also produced by intestinal bacteria
C6:0	Hexanoic acid	Caproic acid	Found in milk triglycerides, named after goats because of its rancid smell
C8:0	Octanoic acid	Caprylic acid	Also named after goats, and found in milk triglycerides
C9:0	Nonanoic acid	Pelargonic acid	Found in meteorites, probably the first membrane lipid ever on this planet
C10:0	Decanoic acid	Capric acid	Also named after goats, found in milk and in coconut oil triglycerides
C12:0	Dodecanoic acid	Lauric acid	Found in laurel and coconut oils, as well as in human breast milk triglycerides
C14:0	Tetradecanoic acid	Myristic acid	Named after the nutmeg *Myristica fragrans*, found in membrane glycerophospholipids
C16:0	Hexadecanoic acid	Palmitic acid	Found in palm oil and tallow triglycerides
C18:0	Octadecanoic acid	Stearic acid	Found in cocoa butter and animal fat
C20:0	Eicosanoic acid	Arachidic acid	Named after the Latin *arachis* (peanut), found in corn oil
C22:0	Docosanoic acid	Behenic acid	Found in behen oil, and in sphingolipids
C24:0	Tetracosanoic acid	Lignoceric acid	Found in wood tar, and in sphingolipids

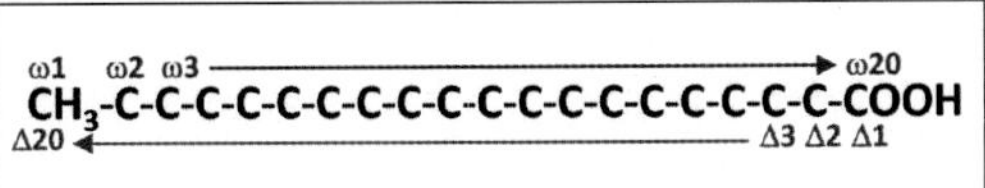

FIGURE 1.6 **Δ and ω numbering systems for arachidonic acid.** Δ starts from the carbon of the carboxylic acid (red numbers) whereas ω starts from the terminal CH_3 group (blue numbers).

1.5.4 Biochemistry of Unsaturated Fatty Acids

If you wish to write the structure of an unsaturated fatty acid, you need to know:

- The number of carbon atoms
- The number of double bonds
- The position of all double bonds in the carbon chain
- The stereochemical configuration of these double bonds.

Let us consider a concrete example: arachidonic acid, an unsaturated fatty acid that plays an important role in neurochemistry as a part of the endocannabinoid neurotransmitter anandamide (see Chapter 5). Arachidonic acid is the polyunsaturated fatty acid containing 20 carbon atoms and 4 double bonds. To write its structure, we will start with a simple chain with 20 carbons. At one end we will put the –COOH group, and the terminal $–CH_3$ group at the other end:

$$CH_3–C–C–C–C–C–C–C–C–C–C–C–C–C–C–C–C–C–C–COOH$$

To position the double bonds we need to number the carbon atoms. Here there are only two possibilities. You can use chemistry rules and number the carbons from the first remarkable function (i.e., the –COOH group). This numbering system is referred to as the Δ nomenclature, according to which the carbon atoms are numbered Δ1 to Δ20 (Fig. 1.6, in red). Unfortunately, this nomenclature does not properly reflect some important biological properties of unsaturated fatty acids. Thus, biochemists often use another numbering, referred to as omega (ω). In the ω system, the C no. 1 corresponds to the terminal methyl group, and whatever the length of the carbon chain it is always numbered ω1 (ω is the last letter of the Greek alphabet). In this case, the 20 carbon atoms of arachidonic acid are numbered from ω1 to ω20 (Fig. 1.6, in blue).

Now we will position the four double bonds of arachidonic acid in the ω system. The first one is between carbons ω6 and ω7 (Fig. 1.7). Thus, we can say that the first double bond is

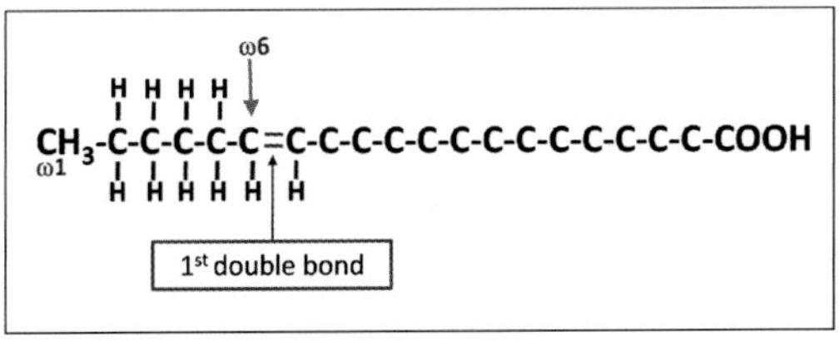

ω6
H H H H
CH3-C-C-C-C-C=C-C-C-C-C-C-C-C-C-C-C-C-C-COOH
ω1
H H H H H H
1st double bond

FIGURE 1.7 **Positioning the first double bond of arachidonic acid.** Count six carbons from the terminal CH_3 group to reach carbon ω6 and position the first double bond (in red) between carbons ω6 and ω7. Note that all carbons between ω2 and ω5 have two hydrogen atoms ($–CH_2$ methylene groups) whereas carbons ω6 and ω7, which form the first double bond, have only one hydrogen.

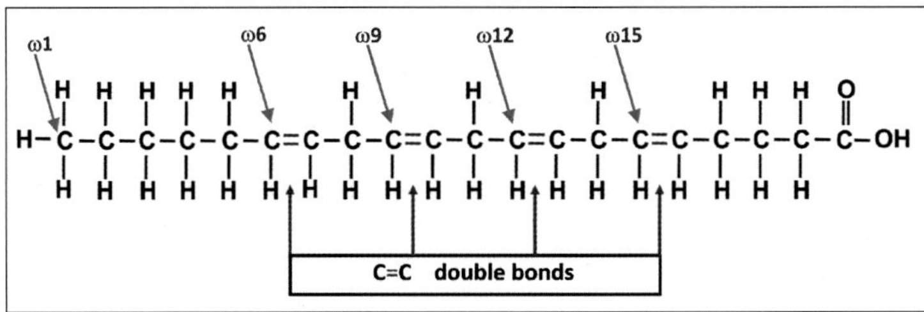

FIGURE 1.8 **Chemical structure of arachidonic acid.** The formula of arachidonic acid is C20:4ω6, indicating this fatty acid has 20 carbon atoms and 4 double bonds. According to this formula, the first double bond is in ω6, which, following the ω + 3 rule, gives the exact position of all other double bonds (thus, ω9, ω12, and ω15).

located on the 6th carbon (designated as ω6), counting from the terminal methyl carbon (designated as ω1) toward the carbon of the carboxylic acid group (ω20).

To position the three other double bonds we will use a simple but universal rule: in polyunsaturated fatty acids, two successive double bonds are separated by a single methylene (–CH_2–) group. Correspondingly, if the first double bond occurs on ω6, the ω7 carbon is also involved in the double bond, the ω8 is the methylene spacer group, and the next double bond is in ω9. In other words, the second double bond is on the ω9 carbon (ω6 + 3 carbons = ω9). Using the same rule we obtain the position of the two remaining double bonds on carbons ω12 (i.e., ω9 + 3) and ω15 (ω12 + 3). This can be summarized in the condensed formula of arachidonic acid: C20:4ω6, which means that arachidonic acid is a polyunsaturated fatty acid with 20 carbons and 4 double bonds. The first double bond occurs on carbon ω6, and the following (not noted in the formula but deduced from the stated universal rule) on carbons ω9, ω12, and ω15 (Fig. 1.8).

Finally, we have to indicate the stereochemistry of the double bonds of arachidonic acid, either *cis* (*Z*) or *trans* (*E*), according to the relative positions of the hydrogen atoms versus chemical groups linked to the carbons (Fig. 1.9).

In natural fatty acids, the configuration of double bonds is generally Z (although not totally licit from a stereochemical point of view, most biologists prefer to use the *cis* terminology). Hence, the formula for arachidonic acid is [C20:4 *cis*-ω6, *cis*-ω9, *cis*-ω12, *cis*-ω15], which can be shortened to C20:4ω6. (Writing it this way indicates the position of the double bonds after ω6 in the *cis* configuration.) An interesting aspect of the omega system is that all unsaturated fatty acids with the first double bond on the ω6 carbon share a common biosynthetic pathway

FIGURE 1.9 ***cis/trans* (or *Z/E*) stereochemistry.** If a = a′ we use the *cis/trans* nomenclature, but if a ≠ a′ (which is the case for fatty acids because the terminus of the a group is the CH_3 group and the terminus of the a′ group is carboxylate), we should more correctly use the Z/E system based on Cahn–Ingold–Prelog priority rules. In practice, it must be noted that most people still use the *cis/trans* system for the stereochemistry of fatty acids.

TABLE 1.3 Some Unsaturated Fatty Acids and Their Role in Brain

Formula in ω system	Formula in Δ system	Common name	Brain function/effect
C18:3ω3	C18:3Δ9	α-Linolenic acid	Reduces stress level and depression
C20:5ω3	C20:5Δ3	Eicosapentaenoic acid, icosapentaenoic acid	Precursor for all eicosanoids (prostaglandin, thromboxane, and leukotriene)
C22:6ω3	C22:6Δ4	Docosahexaenoic acid	Most abundant ω3 fatty acid in the brain
C18:2ω6	C18:2Δ9	Linoleic acid	Crucial for brain development
C18:3ω6	C18:3Δ6	γ-Linolenic acid	Precursor for prostaglandin and thromboxane
C20:4ω6	C20:4Δ5	Arachidonic acid	Involved in endocannabinoid metabolism
C22:4ω6	C22:4Δ 7	Docosatetraenoic acid	Most abundant fatty acids in the early human brain
C18:1ω7	C18:1Δ11	Vaccenic acid	Found in patients with schizophrenia
C18:1ω9	C18:1Δ9	Oleic acid	Abundant in myelin sphingolipids
C24:1ω9	C24:1Δ15	Nervonic acid	Found in brain sphingolipids
C20:3ω9	C20:3Δ5	Mead acid (discovered by James Mead)	Anti-inflammatory effects

and have similar biological properties. For this reason, natural unsaturated fatty acids can be classified into discrete series. Each series has a common precursor, such as α-linolenic acid (C18:3ω3) for the ω3 family and linoleic acid (C18:2ω6) for ω6 fatty acids (Table 1.3). Although the omega nomenclature is widely used in both scientific and public media, the International Union of Pure and Applied Chemistry (IUPAC) recommends replacing the "ω" prefix with "n" as the proper technical abbreviation. In this system, arachidonic acid is noted C20:4*n*–6 (occasionally, $n-6$ can be erroneously identified as "n minus 6," but this is rather confusing because by doing so you indicate the terminal CH_3 as the "*n* minus 1 carbon," which is nonsense). A good exercise before going further is to explain now why in the Δ system arachidonic acid is noted C20:4Δ5. A rapid glance at Table 1.3 will convince you of the usefulness of the ω system versus Δ numbering to classify polyunsaturated fatty acids in homologous series.

1.5.5 Glycerolipids

Glycerolipids are complex lipids formed by the condensation of one, two, or three fatty acid molecules on glycerol, a small compound with three carbon atoms (either numerically numbered C1, C2, C3, or, according to the Greek alphabet, Cα, Cβ, Cα'), with each one bearing a hydroxyl function OH (Fig. 1.10). Glycerol is a symmetrical molecule and its two terminal $-CH_2OH$ groups (referred to as α and α') are stereochemically equivalent. The central carbon C2, which is not asymmetric, is referred to as β.

Each OH function of glycerol can react with the –COOH group of a fatty acid, leading to an ester derivative called glyceride. An ester results from the condensation of a carboxylic acid and an alcohol (Fig. 1.11).

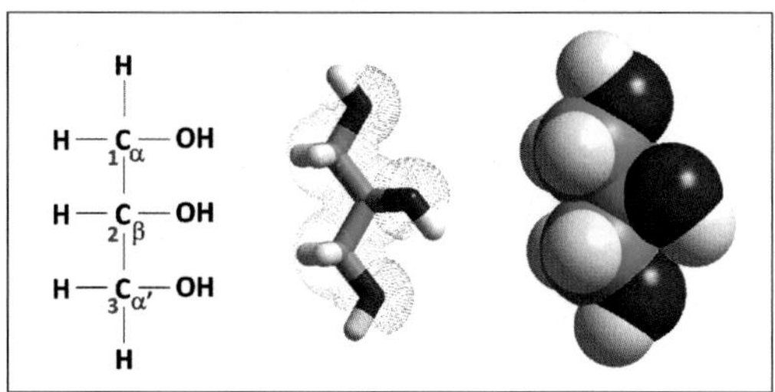

FIGURE 1.10 **Glycerol.** Three representations of glycerol are shown: from left to right, chemical structure, tube model, and sphere model (oxygen in red, carbon in green, nitrogen in blue, and hydrogen in white/gray). The three carbon atoms are numbered numerically and with Greek letters (carbons 1 and 3 are similar, so they are referred to as α and α′).

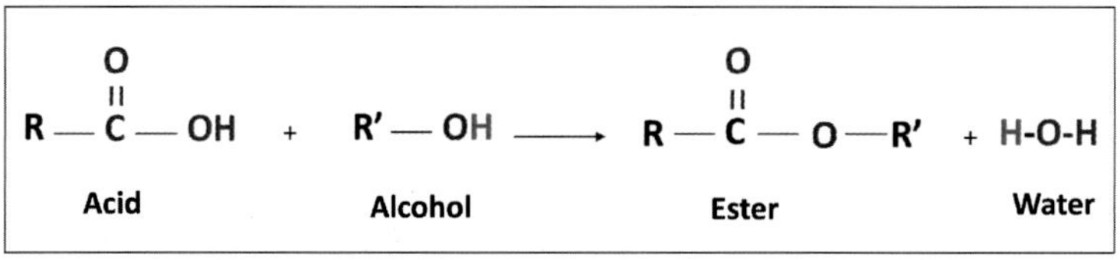

$$\underset{\text{Acid}}{R{-}\overset{O}{\overset{\|}{C}}{-}OH} + \underset{\text{Alcohol}}{R'{-}OH} \longrightarrow \underset{\text{Ester}}{R{-}\overset{O}{\overset{\|}{C}}{-}O{-}R'} + \underset{\text{Water}}{H\text{-}O\text{-}H}$$

FIGURE 1.11 **Esterification reaction.**

Each carbon atom of glycerol can be linked to a fatty acid, which will then become a chemical group by itself referred to as "acyl." If all three carbons of glycerol have reacted with a fatty acid, we will obtain a triacylglycerol (TAG), also named triglyceride. Diacylglycerol (DAG, diglycerides) include either αα′ or αβ derivatives, according to the site of esterification.[5] Similarly, the two categories of monoacylglycerol (MAG, monoglycerides) are α and β. Examples of MAG, DAG, and TAG (mono-, di-, or triacylglycerol) structures are given in Fig. 1.12.

Most membrane glycerolipids are phospholipids. These membrane components are obtained by the condensation of an αβ-DAG with phosphoric acid, leading to a molecule called

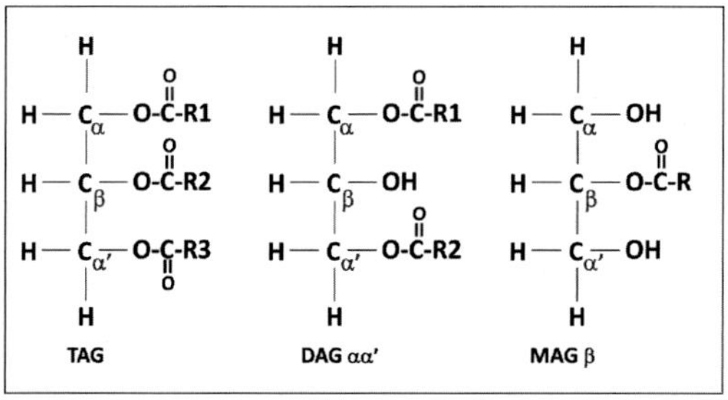

TAG	DAG αα′	MAG β
H	H	H
H—C_{α}—O-C(=O)-R1	H—C_{α}—O-C(=O)-R1	H—C_{α}—OH
H—C_{β}—O-C(=O)-R2	H—C_{β}—OH	H—C_{β}—O-C(=O)-R
H—$C_{\alpha'}$—O-C(=O)-R3	H—$C_{\alpha'}$—O-C(=O)-R2	H—$C_{\alpha'}$—OH
H	H	H

FIGURE 1.12 **Structure of acylglycerol esters.** The aliphatic chains of acyl groups are noted R. The nature of these R chains should be given to write the structure of a particular acylglycerol. In di- and triacylglycerol, R1, R2, and (when concerned) R3 can be either the same or distinct acyl chains. Thus, the terms MAG, DAG, and TAG are generic, and these molecules have a wide biochemical diversity based on their acyl content.

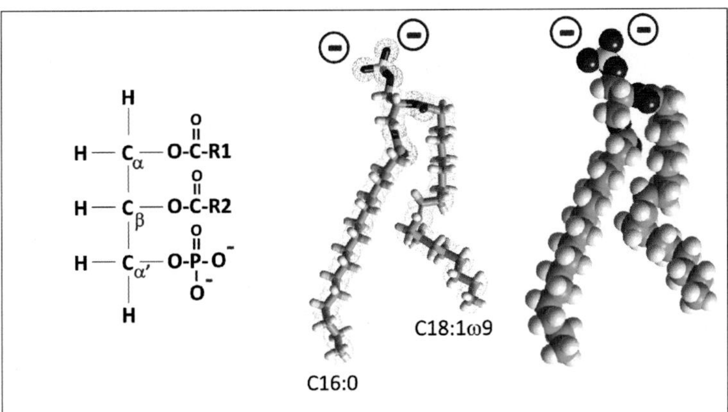

FIGURE 1.13 **Phosphatidic acid.** Carbons α, β, and α′ correspond to C1, C2, and C3, respectively. Note that C2 (Cβ) is asymmetric (R configuration). The two OH of the phosphate can be dissociated at physiological pH, as indicated in the tube and sphere models. The acyl chains represented here are palmitic acid (R1) and oleic acid (R2).

phosphatidic acid (PA) (Fig. 1.13). PA is the precursor of membrane glycerophospholipids. The most important biochemical feature of this class of compounds is the nature of the acyl chain in α and β of glycerol; R1 (in position α) results from the condensation of a saturated fatty acid, whereas R2 (in position β) comes from an unsaturated fatty acid. This unique combination of saturated/unsaturated chains is critical for proper membrane function (see Chapter 2).

PA[6] is a metabolic intermediate in the biosynthetic/degradation pathways of more complex glycerophospholipids. It usually represents less than 1% of total membrane lipids but plays a critical role in signal transduction, due to the unique ionization properties of its phosphate group (see Chapter 3). Membrane glycerophospholipids are derived from PA by condensation with an organic alcohol (general formula X-OH). Phosphatidylcholine (PC), the most abundant membrane lipid, results from the condensation of choline with PA (Fig. 1.14).

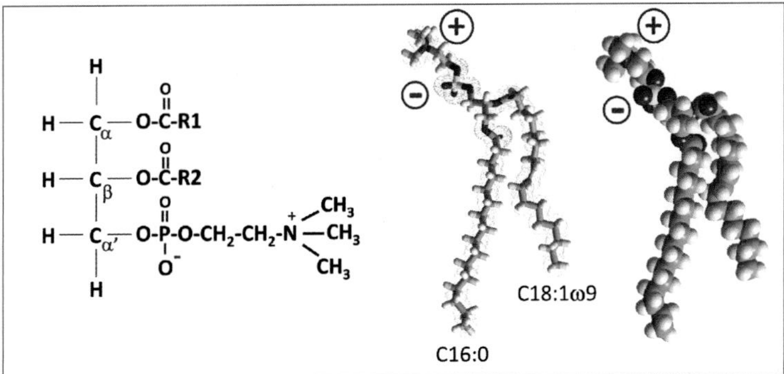

FIGURE 1.14 **Phosphatidylcholine.** This glycerophospholipid is the most abundant lipid of plasma membranes. At pH 7 it is zwitterionic, with a negative charge (in red) on the phosphate group and a positive charge (in blue) on the quaternary nitrogen atom. The acyl chains are palmitic acid (R1) and oleic acid (R2).

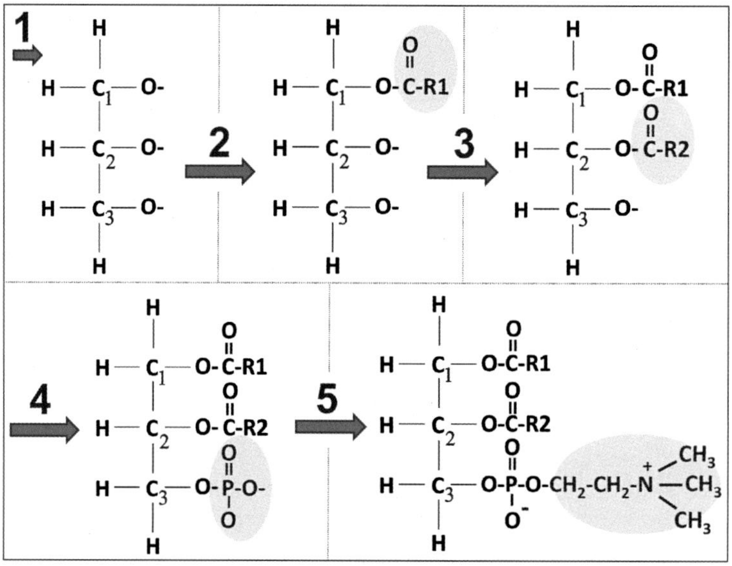

FIGURE 1.15 **How to draw a phosphatidylcholine molecule: a step-by-step procedure. 1.** Write the glycerol backbone structure and omit the hydrogen atoms of hydroxyl groups. **2.** Add the saturated acyl chain R1 on carbon no. 1. **3.** Add the unsaturated acyl chain R2 on carbon no. 2. **4.** Add the phosphate group on carbon no. 3. **5.** Condense with choline. Do not forget to indicate the positive charge on the nitrogen atom and the negative charge on the phosphate group.

How do you write the structure of PC? As a guideline, you may use the step-by-step procedure explained in Fig. 1.15.

Similarly, you can write the structures of phosphatidylethanolamine PE and phosphatidylserine PS (Fig. 1.16). Now you know how to write the structure of the three main membrane glycerophospholipids. At pH 7, both PC and PE are zwitterions (i.e., bear a positive and a negative charge). In contrast, PS has one positive and two negative charges, so that this lipid is globally anionic.[7]

The unique R1/R2 (saturated/unsaturated) acyl chains combination is a key feature that you should keep in mind when studying membrane structure and function (see Chapter 2). Apart from PC, PE, and PS, a few other glycerophospholipids are biologically important and their structures can be easily deduced from the condensation of PA with an organic alcohol (e.g., glycerol or inositol): phosphatidylglycerol (PG), which is present in mitochondrial membranes, and phosphatidylinositol (PI), a minor component of plasma membrane that plays a key role in signal transduction (see Chapter 3).

1.5.6 Sphingolipids

Sphingolipids were discovered by J. L. W. Thudichum, today considered the pioneer in the chemistry of the brain.[8] The common building block of sphingolipids is a long chain base that Thudichum named *sphingosine*, "in commemoration of the many enigmas which it presents to the inquirer" (in reference to the Sphinx enigmas). Indeed, the correct structure of

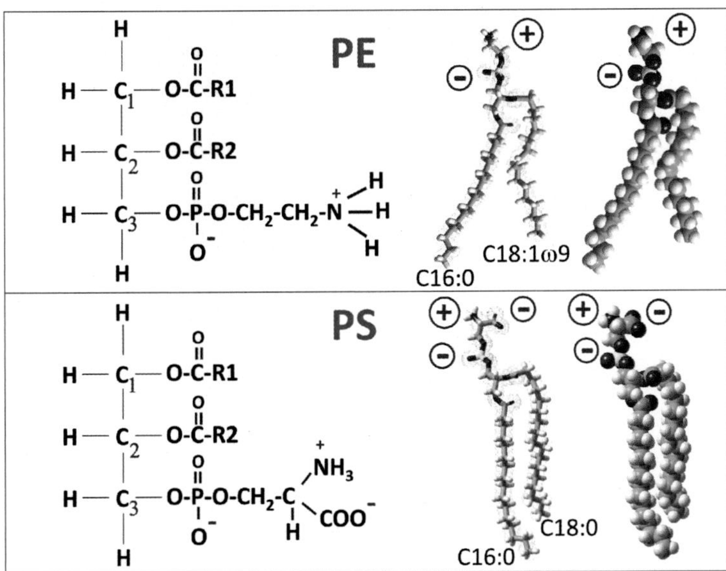

FIGURE 1.16 **Structure of phosphatidylethanolamine (PE) and phosphatidylserine (PS).** At pH 7 PE (top) is zwitterionic but PS (bottom) bears two negative charges and one positive charge so that its mean electric charge is –1. For this reason it belongs to the class of anionic lipids. The acyl chains of PE are palmitic acid (R1) and oleic acid (R2). Note that the PS molecule represented here contains two saturated acyl chains (palmitic acid in R1 and stearic acid in R2) instead of the usual saturated/unsaturated acyl content shown for PE.

sphingosine was established more than a half-century later by Carter et al.[9] In 1958, Shapiro et al. published the total synthesis of sphingosine, confirming its chemical structure as *trans*-D-erythro-1,3-dihydroxy-2-amino-4-octadecene (or, according to the R/S system, *trans*-(2S,3R)-1,3-dihydroxy-2-amino-4-octadecene.[10] In practice, it is a C18 carbon chain (CH_3-$(CH_2)_{16}$-CH_3), with two OH groups, one amino group, and a *trans* (E) double bond (Fig. 1.17).

It is amazing that such a simple molecule has for decades and several generations of talented biochemists resisted definition. Carter et al. have also introduced the term *sphingolipids* to designate lipids derived from the parent base sphingosine.[11] In most textbooks, the way sphingosine is represented makes it difficult to memorize. However, following the step-by-step procedure described in Fig. 1.18 makes it clearer.

Finally, it should be noted that minor chemical variations of the sphingosine molecule can give rise to several distinct sphingoid bases expressed in mammalian cells, including the saturated dihydrosphingosine (sphinganine) and phytosphingosine (a saturated derivative of sphingosine with a third OH group on C4).[12,13]

The amino group of sphingosine can react with a fatty acid to form an amide called ceramide (Fig. 1.19). Once again this generic name indicates that numerous distinct ceramides can be formed, according to the structure of their acyl chain. The acyl chain of ceramide is generally saturated or monounsaturated.

Moreover, the carbon atom linked to the carbonyl group of the acyl chain (named C in α, or Cα) may either display or not display a hydroxyl group (Fig. 1.20). Correspondingly, ceramides can contain an α-hydroxylated fatty acyl (HFA-ceramides) or a standard nonhydroxylated

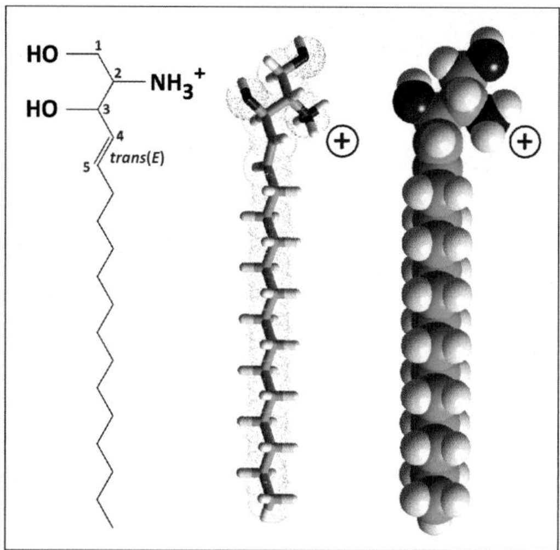

FIGURE 1.17 **Structure of sphingosine.** Three representations of sphingosine are shown: from left to right, chemical structure, tube model, and sphere model. Note the positive charge on the nitrogen atom (at physiological pH values).

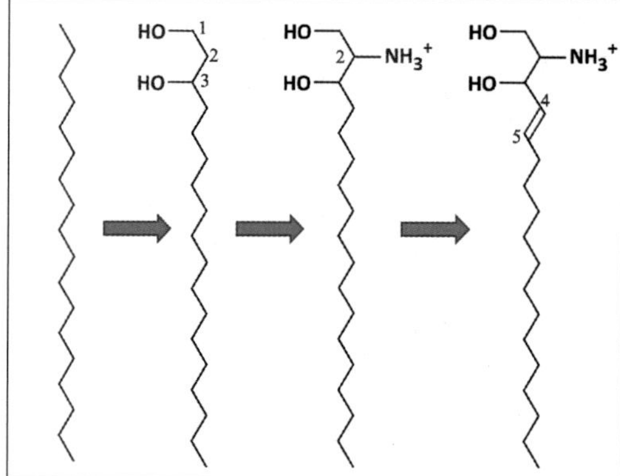

FIGURE 1.18 **A simple method to draw sphingosine.** Draw a C18 chain in zigzag symbol and number the carbons (you can choose any terminus to start the numbering). Then, add the two OH groups on C1 and C3. Add the amino group on C2 (–NH_3^+ at pH 7). Add the *trans* (*E*) double bond between C4 and C5.

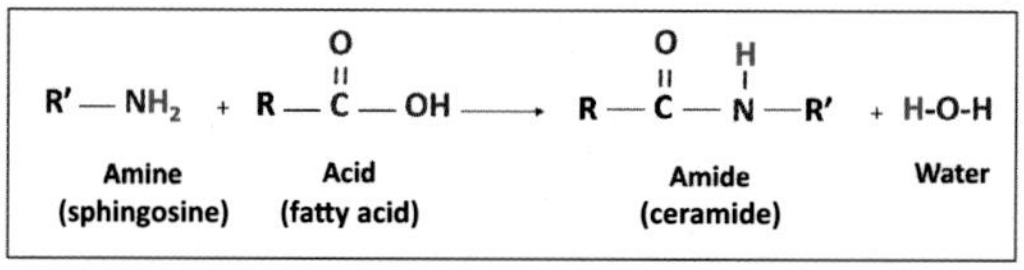

$$\underset{\text{(sphingosine)}}{\underset{\text{Amine}}{R'\!-\!NH_2}} + \underset{\text{(fatty acid)}}{\underset{\text{Acid}}{R\!-\!\overset{\overset{\displaystyle O}{\|}}{C}\!-\!OH}} \longrightarrow \underset{\text{(ceramide)}}{\underset{\text{Amide}}{R\!-\!\overset{\overset{\displaystyle O}{\|}}{C}\!-\!\overset{\overset{\displaystyle H}{|}}{N}\!-\!R'}} + \underset{\text{Water}}{H\text{-}O\text{-}H}$$

FIGURE 1.19 **Ceramide results from the condensation of sphingosine with a fatty acid.**

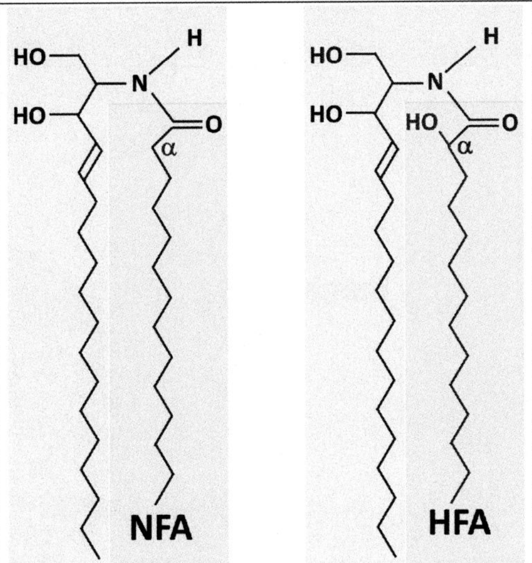

FIGURE 1.20 **α-Hydroxylated and nonhydroxylated ceramides.** α is the first carbon after the carbonyl group of the acyl chain (it corresponds to Δ2 in the Δ numbering system of fatty acids). HFA refers to α-hydroxylated fatty acid, and NFA to nonhydroxylated fatty acid.

fatty acyl (NFA-ceramides).[14,15] This slight chemical modification of ceramides may have a critical effect on ceramide function (see Chapter 3) as well as on the higher incidence of Alzheimer's disease in women versus men (see Chapter 11).

In any case, drawing the structure of a given ceramide is quite easy, provided that both the nature and hydroxylation status of the acyl chain are clearly indicated. In Fig. 1.21 you have

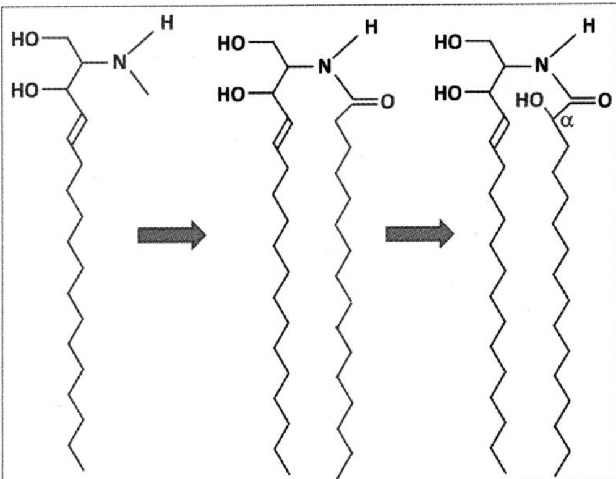

FIGURE 1.21 **How to draw a ceramide molecule.** Draw a sphingosine chain as explained in Fig. 1.18 (just omit one H atom of the $-NH_2$ group). Add the C16:0 acyl chain. If the ceramide is a HFA, add the OH group on the acyl chain (in α of the carbonyl group).

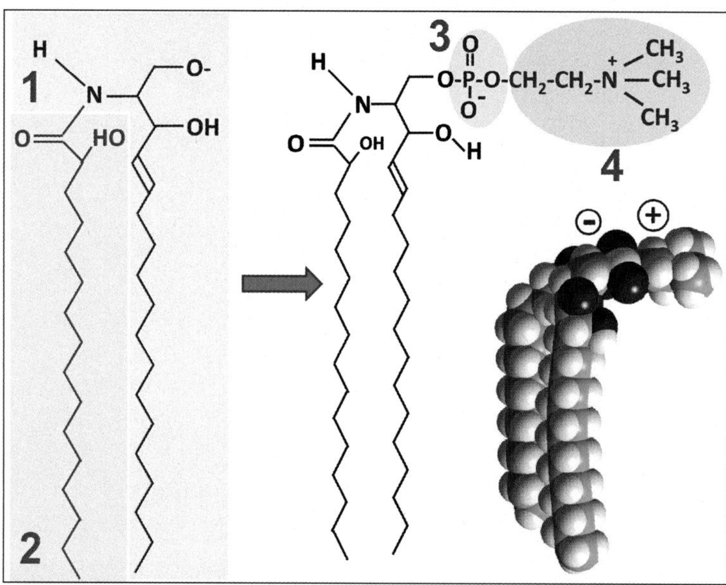

FIGURE 1.22 **How to draw a sphingomyelin molecule. 1.** Draw sphingosine as explained in Fig. 1.21 (do not write the H atom of the C1). **2.** Add the acyl chain (HFA-C16:0 in this example). Now you have completed your ceramide backbone. **3.** Add the phosphate group on carbon C1 of sphingosine. **4.** Finish the sphingomyelin molecule with the choline chain. Note that SM is zwitterionic at pH 7, as indicated in the sphere model.

a tutorial for drawing a ceramide molecule formed by the condensation of sphingosine with the HFA-C16:0 fatty acid.

Once you are familiar with ceramide structure, we can go further and use ceramide as a new starting block for more complex sphingolipids. Such sphingolipids fall into two categories. If ceramide is condensed with phosphorylcholine (not to be confounded with phosphatidylcholine, PC), it forms a sphingomyelin (SM, Fig. 1.22).

If ceramide is condensed with a sugar, it forms a glycosphingolipid (GSL). The glycone part of GSLs can be neutral, sulfated, acidic, or cationic.[16,17] We begin the study of GSLs with the simplest species, the monohexosylceramides. These neutral GSLs are formed by the condensation of a ceramide with either glucose or galactose, leading to glucosylceramide and galactosylceramide, respectively. Glucosylceramide (GlcCer) is the metabolic precursor of most GSLs, whereas galactosylceramide (GalCer) forms a small family consisting essentially of GalCer, its sulfated congener (sulfatide), and its unique ganglioside derivative (GM4). A tutorial for drawing a GalCer molecule is given in Fig. 1.23. Historically, GalCer and GlcCer were referred to as cerebrosides.[18] The names galactocerebroside (for GalCer) and glucocerebroside (for GlcCer) are no longer used. Similarly, the terms *kerasin* (GalCer-NFA) and *phrenosin* (GalCer-HFA) are obsolete.

Complex neutral GSLs (Table 1.4) can be derived from GlcCer by adding one or several neutral sugars, chiefly galactose or *N*-acetylgalactosamine in various combinations. The structure of these sugars is shown in Fig. 1.24.

By combining the information given in Fig. 1.24 and Table 1.4, you can easily draw the structure of any neutral GSL, for example, asialo-GM1 (Fig. 1.25).

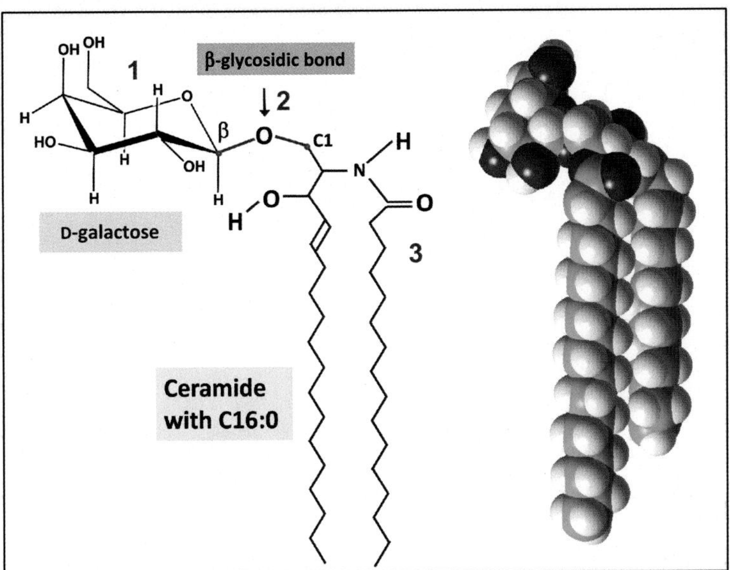

FIGURE 1.23 **How to draw a GalCer molecule.** GalCer is formed by addition of the cyclic form of D-galactose, called D-galactopyranose, onto ceramide. The bond between galactose and ceramide is a β-glycosidic bond. "Glycosidic" means that the reactive OH group of the sugar is the OH beared by the C1; "β" means that at the time the bond is formed, this OH group is over the plane of the sugar ring. Galactose is linked to the C1 of ceramide, so that GalCer is abbreviated Galβ1–1Cer. Use the following method for drawing GalCer. **1.** On the upper left of the paper sheet, draw the galactose molecule. Do not write the H atom of the C1. **2.** Link the oxygen in C1 of galactose (β) to the C1 of sphingosine. **3.** Add the acyl chain (NFA-C16:0 in this example). A sphere model of this GalCer molecule is shown on the right.

In addition to neutral GSLS, three other families of GSLs have glycone part that bears one or several electric charges at pH 7. First, some GSLs contain a negatively charged sulfate group ($-SO_3^-$) generally linked to the C3 of the sugar. They are referred to as sulfated GSLs. The sulfated derivative of GalCer, that is, 3-sulfoGalCer or sulfatide (Fig. 1.26), is the most important of these GSLs. Indeed, GalCer and sulfatides are among the most abundant lipids of the myelin sheath.[19]

TABLE 1.4 Neutral GSLs

Structure	Name
Galβ_1-Cer	Galactosylceramide, GalCer
Glcβ_1-Cer	Glucosylceramide, GlcCer
Galβ_{1-4}Glcβ_1-Cer	Lactosylceramide, LacCer
Galα_{1-4}Galβ_{1-4}Glcβ_1-Cer	Globotriaosylceramide, Gb_3
Galβ_{1-3}GalNAcβ_{1-4}Galβ_{1-4}Glcβ_1-Cer	Asialo-GM1
GalNAcβ_{1-3}Galα_{1-4}Galβ_{1-4}Glcβ_1-Cer	Globotetrahexosylceramide, Gb_4
GalNAcα_{1-3}GalNAcβ_{1-3}Galα_{1-3}Galβ_{1-4}Glcβ_1-Cer	Forssman-like iGb_4

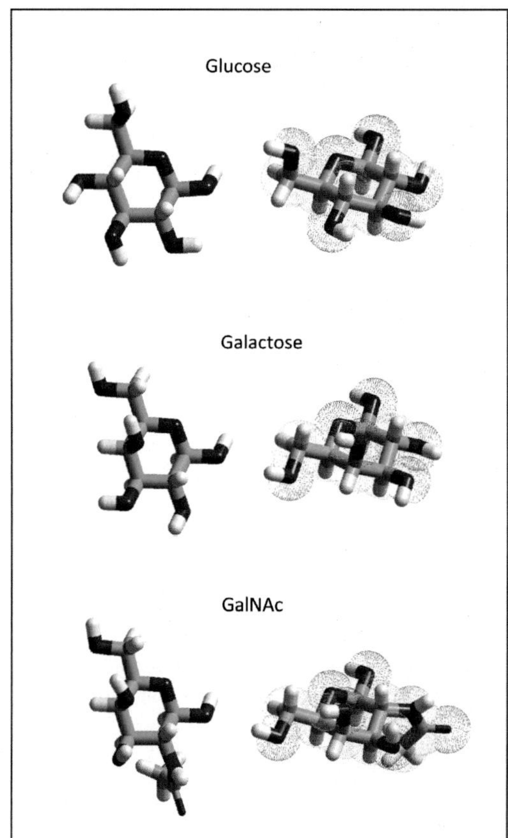

FIGURE 1.24 **The sugars of neutral GSLs.** Two distinct orientations (tube and dotted spheres) of glucose, galactose, and GalNAc are shown.

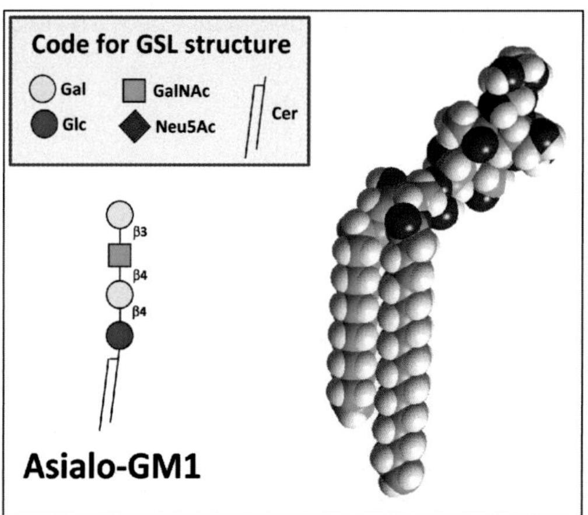

FIGURE 1.25 **Structure of asialo-GM1, a neutral GSL expressed in brain.** Asialo-GM1 is represented in a sphere model (on the right) and in a simplified geometric color code (on the left). This code is particularly useful for drawing simple yet informative structures of complex GSLs.

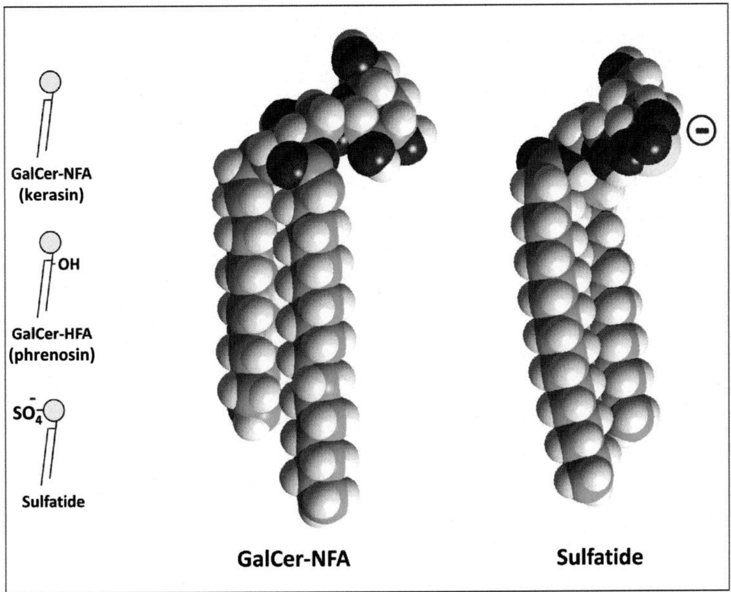

FIGURE 1.26 **GalCer and its sulfated derivative, two main lipid components of myelin.** GalCer-NFA, GalCer-HFA, and sulfatide are represented with geometric symbols (on the left). GalCer-NFA and sulfatide (also with an NFA acyl chain) are represented in sphere model. Note the negative charge of the sulfate group on the head group of sulfatide.

The second class of electrically charged GSLs includes acidic lipids containing at least one sialic acid. These acidic GSLs are called gangliosides.[20] Sialic acids are sugar derivatives with a negatively charged carboxylate group, such as *N*-acetylneuraminic acid (NANA, Fig. 1.27).

Gangliosides can contain one or several sialic acid residues in their carbohydrate part. Figure 1.25 shows the structure of a neutral GSL, asialo-GM1. The name of this GSL means

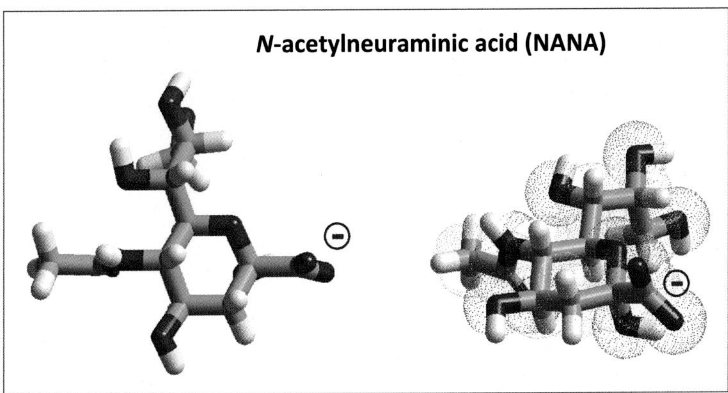

FIGURE 1.27 **Structure of a sialic acid (*N*-acetylneuraminic acid).** Note the negative charge on the carboxylate group.

TABLE 1.5 Major Brain Gangliosides

Structure	Name
NeuAc$\alpha_{2\text{-}3}$Galβ_1-Cer	GM4
NeuAc$\alpha_{2\text{-}3}$Gal$\beta_{1\text{-}4}$Glcβ_1-Cer	GM3
Gal$\beta_{1\text{-}3}$GalNAc$\beta_{1\text{-}4}$(NeuAc$\alpha_{2\text{-}3}$)Gal$\beta_{1\text{-}4}$Glcβ_1-Cer	GM1
NeuAc$\alpha_{2\text{-}3}$Gal$\beta_{1\text{-}3}$GalNAc$\beta_{1\text{-}4}$(NeuAc$\alpha_{2\text{-}3}$)Gal$\beta_{1\text{-}4}$Glcβ_1-Cer	GD1a
Gal$\beta_{1\text{-}3}$GalNAc$\beta_{1\text{-}4}$(NeuAc$\alpha_{2\text{-}8}$NeuAc$\alpha_{2\text{-}3}$)Gal$\beta_{1\text{-}4}$Glcβ_1-Cer	GD1b
NeuAc$\alpha_{2\text{-}8}$NeuAc$\alpha_{2\text{-}3}$ Gal$\beta_{1\text{-}3}$GalNAc$\beta_{1\text{-}4}$(NeuAc$\alpha_{2\text{-}3}$)Gal$\beta_{1\text{-}4}$Glcβ_1-Cer	GT1a
NeuAc$\alpha_{2\text{-}3}$Gal$\beta_{1\text{-}3}$GalNAc$\beta_{1\text{-}4}$(NeuAc$\alpha_{2\text{-}8}$NeuAc$\alpha_{2\text{-}3}$)Gal$\beta_{1\text{-}4}$Glcβ_1-Cer	GT1b

that it is analogous to ganglioside GM1, except that it has lost its sialic acid. Conversely, you can obtain the structure of GM1 by adding a sialic acid to asialo-GM1. You need two pieces of information: the nature of the sialic acid (generally NANA), and the nature of the glycosidic bond linking the sialic acid to asialo-GM1 (α2-3 in this case). The formulas of the most important brain gangliosides are indicated in Table 1.5.

Combining information about the structure of neutral, sulfated, and acidic sugars will help you to draw the structure of any ganglioside, as shown for GM1 in Fig. 1.28.

A rapid glance at Table 1.5 indicates that GD1a is obtained from GM1 by the addition of a sialic acid on the terminal Gal residue. Similarly, GT1a is obtained from GD1a and GT1b from GD1b.[21] To be honest, drawing the complete chemical structure of membrane lipids as complex as these gangliosides will be necessary on only rare occasions (if ever). In any case, the information provided here will allow you to do so, even for a ganglioside that is not described in this book: If you know the sugar sequence, the glycosidic bonds, and the nature of the acyl chain of the ceramide, then you have enough information to draw the structure of any ganglioside. Because you know perfectly how to do that, now we will allow

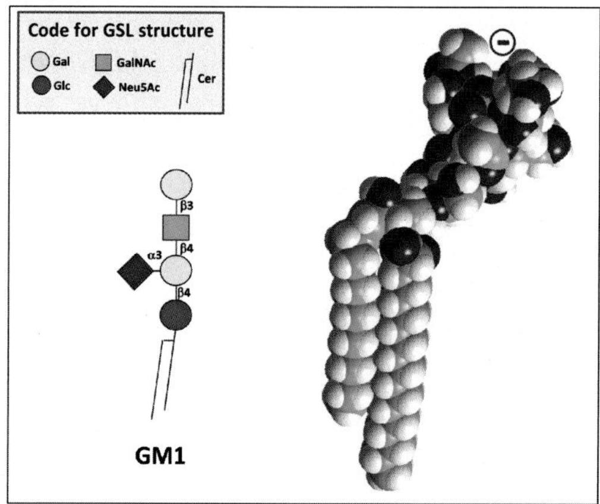

FIGURE 1.28 **Structure of ganglioside GM1.**

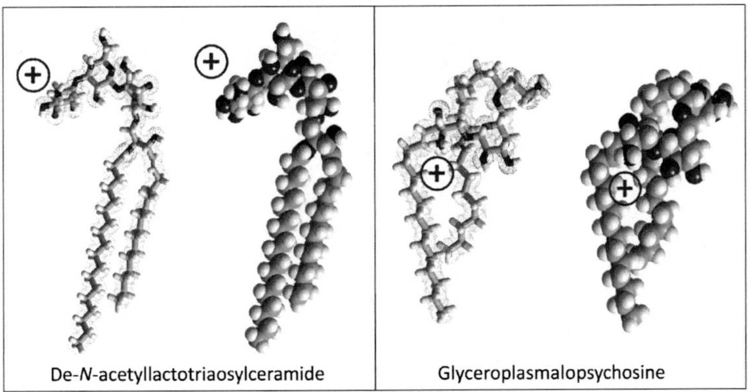

FIGURE 1.29 **Structure of two distinct cationic GSLs purified from brain extracts.** Note the particular shape of glyceroplasmalopsychosine (right panel).

ourselves to simplify the representations of these lipids as much as possible. As you can see in Figs 1.25, 1.26, and 1.28, GSLs can be adequately represented with simple geometric building units and a color code.

Finally, a third class of charged GSLs has been recently discovered in the brain.[22–24] In contrast with sulfated GSLs and gangliosides, these GSLs are cationic because they bear a positively charged $-NH_3^+$ group (Fig. 1.29). This free $-NH_3^+$ group is either the unsubstituted amino group of the sphingosine backbone of the GSL (in the case of glyceroplasmalopsychosine) or a deacetylated *N*-acetyl group of a GlcNAc unit (for De-*N*-acetyllactotriaosylceramide). In all cases, the interaction of such cationic GSLs with negatively charged gangliosides in the plasma membrane of brain cells may allow a fine regulation of membrane function, as proposed by Hakomori and colleagues.[22,24] Hence, further studies on these newly discovered brain GSLs are warranted.

1.5.7 Sterols

In brain membranes, sterol lipids are almost exclusively represented by cholesterol. Although Brown and Goldstein stated that "cholesterol is the most highly decorated small molecule in biology" (1985 Nobel lecture, cited by Nes[25]), drawing its structure is not as difficult as students generally believe. The first thing to put down is a polycyclic compound called sterane (Fig. 1.30). Then you add the polar head of cholesterol, a simple OH group, on carbon C3. The molecule is completed by a double bond between C5 and C6, two methyl groups (on C10 and C13), and an isooctyl chain linked to C17.

If you follow the order of carbon numeration on the first cycle (ring A of sterane) you are moving counterclockwise. The β face of cholesterol is defined according to the system numeration of ring compounds.[26] The methyl and isooctyl groups give a rough topology to this β face, whereas the α face, with no spiking groups, is planar (Fig. 1.31). For this reason, cholesterol has been described as a "bifacial" lipid.[16,27] This unique feature allows a great diversity of protein–lipid and lipid–cholesterol–protein interactions in biological membranes (see Chapter 6).

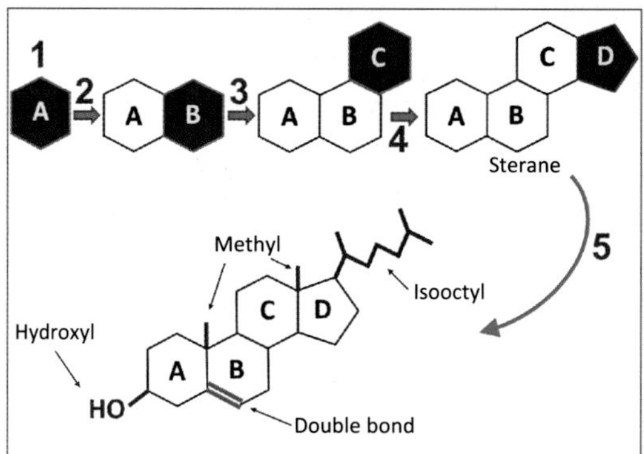

FIGURE 1.30 **A tutorial for drawing cholesterol. 1–4.** Start by drawing the sterane backbone, and sequentially number the four rings (from A to D). **5.** Add the OH group (carbon C3), and the double bond (between C5 and C6). Add a methyl group (the left one is on C10, the right one on C13) and an isooctyl chain (on C17). The numbering of all carbon atoms is indicated in Fig. 1.31.

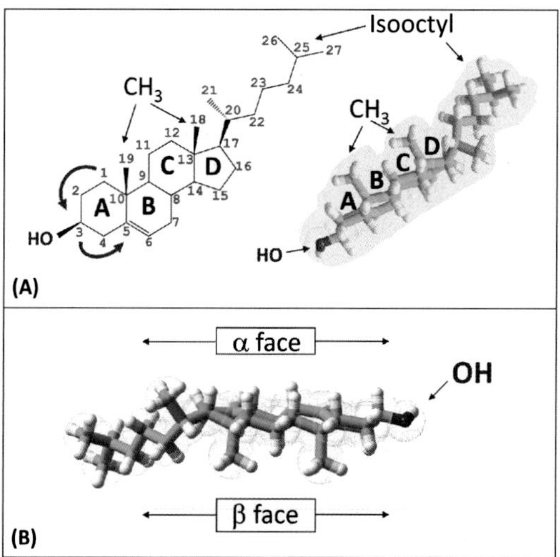

FIGURE 1.31 **Topology of cholesterol, a "bifacial" membrane lipid.** (A) Structure of cholesterol with numbering of the carbon atoms (left panel). If we follow the order of carbon numeration of the first cycle, we are moving in the counterclockwise direction, which defines the β face. The global shape of the cholesterol molecule, in the same orientation, is shown in the right panel. (B) Tube model of cholesterol showing its two topologically distinct faces: the smooth α face is devoid of substituents, whereas the opposite face (referred to as β) has several alkyl groups (two methyl groups and the isooctyl chain) that form a rough terrain.

1.6 BIOCHEMICAL DIVERSITY OF BRAIN LIPIDS

For decades, the lipid components of biological membranes have been considered as a mixture of phospholipids and cholesterol. In best cases, phospholipids were separated into glycerophospholipids (PC, PE, and PS) and sphingomyelin (SM), and GSLs were at least mentioned. Nowadays, we know that the biochemical diversity of membrane lipids has been incredibly underestimated. With the exception of cholesterol, whose structure is unique, membrane lipids may vary in both their protruding polar groups and their membrane-embedded apolar parts. Their relative distribution in both membrane leaflets (intra- and extracellular in the case of the plasma membrane) can be asymmetric. Specific lipids can segregate into discrete domains where they perform specific functions. In addition, some brain lipids may act as neurotransmitters (endocannabinoids), play a role in inflammation (prostaglandins), or control signal transduction pathways (PA). In any case, these functions reflect not only the biochemical structure of these lipids but the multiple facets of their molecular interactions with surrounding lipids or protein receptors. The main purpose of this book is to explain how these lipids ensure a smooth functioning of the brain and, on the opposite side, how they participate in brain dysfunctions.

1.7 A KEY EXPERIMENT: LIPID ANALYSIS BY THIN LAYER CHROMATOGRAPHY [28,29]

In the earliest states of emerging biochemistry at the beginning of the nineteenth century, the isolation of the chemical species that exist in living organisms was developed by French chemist Michel-Eugène Chevreul. For this reason, Chevreul is usually considered as the real founder of biochemical analysis. He called these chemical species "proximate principles" (in French, *principes immédiats*). Chevreul published a series of articles on fats, and in 1817, he coined the term *cholesterine* (derived from the Greek words *khole*, meaning "bile," and *steros*, meaning "solid") to designate the substance discovered in gallstones by Poultier de la Salle, another French chemist, in 1758. Because cholesterine proved to be a secondary alcohol, the compound was eventually named cholesterol. It is fascinating to realize that our knowledge of lipid biochemistry has been acquired by a series of tenacious researchers who struggled for years to decipher complex structures, with the unsophisticated chemical and biochemical approaches that were available in their time. Among these forerunners we should also mention Ludwig Thudichum and Herbert Carter for sphingolipids, and Ernst Klenk for gangliosides.

Klenk coined the name *ganglioside* in 1942 to designate a new category of lipids that he had isolated from brain ganglion cells. He also worked on brain cerebrosides and used thin layer chromatography (TLC) to separate complex GSL mixtures. This technique of separation of molecules was first used in the late 1930s by Nikolai Izmailov and Maria Shraiber who adapted the classical column chromatography for the rapid analysis of pharmaceutical plant extracts.[30] For separating mixtures of GSLs, the lipid extract is spotted onto a glass or an aluminum plate coated with a thin layer of silica gel. This layer can be as thin as 5 μm for high-performance thin layer chromatography (HPTLC). The plate is developed with a solvent system (e.g., chloroform:methanol:water; 60:35:8 for neutral GSLs), and at the end of the migration, a revealing agent is sprayed (orcinol for neutral GSLs, resorcinol for gangliosides). This technique is useful for separating cerebrosides (GalCer and GlcCer) containing NFA and HFA-ceramide, as shown in Fig. 1.32A. Moreover, TLC and HPTLC plates can be incubated with GSL antibodies that can identify a specific GSL in a complex mixture, just like antibody work for proteins in Western blot experiments. The technique can also be used to

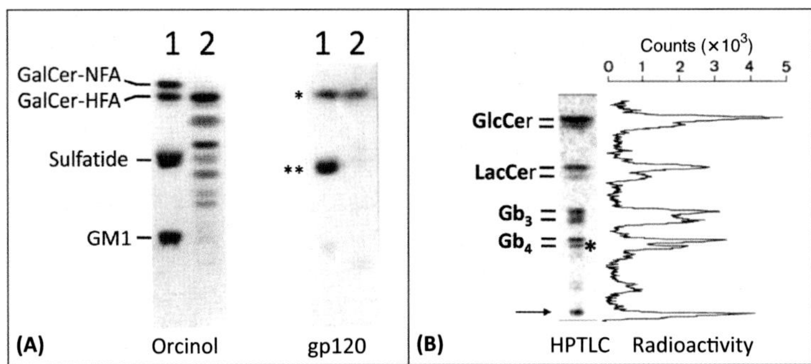

FIGURE 1.32 **Thin layer chromatography of GSLs.** (A) Left panel: orcinol staining of GSLs purified from bovine brain (lane 1) or extracted from colon epithelial cells (lane 2) and separated by HPTLC. Due to its OH group in α of the acyl chain, GalCer-HFA is slightly more polar than its NFA counterpart. Thus, it has a lower migration distance (Rf) in the solvent system used (chloroform:methanol:water, 60:35:8, vol:vol:vol). (B) Right panel: HIV-1 surface envelope glycoprotein gp120 binding detected on the HPTLC plate. Note that gp120 selectively binds to GalCer-HFA (asterisk) and sulfatide (double asterisk), whereas it ignores GalCer-NFA and GM1 (lane 1). In the GSLs extracted from colon cells (lane 2) the only GSL able to bind gp120 is GalCer-HFA. (Reproduced from Fantini et al.[31] Copyright © 1993, National Academy of Sciences.) Right panel: Patterns of [^{14}C]galactose-labeled GSLs expressed by human peripheral blood mononuclear cells. Radioactivity measurements of the HPTLC plate were performed with a phosphoimager. The position of standard GSLs run in parallel and stained with orcinol is shown in the left margin. The arrow indicates the depot spot of the GSLs on the HPTLC plate. The asterisk indicates the detection of small amounts of ganglioside GM3 (NFA species) that co-migrates with the lower Gb_4 band. (*Reproduced from Fantini et al.,[29] with permission.*)

study the binding of microbial proteins (virus glycoproteins such as HIV-1 gp120 or bacterial proteins/toxins) to their GSL receptors.[31] In all these cases, the TLC/HPTLC plates are briefly treated with a solution of poly(isobutyl methacrylate) in *n*-hexane before the incubation with protein ligands.[32] This treatment ensures that the GSLs are correctly oriented on the surface of the plate, so that their sugar part is fully accessible to external ligands, in a topology that mimics the orientation of GLS in a biological membrane.

TLC/HPTLC can also be used for studying the metabolic pathways involved in GSLs biosynthesis and degradation. In this case, GSLs are extracted from cultured cells (or animals) that have been previously treated with radioactive precursors (e.g., ^{3}H or ^{14}C-labeled sugars in the case of GSLs) (Fig. 1.32B). Finally, one should be aware that the way gangliosides are named is derived from their chromatographic mobility. For instance, monosialylated gangliosides are named GM1 to GM4 in reference to their respective migration distance (Rf) when separated on a TLC plate (GM1 has the lower mobility, GM4 the highest). Examples of the separation of gangliosides extracted from various brain cell types are given in Chapter 2.

References

1. Pauling L. *General Chemistry*. 2nd ed. San Francisco: Freeman; 1953.
2. Keutsch FN, Saykally RJ. Water clusters: untangling the mysteries of the liquid, one molecule at a time. *Proc Natl Acad Sci USA*. 2001;98(19):10533–10540.
3. Chaplin MF. Water's hydrogen bond strength. In: Ruth M, Lynden-Bell, Simon Conway M, Barrow JD, Finney JL, Harper CL Jr, eds. *In Water of Life: Counterfactual Chemistry and Fine-Tuning in Biochemistry*. Boca Raton, FL: CRC Press; 2010:69–86.

4. Tanford C. *Ben Franklin Stilled the Waves*. Oxford University Press, Oxford, UK; 2004.
5. Goni FM, Alonso A. Structure and functional properties of diacylglycerols in membranes. *Prog Lipid Res*. 1999;38(1):1–48.
6. Kooijman EE, Burger KN. Biophysics and function of phosphatidic acid: a molecular perspective. *Biochim Biophys Acta*. 2009;1791(9):881–888.
7. Biermann M, Maueröder C, Brauner JM, et al. Surface code – biophysical signals for apoptotic cell clearance. *Phys Biol*. 2013;10(6):065007.
8. Thudichum JLW. Researches on the chemical constitution of the brain. In: *Report of the Medical Officer of the Privy Council and Local Government Board*. Vol. 3; 1874: 113.
9. Carter HE, Glick FJ, Norris WP, Phillips GE. Biochemistry of the sphingolipides III. Structure of sphingosine. *J Biol Chem*. 1947;170(1):285–294.
10. Shapiro D, Segal H, Flowers HM. The total synthesis of sphingosine. *J Am Chem Soc*. 1958;80(5):1194–1197.
11. Carter HE, Haines WJ, Ledyard WE, Norris WP. Biochemistry of the sphingolipides; preparation of sphingolipides from beef brain and spinal cord. *J Biol Chem*. 1947;169(1):77–82.
12. Pruett ST, Bushnev A, Hagedorn K, et al. Biodiversity of sphingoid bases ("sphingosines") and related amino alcohols. *J Lipid Res*. 2008;49(8):1621–1639.
13. Fantini J, Yahi N. Molecular insights into amyloid regulation by membrane cholesterol and sphingolipids: common mechanisms in neurodegenerative diseases. *Expert Rev Mol Med*. 2010;12:e27.
14. Yahi N, Aulas A, Fantini J. How cholesterol constrains glycolipid conformation for optimal recognition of Alzheimer's beta amyloid peptide (Abeta1-40). *PloS One*. 2010;5(2):e9079.
15. Fantini J, Yahi N. Molecular basis for the glycosphingolipid-binding specificity of alpha-synuclein key role of tyrosine 39 in membrane insertion. *J Mol Biol*. 2011;408(4):654–669.
16. Fantini J, Barrantes FJ. Sphingolipid/cholesterol regulation of neurotransmitter receptor conformation and function. *Biochim Biophys Acta*. 2009;1788(11):2345–2361.
17. Fantini J, Garmy N, Mahfoud R, Yahi N. Lipid rafts: structure, function and role in HIV, Alzheimer's and prion diseases. *Expert Rev Mol Med*. 2002;4(27):1–22.
18. Klenk E, Rivera ME. The cerebron fraction of brain cerebrosides and isolation of the individual components. *Hoppe Seylers Z Physiol Chem*. 1969;350(12):1589–1592.
19. Boggs JM, Gao W, Hirahara Y. Myelin glycosphingolipids, galactosylceramide and sulfatide, participate in carbohydrate-carbohydrate interactions between apposed membranes and may form glycosynapses between oligodendrocyte and/or myelin membranes. *Biochim Biophys Acta*. 2008;1780(3):445–455.
20. Klenk E. On the discovery and chemistry of neuraminic acid and gangliosides. *Chem Phys Lipids*. 1970;5(1):193–197.
21. van Echten-Deckert G, Walter J. Sphingolipids: critical players in Alzheimer's disease. *Prog Lipid Res*. 2012;51(4):378–393.
22. Hikita T, Tadano-Aritomi K, Iida-Tanaka N, Anand JK, Ishizuka I, Hakomori S. A novel plasmal conjugate to glycerol and psychosine ("glyceroplasmalopsychosine"): isolation and characterization from bovine brain white matter. *J Biol Chem*. 2001;276(25):23084–23091.
23. Iida-Tanaka N, Hikita T, Hakomori SI, Ishizuka I. Conformational studies of a novel cationic glycolipid, glyceroplasmalopsychosine, from bovine brain by NMR spectroscopy. *Carbohydr Res*. 2002;337(19):1775–1779.
24. Hikita T, Tadano-Aritomi K, Iida-Tanaka N, Ishizuka I, Hakomori S. De-*N*-acetyllactotriaosylceramide as a novel cationic glycosphingolipid of bovine brain white matter: isolation and characterization. *Biochemistry*. 2005;44(27):9555–9562.
25. Nes WD. Biosynthesis of cholesterol and other sterols. *Chem Rev*. 2011;111(10):6423–6451.
26. Rose IA, Hanson KR, Wilkinson KD, Wimmer MJ. A suggestion for naming faces of ring compounds. *Proc Natl Acad Sci USA*. 1980;77(5):2439–2441.
27. Fantini J, Barrantes FJ. How cholesterol interacts with membrane proteins: an exploration of cholesterol-binding sites including CRAC, CARC, and tilted domains. *Front Physiol*. 2013;4:31.
28. Scandroglio F, Loberto N, Valsecchi M, Chigorno V, Prinetti A, Sonnino S. Thin layer chromatography of gangliosides. *Glycoconj J*. 2009;26(8):961–973.
29. Fantini J, Tamalet C, Hammache D, Tourres C, Duclos N, Yahi N. HIV-1-induced perturbations of glycosphingolipid metabolism are cell-specific and can be detected at early stages of HIV-1 infection. *J Acquir Immune Defic Syndr Hum Retrovirol*. 1998;19(3):221–229.
30. Shraiber MS. In: Ettre LS, Zlatkis A, eds. *75 Years of Chromatography: A Historical Dialogue*. Amsterdam: Elsevier; 1979:413–417.

31. Fantini J, Cook DG, Nathanson N, Spitalnik SL, Gonzalez-Scarano F. Infection of colonic epithelial cell lines by type 1 human immunodeficiency virus is associated with cell surface expression of galactosylceramide, a potential alternative gp120 receptor. *Proc Natl Acad Sci USA*. 1993;90(7):2700–2704.
32. Yahi N, Sabatier JM, Baghdiguian S, Gonzalez-Scarano F, Fantini J. Synthetic multimeric peptides derived from the principal neutralization domain (V3 loop) of human immunodeficiency virus type 1 (HIV-1) gp120 bind to galactosylceramide and block HIV-1 infection in a human CD4-negative mucosal epithelial cell line. *J Virol*. 1995;69(1):320–325.

CHAPTER

2

Brain Membranes

Jacques Fantini, Nouara Yahi

OUTLINE

2.1 WHY LIPIDS ARE DIFFERENT FROM ALL OTHER BIOMOLECULES

The famous Francis Crick's adage, "*If you want to understand function, study structure,*" does not apply to lipids. Observe and dissect the structure of phosphatidylcholine (PC) as long as you want, you will not understand how plasma membranes are organized and function. The reason is that lipids do not act alone but cooperate to form ordered molecular assemblies that have acquired specific functional properties. Here the whole is not only more than the sum of its parts, the whole creates the functions the parts totally lack. In first approximation, this unique feature can be compared to a brick (a single lipid molecule) in a wall (here representing the membrane, interestingly historically referred to as the *cell wall*). To begin, we will keep in mind this rough analogy while explaining why lipid molecules can self-assemble into various architectural motifs.

Brain Lipids in Synaptic Function and Neurological Disease. http://dx.doi.org/10.1016/B978-0-12-800111-0.00002-3

2.2 ROLE OF STRUCTURED WATER IN MOLECULAR INTERACTIONS

In biochemistry, when one molecule interacts with another one to form a complex, great attention is given to the chemical groups that are physically involved in the interaction. These chemical groups constitute the so-called "binding site," a three-dimensional domain of a molecule that is specifically devoted to the binding of another one. Such binding sites should be readily accessible to the molecular partner with which it is promised to interact. Let us consider the interaction between two proteins circulating in a biological fluid. Each of these proteins displays a binding site at its periphery. In the water environment in which these proteins circulate, these peripheral binding sites are first occupied by water molecules. When both proteins come in close contact, bound water molecules leave the binding domains so that the proteins can interact through noncovalent bonds (Fig. 2.1).

In this respect, the binding process can be decomposed in two phenomena: (1) the destructuration of numerous water molecules that were initially associated with each binding site through a network of hydrogen bonds, and (2) the formation of a new set of noncovalent interactions between the binding sites of each protein partner.[1] Both mechanisms contribute to the energy of association required to form the complex. We have learned from thermodynamics that the Gibbs free energy change associated with a chemical reaction (i.e., ΔG) takes into account both mechanisms. On one hand, the departure of numerous water molecules from each binding site

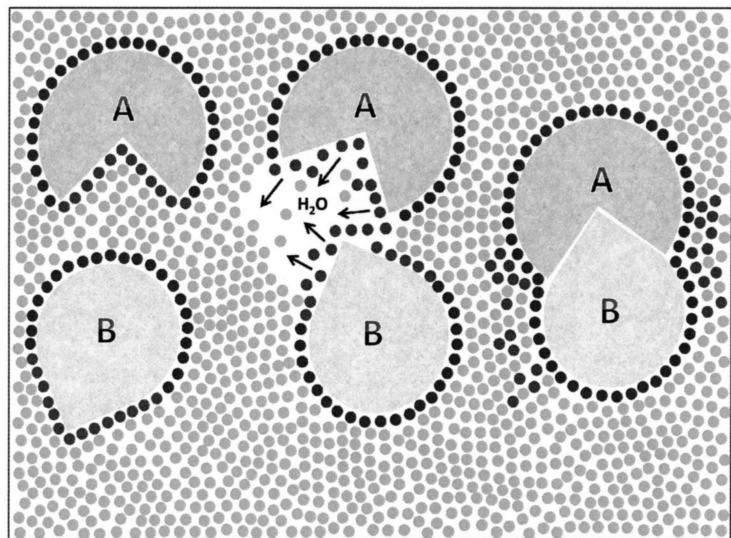

FIGURE 2.1 **Dehydration precedes binding reactions in water.** Consider two proteins, A and B, in solution in water (water molecules of the bulk solvent are represented as blue disks). Each of these proteins has a complementary binding site for the other one, yet initially both binding sites are covered with a layer of bound water (represented as purple disks). The binding reaction is possible only if these water molecules initially interacting with the binding sites are displaced (arrows) and redistributed randomly in the bulk solvent, thereby inducing an increase of entropic disorder. Note that the water molecules bound to regions of proteins A and B not involved in binding (represented as dark blue disks) are not affected by the process.

generates a significant molecular disorder, quantified as an entropy increase (ΔS). Disorder is energy, because going back to the order state requires energy (e.g., teenagers who let the disorder progressively develop in their bedroom know how hard it is to clean it). On the other hand, the noncovalent bonds that are formed between both binding sites contribute to the whole energy of interaction, because you would need energy to break them. This enthalpic contribution of the binding is expressed by the ΔH term of the classical equation $\Delta G = \Delta H - T\Delta S$ where T is the absolute temperature (expressed in Kelvin, allowing ΔG to be expressed in joules per moles or more representatively in kJ mol^{-1}). It is clearly beyond the scope of this book to dwell further on this famous thermodynamic equation. Let us just remark that the free energy change ΔG of any spontaneous reaction must have a negative value. In this respect, the sign of ΔH indicates whether the binding process is chiefly enthalpy-driven or entropy-driven. Schematically, enthalpic binding results from specific molecular interactions, such as hydrogen bonds that require a perfect adjustment of both ligands ($\Delta H < 0$). In contrast, entropic binding is far less specific because it relies primarily on the disorder created in water molecules surrounding the ligands. In practice, "pure" enthalpic or entropic binding does not exist because both contribute to the binding reaction. Indeed, whatever the sign of ΔH (positive or negative), it is clear that a significant value of the entropy change ΔS ensures that ΔG is negative and high, which is particularly critical when ΔH is positive. In this case ΔG will be negative only if the entropic term ΔS is sufficiently high, which will occur if there is enough disorganization of water molecules (Fig. 2.1). Moreover, even for binding reactions with a negative ΔH, it is important that the binding process is associated with the displacement of a maximal number of bound water molecules, which will further increase ΔS and thus ΔG. This information gives a realistic idea of the entropic contribution to binding reactions. Lipid–lipid interactions will not escape this important contribution of water molecules, and as we will see next, water will even do more.

2.3 LIPID SELF-ASSEMBLY, A WATER-DRIVEN PROCESS?

Because lipids are mostly apolar molecules, they are not particularly inclined to accommodate water molecules. However, this statement can be viewed as a tautology because lipids are precisely defined on the basis of their insolubility in water (see Chapter 1). In fact the situation is more complex because in addition to their large apolar, water-insoluble domain, lipids also display a polar head group. Hence, lipid behavior is typically *amphipathic*, their apolar part exhibiting hydrophobic properties that contrast with their hydrophilic polar head group. Before going further, one should understand that the terms *apolar* and *hydrophobic*, and similarly *polar* and *hydrophilic*, cannot be used as synonyms. *Apolar* refers to an intrinsic property of chemical group due to the weak electronegativity of its atoms (chiefly C and H). Because an apolar group cannot form hydrogen bonds with water molecules, it does not interact with water and hence it will have a hydrophobic behavior. On the opposite, a polar group contains electronegative atoms (O, N) often linked to hydrogen (–O–H and –N–H groups), that, due to the simultaneous presence of δ^- and δ^+ partial charges, can establish numerous hydrogen bonds with water molecules. In other word, a polar group is always polar, it is its nature, but it becomes hydrophilic (a molecular behavior) when confronted with water. Do not misuse these different terminologies. As an amphipathic molecule, a lipid can be schematized as shown in Fig. 2.2 for a saturated fatty acid.

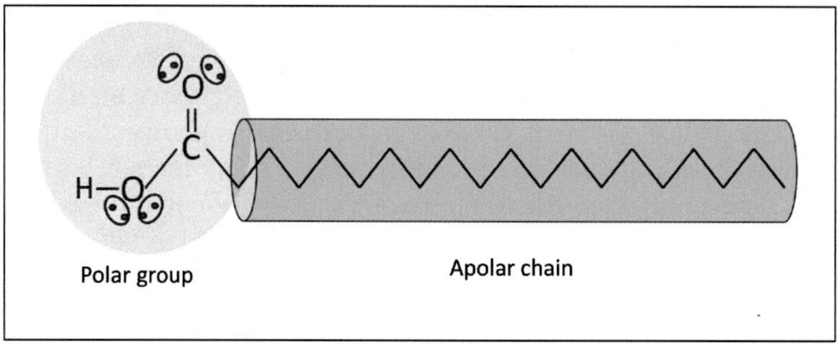

FIGURE 2.2 **How a fatty acid can be schematized as an amphipathic molecule.** The long aliphatic chain is apolar (green cylinder), whereas the carboxylic head group is polar (blue disk). Only the polar head group can form H-bonds with water molecules.

In Chapter 1, we have explained how water molecules interact each other or with polar molecules, on the basis of the formation of intermolecular hydrogen bonds. It is time to study the possibilities of interaction between apolar groups. Let us consider an alkane molecule (e.g., *n*-hexane) consisting of a linear chain with a succession of $-CH_2$ (methylene) groups bordered on each end by a $-CH_3$ (methyl) group (Fig. 2.3). The C–H bonds are mostly nonpolarized, so that this molecule does display any partial charge. However, when two molecules of this alkane are brought together at a short distance, each molecule influences the other one and creates a series of transient dipoles. This phenomenon explains why such neutral molecules initially

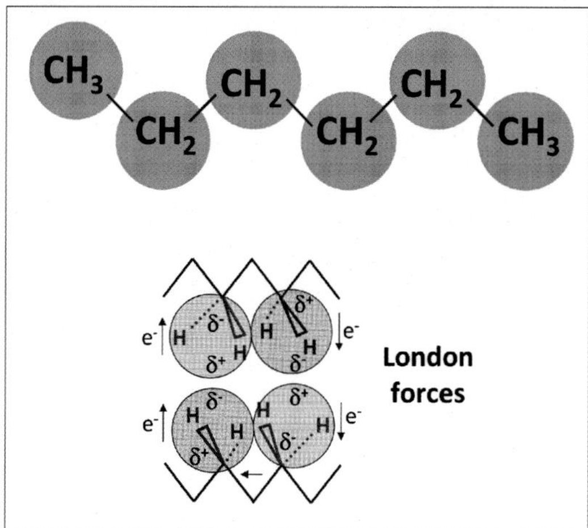

FIGURE 2.3 **London forces stabilize the interaction between neutral alkane molecules.** Single alkane molecules such as *n*-hexane (upper panel) are not polarized because they contain only carbon and hydrogen atoms. However, when two alkane molecules become close (lower panel), slight displacements of the electronic cloud of each C–H group induce a chain reaction resulting in the transient creation of partial δ^- charges (excess of electrons) facing corresponding partial δ^+ charges (poverty of the electronic cloud). Because these partial charges are only transient, the electrons constantly go back and forth, so that all partial δ^- charges become δ^+ for a while, and vice versa (just like windshield wiper movements). The sign of the partial charge changes constantly but the force of attraction remains the same, resulting in an efficient interaction between aliphatic chains.

devoid of partial charges can nevertheless efficiently interact through induced dipoles. The noncovalent forces that stabilize this kind of [induced dipole]–[induced dipole] interactions are called dispersion or London forces. These van der Waals forces have a low energy but they can be especially efficient if their number is high.

Thus, the longer the aliphatic chain is, the stronger the force of interaction between two alkane molecules is (Fig. 2.3). This phenomenon can be quantified by measuring the amount of energy required to separate alkane molecules. The melting temperature (melting point), which corresponds to the quantity of energy needed to obtain the solid-to-liquid phase transition, is an excellent landmark. As expected, it increases with the number of carbons in the alkane series: for example, –95, –57, –30, and +18°C, respectively for hexane, octane, decane, and hexadecane.

The schematic representation of an amphipathic lipid (Fig. 2.2) immediately suggests two possible mechanisms of interaction between two lipids of this kind (i.e., a hydrogen bond between polar head groups and London forces between aliphatic chains). If so, what do water molecules have to do here? The answer is that water may help to bring the apolar parts of lipids together, a step that obviously precedes the establishment of van der Waals interactions. This unique property of water is referred to as the *hydrophobic effect*. Indeed, the interactions between lipid molecules in a water environment can be envisioned in two ways. First, we can say that water exerts a strong repulsive effect on the apolar domain of lipids in a kind of "the enemies of my enemies are my friends" mechanism: the apolar domains of lipids hate water so much that they form a separate apolar milieu from which water molecules are excluded. However, this explanation is not sufficient: the apolar groups need stabilizing forces to remain stuck together (i.e., London forces). Here we find more of a "friends of my friends are my friends" mechanism.

Let us see what happens when a drop of lipid (e.g., palmitic acid, C16:0) is spotted on the surface of water. The drop progressively spreads on water, until it forms a monomolecular film referred to as a lipid monolayer (providing that the number of lipid molecules in the drop is not too high, otherwise multilayers are formed). During spreading, lipid molecules are oriented so that the carboxylic groups interact with water molecules whereas the aliphatic chains extend in the air (Fig. 2.4).

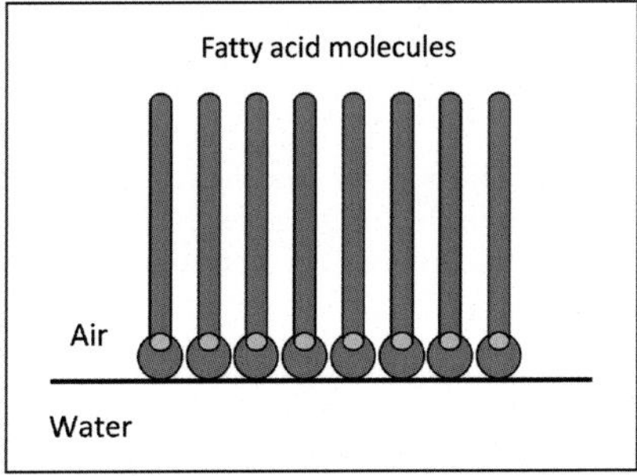

FIGURE 2.4 **Organization of a lipid monolayer at the air–water interface.** All fatty acid molecules are oriented with respect to water so that polar head groups interact with water and apolar chains are rejected in the air. The limit between air and water in such systems is referred to as an air–water interface.

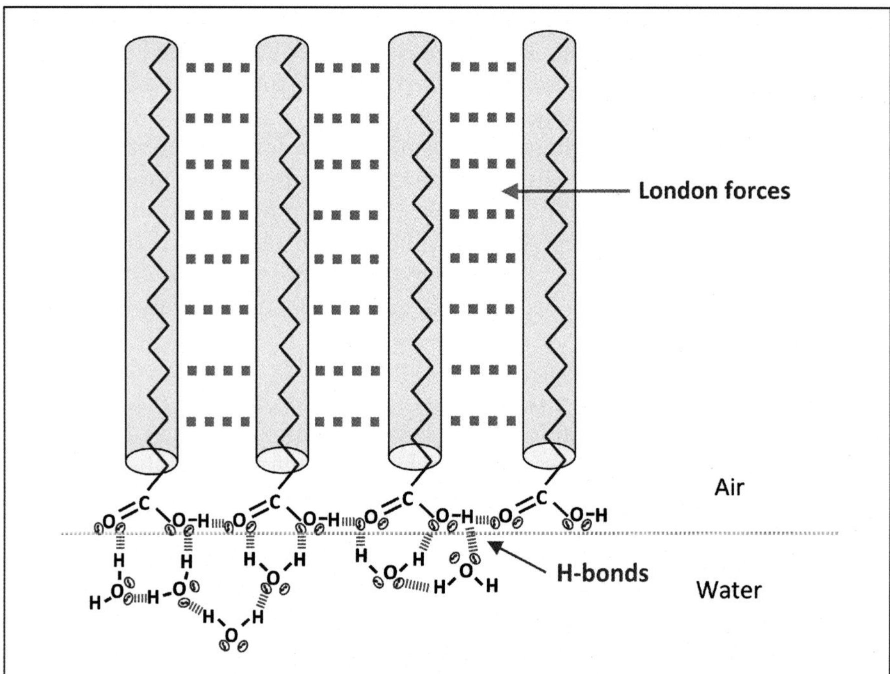

FIGURE 2.5 **Intermolecular forces stabilizing a lipid monolayer at the air–water interface.** The apolar chains interact through London forces. Hydrogen bonds stabilize the interaction of two vicinal carboxylic acid head groups as well as the interaction of water molecules with these polar groups.

At the molecular level, the carboxylic groups form a network of hydrogen bonds with each other and with the last layer of water molecules in contact with the air. The aliphatic chains interact through London forces in the air, which can thus be considered as a hydrophobic milieu (Fig. 2.5).

The molecular mechanisms accounting for the organization of a lipid monolayer have been the subject of an intense debate. Historically, the Nobel Prize recipient Irvin Langmuir thought that the apolar chains of fatty acid have just more affinity for each other than for water. It is interesting to recall his molecular interpretation of the forces involved in the formation of lipid monolayers at the air-water interface:

> When oleic acid is placed on water, it is probable that the carboxyl groups do actually dissolve in water; that is, they combine with the water chemically... The long hydrocarbon chains have too much attraction for each other, however, and too little for water, to be drawn into solution ....[2]

Also obviously true, this historical interpretation of lipid–lipid interactions did not take into account the hydrophobic effect. Conversely, the hydrophobic effect by itself would be ineffective if London forces did not stabilize the interactions between aliphatic chains. A wise position is to consider both. Lipids are brought together first because of the hydrophobic effect, which induces a selective orientation of lipid molecules with their polar part interacting with water and their apolar domain rejected on the opposite side. Then London forces stabilize the interactions between the apolar chains of lipids. We will conclude this brief analysis of the role of water in lipid–lipid interactions with the following remarks.

- First, the hydrophobic effect is not only driving the clustering of lipid molecules in water-free aggregates, but it also increases the strength of the lipid–lipid interaction. This important consequence of the hydrophobic effect is referred to as the hydrophobic interaction, which, according to Israelachvili, describes "the unusually strong attraction between hydrophobic molecules and surfaces in water . . . often stronger than their attraction in free space."[3] This concept is perfectly illustrated by the energy of interaction between two methane molecules *in vacuo* and in water (respectively, -2.5×10^{-21} and -14×10^{-21} J, or a 5.6-fold increase in affinity in water compared with free space). Intuitively, one can consider that water molecules exert a pressure on methane molecules, just like cold weather forces the sheep of a herd to get closer than they would do at higher temperatures. In other words, water really helps lipids to self-aggregate into hydrophobically stabilized clusters.
- Second, the hydrophobic effect does not explain why the lipids dispersed in water do not all eventually form the same supramolecular structures. Indeed, some lipids self-aggregate into micelles, whereas other form bilayers. Understanding the molecular mechanisms controlling the state of organization of lipids in water is one of the main goals of this chapter.
- And finally, neither the hydrophobic effect nor the hydrophobic interaction can explain why a high selectivity of lipid–lipid interactions exists in biological membranes. For instance, plasma membrane cholesterol interacts preferentially with sphingomyelin versus PC. Clearly, something beyond the hydrophobic effect determines the specificity of a given lipid for other lipids. In other words, lipid assemblies are not pure entropic processes (hydrophobic effect) but also rely on enthalpy (binding specificity).To understand this phenomenon, we will refer to the basic principles that govern molecular interactions (i.e., chemical compatibility and geometrical complementarity). All apolar parts of lipids are indeed apolar, but do not necessarily have the same chemical structure. Some apolar chains are saturated, others contain one or several double bonds, and cholesterol will make a choice among these chains. This choice is driven by a higher affinity for certain lipids (e.g., sphingomyelin or, more generally, sphingolipids) than for others (especially PC). The consequence is the segregation of membrane lipids into discrete domains with specific biochemical compositions. At this stage, we can envision the plasma membrane as a heterogeneous bidimensional lipid structure consisting of two types of domains corresponding to cholesterol-rich regions (also containing most sphingolipids) and cholesterol-poor regions (also enriched in PC). This explanation is in line with modern (i.e., post-Singer–Nicolson[4]) representations of the plasma membrane.[5–7]

2.4 LIPID–LIPID INTERACTIONS: WHY SUCH A HIGH SPECIFICITY?

Three key parameters control the interaction of a given lipid with its neighbors in a supramolecular assembly of lipids in water: (1) the melting temperature, which determines in which physical state (solid or liquid) is the lipid at physiological temperature; (2) the global shape of the lipid, here viewed as a LEGO brick, which determines what kind of architecture can be obtained (i.e., either micelle or bilayer); and (3) the chemical structure of the lipid, especially in the apolar part, which plays a critical role in lipid selectivity.

2.4.1 Melting Temperature of Lipids

The melting temperature is the amount of energy required for obtaining the solid-to-liquid phase transition of a lipid. Put a slice of cold butter into a frying pan and heat. When the butter melts, you have furnished enough energy to separate all the lipid molecules of the butter (mostly triglycerides) so that they can freely move, giving to the substance its liquid consistency. This event corresponds to the energy of interaction linking the lipids of the piece of butter in a noncovalent way. Let us consider first the case of a saturated fatty acid (e.g., C18:0, stearic acid). Two forces contribute to the stabilization of stearic acid molecules in a solid phase. The carboxylic groups are linked by hydrogen bonds, and the aliphatic chains interact through London forces. Thus, if one compares two saturated fatty acids of different length (e.g., C14:0 and C18:0), they have the same number of hydrogen bonds per mol, but notably differ in the number of van der Waals interactions that can be formed between their aliphatic chains. As a matter of fact, the number of van der Waals interactions is proportional to the length of the aliphatic chain. Thus, the melting temperature of saturated fatty acids regularly increases with the number of carbon atoms in the chain: 44, 58, 63, and 70°C, respectively, for C12:0, C14:0, C16:0, and C18:0.

Unsaturated fatty acids have a significantly lower melting temperature than their saturated counterpart with the same number of carbons. For instance, the melting temperature of oleic acid (C18:1ω9*cis*) is only 16°C, compared with the 70°C of stearic acid (C18:0). Thus, the presence of a single-double bond is responsible for a 54°C decrease (70–16°C) in the melting temperature. This striking phenomenon results from an architectural constraint due to the *cis* (*Z*) configuration of the double bond in natural fatty acid molecules. Namely, the double bond induces a bending of the chain whose main axis is deflected at approximately 30° (Fig. 2.6).

This rupture of linearity interferes with the establishment of van der Waals interactions between two vicinal chains because the rotation of the chain at the hinge materialized by the double bond generates a revolution cone. This activity efficiently prevents chain packing

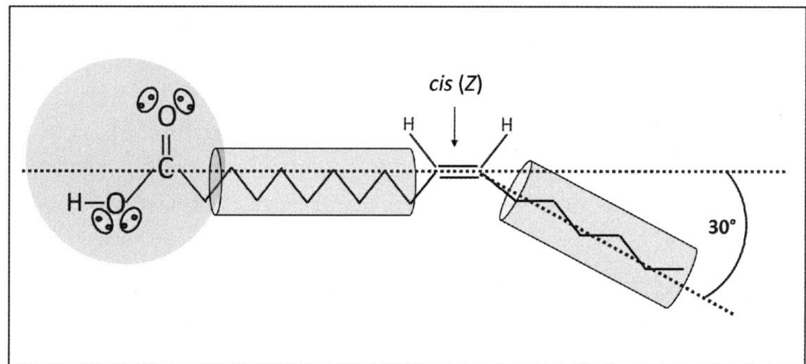

FIGURE 2.6 **The influence of a *cis* double bond in the geometry of a fatty acid.** Each *cis* (*Z*) double bond in the aliphatic chain of a fatty acid induces a kink of 30°. In contrast, a *trans* (*E*) configuration of the double bond would not change the main axis of the aliphatic chain.

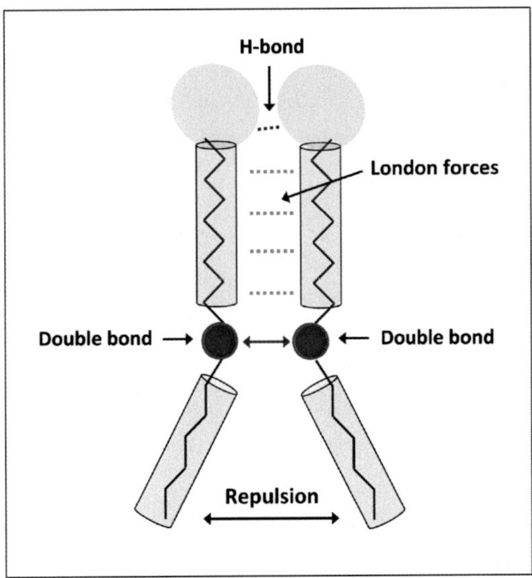

FIGURE 2.7 **Why unsaturated fatty acids have low melting temperatures.** Two vicinal unsaturated fatty acids interact through a combination of hydrogen bonds (between polar head groups) and London forces (between the saturated parts of the acyl chain starting from the head group). The *cis* double bond acts as a hinge, allowing the terminal part of the chain to deviate from the main axis of the chain (30° of angle by double bond). At this level of the chains, no interaction is possible, explaining why unsaturated fatty acids have low melting temperatures compared with saturated fatty acids. Note that the π electron cloud of the double bond also contributes to chain repulsion (double arrow in red).

in the region between the double bond and the terminal methyl (ω1) group (Fig. 2.7). Thus, only the methylene groups located between the carboxyl group and the double bond can participate in van der Waals interaction, which explains the lower value of 16°C for the melting temperature of this C18 fatty acid.

With two double bonds, the deviation is equal to 30 + 30 = 60°, as it is the case for linoleic acid (C18:2ω6), and the melting temperature is –5°C. The record is held by arachidonic acid (C20:4ω6), whose chain adopts a hairpin structure resembling a scorpion tail (30 × 4 = 120°), as shown in Fig. 2.8. This structure further restricts the zone capable of forming van der Waals bonds to three methylene groups (ω17, ω18, and ω19), because the fourth-double bond of this fatty acid is formed between carbons ω15 and ω16. Correspondingly, the melting temperature of arachidonic acid is as low as –49°C. The specific geometry of arachidonic acid induced by its four *cis* double bonds has a high biological significance. As we will see in Chapter 5, arachidonic acid is the precursor of two important endocannabinoid neurotransmitters, anandamide (arachidonoylethanolamide), and 2-AG (2-arachidonoylglycerol).[8,9]

Now that you are more familiar with the melting temperature of lipids, it is time to ask how this notion can help to understand the lipid organization of biological membranes. The frying pan will help us again. Put a slice of cold butter and a spoonful of olive oil in the frying pan. They do not mix, unless you raise the temperature to reach the melting point of butter. However, in biological membranes, the temperature is allowed to vary only in a narrow range above 37°C (for humans). Therefore, if at 37°C one lipid is in a liquid phase and another one in

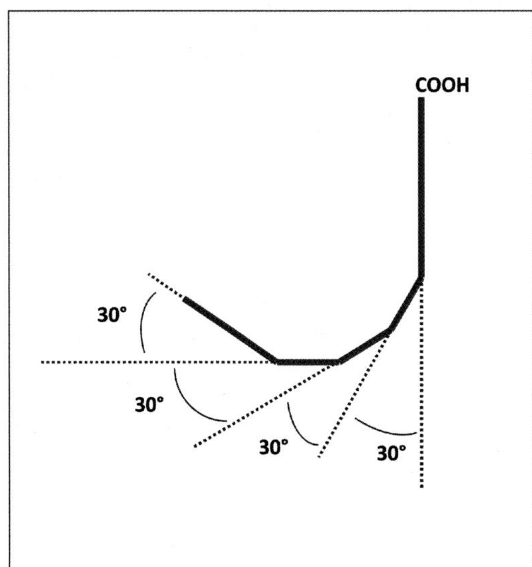

FIGURE 2.8 **Arachidonic acid and its scorpion-like tail.** Each of the four *cis* double bonds of arachidonic acid (C20:4ω6) induces a kink of 30°, leading to a global kink of 30 × 4 = 120° for the hydrocarbon chain of this fatty acid. Hence the chain has a typical scorpion-like shape.

a solid phase, they will not mix. The melting temperature of most sphingolipids is far above 37°C (e.g., 83°C for GalCer purified from bovine brain). In other words, at 37°C, GalCer molecules are densely packed in a paracrystalline, gel-like phase. In contrast, the mean melting temperature of natural PC is –5°C. Thus, at 37°C, PC is liquid. Because they are in two distinct phases, one solid and the other one liquid, sphingolipids and (PC) cannot mix and are thus confined in separate membrane areas.[6,10] Nevertheless, these physicochemical considerations cannot explain by themselves the ability of lipids to adopt different supramolecular structures (e.g., micelles and bilayers). As we will see, this unique lipid polymorphism is chiefly determined by the global molecular shape of individual lipids.

2.4.2 The Molecular Shape of Lipids

Although all membrane lipids have the same apolar/polar type of organization, the respective bulkiness of their polar head groups and apolar tails can greatly affect the overall geometry of the lipid. A thorough analysis of lipid molecular shapes has been published by Israelachvili, who has created and popularized the concept.[11] He assigned a specific packing parameter P calculated as $P = V/a.l$ (where V is the specific volume occupied by the apolar tail (hydrocarbon) domain, a is the area of the polar head group, and l is the effective length of the tail). Accordingly, a cylinder shape is characterized with a value of P close to 1 (Fig. 2.9). If $P < 1$, the lipid has an inverted cone shape due to the oversize of its polar head group compared with the apolar domain. Finally, if $P > 1$, the lipid adopts a cone shape, with a small head group and a large apolar domain.

These shapes will determine what kind of organization hydrophobically driven lipid assemblies will adopt in water. Accordingly, the value of P for each lipid is predictive of a

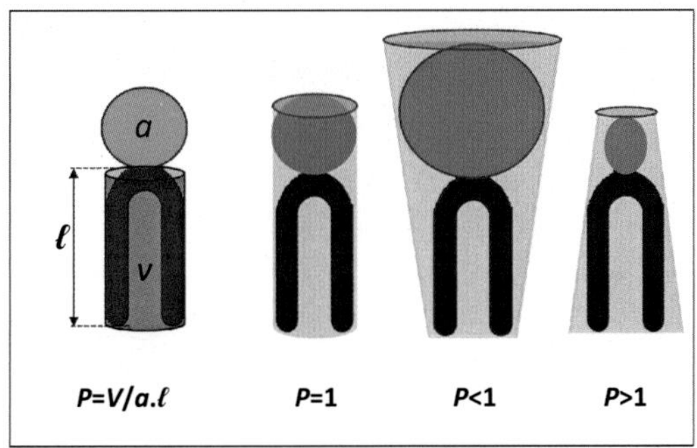

FIGURE 2.9 **The molecular shape of lipids.** Membrane lipids can be considered as simple geometric figures. The packing parameter (P) is calculated according to the formula $P = V/a.l$ (where V is the specific volume occupied by the apolar tail (hydrocarbon) domain, a is the area of the polar head group, and l is the effective length of the tail). Sphingolipids (inverted cones) have P values < 1, cholesterol (with a cone shape) has $P > 1$, and PC (a cylinder) has $P = 1$.

mode of organization. Cylindrical lipids (ideally with $P \sim 1$) such as natural PC (e.g., POPC) will spontaneously form lipid bilayers, whereas inverted cones (e.g., sphingolipids with $0 < P < 1/3$) will form micelles (Fig. 2.10).

In contrast, phosphatidylethanolamine (PE) and cholesterol ($P > 1$) are tapers. However, a more detailed analysis of the packing parameters of natural lipids indicates that the situation is a bit more complex. For instance, lipids with $1/3 < P < 1/2$ will also form micelles, but

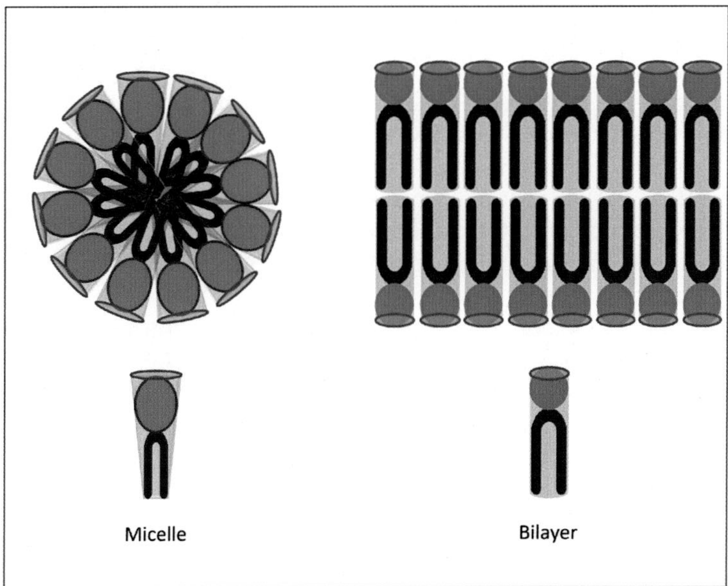

FIGURE 2.10 **Different types of lipid structuration in water: micelles and bilayers.** Sphingolipids (P values < 1) have an inverted cone shape, so that they will not form bilayers but micelles in water (left panel). Bilayer assemblies require cylindrical lipids ($P = 1$) such as POPC (palmytoyl-oleoyl-phosphatidylcholine).

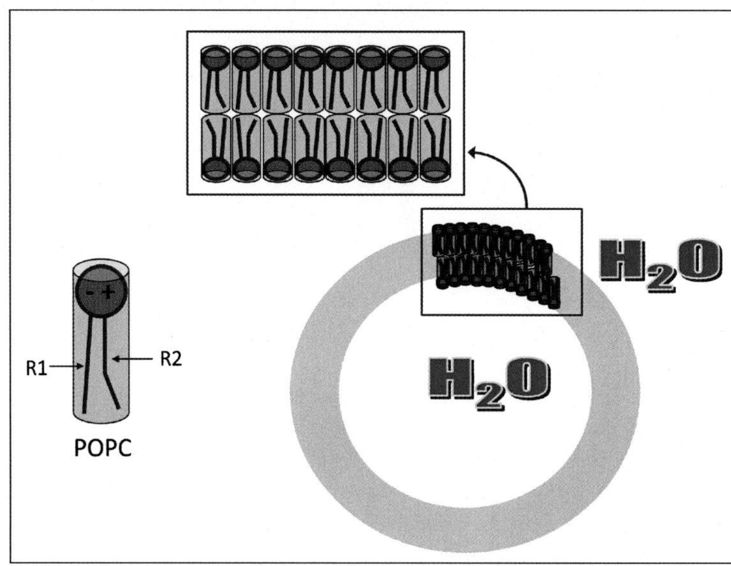

FIGURE 2.11 **A bilayer of phosphatidylcholine.** Palmitoyl-oleoyl-phosphatidylcholine (POPC) molecules spontaneously form bilayer structures in water. These bilayers are closed on themselves so that water molecules are trapped inside the vesicle. Note that the association of a saturated acyl chain in R1 and an unsaturated one in R2 gives a cylinder form to the lipid.

with rodlike shape.[12] Moreover, the formation of bilayer structures tolerates P values ranging from 1/2 to 1. It has also been observed that hydration of polar head groups and their interactions with ions can affect the packing parameters of membrane lipids. In any case, a common parameter of these lipid assemblies is that they hate edges, which would force lipid–water confrontations. For these reasons, lipid structures are closed on themselves. A micelle can be schematized as a sphere whose hydrophobic center totally excludes water molecules (Fig. 2.10). Bilayers close on themselves to prevent the exposure of hydrophobic chains to water molecules (Fig. 2.11). In marked contrast with micelles, which totally exclude water, the vesicles formed by lipid bilayers (also called liposomes) do contain water. From the point of view of lipid organization, a resemblance between the plasma membrane of living cells and liposomes is obvious.

From this discussion, one could logically deduce that the lipids that constitute biological membranes are all cylindrical. Indeed, bilayer structures are generally represented by a regular arrangement of a unique species of lipids (presumably PC) with a typical cylindrical shape (Fig. 2.11). This is a misleading oversimplification.

Although helpful, this schematic representation raises two issues. First, the plasma membrane also contains sphingolipids (inverted cones with $0 < P < 1/3$) and cholesterol (cones with $P = 1.21$), and it is still assumed to adopt a bilayer structure. Second, the lipid composition of each leaflet of the plasma membrane is specific, and this transmembrane asymmetry induces a curvature that would not occur if all lipids were cylindrical. Solving these problems will help us to figure out how a plasma membrane is really organized at the lipid level.

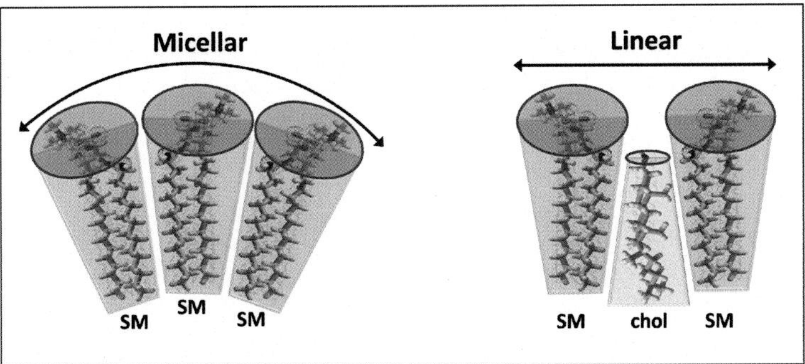

FIGURE 2.12 **Cholesterol forces sphingomyelin to form a bilayer structure.** Sphingomyelin (SM) has not a natural propensity to form bilayer structures except in presence of cholesterol (right panel). In absence of cholesterol, sphingomyelin will form micellar assemblies (left panel).

From a topological point of view, it is obvious that sphingolipids and cholesterol have complementary shapes (respectively, a wedge and a cone), consistent with segregation of both lipids in specific domains separated from PC. In the case of lipid mixtures, the packing parameters is additive. An example of such an additivity has been given by Kumar[12] who studied the bilayer structure formed by an equimolar mixture of lysophosphatidylcholine ($P = 0.397$) and cholesterol ($P = 1.21$). Hence the packing parameter of the mixture could be calculated as $[0.5 \times 0.397] + [0.5 \times 1.21] = 0.804$, a value that falls in the range of the packing parameters for bilayers ($0.5 < P < 1$). From a molecular point of view, the lipid shape theory predicts that cholesterol has more affinity for sphingolipids than for PC. If it has the choice between both of these lipids, cholesterol will thus select sphingolipids over PC. Accordingly, the plasma membrane can schematically be divided in two distinct types of lipid organization, which chiefly differ by their cholesterol content. The areas that contain both sphingolipids and cholesterol form condensed lipid phases that have been referred to as plasma membrane microdomains,[13] lipid rafts,[5] caveolae,[14] or more recently sphingolipid domains.[15] In these domains, cholesterol forces sphingolipids to participate in a counterintuitive bilayer structure (Fig. 2.12).

Cholesterol has been metaphorically compared to a mortar or cement that maintains vertical conic lipids that would otherwise induce an important curvature of the membrane, leading to micelle formation. The presence of cholesterol also prevents the sphingolipids from forming a gel-like structure stabilized by highly efficient van der Waals interactions between their apolar chains. The gel-like structure of pure sphingolipids is referred to as Lβ, but in presence of cholesterol, this Lβ phase is transformed into a slightly more fluid phase called the liquid-ordered phase (Lo). By contrast, the bulk PC regions form, in the presence of small amounts of cholesterol, a liquid crystalline-disordered phase (Ld). Therefore, the Ld phase contains less cholesterol than the Lo phase. A schematic representation of these cholesterol-rich/cholesterol-poor domains of the plasma membrane is shown in Fig. 2.13. This scheme takes into account the respective shapes of membrane lipids, according to the packing parameter defined by Israelachvili.[11] In the exofacial leaflet, lipid raft domains contain a high level of cholesterol together with sphingolipids (sphingomyelin + glycosphingolipids).[16,17] It is important to note that sphingolipids are not found in the cytoplasmic leaflet.[5] Nevertheless,

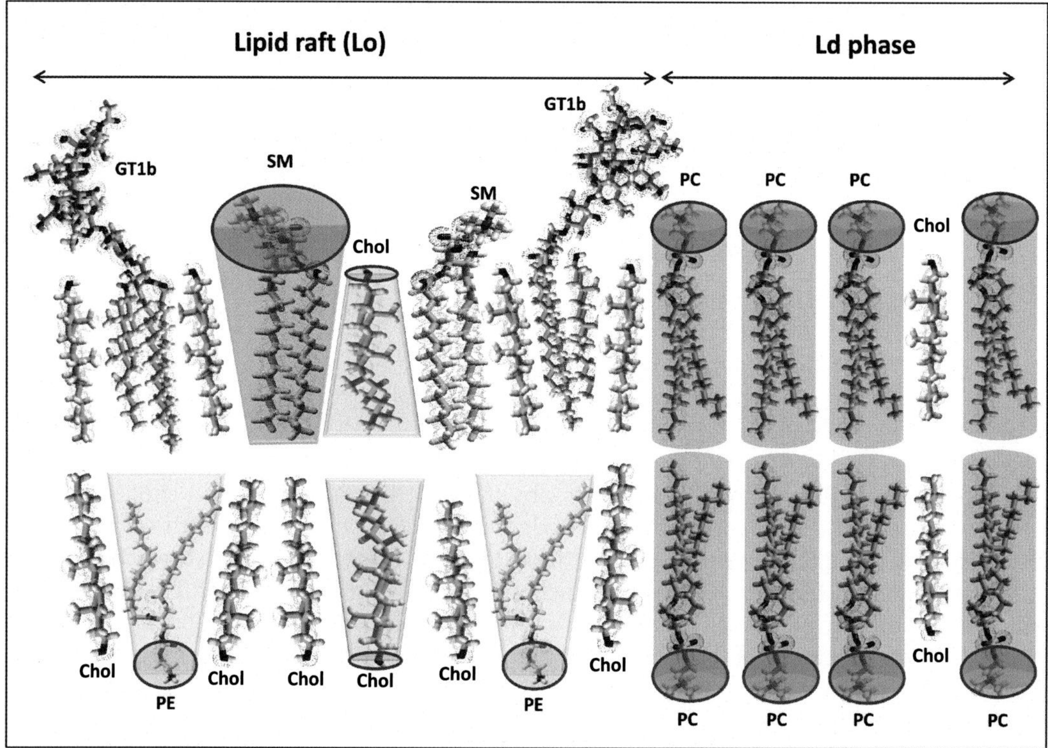

FIGURE 2.13 **Lipid organization of the plasma membrane: a simplified model based on the theoretical shape of membrane lipids.** The upper leaflet corresponds to the exofacial leaflet of the plasma membrane. Lipid raft domains (Lo phase) are enriched in cholesterol (chol) in both leaflets (exofacial and cytoplasmic). Sphingolipids (sphingomyelin SM and glycosphingolipids such as ganglioside GT1b) are located in the exofacial leaflet, whereas phosphatidylethanolamine (PE) is preferentially expressed in the cytoplasmic leaflet. In both lipids, cholesterol ensures the cohesion of the microdomain by filling the voids between highly dissymmetric lipids (SM, GT1b, PE). Note the different modes of interaction of cholesterol molecules: face-to-face in the same leaflet, and transbilayer tail-to-tail topology. The Ld phase is enriched in phosphatidylcholine (PC), but contains less cholesterol than lipid rafts. Other types of lipids found in both Lo and Ld phases have not been represented to improve the clarity of this cartoon. All the lipids are shown in dotted tube rendering (HyperChem software).

this leaflet is enriched in PE, which, like cholesterol, has a taper shape. Cholesterol probably fills the gaps between PE molecules in the cytoplasmic leaflet of lipid rafts, as suggested in Fig. 2.13. Transbilayer tail-to-tail dimers of cholesterol have been evidenced in model membranes.[18–20] Thus, it is likely that such dimers could also exist in natural membranes, thereby acting to stabilize the lipid assemblies found in lipid rafts. At the edge of lipid rafts, the presence of cholesterol could smooth the transition between the Lo and Ld phases,[21] avoiding direct physical contacts between sphingolipids and PC. This cholesterol probably plays a key role in the coalescence of two distinct rafts, a dynamic process that is mandatory for the recruitment of receptor and transducer proteins involved in signal transduction.[21,22] Finally, the schematic rendition of plasma membrane microdomains shown in Fig. 2.13 gives a good idea of the wide lipid heterogeneity of lipid molecules in the Lo phase compared with the more homogeneous content of the Ld phase.

From the preceding discussion, one could conclude that cholesterol is a key organizer of plasma membrane lipids. At this point, we will combine the notions of melting temperature and lipid shape to understand how cholesterol can indeed organize the plasma membrane into distinct lipid domains. Let us consider a theoretical bilayer of pure PC below its melting temperature (T_m) (Fig. 2.14). In this case, the molecules form a densely packed, crystalline gel-like phase (Lβ) structure. At 37°C (i.e., above the T_m of PC), the bilayer forms a fluid liquid-crystalline phase (referred to as Lc or Lα). When cholesterol is added, it abolishes the thermal transition between Lβ and Lc phases, giving membrane properties intermediate between the two phases. The fluid mosaic model of biological membranes[4] is based on this physicochemical feature. In other words, one can consider that cholesterol, in a way, corrects the slight differences in the T_m of the different PC molecules and related glycerophospholipids. In presence of few cholesterol amounts, the Lc phase is transformed into a slightly more rigid phase. In this case, the cholesterol mortar has a rigidifying effect.

So, now what happens to sphingolipids? In marked contrast with PC, sphingolipids have a T_m >37°C. Thus, at 37°C, these lipids form a tightly packed gel phase (Lβ). By interacting preferentially with sphingolipids, cholesterol favors the phase separation between sphingolipids and glycerophospholipids such as PC.[14] Consequently, sphingolipids adopt an intermediate phase referred to as the liquid-ordered (Lo) phase. In the Lo phase, the hydrocarbon chains are tightly packed as in the gel phase, but have a higher degree of mobility owing to the intercalation of cholesterol molecules between sphingolipids. In this case, the cholesterol mortar has a fluidifying effect. It is believed that cholesterol-enriched microdomains (lipid rafts, caveolae, sphingolipid domains) are in the Lo phase.[6] As for the effect of cholesterol on a biological membrane, it can induce either rigidity or fluidity, depending on the specific lipids with which it physically interacts. For a membrane taken as a whole, cholesterol allows the coexistence of distinct phases in an overall fluid regime. In this respect, and for the reasons already explained, cholesterol is the only lipid that can be found at the frontier between Lo and Ld phases (i.e., at the edge of cholesterol-enriched plasma membrane domains).[21]

The second issue that we have to take into account in this discussion of biologically relevant lipid assemblies is membrane asymmetry. In human red blood cells, the outer (extracellular) leaflet contains PC and sphingolipids (sphingomyelin plus glycosphingolipids), whereas the inner (cytoplasmic) leaflet contains phosphatidylethanolamine (PE), phosphatidylserine, and phosphatidylinositol, but no sphingolipids. Cholesterol is the only lipid that, at least locally, may be present in equal amounts in both leaflets. This phenomenon is consistent with model bilayers studies that have shown cholesterol to have a propensity to form tail-to-tail dimers, thereby generating a potential functional link for transmembrane events.[21] Let us consider a bilayer formed by a PC monolayer (mimicking the extracellular leaflet) and a PE monolayer (mimicking the intracellular leaflet). The area of the polar head group of PE is slightly smaller than the one of PC. Thus, PC is more cylindrical than PE, which looks more tapered (Fig. 2.15). Because PE is concentrated in the inner (cytoplasmic) leaflet of the plasma membrane, these small differences in lipid shape affect the topology of the lipid bilayer by inducing a curvature of the structure. A specific enrichment of cholesterol in the cytoplasmic leaflet would also lead to similar curvature effects.

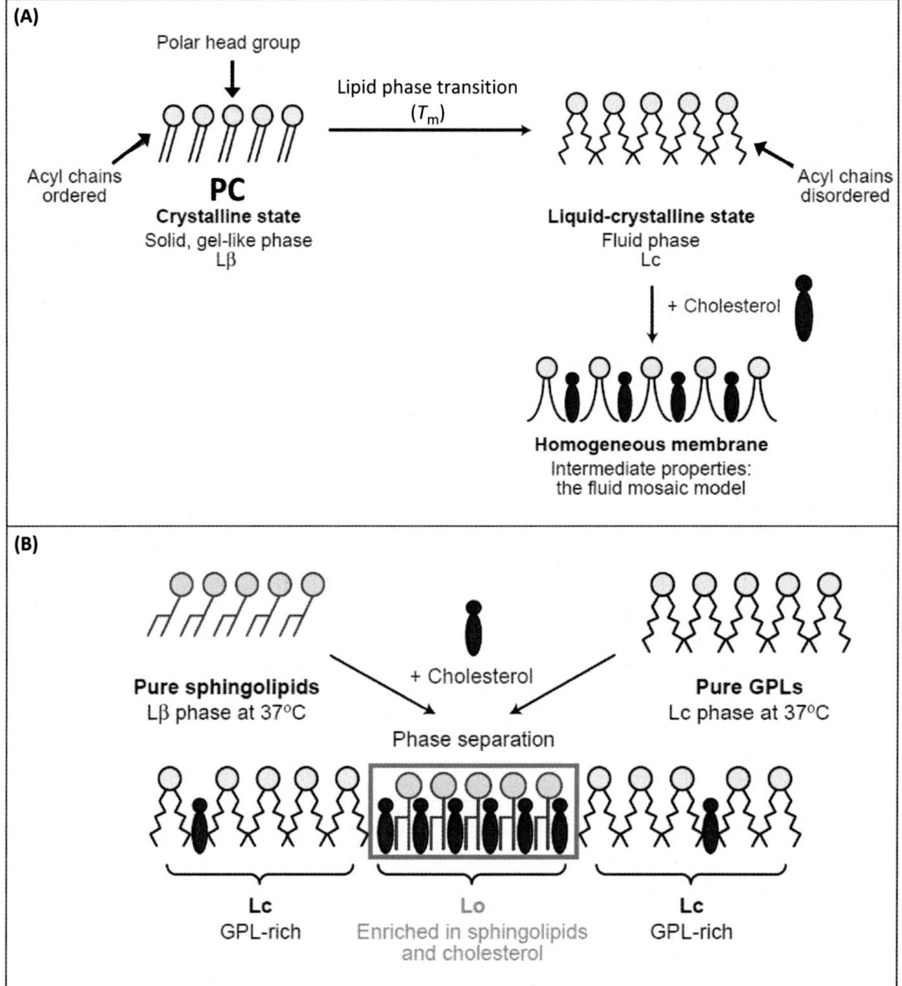

FIGURE 2.14 **Cholesterol favors phase separation of membrane lipids: the origin of microdomain formation.** (A) Below its melting temperature (TM), phosphatidylcholine (PC) forms a solid gel-like phase (Lβ) characterized by a high ordering of acyl chains. At 37°C (thus above the TM), the bilayer converts to a liquid-crystalline, fluid phase (referred to as Lc or Lα). The addition of cholesterol to these PC bilayers abolishes the normal thermal transition between Lβ and Lα phases, resulting in membrane properties intermediate to the two phases. This effect of cholesterol initially suggested that the slight differences in the TM of various glycerophospholipids (GPLs) were "corrected" by cholesterol, resulting in a homogeneous lipid phase at the physiological temperature. (B) Sphingolipids form a gel phase (Lβ) at 37°C, with tight packing of the saturated chains. In mixed sphingolid/PC systems, cholesterol interacts preferentially with sphingolipids, favoring the phase separation between sphingolipids and glycerophospholipids. By extrapolation, it is believed that in the plasma membrane, PC and most other GPLs form a relatively cholesterol-poor Lc phase, whereas sphingolipids form a Lo phase highly enriched in cholesterol. Lipid rafts probably exist in a Lo phase or a state with similar properties. *(Adapted from Fantini et al.,[6] with permission.)*

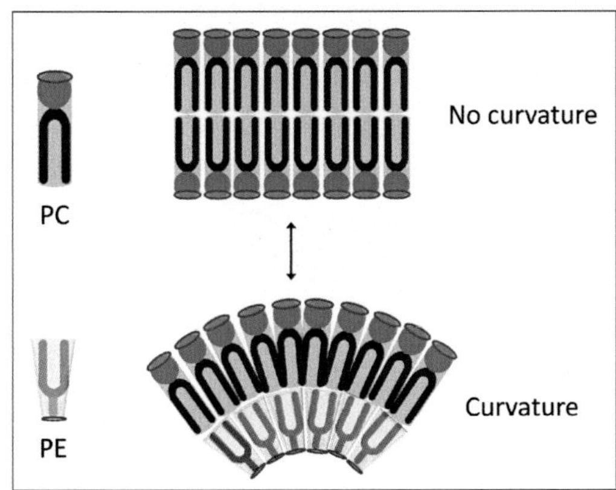

FIGURE 2.15 **Curvature emergence from transmembrane lipid asymmetry.** Membrane domains with phosphatidylcholine (PC) are not curved (upper panel). The emergence of curvature is due to the enrichment of the cytoplasmic leaflet of the plasma membrane with phosphatidylethanolamine (PE).

2.5 NONBILAYER PHASES AND LIPID DYNAMICS

Until now, we have focused our discussion on bilayer structures. Depending in particular on their degree of hydration, some membrane lipids can also adopt hexagonal structures.[23] This tendency is especially true for phosphatidylethanolamine (PE), which can therefore participate in two types of lipid assemblies in biological membranes (i.e., the classical bilayer structure and the hexagonal arrangement referred to as H_{II}).[23,24] The ability of such nonbilayer lipids to adopt different structures has been referred to as *lipid polymorphism*.[23,25,26] This structure is illustrated in Fig. 2.16.

In the H_{II} phase, which can be interpreted as an inverse hexagonal phase, the lipid monolayers form long cylinders with the apolar tails directed toward the periphery of the structure. Thus, several cylinders can make a periodic arrangement of H_{II} phases in which the apolar tails of one cylinder interact with those of its immediate neighbor (Fig. 2.16).

It has been suggested that the hexagonal H_{II} phase plays an important role in membrane fusion.[26–31] More generally, destabilization of lamellar phase and formation of nonbilayer connections can participate in membrane dynamics and renewal.[25] Such transitions from a lamellar to an inverted hexagonal H_{II} phase play a critical role in the exocytosis and recycling of synaptic vesicles.[32] In summary, depending on the global shape of individual lipids, on the hydration status of their head groups, and on subtle changes of the ionic environment, lipid polymorphism may transiently generate phases with a negative curvature (H_{II}) that coexist with lamellar phases (no curvature) and can exhibit, here and there, small areas with a positive curvature.

2.6 THE PLASMA MEMBRANE OF GLIAL CELLS AND NEURONS: THE LIPID PERSPECTIVE

The brain is primarily a lipid machine. Indeed, lipids represent up to 50% of its dry weight,[33] and as much as 80% of the dry weight of myelin. The lipid composition of the whole brain (as determined in adult rats and expressed in μmol g^{-1} of wet weight) is 89 for glycerophospholipids,

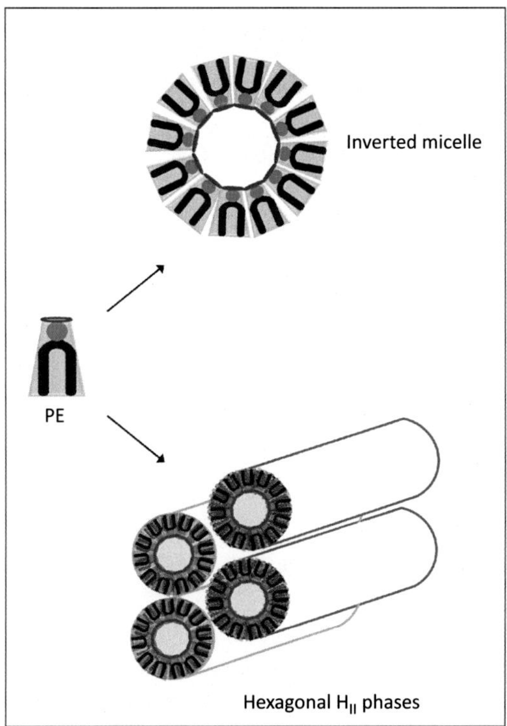

FIGURE 2.16 **Lipid polymorphism of phosphatidylethanolamine: inverted micelles and hexagonal H_{II} phases.** Taper lipids such as phosphatidylethanolamine (PE) have a higher propensity to form micellar rather than bilayer structures. In the plasma membrane, PE can induce curvature effects and hexagonal (H_{II}) phase formation.

69 for cholesterol, and 31 for sphingolipids.[34–36] Beyond these quantitative features, some lipids are overrepresented in the brain, such as GalCer and sulfatide in myelin, or ganglioside GM1 for neurons. Adult mammalian brain gangliosides are dominated by just four species: GM1, GD1a, GD1b, and GT1b, which represent by themselves 97% of gangliosides.[34] All these lipids are not homogeneously distributed in the brain. In fact, glial cells and neurons have specific glycosphingolipid compositions.[37] Thin layer chromatography analysis of the sphingolipid expression in various brain cell types shows how the membranes of these cells are different (Fig. 2.17). In addition to these specific cellular expression patterns, lipid content may also greatly vary in brain areas, as a function of both age and gender, and in neural diseases (see Chapter 4).

One can see that oligodendrocytes, the cells that synthesize myelin, express chiefly GalCer, sulfatide, and GM3. Astrocytes, the cells that are functionally coupled to neurons through metabolic and neurotransmitter transporter networks, express high levels of ganglioside GM3. Finally, neurons have the most complete set of gangliosides, including the four major brain species: GM1, GD1a, GD1b, and GT1b. The other glycosphingolipids detected in these cells (GlCer, LacCer, GM3, and GD3) are metabolic intermediates with low levels of expression in the plasma membrane. Overall, this differential expression of glycosphingolipids

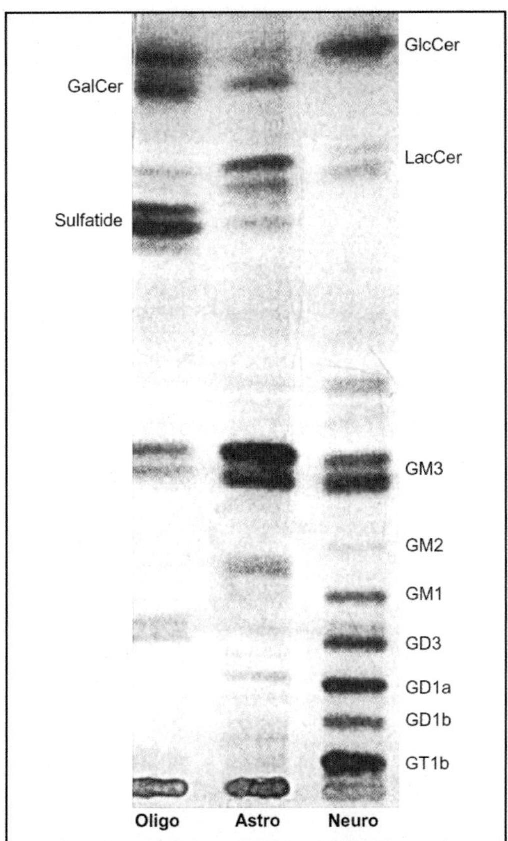

FIGURE 2.17 **Expression of glycosphingolipids in glial cells and neurons.** Thin layer chromatography of glycosphingolipids extracted from oligodendrocytes (oligo), astrocytes (astro), and neurons (neuro) from rat brains. Primary cell cultures were metabolically labeled with ^{14}C galactose prior to extraction and separation. Note that GlcCer and LacCer are metabolic intermediates in the biosynthetic pathways of glycosphingolipid biosynthesis. *(Reprinted from van Echten-Deckert et al.*[37] *Copyright © 2012, with permission from Elsevier.)*

prompted us to study the possible arrangements of sphingolipid microdomains in the plasma membranes of these three brain cell types (Fig. 2.18). To this end, we have merged minimized structures of each relevant glycosphingolipids (obtained with the HyperChem software using the CHARMM force field) and conducted a series of molecular dynamics simulations. In addition to these specific glycosphingolipids, we have incorporated cholesterol in the molecular models of astrocyte, oligodendrocyte, and neuron membranes. The results of this modeling exercise are presented in Fig. 2.18. One can easily visualize that each brain cell type has its own fingerprint of lipid raft structure based on a specific expression of a subset of selected glycosphingolipids. In any case, the high content of brain membranes in glycosphingolipids has a huge influence on the function of neurotransmitter receptors and transporters, as explained in Chapter 5.

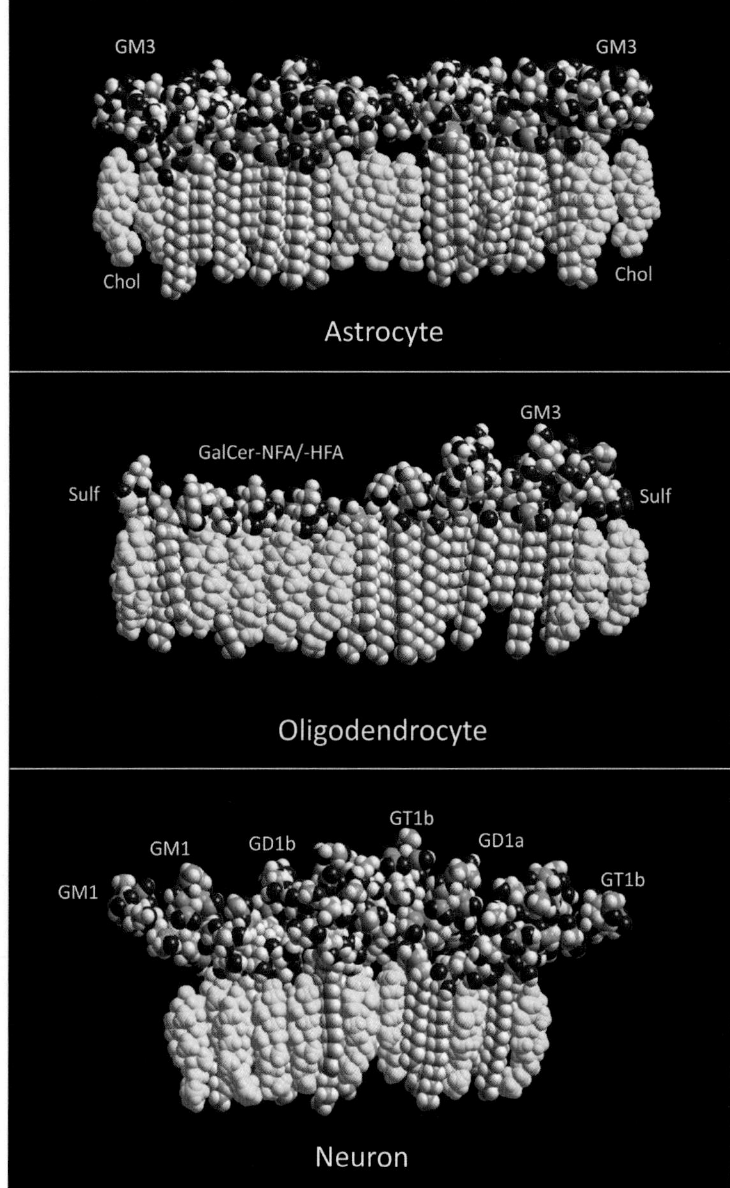

FIGURE 2.18 **Lipid microdomains in astrocytes, oligodendrocytes and neurons: a molecular modeling study.** The model represents the typical glycosphingolipid/cholesterol content of the outer (exofacial) leaflet of a lipid raft present on the plasma membrane of astrocytes, oligodendrocytes, and neurons (obtained by molecular dynamics simulations with the HyperChem software, using the CHARMM force field). Cholesterol is in yellow.

2.7 KEY EXPERIMENTS ON LIPID DENSITY

Two important notions discussed in this chapter merit further emphasis. First, we would like to underscore once again how the fine control of the transition temperature of lipids is crucial for biological systems. The sperm whale has a large head that contains, besides the brain, a wide cavity called the spermaceti organ. This cavity is filled with 3–4 tons of oil (spermaceti oil) consisting of a mixture of triglycerides and waxes. Due to this unique lipid composition, spermaceti oil is liquid at the normal resting body temperature of the whale (i.e., 37°C) as a marine mammal. Lowering the temperature by only few degrees induces the liquid to solid transition of the oil. During a dive, the whale cools its spermaceti oil by inhaling seawater, so that it becomes solid and denser, thus changing the buoyancy of the whale to match the density of seawater. Conversely, during the return to the surface, the whale uses its blood circulation to warm up the frozen spermaceti oil which melts, thereby decreasing its density to match that of the surface water.[38] A fascinating study describing the physical properties of spermaceti oil as a basis for understanding the buoyancy function of the spermaceti organ has been published by M. Clarke.[39] This fine-tuning of oil density around the melting temperature is achieved through a precise adjustment of the lipid content of spermaceti oil.

Interestingly, Marie Curie (double Nobel Prize winner) performed a symmetric experiment during a series of physics lessons taught to children, including her daughter Irene (also a Nobel Prize recipient). In fact, Marie Curie organized a teaching cooperative in which children of famous professors (Jacques Hadamard, Paul Langevin, and Jean Perrin) came to her laboratory for their lessons. During one of these sessions (July 2, 1907), Marie Curie prepared two glasses filled with liquid mixtures. The first contained water and oil: because it is less dense than water, the oil floated on water. The second one contained oil and alcohol: this time oil fell to the bottom of the glass, because it is denser than alcohol. Then, using a poetic metaphor, Marie Curie said: "Since oil swims on the water and drowned in alcohol, one can mix water and alcohol so that the oil does not float nor sinks. You will see that oil will take the form of a ball, and this will be very nice." So did the children, getting beautiful yellow balls suspended in the middle of the liquid. Isabelle Chavannes, one of the children, wrote in her notebook: "Tous les enfants sont ravis" (All the children are delighted). Thanks to her, these wonderful physics lessons taught by Marie Curie have come down to us.[40] Why do we say that this experiment is symmetric to the mechanism used by the whale to control its own density? Obviously the whale cannot change the density of seawater and so has no other issue than changing the density of its oil mass to dive or float. In Marie Curie's experiment, the density of oil is constant; therefore it is the density of the surrounding liquid that changes, according to the water to alcohol ratio. However, in both cases the oil mass (either the whale in the sea or the oil ball in the glass) moves up or down depending on the balance between its own density and the density of the liquid.

References

1. Israelachvili J, Wennerstrom H. Role of hydration and water structure in biological and colloidal interactions. *Nature*. 1996;379(6562):219–225.
2. Langmuir I. The constitution and fundamental properties of solids and liquids. II. Liquids. *J Am Chem Soc*. 1917;39:1848–1906.

3. Israelachvili J. *Intermolecular and Surface Forces*. 3rd ed. USA: Academic Press; 2011.
4. Singer SJ, Nicolson GL. The fluid mosaic model of the structure of cell membranes. *Science*. 1972;175(4023): 720–731.
5. Simons K, Ikonen E. Functional rafts in cell membranes. *Nature*. 1997;387(6633):569–572.
6. Fantini J, Garmy N, Mahfoud R, Yahi N. Lipid rafts: structure, function and role in HIV, Alzheimer's and prion diseases. *Expert Rev Mol Med*. 2002;4(27):1–22.
7. Goni FM. The basic structure and dynamics of cell membranes: an update of the Singer-Nicolson model. *Biochim Biophys Acta*. 2014;1838(6):1467–1476.
8. Di Pasquale E, Chahinian H, Sanchez P, Fantini J. The insertion and transport of anandamide in synthetic lipid membranes are both cholesterol-dependent. *PloS One*. 2009;4(3):e4989.
9. Di Iorio G, Lupi M, Sarchione F, et al. The endocannabinoid system: a putative role in neurodegenerative diseases. *Int J High Risk Behav Addict*. 2013;2(3):100–106.
10. Rietveld A, Simons K. The differential miscibility of lipids as the basis for the formation of functional membrane rafts. *Biochim Biophys Acta*. 1998;1376(3):467–479.
11. Israelachvili JN, Mitchell DJ, Ninham BW. Theory of self-assembly of lipid bilayers and vesicles. *Biochim Biophys Acta*. 1977;470(2):185–201.
12. Kumar VV. Complementary molecular shapes and additivity of the packing parameter of lipids. *Proc Natl Acad Sci USA*. 1991;88(2):444–448.
13. Edidin M. Lipids on the frontier: a century of cell-membrane bilayers. *Nat Rev Mol Cell Biol*. 2003;4(5):414–418.
14. Brown DA, London E. Functions of lipid rafts in biological membranes. *Ann Rev Cell Dev Biol*. 1998;14:111–136.
15. Frisz JF, Lou FK, Klitzing HA, et al. Direct chemical evidence for sphingolipid domains in the plasma membranes of fibroblasts. *Proc Natl Acad Sci USA*. 2013;110(8):E613–E622.
16. Anderson RG, Jacobson K. A role for lipid shells in targeting proteins to caveolae, rafts, and other lipid domains. *Science*. 2002;296(5574):1821–1825.
17. Jacobson K, Mouritsen OG, Anderson RG. Lipid rafts: At a crossroad between cell biology and physics. *Nat Cell Biol*. 2007;9(1):7–14.
18. Harris JS, Epps DE, Davio SR, Kezdy FJ. Evidence for transbilayer, tail-to-tail cholesterol dimers in dipalmitoylglycerophosphocholine liposomes. *Biochemistry*. 1995;34(11):3851–3857.
19. Mukherjee S, Chattopadhyay A. Membrane organization at low cholesterol concentrations: a study using 7-nitrobenz-2-oxa-1,3-diazol-4-yl-labeled cholesterol. *Biochemistry*. 1996;35(4):1311–1322.
20. Rukmini R, Rawat SS, Biswas SC, Chattopadhyay A. Cholesterol organization in membranes at low concentrations: Effects of curvature stress and membrane thickness. *Biophys J*. 2001;81(4):2122–2134.
21. Fantini J, Barrantes FJ. How cholesterol interacts with membrane proteins: an exploration of cholesterol-binding sites including CRAC, CARC, and tilted domains. *Front Physiol*. 2013;4:31.
22. Fantini J, Barrantes FJ. Sphingolipid/cholesterol regulation of neurotransmitter receptor conformation and function. *Biochim Biophys Acta*. 2009;1788(11):2345–2361.
23. Gruner SM, Cullis PR, Hope MJ, Tilcock CP. Lipid polymorphism: The molecular basis of nonbilayer phases. *Ann Rev Biophys Biophys Chem*. 1985;14:211–238.
24. Gruner SM. Intrinsic curvature hypothesis for biomembrane lipid composition: a role for nonbilayer lipids. *Proc Natl Acad Sci USA*. 1985;82(11):3665–3669.
25. Frolov VA, Shnyrova AV, Zimmerberg J. Lipid polymorphisms and membrane shape. *Cold Spring Harb Perspect Biol*. 2011;3(11):a004747.
26. Cullis PR, Hope MJ, Tilcock CP. Lipid polymorphism and the roles of lipids in membranes. *Chem Phys Lipids*. 1986;40(2–4):127–144.
27. Alford D, Ellens H, Bentz J. Fusion of influenza virus with sialic acid-bearing target membranes. *Biochemistry*. 1994;33(8):1977–1987.
28. Allen TM, Hong K, Papahadjopoulos D. Membrane contact, fusion, and hexagonal (HII) transitions in phosphatidylethanolamine liposomes. *Biochemistry*. 1990;29(12):2976–2985.
29. Ellens H, Siegel DP, Alford D, et al. Membrane fusion and inverted phases. *Biochemistry*. 1989;28(9):3692–3703.
30. Kinnunen PK. Fusion of lipid bilayers: a model involving mechanistic connection to HII phase forming lipids. *Chem Phys Lipids*. 1992;63(3):251–258.
31. Siegel DP. Inverted micellar intermediates and the transitions between lamellar, cubic, and inverted hexagonal lipid phases. II. Implications for membrane-membrane interactions and membrane fusion. *Biophys J*. 1986;49(6):1171–1183.

32. Glaser PE, Gross RW. Plasmenylethanolamine facilitates rapid membrane fusion: a stopped-flow kinetic investigation correlating the propensity of a major plasma membrane constituent to adopt an HII phase with its ability to promote membrane fusion. *Biochemistry*. 1994;33(19):5805–5812.
33. Woods AS, Jackson SN. Brain tissue lipidomics: direct probing using matrix-assisted laser desorption/ionization mass spectrometry. *AAPS J*. 2006;8(2):E391–E395.
34. Schnaar RL, Gerardy-Schahn R, Hildebrandt H. Sialic acids in the brain: gangliosides and polysialic acid in nervous system development, stability, disease, and regeneration. *Physiol Rev*. 2014;94(2):461–518.
35. Norton WT, Poduslo SE. Myelination in rat brain: changes in myelin composition during brain maturation. *J Neurochem*. 1973;21(4):759–773.
36. Yu RK, Macala LJ, Taki T, Weinfield HM, Yu FS. Developmental changes in ganglioside composition and synthesis in embryonic rat brain. *J Neurochem*. 1988;50(6):1825–1829.
37. van Echten-Deckert G, Walter J. Sphingolipids: critical players in Alzheimer's disease. *Progr Lipid Res*. 2012;51(4):378–393.
38. Clarke MR. Function of the spermaceti organ of the sperm whale. *Nature*. 1970;228(5274):873–874.
39. Clarke MR. Physical properties of spermaceti oil in the sperm whale. *J Mar Biol Assn UK*. 1978;58:19–25.
40. Curie M, Chavannes I. Leçons de Marie Curie recueillies par Isabelle Chavannes en 1907. Les Ulis France: EDP Sciences; 2003.

CHAPTER

3

Lipid Metabolism and Oxidation in Neurons and Glial Cells

Jacques Fantini, Nouara Yahi

OUTLINE

3.1 GENERAL ASPECTS OF LIPID METABOLISM

Etymologically, metabolism means transformation (from the Greek root *metabole*). This term refers to any chemical reaction occurring in a living organism. Because vital functions occur within organs, tissues, cell, and organelles, transport mechanisms of biomolecules from one location to another are also an important part of metabolism.

Chemical reactions in living organisms are generally coordinated in such a way that the product of a given reaction is the substrate of another one. These chained reactions form a coherent ensemble referred to as a metabolic pathway (Fig. 3.1).

Most of these pathways have been given a name (e.g., glycolysis, Krebs cycle). In a metabolic pathway, a biomolecule, the substrate, is gradually transformed into a product (in some cases several products) through a step-by-step process. Therefore, blocking one reaction of the chain results in the blocking of the whole pathway (e.g., statins inhibit the *de novo* synthesis of

Brain Lipids in Synaptic Function and Neurological Disease. http://dx.doi.org/10.1016/B978-0-12-800111-0.00003-5

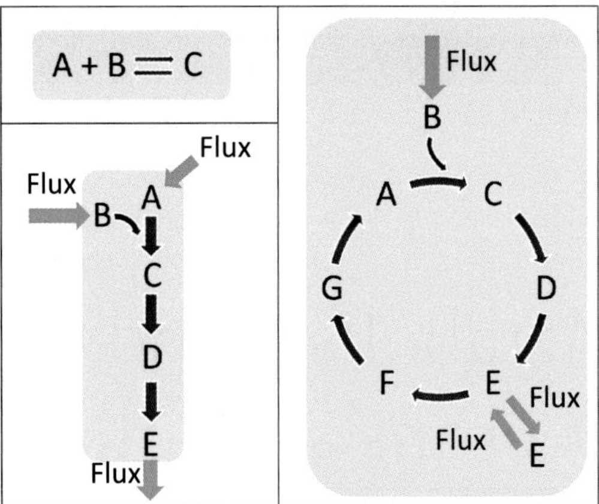

FIGURE 3.1 **Basic principles of metabolism.** The unit of metabolism is the chemical reaction, catalyzed by an enzyme (upper left). Each reaction can be studied individually *in vitro* to determine its mechanism and kinetics parameters. In a linear metabolic pathway (e.g., glycolysis), the reactions are coupled in series (lower left). The rate of the first reaction of the chain (A + B → C) reaction is controlled by the fluxes of A and B. The flux through the whole metabolic pathway is determined by the rate of production of E. The same rules apply for a cyclic metabolic pathway such as Krebs cycle (right panel).

cholesterol by interfering with a single reaction). Each metabolic reaction is catalyzed by its own enzyme, so that characterizing a metabolic pathway implies that all the reactions of the chain are fully elucidated (nature of substrates, products, stoichiometry) and each enzyme identified. However, students should be aware that knowing all the reactions of a metabolic pathway is not sufficient, because this information is static. Dynamic aspects should also been described. The dynamics of metabolism are reflected by the rates of substrate entry into a given metabolic pathway, together with the rate of product exit. This entry–exit corresponds to the metabolic flux. This flux may be extremely variable according to the cell, tissue, or organ in which the transformation occurs. In some cases, several organs need to precisely coordinate their fluxes (e.g., the rate of glucose utilization by the brain should be adjusted to the rate of glucose production by the liver). In addition, metabolic fluxes are subject to periodic variations that reflect the level of cellular activity, just like the intensity of car traffic in a city reveals hectic peaks of human life.

Chemical reactions occurring in aqueous solutions lack landmarks in the three-dimensional space. When two substrates A and B need to interact with each other to form a condensation product C, the probability that A and B form a functional molecular complex in the bulk water is low (Fig. 3.2). Raising the temperature and/or the concentration of A and B in the solution may help, but enzymes are definitely more efficient than these physicochemical parameters (which is a good thing because living organisms cannot act on these parameters as drastically as chemists can in test tubes). To circumvent this difficulty, the enzyme displays specific binding sites for its substrates (Fig. 3.2). Once bound to the enzyme, the substrates are correctly oriented for the reaction which may be further assisted by "catalytic" chemical groups located in the active site of the enzyme.

Overall, the enzyme acts (1) as a concentrating platform for the substrates, which leave the 3D space of the bulk solvent for a solid ground; (2) as an ordering agent that constrains the

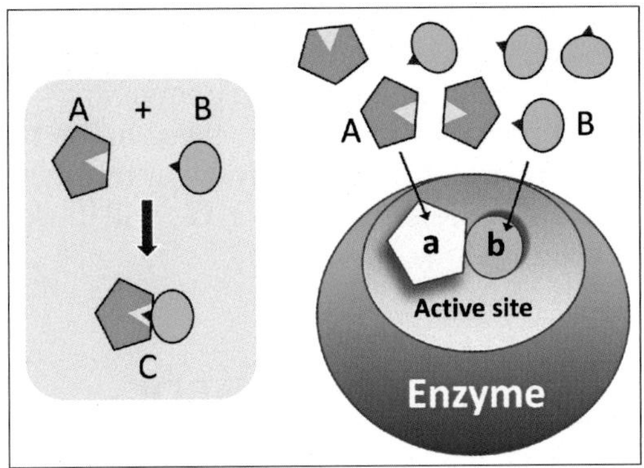

FIGURE 3.2 **Geometric aspects of enzyme function.** Let us consider a condensation reaction A + B → C. To proceed, the reaction requires (1) a physical contact between A and B, and (2) a well-defined orientation of both reactants (left panel). The probability that both events occur simultaneously under physiological conditions of concentration and temperature is low. The enzyme circumvents this difficulty by providing a binding site for each substrate (site a for A, site b for B), so that both A and B are concentrated on a solid phase and, most importantly, are in the correct orientation for the condensation reaction (right panel). The active site of the enzyme may also provide catalytic groups that further accelerate the reaction.

substrates to adopt mutual orientations compatible with the reaction; and (3) as a chemical catalyst that plays an active role in the reaction mechanism. Because lipids are not soluble in water, this general schema of enzymatic reactions cannot be simply extrapolated to lipid metabolism. In particular, substrate orientation is not generally an issue because the lipids are part of highly organized structures (either micelles or bilayers). In fact, most enzymes of the lipid metabolism are associated with biological membranes in which their substrates are trapped in an already metabolic competent topology (Fig. 3.3). In this case, it is not the

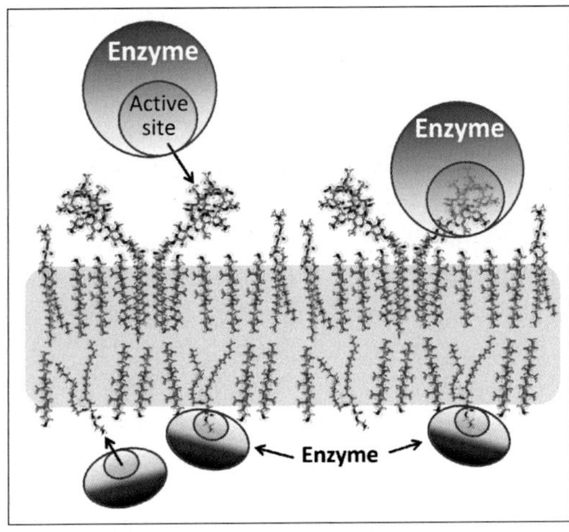

FIGURE 3.3 **Enzymatic action on membrane lipids.** The head groups of membrane lipids have limited mobility due to the dense packing of lipid molecules in the membrane phase. In this case, it is the enzyme that goes to the lipid. This phenomenon occurs on both sides of the membrane.

substrate that goes to the enzyme, but the enzyme that goes to the substrate. For instance, bacterial sphingomyelinases only have to land on the plasma membrane of a host cell to cut the protruding phosphocholine head group of sphingomyelin.[1]

Nevertheless, before being part of a membrane, these lipids have to be synthesized. As for other tissues, lipid biosynthesis in the brain may occur through two main strategies: total *de novo* biosynthesis or recycling of preformed bricks. We will illustrate these different aspects of lipid metabolism in the brain.

3.2 CHOLESTEROL

In the central nervous system cholesterol is chiefly found in myelin sheaths and in the plasma membranes of neurons and astrocytes.[2] Apart from allowing precise adjustments of membrane fluidity and permeability, cholesterol physically interacts with neurotransmitter receptors, controls their conformation, and regulates their function.[3,4] Cholesterol-regulated membrane proteins seem to have coevolved with membrane lipids so that an ideal proportion of approximately 30% of cholesterol is requested for optimal receptor function.[5] Either increasing or decreasing membrane cholesterol content by a few percent can drastically impair the function of membrane proteins, as elegantly shown by Yeagle for Na, K-ATPase.[6] It is probably to prevent such variations of cholesterol content in the plasma membranes of neural cells that brain cholesterol is insulated from circulating pools.[7] As a matter of fact, neither free cholesterol nor cholesterol associated with plasma lipoproteins can traverse the blood–brain barrier.[8–11] Although the details of cholesterol homeostasis in the brain are not fully understood, a schematic view is gradually emerging.[12] In the adult state, neurons synthesize only small amounts of cholesterol and rely heavily on glial cells, mostly astrocytes.[13] Thus, the high energetic costs required for cholesterol synthesis are borne by astrocytes, allowing neurons to focus on electrical activity.[14] Cholesterol is secreted by astrocytes together with apolipoprotein E (apoE) and subsequently delivered to neurons.[7] Finally, cholesterol in excess is converted into oxysterols (e.g., 24S-hydroxycholesterol) that can freely cross the blood–brain barrier.[15]

Because brain cells cannot rely on blood supply for obtaining cholesterol, they have to produce it by *de novo* biosynthesis. The precursor of cholesterol is acetyl-CoA, which is an "activated" form of acetate, a two-carbon building unit. Two acetyl-CoA molecules form acetoacetyl-CoA, which subsequently reacts with a third acetyl-CoA to generate a 6-carbon product called hydroxylmethylglutaryl-CoA (HMG-CoA) (Fig. 3.4).

HMG-CoA is then reduced into mevalonate via a reaction catalyzed by HMG-CoA reductase, a key enzyme of cholesterol biosynthesis. In fact, HMG-CoA reductase is the major rate-limiting enzyme of *de novo* cholesterol synthesis. Consistently, inhibitors of HMG-CoA reductase (statins) block the whole pathway of cholesterol synthesis by preventing the formation of mevalonate. The following steps in the cholesterol synthetic pathway include the ATP-dependent transformation of mevalonate into isopentenyl pyrophosphate (Fig. 3.5).

The condensation and rearrangement of two molecules of isopentenyl pyrophosphate leads to a polyterpenic C15 compound, farnesyl pyrophosphate. Then, two molecules of farnesyl pyrophosphate (2 × C15) form squalene, a C30 compound that is the common precursor of all steroids, including cholesterol. The cyclization of squalene leads to lanosterol

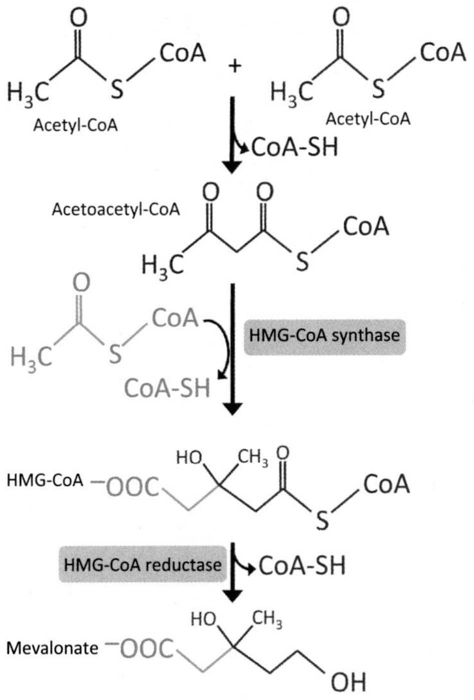

FIGURE 3.4 **Cholesterol biosynthesis from acetyl-CoA to mevalonate.**

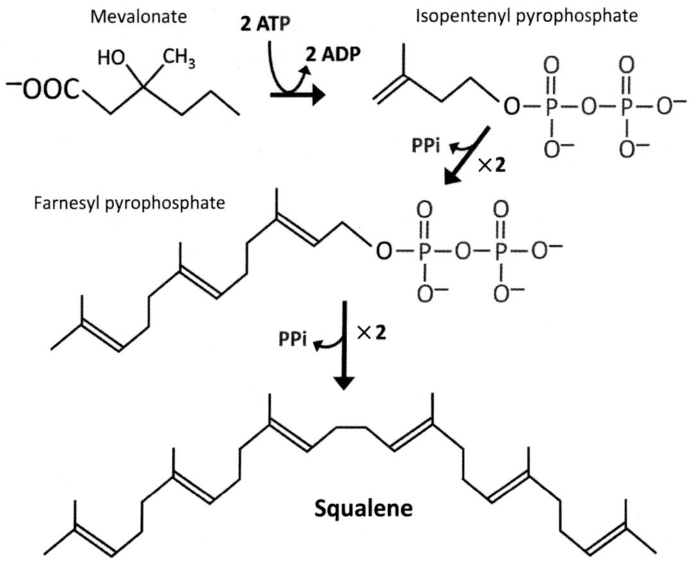

FIGURE 3.5 **Cholesterol biosynthesis from mevalonate to squalene.**

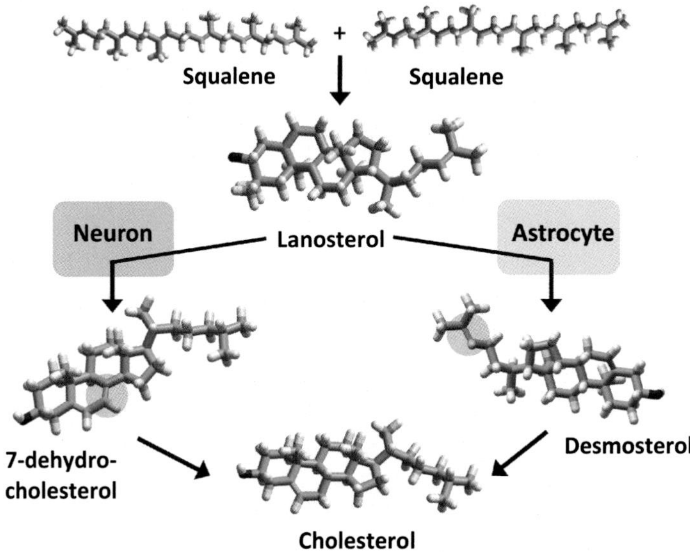

FIGURE 3.6 **Cholesterol biosynthesis in neurons and astrocytes.** Adult neurons synthesize only small amounts of cholesterol, using the Kandutsch–Russell pathway with 7-dehydrocholesterol as the intermediate between lanosterol and cholesterol. Adult astrocytes have a high level of cholesterol biosynthesis, which fulfills neuron needs throughout adult life. In astrocytes, cholesterol is synthesized via the Bloch pathway, with desmosterol as intermediate between lanosterol and cholesterol. Note that lanosterol is obtained via the condensation of two squalene molecules.

(Fig. 3.6). The transformation of lanosterol into cholesterol requires several demethylations and a final reduction step to generate the double bond in the B-cycle of cholesterol (Fig. 3.6).

Between lanosterol and cholesterol, astrocytes and neurons use specific biosynthetic pathways (Fig. 3.6). Let us first examine the case of neurons. Among all brain cell types, neurons have the lowest cholesterol biosynthesis activity.[16] As underscored by Martin et al.[12] cholesterol synthesis has a high energy cost, and neurons have opportunely delegated this function to astrocytes. This transfer gradually takes place during postnatal development of the central nervous system,[17] and it is fully functional in adults.[18] Indeed, postnatal neurons are still able to synthesize cholesterol, but only at a low rate.[13] Because lanosterol accumulation was evident in these cells, their poor cholesterol biosynthesis activity was tentatively attributed to low levels of lanosterol-converting enzymes. In neurons, lanosterol is transformed in 7-dehydrocholesterol, which in turn is reduced into cholesterol by 7-dehydrocholesterol Δ7 reductase, an enzyme that catalyzes the last step in cholesterol biosynthesis (Fig. 3.6). This particular sequence of enzymatic steps from lanosterol to cholesterol is referred to as the Kandutsch–Russell pathway.[19] Cell culture studies have shown that neuronal synthesis of cholesterol is apparently restricted to cell bodies.[20] This restriction has been demonstrated with sympathetic neurons from newborn rats in which the axons and the cell bodies, plated in distinct compartments, can be analyzed separately.[21] The lack of cholesterol synthesis in axons markedly contrasted with other membrane lipids (e.g., phosphatidylcholine and sphingomyelin) whose synthesis was detected in both axons and cell bodies.[20] However, another study suggested that cholesterol synthesized in neuronal cell bodies might be transported to axons.[22] In fact, both the endogenously synthesized cholesterol and cholesterol added exogenously can migrate from cell bodies to the axon. This migration is important because in absence of cholesterol, axonal growth is dramatically impaired.[22]

Neurons and astrocytes have distinct profiles of cholesterol biosynthetic enzymes. Astrocytes contain postlanosterol intermediates that do not belong to the Kandutsch–Russell pathway. In particular, cultured astrocytes display high levels of desmosterol[13,23] (Fig. 3.6). For this reason, it has been suggested that the high desmosterol content of the brain during postnatal development[11,24] was due to cholesterol synthesis by newborn glial cells.[25] The transformation of desmosterol into cholesterol is catalyzed by 24-dehydrocholesterol reductase. The desmosterol route to cholesterol, which is particularly active in astrocytes, is referred to as the Bloch pathway of cholesterol biosynthesis[26] (in reference of Konrad Bloch who deciphered this pathway and obtained the Nobel Prize in 1964). The transfer of cholesterol from astrocytes to neurons is mediated by brain lipoproteins.[27] A lipoprotein is a heterogenous particle containing both lipids and proteins, which allows the transport of lipids through aqueous environments. The lipid-binding proteins that form the core of lipoproteins are named *apoliproteins*. In the brain, apoliproteins are expressed at higher levels in astrocytes than in other cell types.[27] The most abundant apoliprotein produced by astrocytes is apoliprotein E (apoE).[28] The gene coding for the apoE protein (APOE) has three major allelic variants referred to as ε2, ε3, and ε4, whose frequencies in the human population have been estimated respectively to 8, 80, and 12%.[29,30] The two less expressed variants, ε2 and ε4, have been linked to the risk of developing Alzheimer's disease: apoE4 increases the risk of Alzheimer's disease,[31] whereas in contrast apoE2 has neuroprotective anti-Alzheimer effects.[32] The critical role of apoE in brain lipoprotein secretion is supported by *in vitro* studies performed with astrocytes from transgenic animals.[33] In this case, genetically engineered deficiency in the APOE gene abolished lipoprotein secretion by cultured astrocytes. In addition, APOE-deficient rodents displayed several neurological symptoms including memory and learning deficits,[34,35] age-dependent loss of synaptic functions,[36,37] altered synaptic plasticity,[38] and at the subcellular level, a shift in transbilayer distribution of cholesterol in synaptic membranes.[39,40] Taken together, these data strongly support the notion that apoE plays a critical role in the homeostasis of brain cholesterol, and more specifically, in the transcellular transfer of cholesterol from astrocytes to neurons.

The assembly of astrocyte lipoproteins occurs in the endoplasmic reticulum and in the Golgi.[25] The incorporation of cholesterol in these lipoproteins is mediated by ATP-binding cassette (ABC) transporters, such as ABCA1.[41] Indeed, the cholesterol content of astrocyte lipoprotein is associated with the level of expression of the ABCA1 transporter.[42,43] Apart from ABCA1, other members of the ABC transporters (ABCG1 and ABCG4) have been involved in cholesterol efflux from astrocytes.[23] The capacity of several distinct ABC transporters to regulate cholesterol export from astrocytes gives an alternative to the brain in case of defect of one of these transporters.

The uptake of cholesterol-containing lipoprotein by neuronal membranes is mediated by proteins belonging to the low-density lipoprotein receptor family, especially LRP1.[44,45] The journey of cholesterol from the astrocyte to the neuron is summarized in Fig. 3.7.

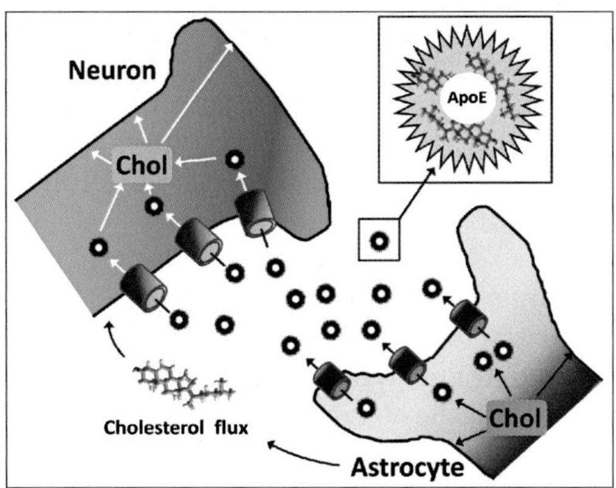

FIGURE 3.7 **Cholesterol flux between astrocytes and neurons.** Because cholesterol cannot travel as a free solute between the astrocyte and the neuron, it is incorporated in apoE-containing lipoproteins (inset). The journey begins in the endoplasmic reticulum of astrocytes, where cholesterol is incorporated in lipoproteins. These lipoproteins are exported via ABC transporters. In neurons, cholesterol-enriched lipoproteins are taken up by lipoprotein receptors (LRP1). Note that the astrocyte has also to fulfill its own cholesterol needs, thus a part of neosynthesized cholesterol is incorporated in astrocytic membranes.

The half-life of cholesterol in the human brain has been estimated at 5 years.[46] This long half-life reflects an efficient recycling of brain cholesterol. Nevertheless, *de novo* synthesis of cholesterol by astrocytes occurs continuously, probably to prevent any deficit in the cholesterol requested by neurons for synaptic functions. To maintain the steady state, excess amounts of brain cholesterol have to be exported into the bloodstream.[7,47] A small proportion of cholesterol associated with apoE lipoproteins can be extracted via the cerebrospinal fluid.[48] Nevertheless, the main mechanism of brain cholesterol efflux involves the enzymatic oxidation of cholesterol into 24S-hydroxycholesterol. To improve its passive diffusion through the blood–brain barrier, brain cholesterol is oxidized into an oxysterol derivative, 24S-hydroxycholesterol.[15] This transforms the amphipathic structure of cholesterol into a roughly symmetric molecule with a polar OH group at both ends (Fig. 3.8). These two opposite OH groups confer to the oxysterol membrane-crossing properties that cholesterol does not display.[49] As a consequence, oxysterol can diffuse through the plasma membrane of endothelial cells[15,46] (Fig. 3.8). This simple and efficient mechanism controls the rate of cholesterol efflux from the brain.

Recent data suggest that brain cholesterol homeostasis might also be controlled by microRNAs.[50] MicroRNAs (miRNAs) are single-stranded short noncoding RNAs that regulate gene expression at the posttranscriptional level.[51,52] For instance, miR-758, a miRNA expressed at high levels in brain cells, has been shown to perturb cholesterol fluxes by decreasing the expression of ABCA1.[53] We are just at the beginning of the elucidation of these new mechanisms of regulation of brain cholesterol homeostasis. Nevertheless, these data support the idea that in addition to classic regulatory mechanisms, miRNAs may play an important role in the control of brain cholesterol metabolism. Incidentally, miRNAs affecting brain cholesterol fluxes may favor the development of several neurodegenerative diseases including Alzheimer's, Huntington's, and Parkinson's diseases.[50] Clearly, the mechanistic

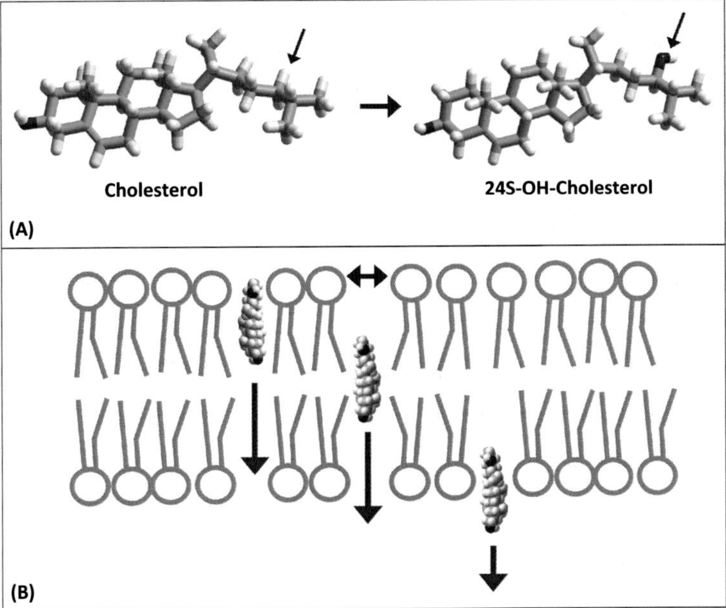

FIGURE 3.8 **Cholesterol has to be oxidized to leave the brain.** Cholesterol cannot pass the blood–brain barrier. Hence, cholesterol in excess is oxidized into 24S–OH–cholesterol (A). The site of oxidation (C24) is indicated by an arrow. This transformation allows cholesterol to find its way through the plasma membranes of endothelial cells by inducing local deformations of the bilayer (B). Indeed, the symmetric OH groups of 24S–OH–cholesterol destabilize the bilayer, so that the oxysterol is gradually excluded from the membrane (red arrows).

links between miRNAs, cholesterol metabolism, and neurodegenerative diseases warrants further investigation.

3.3 SPHINGOLIPIDS

Sphingolipids comprise both constitutive membrane lipids and bioactive lipids involved in signal transduction.[54,55] The term *sphingolipid* was coined in 1947 by Carter for the designation of lipids derived from sphingosine.[56] Sphingosine was first described by Thudichum at the end of the nineteenth century, but it was Carter who explained its chemical structure. Schematically, sphingosine is a long hydrocarbon chain (C18) with one amino and two hydroxyl groups, and a double bond in the *trans* configuration (Fig. 3.9).

It turned out that sphingosine is in fact the most abundant member of a family of structurally related compounds referred to as *sphingoid bases*.[57] The metabolism of sphingolipids is particularly complex because two pathways, one for the synthesis of constitutive membrane sphingolipids and a second for the generation of bioactive sphingolipids, coexist in brain cells. Schematically, the biosynthetic pathway of sphingolipids first leads to ceramide, which in turn generates sphingomyelin and glycosphingolipids (Fig. 3.9). The processing of these complex sphingolipids may then give rise to a series of bioactive sphingolipids including ceramide, ceramide-1-phosphate, sphingosine, and sphingosine-1-phosphate.[54]

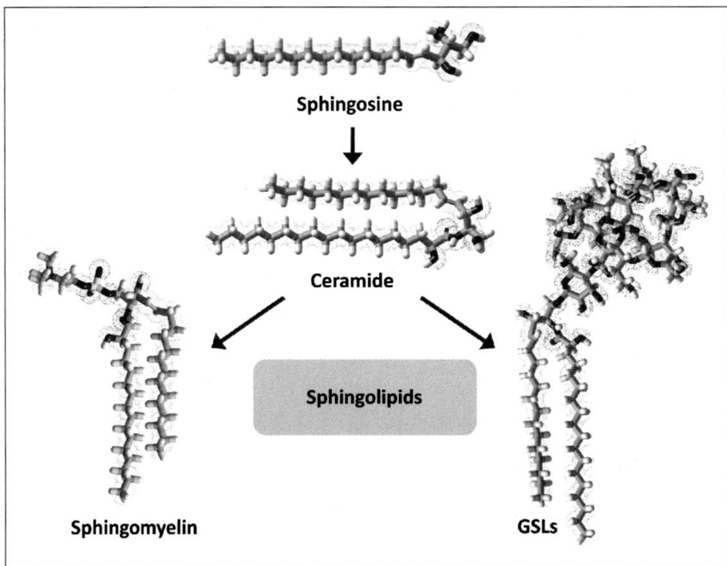

FIGURE 3.9 **Key structures in sphingolipid biochemistry.** Sphingosine is the common building unit of all sphingolipids, including ceramides, sphingomyelins, and glycosphingolipids (GSLs). However, sphingosine is not an intermediary metabolite in the biosynthetic pathways of complex sphingolipids, but a bioactive lipid messenger generated from the degradation of some of these sphingolipids.

3.3.1 Biosynthesis of Ceramide in the Endoplasmic Reticulum

Ceramide is generated *de novo* in the endoplasmic reticulum.[58] The first step in the biosynthesis of ceramide is the condensation of L-serine and palmitoyl-CoA, a reaction catalyzed by serine palmitoyltransferase. The condensation product obtained is 3-ketosphinganine (Fig. 3.10), and this reaction is considered as the rate-limiting enzyme in sphingolipid synthesis.[59]

The next step is the reduction of the ketone group of 3-ketosphinganine catalyzed by 3-ketosphinganine reductase. The resulting product, sphinganine, is then acylated on the amino group (N-acylation) by an enzyme improperly referred to as "ceramide synthase." In fact, the N-acylation of sphinganine does not generate ceramide but dihydroceramide (Fig. 3.10). Hence, as noted by Tidhar and Futerman, this enzyme should be logically known as "dihydroceramide synthase."[58] Nevertheless, the designation of this enzyme as ceramide synthase has been established by usage. Finally, dihydroceramide is converted into ceramide by a desaturase which catalyzes the formation of a *trans* double bonds between carbons 4 and 5. Ceramide can be converted to sphingomyelin, sphingosine, or various glycosphingolipids, which are ubiquitous constituents of membrane lipids and may also be involved in various cellular events, including signal transduction, proliferation, differentiation, apoptosis, and the maintenance of neuronal tissues and cells.

3.3.2 Biosynthesis of Sphingomyelin in the Golgi

The conversion of ceramide into sphingomyelin occurs on the lumen of Golgi membranes.[60] It requires the transport of ceramide from the reticulum to the Golgi, a process involving a

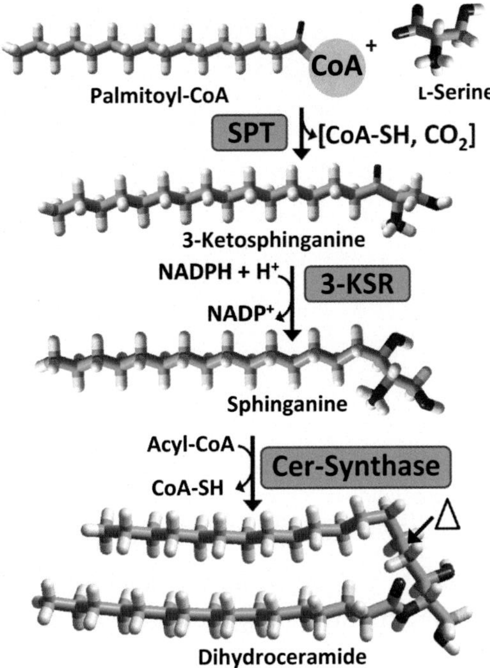

FIGURE 3.10 **Biosynthetic pathway of ceramide.** SPT, serine palmitoyltransferase; 3-KSR, 3-ketosphinganine reductase; Cer-Synthase, ceramide synthase. In fact, the reaction catalyzed by ceramide synthase leads to dihydroceramide. Ceramide is obtained by the secondary action of a desaturase that creates a *trans* double bond between carbons C4 and C5 of the sphingosine moiety (**Δ**).

transporter protein (CERT).[58] The synthesis of sphingomyelin from ceramide is catalyzed by sphingomyelin synthase (an enzyme expressed in human brain) that transfers a phosphocholine residue from phosphatidylcholine to the C1–OH group of ceramide, thereby producing sphingomyelin and diacylglycerol[61] (Fig. 3.11).

In addition to the classic synthesis of sphingomyelin in the *cis*-Golgi, oligodendrocytes may produce as much as 50% of sphingomyelin at the plasma membrane and in the myelin sheath.[62] This unique property of oligodendrocytes is probably due to their capacity to synthesize large amounts of myelinating membranes.

3.3.3 Glycosphingolipids

In the brain, two distinct pathways of glycosphingolipid biosynthesis have been characterized (Fig. 3.12). The first one leads to myelin galactolipids (GalCer, sulfatide, and GM4) (Fig. 3.13), whereas the second one includes all other glycosphingolipids (both neutral and anionic) derived from glucosylceramide (GlcCer) (Fig. 3.14).

GlcCer is synthesized in the Golgi apparatus via glycosylation of ceramide in a reaction catalyzed by GlcCer synthase (UDP-glucose:ceramide glucosyltransferase, GlcT-1)[63,64]: ceramide + UDP-Glc → GlcCer + UDP. GlcCer is the key precursor lipid for the synthesis of

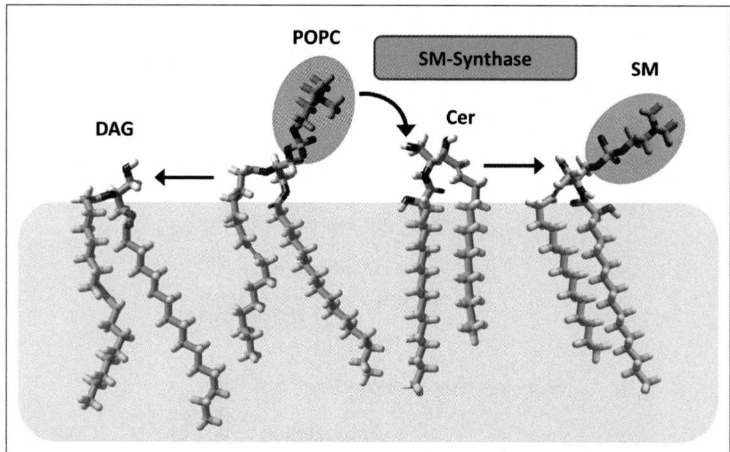

FIGURE 3.11 **Biosynthesis of sphingomyelin.** Sphingomyelin (SM) is synthesized by SM synthase, an enzyme that catalyzes the transfer of a phosphocholine unit from phosphatidylcholine (POPC) onto the C1 of ceramide (Cer). The reaction generates diacylglycerol (DAG). This is typically an "all-membrane" process.

more than 400 glycosphingolipid species, including major brain gangliosides.[65] The biosynthetic pathway of the most important brain glycosphingolipids is summarized in Figs 3.13 and 3.14.

Sialic acid units are grafted onto neutral glycosphingolipid carbohydrates to produce gangliosides. The donor is an activated nucleotide sugar (CMP-sialic acid) that is transferred to an oligosaccharide chain by a sialyltransferase.[66] Conversely, sialic acids can be removed from

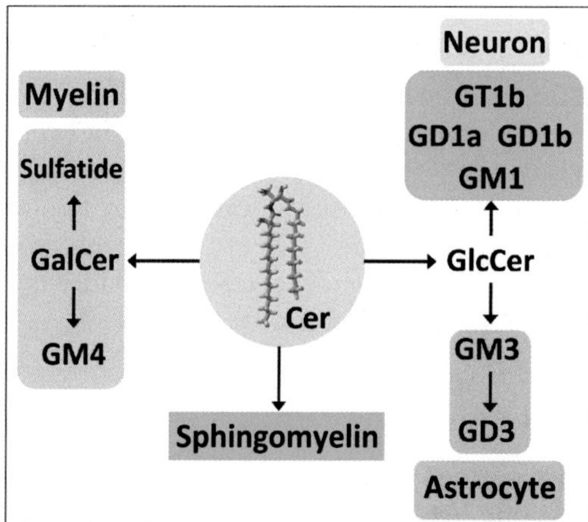

FIGURE 3.12 **Overview of sphingolipid metabolism in the brain.** Neurons synthesize chiefly gangliosides GM1, GD1a, GD1b, and GT1b. Astrocytes synthesize GM3 and, in lower amounts, GD3. Oligodendrocytes synthesize the myelin sphingolipids GalCer and sulfatide. GM4 is selectively expressed in human myelin and glial cells. All brain cells express high-levels of sphingomyelin.

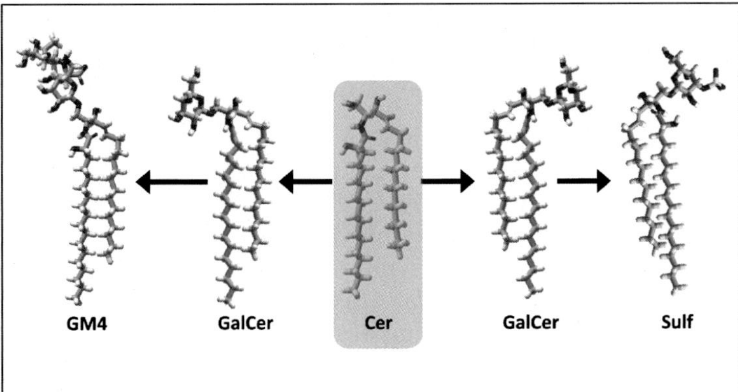

FIGURE 3.13 **Biosynthesis of myelin galactosphingolipids.** Ceramide (Cer) is the precursor of GalCer. GalCer can be sulfated (Sulf) or sialylated (ganglioside GM4). GalCer, sulfatide, and GM4 are major component of the human myelin.

gangliosides by the enzymatic action of sialidases (also referred to as neuraminidase in reference of the most common sialic acid, *N*-acetylneuraminic acid). Sialyltransferases are located in the Golgi apparatus,[66] where gangliosides are synthesized. Sialidases have a variety of subcellular distributions, including intracellular compartments and plasma membranes.[66,67] An intriguing aspect of the synaptic function is the postulated existence of desialylation/ sialylation cycle at the presynaptic membrane[68] (see Chapter 5). The topology of membrane-bound gangliosides renders possible the existence of such cycles because the desialyated

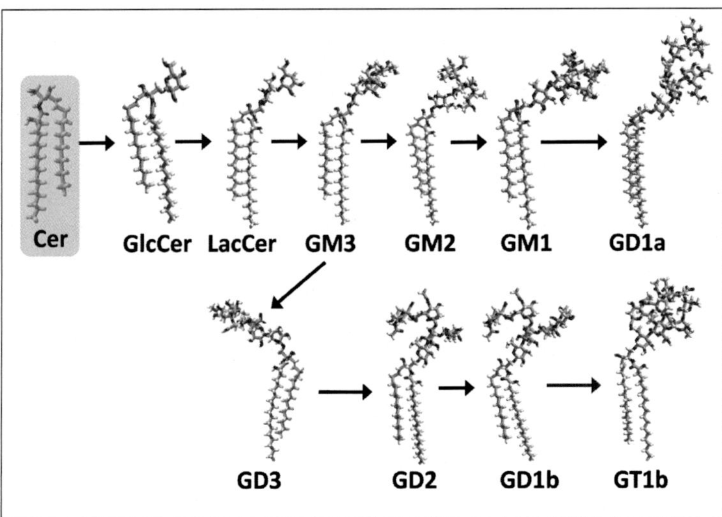

FIGURE 3.14 **Metabolic connections between GlcCer-derived gangliosides.** The neutral glycosphingolipids GlcCer and LacCer are intermediate metabolites in the biosynthetic pathway of gangliosides. Adult neurons express chiefly GM1, GD1a, GD1b, and GT1b, whereas astrocytes express GM3 and GD3.

gangliosides remain bound to the membrane and are not relocalized after the enzymatic reaction. Thus, the resialylation of these glycosphingolipids restores the initial conditions. As discussed in the first part of this chapter, this process would not be possible for water-soluble substrates.

GalCer and sulfatide are the most abundant glycosphingolipids of the myelin sheath.[69] These galactolipids contribute to the insulating properties of the myelin membrane.[70] GalCer is synthesized in oligodendrocytes by the enzyme UDP-Gal:ceramide galactosyltransferase (GalCer synthase) according to reaction: ceramide + UDP-Gal → GalCer + UDP[70] (Fig. 3.13). Cerebroside sulfotransferase (CST) catalyzes the synthesis of sulfatide (3′-sulfo-GalCer) in the luminal side of the Golgi apparatus.[71] The sulfate group is transferred from 3′-phosphoadenosine 5′-phosphosulfate (PAPS) to the 3′OH group of GalCer. GalCer also serves as a precursor for the synthesis of ganglioside GM4 (Fig. 3.13). The sialylation of GalCer is catalyzed by GM3 synthase, a silalyltransferase that is also involved in the generation of ganglioside GM3 from LacCer.[72] Thus, there is no specific GM4 synthase but a GM3/GM4 synthase activity carried by a unique protein. Although GM4 expression has been detected in oligodendrocytes and astrocytes,[73,74] its physiological function remains unclear. *In vitro* studies have shown that the Parkinson's disease-associated protein α−synuclein can interact with GM4,[75] suggesting that this ganglioside could be linked to the pathogenesis of neurodegenerative diseases. Moreover, it has been reported that the HIV-1 surface envelope glycoprotein gp120 binds to GalCer, sulfatide, and GM4 extracted from the human brain.[76]

3.3.4 Sphingolipid Messengers

Glycosphingolipids and sphingomyelin can be enzymatically processed to generate ceramide (Fig. 3.15). Sphingomyelinases catalyze the hydrolysis of the phosphocholine head group of sphingomyelin.[77] Neutral sphingomyelinases are expressed at high levels in brain.[78,79] Most sphingomyelinases have a large tissue distribution, yet a brain-specific, Mg^{2+}-dependent neutral sphingomyelinase (nSMase2) has been characterized.[80] Immunofluorescence studies with anti-nSMase2 antibodies revealed a pattern of expression restricted to neurons. This enzyme has been involved in various types of neurodegeneration, including age-related cognitive decline, Alzheimer's, Parkinson's, and Huntington's diseases.[77]

Apart from sphingomyelinase, the other degradative pathway leading to ceramide is catalyzed by a lysosomal enzyme, glucosylceramidase (glucocerebrosidase).[81] Mutation-associated glucocerebrosidase deficiency results in Gaucher disease, a rare lysosomal storage disorder due to the accumulation of GlcCer inside the lysosomes of cells composing the reticuloendothelial system.[82] In addition, an intriguing association occurs between mutations in the glucocerebrosidase gene and the development of Parkinson's disease and other α−synucleinopathies.[81]

Ceramides exert a wide range of critical biological functions linked to cellular growth, differentiation, and apoptosis.[83] In addition, it is the precursor of three other bioactive sphingolipids: ceramide-1-phosphate, sphingosine, and sphingosine-1-phosphate, which involve respectively, ceramide kinase, ceramidase, and sphingosine kinase[54] (Fig. 3.15). Finally, sphingosine-1-phosphate is degraded into ethanolamine phosphate and a fatty aldehyde via a reaction catalyzed by sphingosine-1-phosphate lyase.[84] All these metabolites contribute to the turnover of sphingolipids in response to a broad range of agonists such as growth factors. According to the

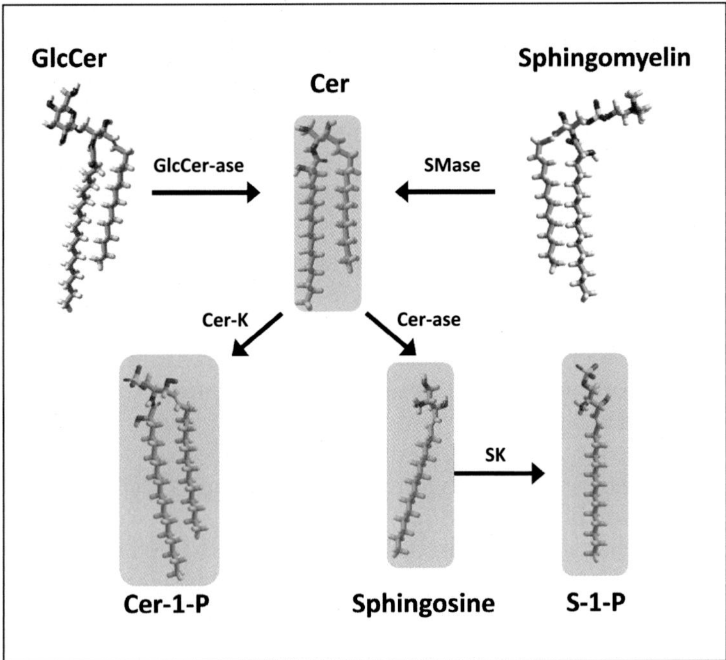

FIGURE 3.15 **Generation of sphingolipid messengers.** Ceramide can be obtained via sphingomyelinase (SMase) processing of sphingomyelin or by deglycosylation of GlcCer by glucocerebrosidase (GlcCer-ase). Ceramide is phosphorylated into ceramide-1-phosphate (Cer-1-P) by ceramide kinase (Cer-K). It can also be degraded into sphingosine (+ a fatty acid) by ceramidase (Cer-ase). Sphingosine is phosphorylated in sphingosine-1-phosphate (S-1-P) by sphingosine kinase (SK).

type of sphingolipid generated under this type of stimulation, the response may then be either growth inhibitory/proapoptotic or growth stimulatory/antiapoptotic.[54] Indeed, sphingosine-1-phosphate promotes cell proliferation and prevents apoptosis,[85] whereas sphingosine is rather associated with growth arrest and proapoptotic effects.[83] Similarly, ceramide is proapoptotic,[86] whereas its phosphorylated derivative ceramide-1-phosphate has antiapoptotic effects and favors cell survival.[87] As emphasized by Mencarelli and Martinez-Martinez,[83] phosphorylation/dephosphorylation of sphingolipids leads to the easily reversible generation of compounds with diametrically different biological effects. All these lipid second messengers exert important functions in brain physiology and are also involved in age-related cognitive deficits as well as in various neurodegenerative diseases.[77,83]

An important issue to solve concerns the mechanisms controlling the cellular effects of sphingolipid messengers. The first level of analysis is to compare the biochemical structure of ceramide, ceramide-1-phosphate, and sphingosine, sphingosine-1-phosphate (Fig. 3.15) and analyze the impact of these structures on the physicochemical properties of all these sphingolipids.[88] Ceramide has intriguing amphipathic properties that can be evidenced by measuring the surface tension of water upon addition of this lipid.[89] When an amphipathic compound is injected in water, it readily goes to the interface where it forms a monolayer.[90,91] The presence of chemical groups able to form hydrogen bonds with water molecules at the interface decreases the surface tension of water. However, when a long-chain ceramide in organic solution is injected in the aqueous phase, the surface tension of water remains unchanged for several hours,[89] indicating a high hydrophobicity for ceramides. When injected in water, ceramides separate from the organic solvent in which they were dissolved and massively aggregate in the trough. It is possible to force ceramide to form a monolayer by carefully depositing a drop of organic ceramide solution at the water surface. These data indicate that ceramide can still form monolayers, which is consistent with its presence in biological membranes. Nevertheless, compared with its membrane lipid precursors (e.g., sphingomyelin), ceramide has an extreme hydrophobicity and a more cylindrical shape (Fig. 3.15), which has two main consequences. First, ceramide molecules form condensed membrane platforms[92] stabilized by van der Waals interactions between their hydrophobic chains and hydrogen bonds between their polar head groups.[83,93] Second, the generation of ceramide from sphingomyelin induces a significant redistribution of cholesterol. The tight association of sphingomyelin and cholesterol is generally considered as the driving force of raft domain formation,[94,95] because in ternary lipid systems, cholesterol interacts preferentially with sphingomyelin rather than phosphatidylcholine.[96–98] The cholesterol–sphingomyelin interaction is facilitated by the remarkable complementary shapes of both lipids (Fig. 3.16).

However, ceramide has a cylindrical shape that does not fit well with cholesterol, so that in mixed ceramide/cholesterol systems, ceramide self-aggregates into highly ordered domains from which cholesterol is excluded.[99] As noticed by Megha and London,[99] cholesterol and ceramides have small polar head groups, so that both may compete for binding to raft lipids with large head groups in a geometrically driven mechanism (Fig. 3.16). In addition, the small size of the head group results in a minimal exposure to water, allowing ceramides to self-associate into packed assemblies.[100] Taken together, these data indicate that the displacement of raft cholesterol might be a potent mechanism of regulation controlled by membrane ceramide.

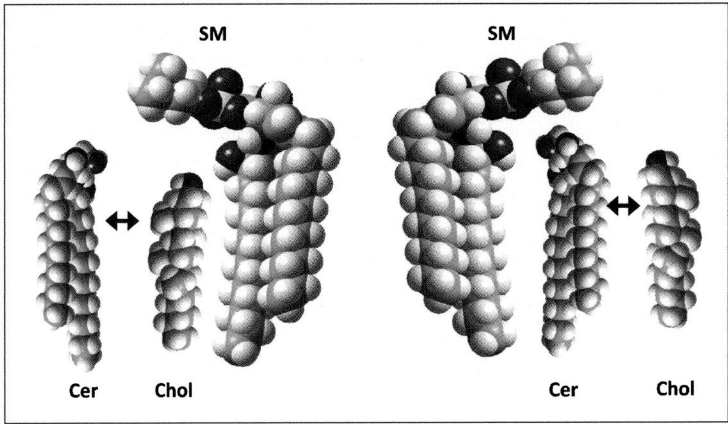

FIGURE 3.16 **Competition between cholesterol and ceramide for binding to sphingomyelin.** The polar head groups of both ceramide (Cer) and cholesterol (Chol) occupy a small volume that fits very well with the large head group of sphingomyelin (SM). Sphingomyelin can interact with either cholesterol (left panel) or ceramide (right panel). Thus ceramide and cholesterol are in competition for binding to sphingomyelin (arrows).

Apart from these physicochemical effects, ceramide may affect neuronal apoptosis through direct effects on protein kinases and phosphatases.[101–105] Indeed, both a ceramide-activated kinase (CAPK) and ceramide-activated protein phosphatases (CAPPs) have been characterized.[101,102]

Ceramide-1-phosphate is produced via the ATP-dependent conversion of ceramide by ceramide kinase.[106] This enzyme is a calcium-stimulated lipid kinase that copurified with brain synaptic vesicles.[107] These data indicated that either the decrease of ceramide or the presence of ceramide-1-phosphate (or both) in synaptic vesicles is important for neurotransmitter exocytosis. Unfortunately, little is known about the role of ceramide and ceramide-1-phosphate in the function of synaptic vesicles. In contrast with ceramide, ceramide-1-phosphate does not induce domain formation in phosphatidylcholine bilayers.[108] The biophysical properties of ceramide-1-phosphate have been studied by Kooijman et al.[109] These authors observed that in contrast to ceramide, ceramide-1-phosphate forms bilayers in aqueous media. However, when mixed with glycerophospholipids, both ceramide and ceramide-1-phosphate may induce a negative curvature in lipid membranes and facilitate the transition from lamellar to inverted hexagonal phases.[88] In this respect, there is no significant functional difference between ceramide and its phosphorylated metabolite, and both may exert similar stimulatory effects on the fusion of synaptic vesicles. However, an important point should be underscored. Ceramide-1-phosphate is structurally related to phosphatidic acid (Fig. 3.17). Interestingly,

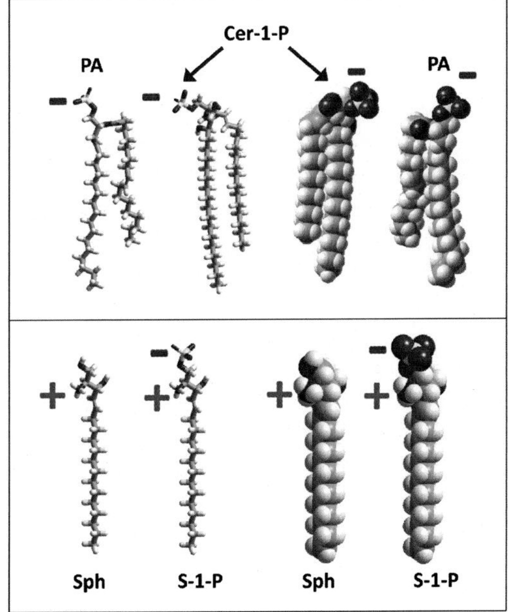

FIGURE 3.17 **Electric charges on sphingolipid messengers: why it matters.** In the upper panel one can see that ceramide-1-phosphate (Cer-1-P) and phosphatidic acid (PA) share interesting structural analogy due to their common phosphate head group bearing one negative charge. In the lower panel, one can see that sphingosine (Sph) is cationic, whereas its phosphorylated derivative, sphingosine-1-phosphate (S-1-P) is zwitterionic (one positive and one negative charge). These electric charges have a dramatic influence on the topology, biophysical properties, and thus biological activity of sphingolipid messengers.

an accumulation of phosphatidic acid has been consistently observed at the active sites of vesicle exocytosis, suggesting that phosphatidic acid has a direct role in membrane fusion.[110] In this case, one could hypothesize that the phosphate group of both ceramide-1-phosphate and phosphatidic acid confer specific membrane curvature effects that improve the fusion properties of membrane lipids.

The last two sphingolipid metabolites are sphingosine and sphingosine-1-phosphate. The most obvious difference between these sphingolipids is their electric charge (Fig. 3.17). At physiological pH, the amino group of sphingosine is protonated, so that sphingosine should be considered as a cationic amphiphile.[88] Correspondingly, sphingosine has the capacity to permeabilize membrane bilayers.[111] In contrast, sphingosine-1-phosphate is zwitterionic (one positive charge on the amino group and one negative charge on the phosphate) (Fig. 3.17). Among bioactive sphingolipids, sphingosine-1-phosphate is quite unique because it may exist as a dispersed solution in the cytosol.[112] In particular, sphingosine-1-phosphate displays a monomer-micelle transition in the micromolar concentration range. Consequently, sphingosine-1-phosphate is in equilibrium between the cytosolic and membranous compartments. In the plasma membrane, the zwitterionic nature of sphingosine-1-phosphate stabilizes the bilayer structure.[112]

Overall the distinctive physicochemical properties of bioactive sphingolipid metabolites account for various regulatory effects on a broad range of brain cells, including neurons, oligodendrocytes, and astrocytes.[113]

3.4 PHOSPHOINOSITIDES

Bioactive sphingolipids represent only a part of the intricate web of bioactive lipids involved in the control of brain functions.[114] Other lipid second messengers, including phosphatidylinositol phosphate (and related phosphoinositides) and diacylglycerol also play important regulatory functions (Fig. 3.18). The case of phosphatidylinositol diphosphate (PIP_2), a membrane lipid located in the inner leaflet of the plasma membrane, is particularly relevant for the regulation of synaptic functions. PIP_2 is cleaved by phospholipase C into diacylglycerol (which remains in the membrane) and inositol triphosphate (IP_3) that is released in the cytosol and binds to the IP_3 receptor, inducing Ca^{2+} release from the endoplasmic reticulum[115] (Fig. 3.18).

As detailed in Chapter 5, glutamate receptors can be divided in two classes: ionotropic (iGluR) and metabotropic (mGluR). The latter are seven transmembrane domain receptors coupled to a transducer G protein (GPCR). The three subfamilies of mGluR that differ in the second messenger systems are activated upon glutamate stimulation.[116] One of these subgroups (referred to as group I) contains mGluR1 and mGluR5, which transmit signaling through cleavage of PIP_2 and mobilize $Ca^{2+}$from intracellular stores. In the brain, mGluR1 receptors play important regulatory functions pertaining to neuron excitability and synaptic plasticity.[117] Correspondingly, mGluR1 receptors have a broad expression in various brain areas,[118] including neurons of the olfactory system,[118] hypothalamus,[119] thalamus,[120] basal ganglia,[118,121] hippocampus,[122] and cerebellum.[123,124]

From a functional point of view, mGluR1 receptors are primarily coupled to transducer proteins of the Gq family, which activates phospholipase Cβ (PLCβ).[117] PLCβ cleaves phosphatidylinositol-4,5-bisphosphate (PIP_2) into inositol-1,4,5-trisphosphate (IP_3) and diacylglycerol (DAG) (Fig. 3.19). In the classic pathway, IP_3 activates IP_3 and ryanodine

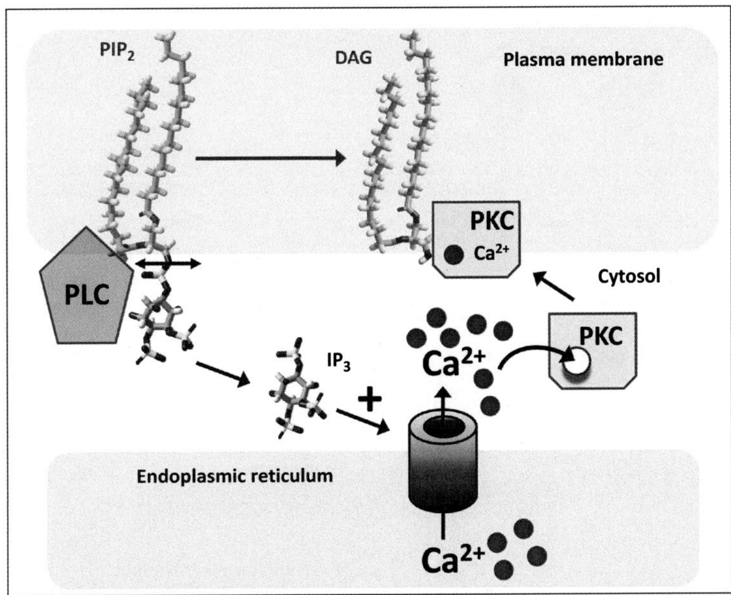

FIGURE 3.18 **Overview of the PIP_2/IP_3/DAG pathway of signal transduction.** DAG, diacylglycerol; PLC, phospholipase C; PKC, protein kinase C. The sequence of events begins with the activation of phospholipase C that cleaves PIP_2 into DAG and IP_3. IP_3 interacts with IP_3/ryanodine receptors on endoplasmic reticulum membranes. The activation of these Ca^{2+} channel receptors evokes the release of Ca^{2+} into the cytosol. The binding of Ca^{2+} to protein kinase C induces its translocation toward the plasma membrane where it interacts with DAG, which triggers the activation of protein kinase C.

receptors on the surface of the endoplasmic reticulum.[125,126] This process induces the release of Ca^{2+} from intracellular stores, which in turn results in the activation of protein kinase C, thereby triggering a signaling cascade involving Akt, ERK, and MAP kinases[127] (Fig. 3.19).

Besides this canonical effect, the activation of mGluR1 receptors can also modulate intracellular Ca^{2+} concentrations via the control of voltage-sensitive calcium channels.[128] In cerebellar granule cells, a functional coupling among mGluR1, ryanodine receptors, and L-type Ca^{2+} channels was demonstrated.[129] Overall, mGluR1 has been associated with various signaling pathways controlling synaptic functions, neuronal survival, and plasticity.[130] The signal transduction pathways activated by mGluR1 also include phospholipase D, which produces the second messenger phosphatidic acid,[131] and phospholipase A2, which generates arachidonic acid.[132] Overall, the activation of mGluR1 leads to a massive lipid redistribution and turnover in neuronal membranes, and to the production of lipid messengers that allow a fine-tuning of synaptic transmission through modulation of synaptic ion channels.[133]

3.5 PHOSPHATIDIC ACID

Phosphatidic acid is a key intermediate metabolite in the synthesis pathways of all membrane glycerophospholipids, including phosphatidylcholine (PC), phosphatidylethanolamine (PE), phosphatidylserine (PS), as well as phosphatidylinositol (PI) and its phosphorylated

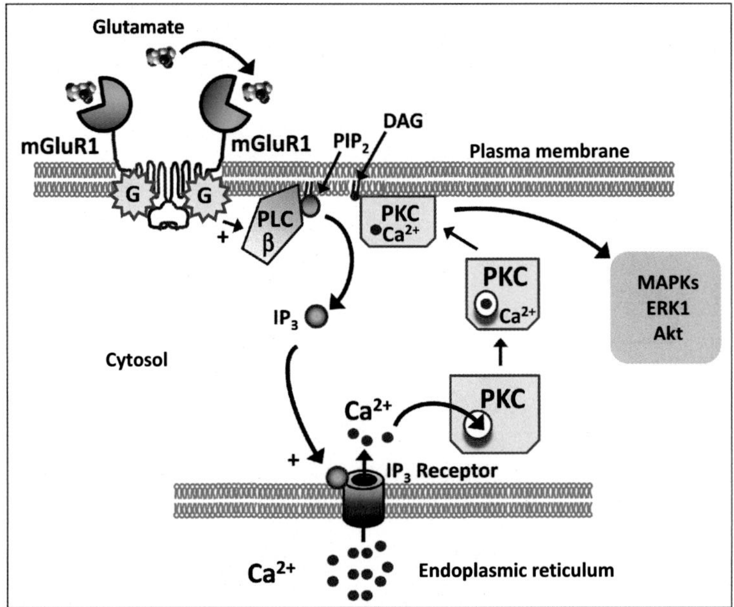

FIGURE 3.19 **Activation of the PIP_2/IP_3/DAG pathway by mGluR1.** In the plasma membrane of neurons, metabotropic glutamate receptors mGluR1 are expressed as functional dimeric units coupled to a transducer protein of the Gq family (G, in yellow). When glutamate binds to mGluR1 in the extracellular space, a conformational change in the receptor activates the transducer, which in turn activates phospholipase Cβ (PLCβ). As illustrated in Fig. 3.18, PLCβ cleaves PIP_2 into diacylglycerol (DAG) and IP_3. IP_3 activates receptors/Ca^{2+} channels on the reticulum membrane, which triggers the release of Ca^{2+} in the cytosol. Protein kinase C (PKC) is activated by Ca^{2+} (translocation) and DAG. PKC triggers a cascade of activation, which includes Akt, ERK1, and MAP kinases.

derivatives (e.g., PIP_2).[134] Besides this metabolic contribution to membrane biogenesis, phosphatidic acid is an important signaling lipid, especially in the central nervous system.[135,136] Phosphatidic acid can be formed by phosphorylation of diacylglycerol, a reaction catalyzed by diacylglycerol kinase[137] (Fig. 3.20).

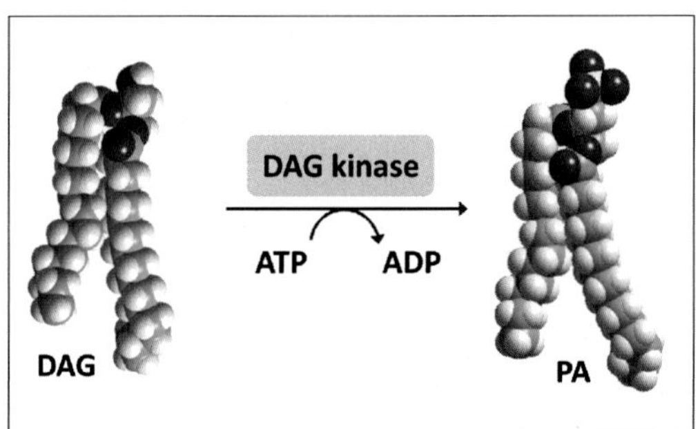

FIGURE 3.20 **Biosynthesis of phosphatidic acid.** Diacylglycerol (DAG) is phosphorylated into phosphatidic acid (PA). In this reaction, the donor of phosphate is ATP.

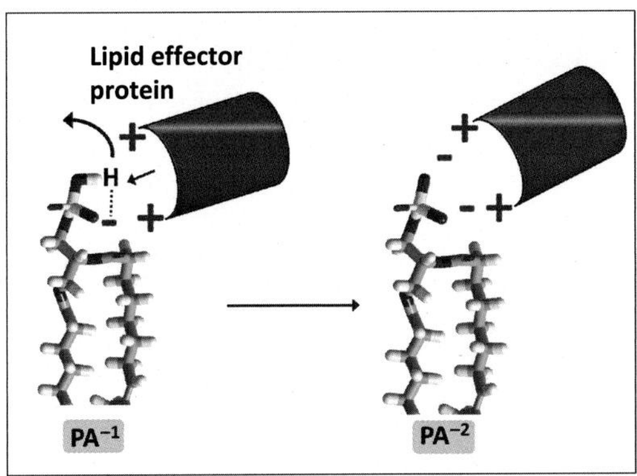

FIGURE 3.21 **The electrostatic/hydrogen bond switching mechanism.** Once the first proton has left the phosphate group of phosphatidic acid, the second one establishes an intramolecular hydrogen bond that stabilizes the negative charge (this form of phosphatidic acid has one negative charge and it is noted PA^{-1}). When a lipid effector protein binds to the head group PA^{-1}, its cationic amino acids (Lys, Arg) destabilize the remaining proton so that PA then has two negative charges (PA^{-2}). At this stage, the PA–protein complex is stabilized by two electrostatic bridges. This unique mechanism explains why PA is a target of choice for promoting the membrane insertion of proteins that bind to anionic lipids.

The other possible pathway is the cleavage of membrane glycerophospholipids by phospholipase D. Total *de novo* synthesis of phosphatidic acid from dihydroxyacetone phosphate allows mammalian cells to counterbalance the conversion of phosphatidic acid into membrane glycerophospholipids.[138] A key feature of phosphatidic acid is its phosphomonoester (phosphate) group, which displays peculiar ionization properties[139] (Fig. 3.21). In particular, the phosphate group of phosphatidic acid may form an intramolecular hydrogen bond upon initial deprotonation (i.e., when its charge is –1), which stabilizes the second proton against dissociation[140] (Fig. 3.21).

Most importantly, lysine and arginine residues in proteins were found to increase the charge of phosphatidic acid by interacting with its phosphate group. An elegant model describing these properties, referred to as the "electrostatic/hydrogen-bond switch mechanism" has been developed by Kooijman et al.[141] (Fig. 3.21). This model explains why phosphatidic acid is particularly suited for promoting the interfacial insertion of proteins such as dynamin, a vesicle-binding protein involved in cellular endocytosis.[139] Indeed, dynamin binds to negatively charged membranes but penetrates more efficiently in membranes containing phosphatidic acid.[142] According to this model, a lipid effector protein is initially attracted by the head group of a protonated phosphatidic acid molecule whose phosphate group displays one negative charge (Fig. 3.21). Then, basic amino acids of the protein interact with the phosphate group via a hydrogen bond. This interaction induces the dissociation of the remaining proton of phosphatidic acid that has now two negative charges that reinforce the association of the protein to the lipid through electrostatic interactions. In addition, to this "switch" mechanism, phosphatidic acid has a pronounced cone shape that facilitates protein insertion by reducing the packing capacity of lipid head groups in the bilayer.[143] Overall, the generation of phosphatidic acid in synaptic membranes is a means to control key synaptic functions such as neuronal signal transduction,[144] synaptic vesicle traffic,[145] and dendritic spine stability.[146] Finally, Shin and Loewen have suggested that phosphatidic acid could be a pH sensor lipid capable of coupling changes in pH with signal transduction pathways.[134] This function of phosphatidic acid could participate in the Na^+/H^+ exchanger-mediated regulation of pH during neurite morphogenesis and dendritic spine growth.[147,148]

3.6 ENDOCANNABINOIDS

The major psychoactive component of cannabis, Δ9-tetrahydrocannabinol (THC), was characterized in 1964.[149] A first receptor for THC expressed in the brain (CB1) was discovered in 1990,[150] and a second one expressed in peripheral tissues (CB2) 2 years later.[151] The postulated endogenous ligands of these receptors were referred to as *endocannabinoids*.[152] The first discovered endocannabinoid, *N*-arachidonoylethanolamide, was isolated from a pig brain and named *anandamide* based on the Sanskrit word *ananda* that means "bliss," together with the suffix *amide* in reference to its chemical nature.[153] Another major endocannabinoid, 2'arachidonoylglycerol (2-AG) was isolated in brain tissue.[154] 2-AG is expressed in the brain at higher concentrations than anandamide.[155] However, this higher level does not mean that 2-AG is a more important neurotransmitter than anandamide. Indeed, it should be noted that due to its larger head group, 2-AG, but not anandamide, may form micellar structures (Fig. 3.22). Thus, the extracellular transport of 2-AG could require significant amounts of neurotransmitter to

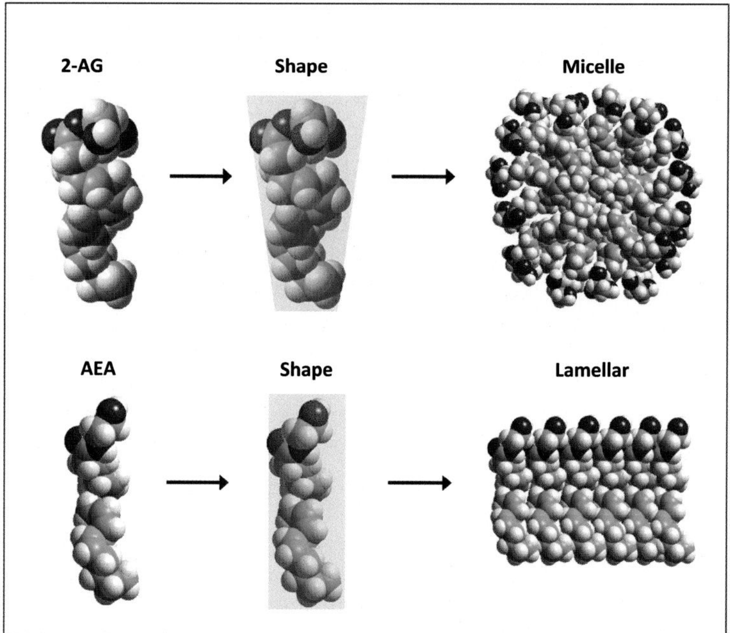

FIGURE 3.22 **Molecular shape and topology of endocannabinoids.** 2-Arachidonoylglycerol (2-AG) has a conic shape that is compatible with a micellar organization. In contrast, anandamide (AEA) has a more cylindrical shape that prevents micelle formation and favors lamellar organization (see Chapter 2 for a discussion on lipid shape).

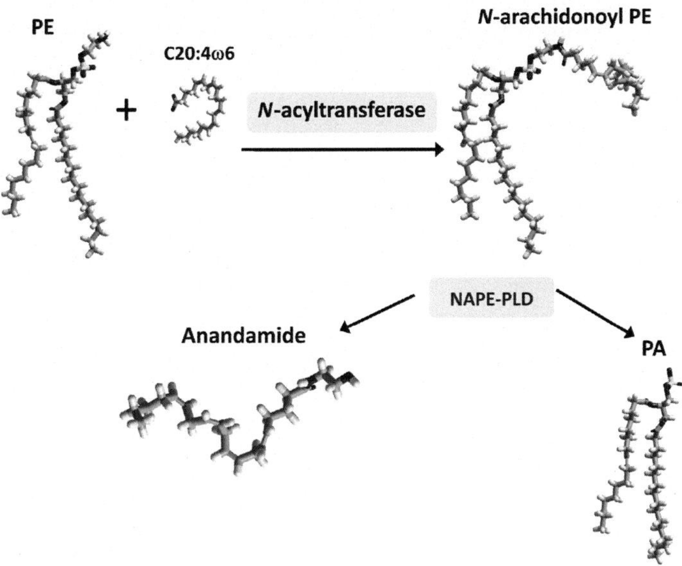

FIGURE 3.23 **Biosynthesis of anandamide.** Phosphatidylethanolamine (PE) is condensed with an arachidonate (C20:4ω6) group given by a glycerophospholipid (not shown) in a reaction catalyzed by *N*-acyltransferase. This results in the synthesis of *N*-arachidonoyl PE, the precursor of anandamide. Then, this precursor is cleaved into anandamide and phosphatidic acid (PA) by *N*-acylphosphatidylethanolamine-hydrolyzing phospholipase D (NAPE-PLD).

fill the micelles, whereas in contrast anandamide could be transported as a single molecule bound to a lipid-binding protein (see Chapter 5 for further discussion of this issue).

Anandamide is produced through Ca^{2+}-stimulated hydrolysis of the phosphatidylethanolamine (PE) derivative *N*-arachidonoyl PE (Fig. 3.23). The *N*-arachidonoyl PE precursor is synthesized in the brain by an *N*-acyltransferase that catalyzes the transfer of an arachidonate group from a glycerophospholipid to the amino group of PE.[156,157] The second step of anandamide synthesis is the hydrolysis of *N*-arachidonoyl PE to anandamide and phosphatidic acid, a reaction catalyzed by *N*-acyl-phosphatidylethanolamine-hydrolyzing phospholipase D (NAPE-PLD).[158]

Several metabolic pathways may produce 2-AG.[152] The principal route is Ca^{2+}-dependent and is mediated by two enzymes, phospholipase C and diacylglycerol lipase.[152] Phospholipase C hydrolyzes an arachidonic acid–containing membrane phospholipid (e.g., phosphatidylinositol) and produces an arachidonic acid–containing diacylglycerol. Then, 2-AG is produced from this diacylglycerol by the action of diacylglycerol lipase.

Both anandamide and 2-AG can be degraded by hydrolysis or oxidation.[159] The hydrolysis of endocannabinoids is catalyzed by fatty acid amide hydrolase (FAAH), an enzyme which cleaves the amide bond of anandamide and the ester bond of 2-AG.[160] Molecular details of the binding of endocannabinoids to FAAH are given in Chapter 5. Nevertheless, although FAAH may hydrolyze 2-AG *in vitro*, it turned out that in the brain, this enzyme is not involved in 2-AG degradation. *In vivo*, more than 80% of 2-AG is hydrolyzed by another enzyme, monoacylglycerol lipase (MGL).[161] The oxidation pathway of endocannabinoids involves cyclooxygenase (COX) and lipoxygenase (LOX) enzymes, as well as several cytochrome P450 enzymes.[162,163] In particular, the inflammation-induced COX-2 enzyme generates several bioactive derivatives of endocannabinoids, including prostaglandin, thromboxane, and prostacyclin ethanolamides.[164,165] Therefore, the metabolisms of endocannabonoids and inflammatory lipid messengers are functionally interconnected.

The elucidation of synaptic mechanisms involving endocannabinoids has revealed some surprises. Indeed, endocannabinoids are released in the extracellular space from postsynaptic neurons in response to a rise in intracellular Ca^{2+}. Then, endocannabinoid neurotransmitters travel backward across synapses, and eventually stimulate CB1 receptor on the presynaptic neuron.[166] The main effect of endocannabinoids on presynaptic neurons is to decrease the release of either the inhibitory γ-aminobutyric acid (GABA) or the excitatory glutamate neurotransmitters,[167] resulting in a control of a broad range of physiological functions including food intake,[168] fear, [169] and anxiety.[170]

3.7 LIPID PEROXIDATION

Membrane lipids are particularly vulnerable to oxidation due to their high content in polyunsaturated acyl chains.[171] The oxidation of membrane lipids generates important changes in their physicochemical properties, thereby significantly affecting membrane behavior. Two main pathways of oxidation of polyunsaturated acyl chains include enzymatic and nonenzymatic mechanisms.[172] Membrane cholesterol also provides an oxidation pathway.[172] Oxidation of membrane lipids occurs upon aging and in neurodegenerative diseases.[173] Indeed, postmortem analysis has indicated that the extent of lipid oxidation in brains affected by Alzheimer's disease exceeds that in age-matched control individuals.[174,175] The main oxidized forms of cholesterol found in brain tissue are listed in Fig. 3.24.

The oxidation of ω3 and ω6 polyunsaturated fatty acids in membrane phospholipids results in the addition of numerous polar oxygen atoms that perturb the packing properties of these lipids and modifies their membrane topology. The consequence is that previously hydrophobic acyl chains buried in the membrane may gain enough polarity to project into the aqueous environment[176] (Fig. 3.25). This phenomenon has been metaphorically referred to as the lipid whisker model.[176,177] Indeed, cell membranes "grow whiskers" as glycerophospholipids undergo peroxidation, and many of their oxidized acyl chains protrude at the surface.

3.8 KEY EXPERIMENT: ALZHEIMER'S DISEASE, CHOLESTEROL, AND STATINS: WHERE IS THE LINK?

Although cholesterol is considered as a hazardous cofactor in Alzheimer's disease, the mechanisms linking the disease to cholesterol are still debated.[178] A number of studies indicate that high cholesterol, especially after midlife, significantly increases the risk of developing Alzheimer's disease,[179,180] but this conclusion has been recently challenged.[178] In parallel, the effectiveness of cholesterol-lowering treatments for the prevention and/or the treatment

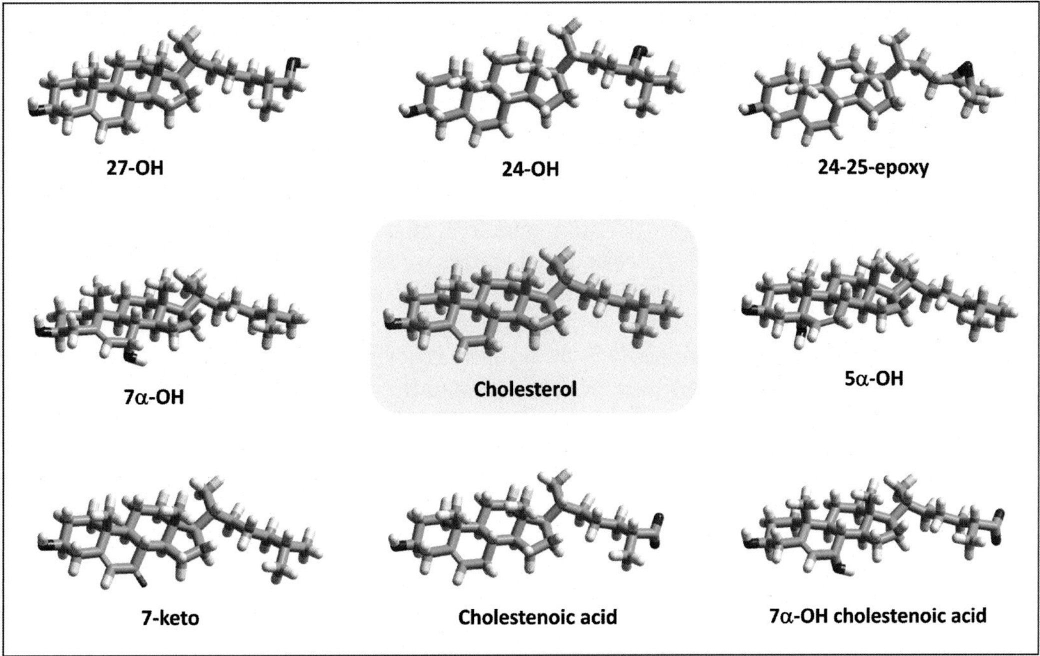

FIGURE 3.24 **Structure of oxysterols detected in the brain.**

of Alzheimer's disease has been seriously questioned.[181] This interrogation contrasts with the rather clear-cut data obtained in animal models and cell culture systems.[182] As a matter of fact, increasing cholesterol levels in neural cells was consistently associated with an increased production and neurotoxicity of Alzheimer's β-amyloid peptides,[183] whereas decreasing plasma membrane cholesterol levels had global neuroprotective effects.[178] In face of such spectacular

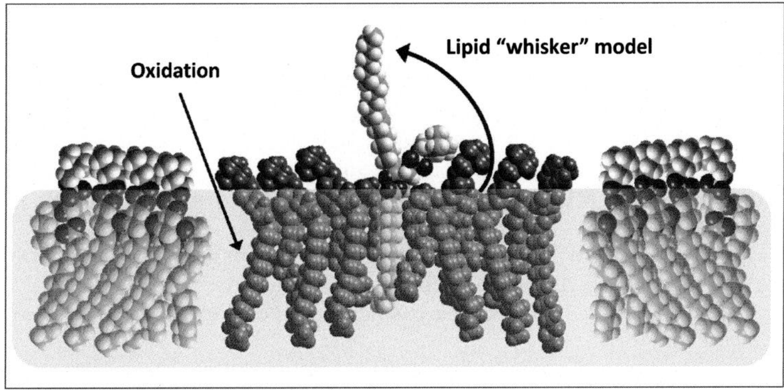

FIGURE 3.25 **The lipid whisker model.** The oxidation of the polyunsaturated acyl chain in membrane glycerophospholipids increases its polarity so that it can be excluded from the apolar phase of the membrane. The other acyl chain, which is saturated, maintains the glycerophospholipid anchored in the membrane.

results, it is somewhat frustrating to have to admit that statins are not the definitive cure for Alzheimer's disease.

Can we explain the reasons for this failure? First, we have to consider that estimating total cholesterol levels in the brain may not reflect subtle changes in the cholesterol content of lipid rafts (e.g., transbilayer redistribution in the exofacial leaflet of the plasma membrane that gradually occurs upon aging).[39,184] Second, in addition to cholesterol, lipid rafts are also enriched in statin-insensitive lipids (sphingolipids) that have been incriminated in the pathogenesis of Alzheimer's disease.[185,186] Third, and this might well be the chief reason of their failure as a cure for Alzheimer's disease, most statins do not cross the blood–brain barrier, so that they have little, if any, effect on the homeostasis of brain cholesterol.[47] Therefore, if statins exert any protective or curative effect in Alzheimer's disease patients, this has probably little to do with brain cholesterol levels.[187,188] As antiatherosclerosis agents, statins may improve blood circulation and thus ameliorate brain oxygenation.[47] These global effects could greatly vary among patients, explaining some inconsistent results obtained in clinical trials. In summary, although it is clear that cholesterol depletion is a good *in vitro* approach to cure neural cells "infected" by Alzheimer's β-amyloid peptides, it is not yet possible to use this strategy in patients with neurodegenerative diseases.

References

1. Openshaw AE, Race PR, Monzo HJ, Vazquez-Boland J, Banfield M. Crystal structure of SmcL, a bacterial neutral sphingomyelinase C from Listeria. *J Biol Chem*. 2005;280(41):35011–35017.
2. Snipes GJ, Suter U. Cholesterol and myelin. *Subcell Biochem*. 1997;28:173–204.
3. Fantini J, Barrantes FJ. Sphingolipid/cholesterol regulation of neurotransmitter receptor conformation and function. *Biochim Biophys Acta*. 2009;1788(11):2345–2361.
4. Fantini J, Barrantes FJ. How cholesterol interacts with membrane proteins: an exploration of cholesterol-binding sites including CRAC, CARC, and tilted domains. *Front Physiol*. 2013;4:31.
5. Haines TH. Do sterols reduce proton and sodium leaks through lipid bilayers? *Prog Lipid Res*. 2001;40(4): 299–324.
6. Yeagle PL. Modulation of membrane function by cholesterol. *Biochimie*. 1991;73(10):1303–1310.
7. Bjorkhem I, Meaney S. Brain cholesterol: long secret life behind a barrier. *Arterioscler Thromb Vasc Biol*. 2004;24(5):806–815.
8. Edmond J, Korsak RA, Morrow JW, Torok-Both G, Catlin DH. Dietary cholesterol and the origin of cholesterol in the brain of developing rats. *J Nutr*. 1991;121(9):1323–1330.
9. Danik M, Champagne D, Petit-Turcotte C, Beffert U, Poirier J. Brain lipoprotein metabolism and its relation to neurodegenerative disease. *Crit Rev Neurobiol*. 1999;13(4):357–407.
10. Dietschy JM, Turley SD. Cholesterol metabolism in the brain. *Curr Opin Lipidol*. 2001;12(2):105–112.
11. Quan G, Xie C, Dietschy JM, Turley SD. Ontogenesis and regulation of cholesterol metabolism in the central nervous system of the mouse. *Brain Res Dev Brain Res*. 2003;146(1–2):87–98.
12. Martin M, Dotti CG, Ledesma MD. Brain cholesterol in normal and pathological aging. *Biochim Biophys Acta*. 2010;1801(8):934–944.
13. Nieweg K, Schaller H, Pfrieger FW. Marked differences in cholesterol synthesis between neurons and glial cells from postnatal rats. *J Neurochem*. 2009;109(1):125–134.
14. Pfrieger FW. Outsourcing in the brain: do neurons depend on cholesterol delivery by astrocytes? *BioEssays*. 2003;25(1):72–78.
15. Lutjohann D, Breuer O, Ahlborg G, et al. Cholesterol homeostasis in human brain: evidence for an age-dependent flux of 24S-hydroxycholesterol from the brain into the circulation. *Proc Natl Acad Sci USA*. 1996;93(18):9799–9804.
16. Saito M, Benson EP, Saito M, Rosenberg A. Metabolism of cholesterol and triacylglycerol in cultured chick neuronal cells, glial cells, and fibroblasts: accumulation of esterified cholesterol in serum-free culture. *J Neurosci Res*. 1987;18(2):319–325.

17. Saito K, Dubreuil V, Arai Y, et al. Ablation of cholesterol biosynthesis in neural stem cells increases their VEGF expression and angiogenesis but causes neuron apoptosis. *Proc Natl Acad Sci USA*. 2009;106(20):8350–8355.
18. Funfschilling U, Saher G, Xiao L, Mobius W, Nave KA. Survival of adult neurons lacking cholesterol synthesis *in vivo*. *BMC Neurosci*. 2007;8:1.
19. Kandutsch AA, Russell AE. Preputial gland tumor sterols. 3. A metabolic pathway from lanosterol to cholesterol. *J Biol Chem*. 1960;235:2256–2261.
20. Vance JE, Pan D, Campenot RB, Bussiere M, Vance DE. Evidence that the major membrane lipids, except cholesterol, are made in axons of cultured rat sympathetic neurons. *J Neurochem*. 1994;62(1):329–337.
21. Campenot RB. Development of sympathetic neurons in compartmentalized cultures. Il Local control of neurite growth by nerve growth factor. *Dev Biol*. 1982;93(1):1–12.
22. de Chaves EI, Rusinol AE, Vance DE, Campenot RB, Vance JE. Role of lipoproteins in the delivery of lipids to axons during axonal regeneration. *J Biol Chem*. 1997;272(49):30766–30773.
23. Wang N, Yvan-Charvet L, Lutjohann D, et al. ATP-binding cassette transporters G1 and G4 mediate cholesterol and desmosterol efflux to HDL and regulate sterol accumulation in the brain. *FASEB J*. 2008;22(4):1073–1082.
24. Hinse CH, Shah SN. The desmosterol reductase activity of rat brain during development. *J Neurochem*. 1971;18(10):1989–1998.
25. Pfrieger FW, Ungerer N. Cholesterol metabolism in neurons and astrocytes. *Prog Lipid Res*. 2011;50(4):357–371.
26. Bloch K. The biological synthesis of cholesterol. *Science*. 1965;150(3692):19–28.
27. Wang H, Eckel RH. What are lipoproteins doing in the brain? *Trends Endocrinol Metab*. 2014;25(1):8–14.
28. Boyles JK, Pitas RE, Wilson E, Mahley RW, Taylor JM. Apolipoprotein E associated with astrocytic glia of the central nervous system and with nonmyelinating glia of the peripheral nervous system. *J Clin Invest*. 1985;76(4):1501–1513.
29. Zannis VI, Just PW, Breslow JL. Human apolipoprotein E isoprotein subclasses are genetically determined. *Am J Hum Genet*. 1981;33(1):11–24.
30. Myers RH, Schaefer EJ, Wilson PW, et al. Apolipoprotein E epsilon4 association with dementia in a population-based study: the Framingham study. *Neurology*. 1996;46(3):673–677.
31. Corder EH, Saunders AM, Strittmatter WJ, et al. Gene dose of apolipoprotein E type 4 allele and the risk of Alzheimer's disease in late onset families. *Science*. 1993;261(5123):921–923.
32. Suri S, Heise V, Trachtenberg AJ, Mackay CE. The forgotten APOE allele: a review of the evidence and suggested mechanisms for the protective effect of APOE varepsilon2. *Neurosci Biobehav Rev*. 2013;37(10 Pt 2): 2878–2886.
33. Fagan AM, Holtzman DM, Munson G, et al. Unique lipoproteins secreted by primary astrocytes from wild type, apoE (−/−), and human apoE transgenic mice. *J Biol Chem*. 1999;274(42):30001–30007.
34. Oitzl MS, Mulder M, Lucassen PJ, Havekes LM, Grootendorst J, de Kloet ER. Severe learning deficits in apolipoprotein E-knockout mice in a water maze task. *Brain Res*. 1997;752(1–2):189–196.
35. Gordon I, Grauer E, Genis I, Sehayek E, Michaelson DM. Memory deficits and cholinergic impairments in apolipoprotein E-deficient mice. *Neurosci Lett*. 1995;199(1):1–4.
36. Masliah E, Mallory M, Ge N, Alford M, Veinbergs I, Roses AD. Neurodegeneration in the central nervous system of apoE-deficient mice. *Exp Neurol*. 1995;136(2):107–122.
37. Buttini M, Orth M, Bellosta S, et al. Expression of human apolipoprotein E3 or E4 in the brains of Apoe−/− mice: isoform-specific effects on neurodegeneration. *J Neurosci*. 1999;19(12):4867–4880.
38. Krugers HJ, Mulder M, Korf J, Havekes L, de Kloet ER, Joels M. Altered synaptic plasticity in hippocampal CA1 area of apolipoprotein E deficient mice. *Neuroreport*. 1997;8(11):2505–2510.
39. Igbavboa U, Avdulov NA, Chochina SV, Wood WG. Transbilayer distribution of cholesterol is modified in brain synaptic plasma membranes of knockout mice deficient in the low-density lipoprotein receptor, apolipoprotein E, or both proteins. *J Neurochem*. 1997;69(4):1661–1667.
40. Hayashi H, Igbavboa U, Hamanaka H, et al. Cholesterol is increased in the exofacial leaflet of synaptic plasma membranes of human apolipoprotein E4 knock-in mice. *Neuroreport*. 2002;13(4):383–386.
41. Kim WS, Guillemin GJ, Glaros EN, Lim CK, Garner B. Quantitation of ATP-binding cassette subfamily-A transporter gene expression in primary human brain cells. *Neuroreport*. 2006;17(9):891–896.
42. Wahrle SE, Jiang H, Parsadanian M, et al. ABCA1 is required for normal central nervous system ApoE levels and for lipidation of astrocyte-secreted apoE. *J Biol Chem*. 2004;279(39):40987–40993.
43. Hirsch-Reinshagen V, Zhou S, Burgess BL, et al. Deficiency of ABCA1 impairs apolipoprotein E metabolism in brain. *J Biol Chem*. 2004;279(39):41197–41207.

44. Bu G, Maksymovitch EA, Nerbonne JM, Schwartz AL. Expression and function of the low density lipoprotein receptor-related protein (LRP) in mammalian central neurons. *J Biol Chem*. 1994;269(28):18521–18528.
45. Ishiguro M, Imai Y, Kohsaka S. Expression and distribution of low density lipoprotein receptor-related protein mRNA in the rat central nervous system. *Brain Res*. 1995;33(1):37–46.
46. Bjorkhem I, Lutjohann D, Diczfalusy U, Stahle L, Ahlborg G, Wahren J. Cholesterol homeostasis in human brain: turnover of 24S-hydroxycholesterol and evidence for a cerebral origin of most of this oxysterol in the circulation. *J Lipid Res*. 1998;39(8):1594–1600.
47. Martins IJ, Berger T, Sharman MJ, Verdile G, Fuller SJ, Martins RN. Cholesterol metabolism and transport in the pathogenesis of Alzheimer's disease. *J Neurochem*. 2009;111(6):1275–1308.
48. Pitas RE, Boyles JK, Lee SH, Foss D, Mahley RW. Astrocytes synthesize apolipoprotein E and metabolize apolipoprotein E-containing lipoproteins. *Biochim Biophys Acta*. 1987;917(1):148–161.
49. Kessel A, Ben-Tal N, May S. Interactions of cholesterol with lipid bilayers: the preferred configuration and fluctuations. *Biophys J*. 2001;81(2):643–658.
50. Goedeke L, Fernandez-Hernando C. MicroRNAs: a connection between cholesterol metabolism and neurodegeneration. *Neurobiol Dis*. 2014;72PA:48–53.
51. Ambros V. The functions of animal microRNAs. *Nature*. 2004;431(7006):350–355.
52. Bartel DP. MicroRNAs: genomics, biogenesis, mechanism, and function. *Cell*. 2004;116(2):281–297.
53. Ramirez CM, Davalos A, Goedeke L, et al. MicroRNA-758 regulates cholesterol efflux through posttranscriptional repression of ATP-binding cassette transporter A1. *Arterioscler Thromb Vasc Biol*. 2011;31(11):2707–2714.
54. Merrill Jr AH, Sullards MC, Wang E, Voss KA, Riley RT. Sphingolipid metabolism: roles in signal transduction and disruption by fumonisins. *Environ Health Perspect*. 2001;109(suppl 2):283–289.
55. Merrill Jr AH. Sphingolipid and glycosphingolipid metabolic pathways in the era of sphingolipidomics. *Chem Rev*. 2011;111(10):6387–6422.
56. Carter HE, Glick FJ, Norris WP, Phillips GE. Biochemistry of the sphingolipides III. Structure of sphingosine. *J Biol Chem*. 1947;170:285.
57. Pruett ST, Bushnev A, Hagedorn K, et al. Biodiversity of sphingoid bases ("sphingosines") and related amino alcohols. *J Lipid Res*. 2008;49(8):1621–1639.
58. Tidhar R, Futerman AH. The complexity of sphingolipid biosynthesis in the endoplasmic reticulum. *Biochim Biophys Acta*. 2013;1833(11):2511–2518.
59. Hanada K. Serine palmitoyltransferase, a key enzyme of sphingolipid metabolism. *Biochim Biophys Acta*. 2003;1632(1–3):16–30.
60. Futerman AH, Stieger B, Hubbard AL, Pagano RE. Sphingomyelin synthesis in rat liver occurs predominantly at the cis and medial cisternae of the Golgi apparatus. *J Biol Chem*. 1990;265(15):8650–8657.
61. Rozhkova AV, Dmitrieva VG, Zhapparova ON, et al. Human sphingomyelin synthase 1 gene (SMS1): organization, multiple mRNA splice variants and expression in adult tissues. *Gene*. 2011;481(2):65–75.
62. Vos JP, Giudici ML, van der Bijl P, et al. Sphingomyelin is synthesized at the plasma membrane of oligodendrocytes and by purified myelin membranes: a study with fluorescent- and radio-labelled ceramide analogues. *FEBS Lett*. 1995;368(2):393–396.
63. Hirabayashi Y. A world of sphingolipids and glycolipids in the brain – novel functions of simple lipids modified with glucose. *Proc Japan Acad Ser B Phys Biol Sci*. 2012;88(4):129–143.
64. Ishibashi Y, Kohyama-Koganeya A, Hirabayashi Y. New insights on glucosylated lipids: metabolism and functions. *Biochimica Biophysica Acta*. 2013;1831(9):1475–1485.
65. Hirabayashi Y. A world of sphingolipids and glycolipids in the brain–novel functions of simple lipids modified with glucose. *Proc Japan Acad Ser B Phys Biol Sci*. 2012;88(4):129–143.
66. Schnaar RL, Gerardy-Schahn R, Hildebrandt H. Sialic acids in the brain: gangliosides and polysialic acid in nervous system development, stability, disease, and regeneration. *Physiol Rev*. 2014;94(2):461–518.
67. Da Silva JS, Hasegawa T, Miyagi T, Dotti CG, Abad-Rodriguez J. Asymmetric membrane ganglioside sialidase activity specifies axonal fate. *Nat Neurosci*. 2005;8(5):606–615.
68. Thomas PD, Brewer GJ. Gangliosides and synaptic transmission. *Biochim Biophys Acta*. 1990;1031(3):277–289.
69. Boggs JM, Gao W, Hirahara Y. Myelin glycosphingolipids, galactosylceramide and sulfatide, participate in carbohydrate-carbohydrate interactions between apposed membranes and may form glycosynapses between oligodendrocyte and/or myelin membranes. *Biochim Biophys Acta*. 2008;1780(3):445–455.
70. Schulte S, Stoffel W. Ceramide UDPgalactosyltransferase from myelinating rat brain: purification, cloning, and expression. *Proc Natl Acad Sci USA*. 1993;90(21):10265–10269.

71. Tennekoon G, Zaruba M, Wolinsky J. Topography of cerebroside sulfotransferase in Golgi-enriched vesicles from rat brain. *J Cell Biol*. 1983;97(4):1107–1112.
72. Uemura S, Go S, Shishido F, Inokuchi J. Expression machinery of GM4: the excess amounts of GM3/GM4S synthase (ST3GAL5) are necessary for GM4 synthesis in mammalian cells. *Glycoconj J*. 2014;31(2):101–108.
73. Ueno K, Ando S, Yu RK. Gangliosides of human, cat, and rabbit spinal cords and cord myelin. *J Lipid Res*. 1978;19(7):863–871.
74. Chiba A, Kusunoki S, Obata H, Machinami R, Kanazawa I. Ganglioside composition of the human cranial nerves, with special reference to pathophysiology of Miller Fisher syndrome. *Brain Res*. 1997;745(1–2):32–36.
75. Fantini J, Yahi N. Molecular basis for the glycosphingolipid-binding specificity of alpha-synuclein: key role of tyrosine 39 in membrane insertion. *J Mol Biol*. 2011;408(4):654–669.
76. Bhat S, Spitalnik SL, Gonzalez-Scarano F, Silberberg DH. Galactosyl ceramide or a derivative is an essential component of the neural receptor for human immunodeficiency virus type 1 envelope glycoprotein gp120. *Proc Natl Acad Sci USA*. 1991;88(16):7131–7134.
77. Horres CR, Hannun YA. The roles of neutral sphingomyelinases in neurological pathologies. *Neurochem Res*. 2012;37(6):1137–1149.
78. Das DV, Cook HW, Spence MW. Evidence that neutral sphingomyelinase of cultured murine neuroblastoma cells is oriented externally on the plasma membrane. *Biochim Biophys Acta*. 1984;777(2):339–342.
79. Spence MW, Burgess JK. Acid and neutral sphingomyelinases of rat brain. Activity in developing brain and regional distribution in adult brain. *J Neurochem*. 1978;30(4):917–919.
80. Hofmann K, Tomiuk S, Wolff G, Stoffel W. Cloning and characterization of the mammalian brain-specific, Mg^{2+}-dependent neutral sphingomyelinase. *Proc Natl Acad Sci USA*. 2000;97(11):5895–5900.
81. Siebert M, Sidransky E, Westbroek W. Glucocerebrosidase is shaking up the synucleinopathies. *Brain*. 2014;137(Pt 5):1304–1322.
82. Beutler E, Nguyen NJ, Henneberger MW, et al. Gaucher disease: gene frequencies in the Ashkenazi Jewish population. *Am J Hum Genet*. 1993;52(1):85–88.
83. Mencarelli C, Martinez-Martinez P. Ceramide function in the brain: when a slight tilt is enough. *Cell Mol Life Sci*. 2013;70(2):181–203.
84. Stoffel W, Assmann G. Metabolism of sphingosine bases. XV. Enzymatic degradation of 4t-sphingenine 1-phosphate (sphingosine 1-phosphate) to 2t-hexadecen-1-al and ethanolamine phosphate. *Hoppe Seyler Z Physiol Chem*. 1970;351(8):1041–1049.
85. Olivera A, Kohama T, Edsall L, et al. Sphingosine kinase expression increases intracellular sphingosine-1-phosphate and promotes cell growth and survival. *J Cell Biol*. 1999;147(3):545–558.
86. Obeid LM, Linardic CM, Karolak LA, Hannun YA. Programmed cell death induced by ceramide. *Science*. 1993;259(5102):1769–1771.
87. Gomez-Munoz A, Kong JY, Salh B, Steinbrecher UP. Ceramide-1-phosphate blocks apoptosis through inhibition of acid sphingomyelinase in macrophages. *J Lipid Res*. 2004;45(1):99–105.
88. Goni FM, Alonso A. Biophysics of sphingolipids I. Membrane properties of sphingosine, ceramides and other simple sphingolipids. *Biochim Biophys Acta*. 2006;1758(12):1902–1921.
89. Sot J, Goni FM, Alonso A. Molecular associations and surface-active properties of short- and long-*N*-acyl chain ceramides. *Biochim Biophys Acta*. 2005;1711(1):12–19.
90. Di Scala C, Chahinian H, Yahi N, Garmy N, Fantini J. Interaction of Alzheimer's beta-amyloid peptides with cholesterol: mechanistic insights into amyloid pore formation. *Biochemistry*. 2014;53(28):4489–4502.
91. Hammache D, Pieroni G, Maresca M, Ivaldi S, Yahi N, Fantini J. Reconstitution of sphingolipid-cholesterol plasma membrane microdomains for studies of virus-glycolipid interactions. *Methods Enzymol*. 2000;312:495–506.
92. Kolesnick RN, Goni FM, Alonso A. Compartmentalization of ceramide signaling: physical foundations and biological effects. *J Cell Physiol*. 2000;184(3):285–300.
93. Holopainen JM, Subramanian M, Kinnunen PK. Sphingomyelinase induces lipid microdomain formation in a fluid phosphatidylcholine/sphingomyelin membrane. *Biochemistry*. 1998;37(50):17562–17570.
94. Simons K, Ikonen E. Functional rafts in cell membranes. *Nature*. 1997;387(6633):569–572.
95. Fantini J, Garmy N, Mahfoud R, Yahi N. Lipid rafts: structure, function and role in HIV, Alzheimer's and prion diseases. *Exp Rev Mol Med*. 2002;4(27):1–22.
96. Brown DA, London E. Functions of lipid rafts in biological membranes. *Annu Rev Cell Dev Biol*. 1998;14:111–136.

97. Brown RE. Sphingolipid organization in biomembranes: what physical studies of model membranes reveal. *J Cell Sci*. 1998;111(Pt 1):1–9.
98. Mattjus P, Slotte JP. Does cholesterol discriminate between sphingomyelin and phosphatidylcholine in mixed monolayers containing both phospholipids? *Chem Phys Lipids*. 1996;81(1):69–80.
99. Megha, London E. Ceramide selectively displaces cholesterol from ordered lipid domains (rafts): implications for lipid raft structure and function. *J Biol Chem*. 2004;279(11):9997–10004.
100. Megha, Sawatzki P, Kolter T, Bittman R, London E. Effect of ceramide *N*-acyl chain and polar headgroup structure on the properties of ordered lipid domains (lipid rafts). *Biochim Biophys Acta*. 2007;1768(9):2205–2212.
101. Mathias S, Dressler KA, Kolesnick RN. Characterization of a ceramide-activated protein kinase: stimulation by tumor necrosis factor alpha. *Proc Natl Acad Sci USA*. 1991;88(22):10009–10013.
102. Galadari S, Kishikawa K, Kamibayashi C, Mumby MC, Hannun YA. Purification and characterization of ceramide-activated protein phosphatases. *Biochemistry*. 1998;37(32):11232–11238.
103. Dobrowsky RT, Hannun YA. Ceramide-activated protein phosphatase: partial purification and relationship to protein phosphatase 2A. *Adv Lipid Res*. 1993;25:91–104.
104. Liu G, Kleine L, Hebert RL. Advances in the signal transduction of ceramide and related sphingolipids. *Crit Rev Clin Lab Sci*. 1999;36(6):511–573.
105. Stoica BA, Movsesyan VA, Knoblach SM, Faden AI. Ceramide induces neuronal apoptosis through mitogen-activated protein kinases and causes release of multiple mitochondrial proteins. *Mol Cell Neurosci*. 2005;29(3): 355–371.
106. Hoeferlin LA, Wijesinghe DS, Chalfant CE. The role of ceramide-1-phosphate in biological functions. *Handb Exp Pharmacol*. 2013;215:153–166.
107. Bajjalieh SM, Martin TF, Floor E. Synaptic vesicle ceramide kinase. A calcium-stimulated lipid kinase that co-purifies with brain synaptic vesicles. *J Biol Chem*. 1989;264(24):14354–14360.
108. Morrow MR, Helle A, Perry J, Vattulainen I, Wiedmer SK, Holopainen JM. Ceramide-1-phosphate, in contrast to ceramide, is not segregated into lateral lipid domains in phosphatidylcholine bilayers. *Biophys J*. 2009;96(6):2216–2226.
109. Kooijman EE, Sot J, Montes LR, et al. Membrane organization and ionization behavior of the minor but crucial lipid ceramide-1-phosphate. *Biophys J*. 2008;94(11):4320–4330.
110. Zeniou-Meyer M, Zabari N, Ashery U, et al. Phospholipase D1 production of phosphatidic acid at the plasma membrane promotes exocytosis of large dense-core granules at a late stage. *J Biol Chem*. 2007;282(30): 21746–21757.
111. Siskind LJ, Fluss S, Bui M, Colombini M. Sphingosine forms channels in membranes that differ greatly from those formed by ceramide. *J Bioenerg Biomembr*. 2005;37(4):227–236.
112. Garcia-Pacios M, Collado MI, Busto JV, et al. Sphingosine-1-phosphate as an amphipathic metabolite: its properties in aqueous and membrane environments. *Biophys J*. 2009;97(5):1398–1407.
113. Posse de Chaves EI. Sphingolipids in apoptosis, survival, and regeneration in the nervous system. *Biochim Biophys Acta*. 2006;1758(12):1995–2015.
114. Bieberich E. It's a lipid's world: bioactive lipid metabolism and signaling in neural stem cell differentiation. *Neurochem Res*. 2012;37(6):1208–1229.
115. Berridge MJ, Irvine RF. Inositol trisphosphate, a novel second messenger in cellular signal transduction. *Nature*. 1984;312(5992):315–321.
116. Willard SS, Koochekpour S. Glutamate signaling in benign and malignant disorders: current status, future perspectives, and therapeutic implications. *Int J Biol Sci*. 2013;9(7):728–742.
117. Ferraguti F, Crepaldi L, Nicoletti F. Metabotropic glutamate 1 receptor: current concepts and perspectives. *Pharmacol Rev*. 2008;60(4):536–581.
118. Shigemoto R, Nakanishi S, Mizuno N. Distribution of the mRNA for a metabotropic glutamate receptor (mGluR1) in the central nervous system: an *in situ* hybridization study in adult and developing rat. *J Comp Neurol*. 1992;322(1):121–135.
119. Park YK, Galik J, Ryu PD, Randic M. Activation of presynaptic group I metabotropic glutamate receptors enhances glutamate release in the rat spinal cord substantia gelatinosa. *Neurosci Lett*. 2004;361(1–3):220–224.
120. Petralia RS, Wang YX, Singh S, et al. A monoclonal antibody shows discrete cellular and subcellular localizations of mGluR1 alpha metabotropic glutamate receptors. *J Chem Neuroanat*. 1997;13(2):77–93.
121. Testa CM, Standaert DG, Young AB, Penney Jr JB. Metabotropic glutamate receptor mRNA expression in the basal ganglia of the rat. *J Neurosci*. 1994;14(5 Pt 2):3005–3018.

122. Yanovsky Y, Sergeeva OA, Freund TF, Haas HL. Activation of interneurons at the stratum oriens/alveus border suppresses excitatory transmission to apical dendrites in the CA1 area of the mouse hippocampus. *Neuroscience*. 1997;77(1):87–96.
123. Baude A, Nusser Z, Roberts JD, Mulvihill E, McIlhinney RA, Somogyi P. The metabotropic glutamate receptor (mGluR1 alpha) is concentrated at perisynaptic membrane of neuronal subpopulations as detected by immunogold reaction. *Neuron*. 1993;11(4):771–787.
124. Fotuhi M, Sharp AH, Glatt CE, et al. Differential localization of phosphoinositide-linked metabotropic glutamate receptor (mGluR1) and the inositol 1,4,5-trisphosphate receptor in rat brain. *J Neurosci*. 1993;13(5): 2001–2012.
125. del Rio E, McLaughlin M, Downes CP, Nicholls DG. Differential coupling of G-protein-linked receptors to Ca^{2+} mobilization through inositol(1,4,5)trisphosphate or ryanodine receptors in cerebellar granule cells in primary culture. *Eur J Neurosci*. 1999;11(9):3015–3022.
126. Fagni L, Chavis P, Ango F, Bockaert J. Complex interactions between mGluRs, intracellular Ca^{2+} stores and ion channels in neurons. *Trends Neurosci*. 2000;23(2):80–88.
127. Hermans E, Challiss RA. Structural, signalling, and regulatory properties of the group I metabotropic glutamate receptors: prototypic family C G-protein-coupled receptors. *Biochem J*. 2001;359(Pt 3):465–484.
128. Choi S, Lovinger DM. Metabotropic glutamate receptor modulation of voltage-gated Ca^{2+} channels involves multiple receptor subtypes in cortical neurons. *J Neurosci*. 1996;16(1):36–45.
129. Chavis P, Fagni L, Lansman JB, Bockaert J. Functional coupling between ryanodine receptors and L-type calcium channels in neurons. *Nature*. 1996;382(6593):719–722.
130. Ribeiro FM, Paquet M, Cregan SP, Ferguson SS. Group I metabotropic glutamate receptor signalling and its implication in neurological disease. *CNS Neurol Disord Drug Targets*. 2010;9(5):574–595.
131. Servitja JM, Masgrau R, Pardo R, et al. Metabotropic glutamate receptors activate phospholipase D in astrocytes through a protein kinase C-dependent and Rho-independent pathway. *Neuropharmacology*. 2003;44(2):171–180.
132. Aramori I, Nakanishi S. Signal transduction and pharmacological characteristics of a metabotropic glutamate receptor, mGluR1, in transfected CHO cells. *Neuron*. 1992;8(4):757–765.
133. Carta M, Lanore F, Rebola N, et al. Membrane lipids tune synaptic transmission by direct modulation of presynaptic potassium channels. *Neuron*. 2014;81(4):787–799.
134. Shin JJ, Loewen CJ. Putting the pH into phosphatidic acid signaling. *BMC Biol*. 2011;9:85.
135. Goto K, Kondo H. Molecular cloning and expression of a 90-kDa diacylglycerol kinase that predominantly localizes in neurons. *Proc Natl Acad Sci USA*. 1993;90(16):7598–7602.
136. Ishisaka M, Hara H. The roles of diacylglycerol kinases in the central nervous system: review of genetic studies in mice. *J Pharmacol Sci*. 2014;124(3):336–343.
137. Tu-Sekine B, Raben DM. Regulation and roles of neuronal diacylglycerol kinases: a lipid perspective. *Crit Rev Biochem Mol Biol*. 2011;46(5):353–364.
138. Athenstaedt K, Daum G. Phosphatidic acid, a key intermediate in lipid metabolism. *Eur J Biochem*. 1999;266(1): 1–16.
139. Kooijman EE, Burger KN. Biophysics and function of phosphatidic acid: a molecular perspective. *Biochim Biophys Acta*. 2009;1791(9):881–888.
140. Kooijman EE, Carter KM, van Laar EG, Chupin V, Burger KN, de Kruijff B. What makes the bioactive lipids phosphatidic acid and lysophosphatidic acid so special? *Biochemistry*. 2005;44(51):17007–17015.
141. Kooijman EE, Tieleman DP, Testerink C, et al. An electrostatic/hydrogen bond switch as the basis for the specific interaction of phosphatidic acid with proteins. *J Biol Chem*. 2007;282(15):11356–11364.
142. Burger KN, Demel RA, Schmid SL, de Kruijff B. Dynamin is membrane-active: lipid insertion is induced by phosphoinositides and phosphatidic acid. *Biochemistry*. 2000;39(40):12485–12493.
143. van den Brink-van der Laan E, Killian JA, de Kruijff B. Nonbilayer lipids affect peripheral and integral membrane proteins via changes in the lateral pressure profile. *Biochim Biophys Acta*. 2004;1666(1–2):275–288.
144. Hozumi Y, Goto K. Diacylglycerol kinase beta in neurons: functional implications at the synapse and in disease. *Adv Biol Reg*. 2012;52(2):315–325.
145. Huttner WB, Schmidt A. Lipids, lipid modification, and lipid-protein interaction in membrane budding and fission—insights from the roles of endophilin A1 and synaptophysin in synaptic vesicle endocytosis. *Curr Opin Neurobiol*. 2000;10(5):543–551.
146. Kim K, Yang J, Kim E. Diacylglycerol kinases in the regulation of dendritic spines. *J Neurochem*. 2010;112(3): 577–587.

147. Sin WC, Moniz DM, Ozog MA, Tyler JE, Numata M, Church J. Regulation of early neurite morphogenesis by the Na^+/H^+ exchanger NHE1. *J Neurosci*. 2009;29(28):8946–8959.
148. Diering GH, Mills F, Bamji SX, Numata M. Regulation of dendritic spine growth through activity-dependent recruitment of the brain-enriched Na(+)/H(+) exchanger NHE5. *Mol Biol Cell*. 2011;22(13):2246–2257.
149. Gaoni Y, Mechoulam R. Isolation, structure, and partial synthesis of an active constituent of hashish. *J Am Chem Soc*. 1964;86(8):1646–1647.
150. Matsuda LA, Lolait SJ, Brownstein MJ, Young AC, Bonner TI. Structure of a cannabinoid receptor and functional expression of the cloned cDNA. *Nature*. 1990;346(6284):561–564.
151. Munro S, Thomas KL, Abu-Shaar M. Molecular characterization of a peripheral receptor for cannabinoids. *Nature*. 1993;365(6441):61–65.
152. Kano M, Ohno-Shosaku T, Hashimotodani Y, Uchigashima M, Watanabe M. Endocannabinoid-mediated control of synaptic transmission. *Physiol Rev*. 2009;89(1):309–380.
153. Devane WA, Hanus L, Breuer A, et al. Isolation and structure of a brain constituent that binds to the cannabinoid receptor. *Science*. 1992;258(5090):1946–1949.
154. Sugiura T, Kondo S, Sukagawa A, et al. 2-Arachidonoylglycerol: a possible endogenous cannabinoid receptor ligand in brain. *Biochem Biophys Res Comm*. 1995;215(1):89–97.
155. Sugiura T, Kishimoto S, Oka S, Gokoh M. Biochemistry, pharmacology, and physiology of 2-arachidonoylglycerol, an endogenous cannabinoid receptor ligand. *Prog Lipid Res*. 2006;45(5):405–446.
156. Di Marzo V, Fontana A, Cadas H, et al. Formation and inactivation of endogenous cannabinoid anandamide in central neurons. *Nature*. 1994;372(6507):686–691.
157. Cadas H, di Tomaso E, Piomelli D. Occurrence and biosynthesis of endogenous cannabinoid precursor, *N*-arachidonoyl phosphatidylethanolamine, in rat brain. *J Neurosci*. 1997;17(4):1226–1242.
158. Okamoto Y, Morishita J, Tsuboi K, Tonai T, Ueda N. Molecular characterization of a phospholipase D generating anandamide and its congeners. *J Biol Chem*. 2004;279(7):5298–5305.
159. Vandevoorde S, Lambert DM. The multiple pathways of endocannabinoid metabolism: a zoom out. *Chem Biodivers*. 2007;4(8):1858–1881.
160. Cravatt BF, Giang DK, Mayfield SP, Boger DL, Lerner RA, Gilula NB. Molecular characterization of an enzyme that degrades neuromodulatory fatty-acid amides. *Nature*. 1996;384(6604):83–87.
161. Blankman JL, Simon GM, Cravatt BF. A comprehensive profile of brain enzymes that hydrolyze the endocannabinoid 2-arachidonoylglycerol. *Chem Biol*. 2007;14(12):1347–1356.
162. Kozak KR, Prusakiewicz JJ, Marnett LJ. Oxidative metabolism of endocannabinoids by COX-2. *Curr Pharm Des*. 2004;10(6):659–667.
163. Snider NT, Walker VJ, Hollenberg PF. Oxidation of the endogenous cannabinoid arachidonoyl ethanolamide by the cytochrome P450 monooxygenases: physiological and pharmacological implications. *Pharmacol Rev*. 2010;62(1):136–154.
164. Yu M, Ives D, Ramesha CS. Synthesis of prostaglandin E2 ethanolamide from anandamide by cyclooxygenase-2. *J Biol Chem*. 1997;272(34):21181–21186.
165. Kozak KR, Crews BC, Morrow JD, et al. Metabolism of the endocannabinoids, 2-arachidonylglycerol and anandamide, into prostaglandin, thromboxane, and prostacyclin glycerol esters and ethanolamides. *J Biol Chem*. 2002;277(47):44877–44885.
166. Alger BE. Endocannabinoids: getting the message across. *Proc Natl Acad Sci USA*. 2004;101(23):8512–8513.
167. Alger BE. Retrograde signaling in the regulation of synaptic transmission: focus on endocannabinoids. *Prog Neurobiol*. 2002;68(4):247–286.
168. Di Marzo V, Goparaju SK, Wang L, et al. Leptin-regulated endocannabinoids are involved in maintaining food intake. *Nature*. 2001;410(6830):822–825.
169. Marsicano G, Wotjak CT, Azad SC, et al. The endogenous cannabinoid system controls extinction of aversive memories. *Nature*. 2002;418(6897):530–534.
170. Kathuria S, Gaetani S, Fegley D, et al. Modulation of anxiety through blockade of anandamide hydrolysis. *Nat Med*. 2003;9(1):76–81.
171. Catala A. An overview of lipid peroxidation with emphasis in outer segments of photoreceptors and the chemiluminescence assay. *Int J Biochem Cell Biol*. 2006;38(9):1482–1495.
172. Niki E, Yoshida Y, Saito Y, Noguchi N. Lipid peroxidation: mechanisms, inhibition, and biological effects. *Biochem Biophys Res Comm*. 2005;338(1):668–676.
173. Montine TJ, Neely MD, Quinn JF, et al. Lipid peroxidation in aging brain and Alzheimer's disease. *Free Radic Biol Med*. 2002;33(5):620–626.

174. Markesbery WR, Lovell MA. Four-hydroxynonenal, a product of lipid peroxidation, is increased in the brain in Alzheimer's disease. *Neurobiol Aging*. 1998;19(1):33–36.
175. Sayre LM, Zelasko DA, Harris PL, Perry G, Salomon RG, Smith MA. 4-Hydroxynonenal-derived advanced lipid peroxidation end products are increased in Alzheimer's disease. *J Neurochem*. 1997;68(5):2092–2097.
176. Greenberg ME, Li XM, Gugiu BG, et al. The lipid whisker model of the structure of oxidized cell membranes. *J Biol Chem*. 2008;283(4):2385–2396.
177. Catala A. Lipid peroxidation modifies the picture of membranes from the "Fluid Mosaic Model" to the "Lipid Whisker Model". *Biochimie*. 2012;94(1):101–109.
178. Wood WG, Li L, Muller WE, Eckert GP. Cholesterol as a causative factor in Alzheimer's disease: a debatable hypothesis. *J Neurochem*. 2013;129(4):559–572.
179. Kivipelto M, Helkala EL, Laakso MP, et al. Midlife vascular risk factors and Alzheimer's disease in later life: longitudinal, population based study. *BMJ*. 2001;322(7300):1447–1451.
180. Pappolla MA, Bryant-Thomas TK, Herbert D, et al. Mild hypercholesterolemia is an early risk factor for the development of Alzheimer amyloid pathology. *Neurology*. 2003;61(2):199–205.
181. McGuinness B, O'Hare J, Craig D, Bullock R, Malouf R, Passmore P. Cochrane review on Statins for the treatment of dementia. *Int J Geriatr Psychiatry*. 2013;28(2):119–126.
182. Simons M, Keller P, Dichgans J, Schulz JB. Cholesterol and Alzheimer's disease: is there a link? *Neurology*. 2001;57(6):1089–1093.
183. Abramov AY, Ionov M, Pavlov E, Duchen MR. Membrane cholesterol content plays a key role in the neurotoxicity of β-amyloid: implications for Alzheimer's disease. *Aging Cell*. 2011;10(4):595–603.
184. Igbavboa U, Avdulov NA, Schroeder F, Wood WG. Increasing age alters transbilayer fluidity and cholesterol asymmetry in synaptic plasma membranes of mice. *J Neurochem*. 1996;66(4):1717–1725.
185. Posse de Chaves E, Sipione S. Sphingolipids and gangliosides of the nervous system in membrane function and dysfunction. *FEBS Lett*. 2010;584(9):1748–1759.
186. van Echten-Deckert G, Walter J. Sphingolipids: critical players in Alzheimer's disease. *Prog Lipid Res*. 2012;51(4):378–393.
187. Laufs U, Gertz K, Dirnagl U, Bohm M, Nickenig G, Endres M. Rosuvastatin, a new HMG-CoA reductase inhibitor, upregulates endothelial nitric oxide synthase and protects from ischemic stroke in mice. *Brain Res*. 2002;942(1–2):23–30.
188. Mohmmad Abdul H, Wenk GL, Gramling M, Hauss-Wegrzyniak B, Butterfield DA. APP and PS-1 mutations induce brain oxidative stress independent of dietary cholesterol: implications for Alzheimer's disease. *Neurosci Lett*. 2004;368(2):148–150.

CHAPTER

4

Variations of Brain Lipid Content

Jacques Fantini, Nouara Yahi

OUTLINE

4.1 BRAIN LIPIDS: HOW TO BRING ORDER TO THE GALAXY

As repeatedly stated in this book, the brain is a lipid machine with approximately 50% of its organic matter composed of lipids. To complicate the problem, lipids display broad biochemical diversity, allowing endless variations, from the more subtle (e.g., α-hydroxylation of the acyl chain of GalCer) to the more extreme (compare the structure of GT1b with that of cholesterol). To get a better idea of the situation, representative examples of closely related and totally unrelated lipid structures are given in Fig. 4.1.

More than 37,000 lipid structures are stored in the LIPID MAPS Structure Database (LMSD), a relational database encompassing structures and annotations of biologically relevant lipids.[1] Lipidomic analysis of the brain revealed hundreds of distinct lipid species.[2–4] The pattern of lipid expression was significantly modified following experimental traumatic brain injury in animal models.[5] Aging, gender, and disease factors have also been correlated with specific changes in brain lipid expression.[4,6]

At first glance, establishing the lipid composition of the brain could be compared to counting the stars in a galaxy. This metaphor is not particularly stimulating, but it is not totally incorrect

Brain Lipids in Synaptic Function and Neurological Disease. http://dx.doi.org/10.1016/B978-0-12-800111-0.00004-7

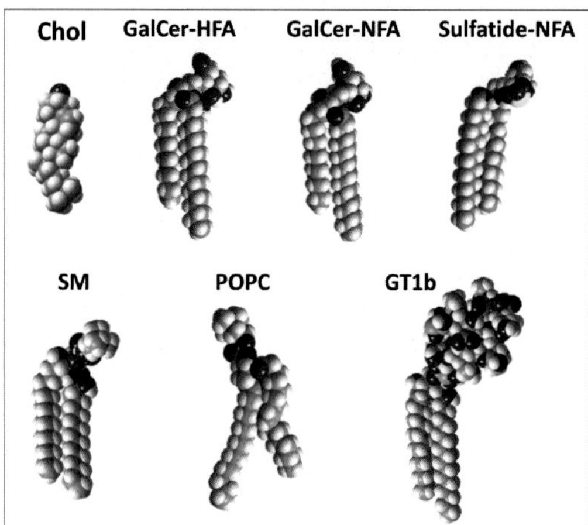

FIGURE 4.1 **Diversity of lipid structures.** Diverse lipid structures, from cholesterol (chol) to GT1b, a trisialylated ganglioside, are shown in sphere rendering (carbon green, oxygen red, nitrogen blue, hydrogen white). Sphingomyelin (SM) and phosphatidylcholine (POPC for palmitoyl oleoyl phosphatidylcholine) share the same polar head group (i.e., phosphorylcholine) but have significantly distinct shapes. GalCer-HFA and GalCer-NFA differ by only an OH group in the acyl chain of the ceramide moiety. Sulfatide is a sulfated derivative of GalCer.

either. The task is colossal, and we might well be lost before discovering anything. If we are too exhaustive, or on the opposite if we neglect a key lipid, we miss the point. In such cases, it is always good to break the problem into slices, or, in a more figurative representation, in concentric circles (Fig. 4.2). The central position is occupied by cholesterol, which is the major brain lipid involved in virtually all brain functions, starting with synaptic transmission. In marked contrast with all other lipids, cholesterol has a unique invariable structure. Hence, its variations

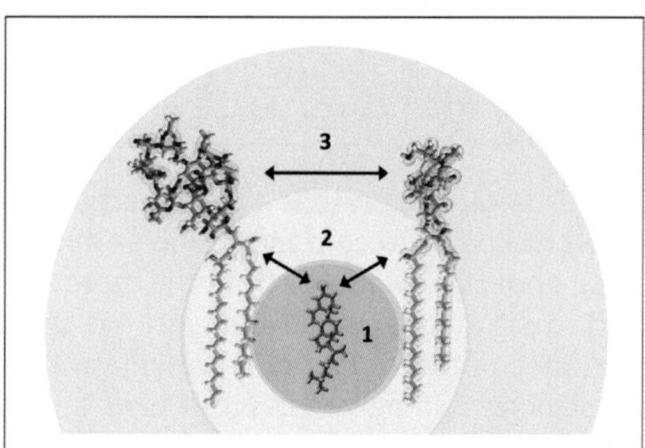

FIGURE 4.2 **Variability of lipid expression in the brain: a rational approach.** Most studies on lipid variability in the brain are focused on raft lipids, chiefly cholesterol and gangliosides. The different levels of investigation of brain lipid variations can be illustrated by concentric circles. Cholesterol is at the center of this multicircular system (rose disk, **1**). It may vary only in quantities and/or geographical localization. The ceramide moiety of gangliosides is represented in the second circle (yellow disk, **2**): it may vary in the structure of its two hydrophobic chains (type of sphingoid base, hydroxylation, and/or length of the acyl chain). Physical interactions between ceramide and cholesterol may finely tune ganglioside conformation (arrows). The third and last level of variability concerns the sugar head group of gangliosides (blue disk, **3**). Sialyltranferases and sialidases (neuraminidases) catalyze the interconversion of brain gangliosides (arrows), resulting in various levels of expression during development, aging, and neurodegenerative diseases.

are limited to changes in concentration and/or geographical localization. Correspondingly, one can establish the concentration of cholesterol in a brain region, in cell type, and in each leaflet of the plasma membrane. Thus, studying these contents as a function of time, gender, disease, or trauma is conceptually equivalent to solving a linear equation with one unknown.

The situation is more complex in the next concentric circles. In the brain, the second most important biologically active lipids are gangliosides. A ganglioside consists of two parts (glycone and ceramide) that are both subject to variability.[7,8] Changes in the sugar head group (glycone part) lead to a different ganglioside. For instance, adding a sialic acid to GD1b generates GT1b. In contrast, changes in the ceramide moiety do not modify the name of the ganglioside. Thus, the same ganglioside name (e.g., GM1) may represent a collection of gangliosides with the same sugar head group but with distinct ceramide parts. Ceramide has two hydrophobic chains: sphingosine and an acyl chain. Several types of sphingosine-like compounds (sphingoid bases) differing by their chain length, number of double bonds, and degree of hydroxylation have been described in mammalian cells. The acyl chain of ceramide may also vary in length, number of double bonds, and hydroxylation. Overall, what we call "GM1" is a family of gangliosides that may number more than 20 members, each with a unique biochemical structure conferring specific physicochemical properties. At this level, if we consider the acyl chain and the sphingoid base of ceramide as the unknown terms of the equation, we are solving a quadratic equation with two unknowns.

For this reason, we have chosen concentric circles to represent brain lipidomics approaches. The first circle after cholesterol represents the variations of the ceramide part of the ganglioside. The structure of ceramide influences the physical association of ganglioside with cholesterol.[9] In particular, the absence or presence of an α–OH group in the acyl chain has a major impact on ganglioside–cholesterol interactions.[9–11] On the other hand, the next circle represents the variations in the sugar head group of gangliosides. The second and third circles are functionally linked by metabolic enzymes (neuraminidases and sialyltransferases) that control the interconversion of gangliosides in biosynthetic and degradative pathways (see Chapter 3). However, the third circle is not connected to the central one (cholesterol) because the physical interaction of gangliosides with cholesterol is chiefly determined by the ceramide and the ceramide–sugar linkage. Hence, the nature of the sugar head group has little (if any) influence on ganglioside–cholesterol interactions. However, both the composition of the ceramide part and the structure of the glycone domain have an impact on the biological activity of gangliosides, which further justifies the position of the second and third circles in our representation. Now, the three unknowns in the equation include two for the ceramide part and one for the sugar head group. At this stage, lipidomics continues to give us this precise structural information (complete biochemical structure of gangliosides), but it becomes difficult to interpret these data at the molecular level. In the following paragraphs, we will try to remain as schematic as possible and, whenever possible, describe the functional consequences of the variation in brain lipid content.

4.2 VARIATIONS IN BRAIN CHOLESTEROL CONTENT

4.2.1 Cholesterol and Aging

As explained, the structure of cholesterol is immutable. Cholesterol may be oxidized, yet oxidized cholesterol is no longer cholesterol. Thus, cholesterol may only vary in localization and quantities. In this respect, global estimations of brains cholesterol could be considered as intrinsically biased because cholesterol content may greatly vary from one region of the brain to another. Nevertheless, comparative studies of brain lipid contents during the span of adult human life have given interesting information on the physiological evolution of cholesterol content. Most importantly, these studies have established that the brain weight begins to diminish at 20 years of age.[12] One can consider that a centenarian has lost about 20% of his brain mass.[12] During the same period (20–100 years), brain cholesterol decreases 20–50%.[12,13] In fact, a regular decrease of cholesterol content occurs as early as 20 years of age in both frontal and temporal lobes.[13] Before this comprehensive study, it was generally admitted that the loss of membrane lipids, including cholesterol, began in the white matter. One important finding of the study published by Svennerholm et al.[12] is that a significant loss of cholesterol occurs in the brain cortex. In line with these data, Martin et al. observed a moderate but significant loss of cholesterol levels in the membrane of aged rodent hippocampal neurons.[14] Moreover, decreased cholesterol contents were also evidenced in the cerebellum.[15] Overall, these data could be logically explained by a decrease in cholesterol biosynthesis upon aging,[16] and by increased levels of cholesterol removal from the brain.[17]

However, it should be mentioned that some studies failed to detect any significant age-related changes in cholesterol levels of certain brain areas.[15] It does not mean that cholesterol homeostasis is not affected at all in these regions. In fact, more subtle variations of cholesterol concentrations may occur locally. Igbavboa et al.[18] studied the respective cholesterol content of the exofacial and cytoplasmic leaflets of brain synaptic plasma membranes isolated from young and old mice. In young animals, the exofacial leaflet contained substantially less cholesterol than did the cytofacial leaflet (13% vs. 87%, respectively). However, exofacial leaflet cholesterol in the oldest group increased twofold over the youngest age group. This age-dependent modification of the asymmetric content of cholesterol in both leaflets of the plasma membrane had a marked effect on membrane fluidity. It is important to note that in this study, cholesterol and the cholesterol-to-phospholipid molar ratio did not differ among the different age groups of the studied mice. How could the redistribution of cholesterol in plasma membrane leaflets affect the fluidity properties of the membrane? An explanation of this phenomenon is given in Fig. 4.3.

In the plasma membrane of young animals, cholesterol is unevenly distributed. In the exofacial leaflet, it is concentrated in lipid raft domains where it interacts with sphingolipids. The other regions of the membrane are enriched in phosphatidylcholine and have lower cholesterol contents. Because sphingolipids are not present in the inner leaflet, the cytosolic face of lipid rafts has to compensate for this lack by specific phospholipids (e.g., phosphatidylserine) and by an increased content in cholesterol (Fig. 4.3A). Due to its higher content in cholesterol, the cytoplasmic leaflet is more rigid than the exofacial leaflet. Upon aging, the twofold cholesterol enrichment of the exofacial leaflet cannot be absorbed by lipid rafts, so the excess of cholesterol is spilled in the liquid disordered (Ld) phase of the membrane (Fig. 4.3B). Increasing the cholesterol content of the Ld phase will have a significant rigidifying effect.[7] Thus, the large difference in fluidity between the two leaflets observed in young mice is probably gradually abolished upon aging.

In addition, cholesterol is required for the fusion of synaptic vesicles with the presynaptic membrane.[19] Consistently, reducing cholesterol levels in cultured hippocampal neurons impaired vesicles exocytosis, and the effect was reversed by cholesterol reloading.[19] The

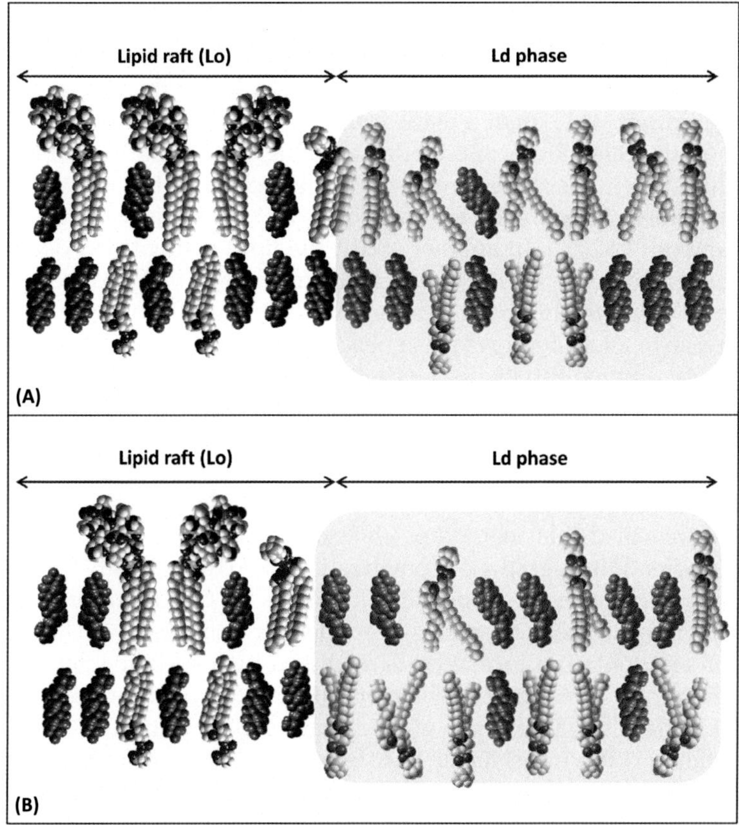

FIGURE 4.3 **Transbilayer variations of cholesterol content in brain membranes.** (A) In young individuals, cholesterol is enriched in the cytoplasmic leaflet of the plasma membrane, both in lipid rafts (Lo phase) and in the bulk membrane (Ld phase). (B) In older subjects, cholesterol is less abundant in the cytosolic leaflet and occurs in higher amounts in the exofacial leaflet, especially in the Ld phase where it may exert a rigidifying effect.

involvement of cholesterol in the fusion of synaptic vesicles is associated with both global and specific mechanisms. Cholesterol forces the vesicles to adopt an intrinsic curvature that is compatible with fusion.[20,21] In addition, in the vesicle membrane, cholesterol interacts specifically with the proteins of the fusion machinery (v-SNARE). Upon cholesterol binding, these proteins reorient their transmembrane domains in a functional parallel topology that is requested for the assembly of the fusion complex.[22] Therefore, it is likely that age-related decreases in cholesterol levels might compromise the curvature capacity of synaptic vesicles and synaptic plasma membranes.[23] The redistribution of cholesterol in the exofacial leaflet of synaptic membranes upon aging[18,24] could also perturb synaptic transmission. Finally, cholesterol interacts with a broad range of neurotransmitter receptors[25] (the molecular aspects of these interactions are treated in Chapter 7). In this respect, the age-associated changes of cholesterol amount and localization in synaptic membranes may progressively induce chaos in the well-oiled synapse machinery.

4.2.2 Brain Cholesterol and Neurodegenerative Diseases

The lipid composition has been carefully analyzed in different regions of the brain in patients with Alzheimer's disease.[26] The regions studied included the frontal, temporal, precentral, and occipital cortex, white matter in the frontal lobe area, the caudate nucleus, the hippocampus, the pons, the cerebellum, and medulla oblongata. The authors of this study did not detect any modification in the cholesterol levels of the frontal cortex, white matter, or medulla oblongata. In the other regions, they found a small, but not statistically significant, decrease in cholesterol amount (<20%). In another study, the lipid content was determined in brain tissue from patients with early onset (type I) and late onset (type II) Alzheimer's disease.[27] In the brains of type I patients, cholesterol content was only significantly reduced in the caudate nucleus. A decrease in cholesterol was also observed in frontal lobe white matter, especially in Alzheimer's disease type II patients. Overall, these data indicated that moderate changes of whole cholesterol may occur in the brain of Alzheimer's disease patients. However, no definitive model has emerged from these studies. In fact, it is possible that more subtle alterations such as a progressive increase of cholesterol in the exofacial leaflet of neuronal membranes could be of importance in Alzheimer's disease.[28] Indeed, Ca^{2+}-imaging studies in live neural cells have revealed that membrane cholesterol facilitated amyloid pore formation.[29] These pores are formed through the oligomerization of Alzheimer's β-amyloid peptides.[30–33] In fact, these oligomeric pores are responsible for increased Ca^{2+} fluxes in neural cells and are considered as the most neurotoxic species in Alzheimer's disease.[34–38] The stimulatory effect of cholesterol on amyloid pore formation is mediated by a specific interaction between the C-terminal hydrophobic domain of Alzheimer's β-amyloid peptide and cholesterol located in the exofacial leaflet.[29,39,40] Two types of membrane redistribution of cholesterol could favor this interaction: (1) the transbilayer transfer of cholesterol from the inner leaflet to the exofacial leaflet,[28] and (2) the lateral transfer of exofacial cholesterol from lipid raft domains to the bulk membrane. Interestingly, detergent-resistant membranes (DRM) prepared from temporal cortex of Alzheimer's disease brains showed strikingly low cholesterol contents.[41] Given that these membrane fractions are supposed to be derived from lipid raft domains,[42] they were expected to be enriched in cholesterol. Although the reliability of this method of preparation of membrane microdomains has been discussed,[43,44] it remains that the DRM from Alzheimer's disease brains were significantly depleted in cholesterol.[41] It could mean that cholesterol is excluded from these membrane domains or that the cholesterol present in these membranes has distinctive solubilization properties. Indeed, a concentration of cholesterol in the exofacial leaflet outside lipid raft domains could perfectly explain these results. If this proved to be the case, then we could draw a parallel between the transbilayer redistribution of cholesterol observed upon aging and Alzheimer's disease-associated changes in brain cholesterol. Consistently, a phenomenon referred to as "lipid raft aging," and characterized by important changes in the lipid content of lipid rafts (including variations of cholesterol), is exacerbated in a mouse model of Alzheimer's disease.[45]

Alterations in cholesterol homeostasis have also been associated with Huntington's disease.[46] In particular, both the biosynthesis and transport of cholesterol from astrocytes to neurons are affected.[47] Moreover, the expression of genes involved in the cholesterol biosynthetic pathway is reduced in the brain of several animal models of Huntington's disease.[48] Concomitantly, an accumulation of cholesterol, potentially linked to enhanced excitotoxicity,

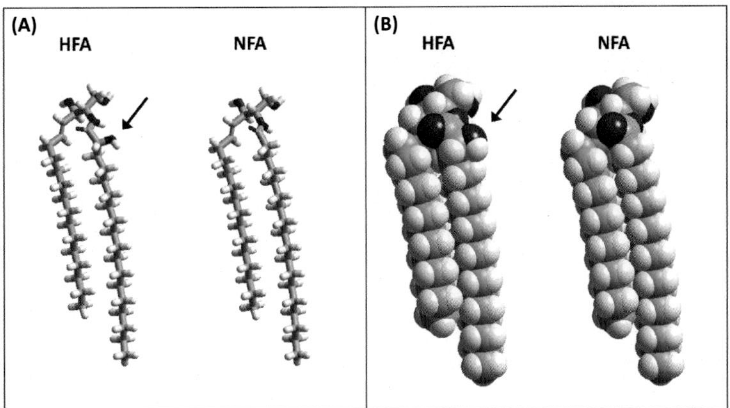

FIGURE 4.4 **Two versions of ceramides coexist in the brain.** HFA, α-hydroxylated acyl chain (ceramide-HFA for α-hydroxylated fatty acid); NFA, nonhydroxylated acyl chain (ceramide-NFA for nonhydroxylated fatty acid). These two distinct types of ceramides are represented in tube (A) or spheres (B). The αOH group of ceramide-HFA is indicated by an arrow.

has been reported in Huntington's disease.[49] Taken together, these data strongly suggest that like other neurodegenerative diseases, Huntington's disease is closely associated with an important dysregulation of cholesterol homeostasis.[50]

4.3 VARIATIONS IN BRAIN GANGLIOSIDE CONTENT

4.3.1 Ceramide Variability

Ceramide is the metabolic precursor of all glycosphingolipids, including gangliosides. In the plasma, it is the anchor that interacts with the apolar region of the membrane. It consists of two hydrocarbon chains, one corresponding to sphingosine, the other to an acyl group derived from a fatty acid. The biochemical diversity of brain ceramides chiefly arises from subtle variations in this acyl chain. In particular, the carbon atom linked to the carbonyl group (referred to as Cα) may carry an OH group (Fig. 4.4). Such α-hydroxylated ceramides are referred to as ceramide-HFA (for α-hydroxylated fatty acid) as explained in Chapter 1. Otherwise, the ceramide is nonhydroxylated (ceramide-NFA for nonhydroxylated fatty acid).

Ceramide-HFA and ceramide-NFA have significantly distinct physicochemical properties in lipid membranes.[51] Because ceramide is the common precursor of the glycosphingolipid biosynthetic pathway,[52] its hydroxylation status may influence the behavior of all glycosphingolipids independently of the nature of their glycone moiety.[53] This feature provides a unique opportunity for the brain to regulate glycosphingolipid function during development.[54] This is particularly important for galactosylceramide (GalCer) and its sulfated derivative (sulfatide) that are especially abundant in myelin.[55] Indeed, it has been shown that during the first postnatal month, the ratio of NFA/HFA of rat cerebral GalCer progressively declined, whereas this ratio increased for sulfatide.[56] The hydroxylation status has a dramatic influence on the conformation of GalCer[57] and its interaction with cholesterol.[10] In absence of cholesterol, Alzheimer's β-amyloid peptide (Aβ1–40) recognizes GalCer-HFA, but not GalCer-NFA monolayers.[10] Moreover, cholesterol inhibited the interaction of Aβ1–40 with GalCer-HFA, and, on the opposite, allowed Aβ1–40 to interact with GalCer-NFA. This dual effect was explained by a conformational tuning of cholesterol on GalCer conformation (see Chapter 11 for a full description of this conformational effect). Interestingly, gender-associated variations in ceramide NFA/HFA ratio have been reported[58] in a mouse model of Alzheimer's disease (the APP^{SL}/PS1Ki).[59] A strong increase in ceramide-HFA species occurred in female mice, whereas the males showed an increase in ceramide-NFA species.[58] Given that significant elevations of ceramide levels were detected in the brain of Alzheimer's disease patients,[60] it is tempting to speculate that (1) dysregulations of sphingolipid metabolism may contribute to the pathogenesis of Alzheimer's disease,[61] and (2) the gender-dependent accumulation of ceramide-HFA in the cortex of APP^{SL}/PS1Ki mice[58] may be mechanistically linked to the higher incidence and prevalence of Alzheimer's disease in women.[62] Finally one should note that the regulatory effect of cholesterol on the conformation of glycosphingolipids containing a NFA ceramide and the resulting stimulation of Aβ binding to these glycolipids has also been demonstrated for gangliosides GM1 and GM3.[9,10]

The notion that the same ganglioside may in fact represent a collection of molecules sharing the same carbohydrate head group but differing in their ceramide moiety is often overlooked. Indeed, reliable techniques for studying their expression and their possible age- and/or disease-associated evolution in brain tissues have not been available until recently. In this respect, imaging mass spectrometry of biological tissues has allowed a major breakthrough in both fundamental and clinical neurosciences.[63] This technique has formally established that the major ganglioside species in the mouse brain have either a C18 or a C20 chain in their ceramide part.[64] Moreover, these ceramide-distinct types of gangliosides have a specific distribution in the mouse brain. In particular, the C-20 species were selectively localized along the entorhinal-hippocampus projections, whereas C-18 gangliosides were widely distributed throughout the frontal brain.[64] These data are in line with the concept that variations in the acyl chain length of gangliosides may be associated with differential expression and distinct roles in brain functions.[65] It remains to determine how subtle changes in acyl chain length may affect the physicochemical properties of gangliosides, their capacity to interact with membrane lipids (especially cholesterol) and proteins (neurotransmitter receptors), and their role in the synapse.

Changes in ceramide structure have also been reported in Parkinson's disease.[66] The comparison of the major ceramide species showed important variations of acyl chain length in brain samples from normal individuals and Parkinson's disease patients.[66] In the anterior cingulate cortex from Parkinson's disease brains, total ceramide levels were reduced from control levels by 53%. Moreover, a significant shift toward shorter acyl length was noted. In parallel, an upregulation of the ceramide synthase gene was interpreted as a response to the decrease in ceramide levels.[66] Ceramide is the precursor of glycosphingolipids but also of signaling lipids such as ceramide-1-phosphate and sphingosine-1-phosphate. Hence, contradictory perturbations of ceramide metabolism (decreased content together with increased expression of ceramide synthase) may profoundly affect neuronal functions.[60] Moreover, it has been reported that in Alzheimer's disease, apoptosis could be mediated by ceramide-enriched exosomes secreted by astrocytes.[67] Overall, it is not surprising that both signaling sphingolipids and gangliosides have been recognized as critical players in neurodegenerative diseases.[61,68–71]

4.3.2 Sugar Head Group Variability

Cholesterol quantities can be measured by an enzymatic assay.[72] Ceramide amounts can be estimated by quantitative scanning of high-performance thin layer chromatography (HPTLC)[58] or by high-performance liquid chromatography (HPLC).[66] In both cases a lipid extract of the tissue has to be done prior the quantitation. Identification and quantification of specific gangliosides by density scanning or autoradiography of HPTLC plates is also possible.[61,73] In addition, the sugar head groups of gangliosides can be identified *in situ* by immunofluorescence and/or immunohistochemistry approaches.[74] This requires the obtention of antiganglioside antibodies, which is not an easy task because gangliosides are weak immunogens.[75,76] Nevertheless, a quite complete panel of antiganglioside antibodies, either polyclonal or monoclonal, has been progressively generated during the last 50 years.[74,77,78] Although these antibodies represent valuable tools for studying the expression of gangliosides in brain tissues,[75,76] two important artifacts should be mentioned: (1) the fixation method may not preserve the gangliosides (e.g., ethanol fixation should not be used because gangliosides are solubilized by alcohol; for the same reason, acetone, and detergent should be avoided)[79]; (2) another important issue is the cryptic behavior of the carbohydrate part of gangliosides that can be masked by cell surface proteins.[11,80–82] Therefore, the absence of labeling does not always reflect a lack of expression, which renders the interpretation of immunofluorescence images somewhat puzzling. In addition, the "cryptic behavior" may also be triggered by intramolecular interactions between the ceramide and the sugar head group of the ganglioside.[10,81] In this case, the structural variability of the acyl chain (α–OH, chain length) in different brain areas could give rise to an apparent yet artifactual geographical restriction of ganglioside expression.

The literature on the brain localization of gangliosides is plethoric. In 1996, Schwarz and Futerman[74] conducted a meta-analysis of about 40 studies using antiganglioside antibodies. Most of these data concerned the brain of rodents, yet some studies were also performed with human tissue.[13,27,83] To remain as simple as possible, we will first present the overall distribution of gangliosides in the adult brain. Then we will analyze the evolution of ganglioside expression upon aging (including embryonic development). In all cases, we will try to interpret these changes in the context of the development and aging of synaptic functions. Then we will move to the cellular scale and analyze the differential expression of gangliosides in neurons and glia. Finally we will discuss the impact of gangliosides expression in neurodegenerative diseases.

Between 20 and 40 years of age, a human brain's four most abundant gangliosides are GM1, GD1a, GD1b, and GT1b.[13,84,85] As explained in Chapter 3, these four gangliosides form a small cluster of metabolically linked structures.[8] The structure of these gangliosides is shown in Fig. 4.5. All have a voluminous glycone part displaying at least one (GM1), two (GD1a and GD1b), or three negative charges (GT1b).

Interestingly, a net predominance of di- and trisialylated gangliosides (69% of total brain gangliosides) over the monosialylated GM1 (28% of total brain gangliosides) was observed. This distribution reflects an intriguing aspect of synaptic function. The ganglioside charges form an electronegative screen that may push away negatively charged compounds (e.g., glutamate) but in contrast attract cationic molecules (e.g., monoamine neurotransmitters). As explained in Chapter 5 this screen controls the flux of neurotransmitters across the synaptic

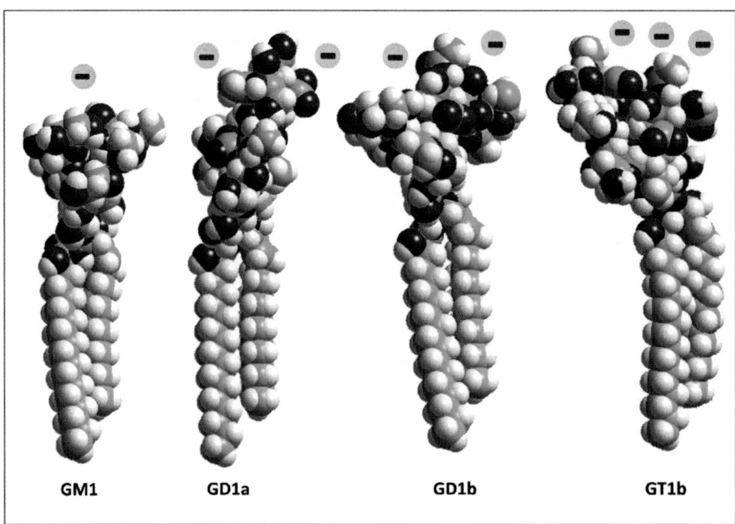

FIGURE 4.5 **Ganglioside structure, with special emphasis on the number of electrical negative charges.** Monosialylated gangliosides such as GM1 have one sialic acid, and thus one negative charge. Disialylated gangliosides (GD1a, GD1b) have two negative charges, and trisialylated gangliosides (GT1b) three. Otherwise, all these gangliosides have a large carbohydrate head group.

cleft. The notion that, at least for some functions, the charge of complex gangliosides might be more important than their precise structure is supported by experiments with a series of knockout mice.[8] These mice with disrupted GM2/GD2 synthase gene lacked complex gangliosides but did not show any major defects in their nervous systems.[86] In fact, they expressed higher levels of GM3 and GD3 in the brain that could apparently compensate for the lack of more complex gangliosides. In another study, the authors wanted to identify, among previously characterized α2,3-sialyltransferases, which of these enzyme(s) were responsible for adding the terminal α2,3 sialic acid to GM1 and GD1b.[87] It turned out that mice lacking GD1a and GT1b expressed increased levels of GM1 and GD1b and had no evident nervous system deficit. Finally, ganglioside-null mice lacking all ganglio-series gangliosides (including the four major brain gangliosides GM1, GD1a, GD1b, and GT1b) developed a severe neurodegenerative disease that resulted in early death.[88] The brains of the these mice showed a progressive decrease in size between 2 and 3 months of age, indicating that complex gangliosides are required right after birth for ensuring the stability of the maturing nervous system.

In line with these data, the development stages of the nervous system are associated with a coordinated expression of gangliosides.[89] In the mouse embryo, the core structure of all glycosphingolipids (i.e., ceramide) appears shortly after the first cell division (2–4 cell stage), that is, before neural tube formation.[90] Then the embryo successively expresses the globo-, lacto-, and ganglio-series of glycosphingolipids.[91] Gangliosides GM3 and GD3 appear at day E12 and remain predominant in the central nervous system until E16. At this stage, embryonic brain ganglioside expression shifts: GM1, GD1a, GD1b, and GT1b become the most abundant species whereas expression of GM3 and GD3 start to decline[54] (Fig. 4.6).

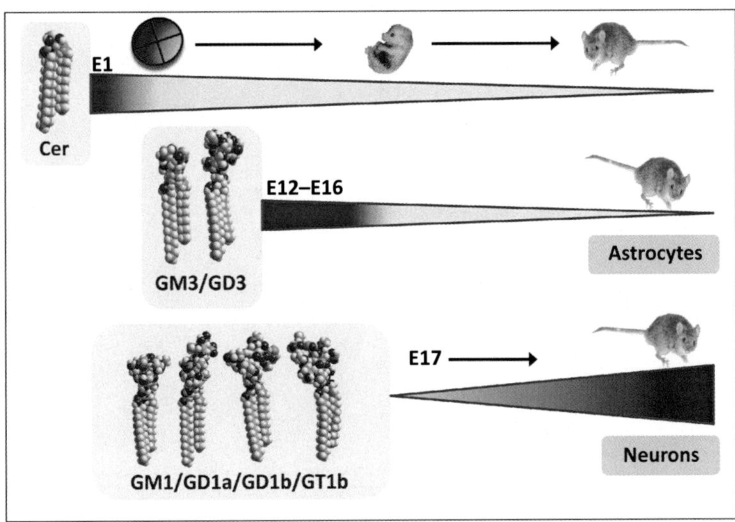

FIGURE 4.6 **Ceramide and ganglioside expression during mouse development.** This cartoon gives a schematic rendition of the evolution of ceramide and ganglioside expression during mouse development. Ceramide (Cer) is expressed early (2–4 cell stage of the mouse embryo). Then GM3 and GD3 appear at day E12 and remain predominant in the developing brain until day E16. At this stage a shift occurs in ganglioside expression, and the most abundant species become GM1, GD1a, GD1b, and GT1b. In the adult mouse, GM3 and GD3 are expressed by astrocytes, whereas the other gangliosides are expressed by neurons.

An increase of complex gangliosides (especially GD1a) was also observed during the prenatal development of the human cortex.[83] The level of the four major gangliosides is maximal in human adults between 20 and 40 years.[13] Then, a regular decrease occurs upon aging in both the gray and white matter. The most important change was a decrease in GD1a and GM1, whereas GD1b, GM3, and GD3 content were significantly increased.[13] Because GM1/GD1a are expressed in neurons and GM3/GD3 in astrocytes,[61,73] increases in GM3/GD3 cannot be interpreted as compensatory mechanism for the loss of GM1/GD1a from synaptic membranes. As suggested by Svennerholm et al.,[13] this change may rather indicate increased glial activity with aging.

The differential expression of gangliosides in neurons and glial cells has high biological significance. The main gangliosides expressed by astrocytes are GM3 and, yet in much lower amounts, GD3.[73,92] Together with the lack of complex gangliosides such as GT1b, the enrichment of astrocytes in GM3 could be interpreted as a remarkable adaptation to the glutamate-clearance function of these cells. Indeed, the sugar head group of GM3 is relatively small and this ganglioside has only one negative charge that is flush with the membrane[11] (Fig. 4.7).

Therefore, GM3 is not able to push away glutamate as complex gangliosides of the postsynaptic membrane can do. In fact, GT1b-enriched neuronal membranes act as a repulsive electronegative screen for glutamate (see Chapter 5). The presence of this screen implies that the glutamate binding site on its receptors has to be located in an extracellular domain, far from the membrane surface. Moreover, guided by this conductive surface that generates a continuous electronegative flux, glutamate is expulsed from the synaptic cleft and taken up by glutamate transporters expressed on the astrocyte membrane. In this respect, the differential

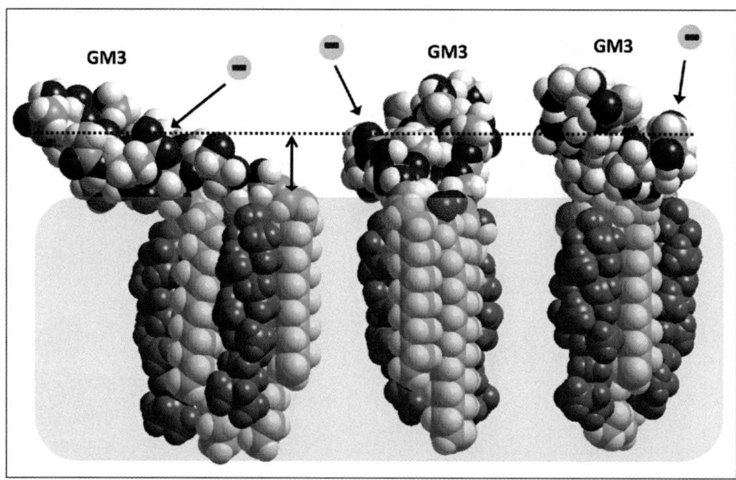

FIGURE 4.7 **Membrane topology of GM3 gangliosides.** GM3 molecules are represented in atom colors. Each GM3 interacts with a cholesterol molecule (in purple). The mean position of the carboxylate group of sialic acid (negative electric charge, arrows) is represented by a dotted line. It stands at a mean distance of 6 Å from the membrane surface (double arrow).

expression of gangliosides in neuronal and astrocytic membranes can be considered an intrinsic feature of the tripartite synapse.

Can any difference be noted between the gangliosides of the pre- and postsynaptic membranes? The analysis of membrane fractions extracted from *Torpedo marmorata* electric organ revealed an interesting asymmetric content.[93] In this study, presynaptic membranes were characterized by a low content of monosialylated gangliosides and a total lack of GM1. In contrast, the postsynaptic membrane contained high amounts of GM1. Otherwise, both pre- and postsynaptic membranes had a similar pattern of di- and trisialylated gangliosides. The question that immediately arises is whether the absence of GM1 from presynaptic membranes is a general feature that can be extrapolated to all presynaptic membranes. It is difficult to answer to this question because the ganglioside content of a given membrane area may greatly vary according to the brain region and upon aging. In this respect, an age-dependent GM1 accumulation, associated with high-density clustering, has been evidenced in synaptosomes prepared from mouse brains.[94] Therefore, it might be possible that presynaptic membranes could be gradually enriched in GM1 upon aging. Together with the redistribution of cholesterol in the exofacial leaflet of neuronal membrane,[18] GM1 enrichment might render synaptic transmission in the aged brain more chaotic. Conversely, the existence of an asymmetric ganglioside composition between the pre- and postsynaptic membranes would be logical if one considers the synapse as vectorial transmission machinery. (See Chapter 5 for further discussion of this issue.)

Changes in brain ganglioside expression have been documented in patients with Alzheimer's disease.[27,95] Compared with control brains, patients with early onset (type I) Alzheimer's disease exhibited an increase in GD3 and GM1 and a decrease in GD1a and GT1b.[27] Impaired ganglioside synthesis has also been noted with Huntington's disease.[96] It has been suggested that decreased GM1 levels could increase the apoptotic susceptibility of brain cells in patients with Huntington's disease. In favor of this hypothesis, inhibition of ganglioside synthesis in wild-type striatal cells induced a Huntington's disease-like phenotype that was reversed upon exogenous GM1 administration.[96] These data suggested that decreased brain GM1 levels might be linked to Huntington's disease pathogenesis. Similarly, a significant GM1 deficiency was observed in nigral dopaminergic neurons in patients with Parkinson's disease.[97] The mechanistic link between GM1 expression and Parkinson's disease was confirmed with the demonstration that genetically engineered mice devoid of GM1 exhibit typical Parkinson's disease symptoms, including motor impairment and depletion of dopamine in the striatum.[97] In Creutzfeldt–Jakob's disease, a decrease in complex gangliosides (GD1b, GT1b, and GQ1b) has been reported, together with an increase of GM1 and GD3.[98] The increased content of GD3 was attributed to the proliferation of astrocytes. The increase in GM1 might be related to the GM1-binding properties of the PrP protein.[99] Interestingly, abnormalities in acyl chain composition of gangliosides (lower expression of C20 species) have been evidenced in the brain of patients with Creutzfeldt–Jakob disease.[100] In remarkable convergence with these data obtained in various human neurodegenerative contexts, neural degeneration could be induced by disruption of ganglioside synthase genes in mice.[101] It cannot be a better argument to support the concept that gangliosides exert crucial influence on brain functions.

4.4 VARIATIONS IN MYELIN LIPIDS

In addition to cholesterol, a decrease in myelin-associated glycosphingolipids (i.e., GalCer and sulfatide, or galactosylsulfatide), occurs upon aging, especially after 50 years.[102] Indeed, the morphological changes in nerve cells during normal aging include a reorganization of myelin sheaths following segmental demyelination.[103] These anatomical alterations of myelin are associated with reduced speeds of information processing and axon transmission rates.[104] Myelin damage is also an important component of Alzheimer's disease.[105]

In the central nervous system, myelin is formed by oligodendrocytes. These cells express chiefly GalCer and sulfatide.[61] In mouse embryos, GalCer and sulfatide are the most lately expressed glycosphingolipids, becoming detectable shortly before birth at E17.[106] In the myelin sheath, these galactosphingolipids interact with each other by "*trans*" carbohydrate–carbohydrate interactions.[107] These interactions are topologically possible because the myelin sheath is a multilayered membrane system in which the successive layers face each other (Fig. 4.8). The strong interaction between GalCer and sulfatide is an important mechanism of stabilization of the myelin sheath.[107] Consequently, decreasing the expression of oligodendrocyte galactosphingolipids (as it occurs upon aging) may affect the functional organization of myelin.[108,109] An important aspect of *trans* GalCer–sulfatide interactions in myelin should be noted. At physiological pH, sulfatide bears a negative charge (due to its dissociated sulfonate group) that can bind electrostatically to positively charged lysine and arginine side chains of surrounding proteins. Such electrostatic interactions drive the association of sulfatide with a broad range of basic proteins, including laminin, an extracellular matrix protein.[110] Because the sulfonate group of sulfatide is not involved in *trans* GalCer–sulfatide interactions,[111] galactolipid-mediated interactions in the apposed myelin membranes are still compatible with the lateral organization of myelin proteins controlled by the extracellular matrix.[112]

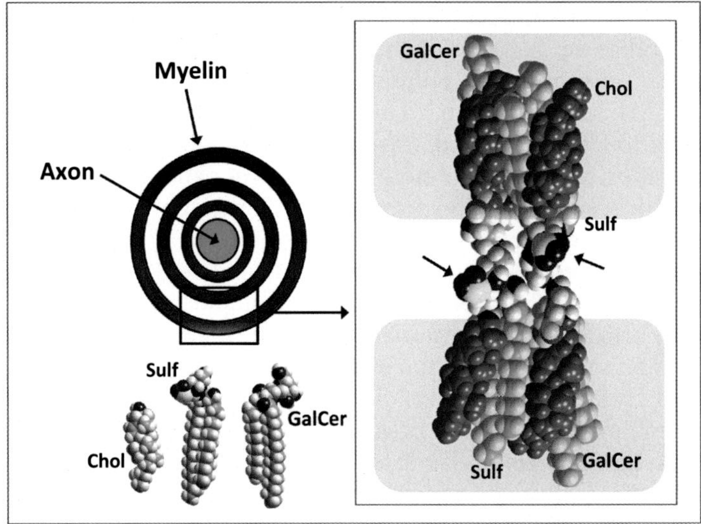

FIGURE 4.8 **GalCer–sulfatide interactions in myelin sheath.** Myelin is formed by a differentiation of oligodendrocyte plasma membranes that wrap around the axon in several concentric layers (left panel). The main lipids of myelin are cholesterol, GalCer, and sulfatide (the sulfated derivative of GalCer). *Trans* carbohydrate–carbohydrate interactions between GalCer and sulfatide on apposed membranes contribute to stabilize the structure of myelin (right panel). Note that in these interactions, the negatively charged sulfonate group of sulfatide remains fully accessible (arrows). It can thus interact with basic proteins that may further stabilize the myelinated building.

Pathological myelin changes were observed in the hippocampal complex of a mouse model of Alzheimer's disease.[113] Interestingly, *in vitro* studies have shown that Alzheimer's β-amyloid peptide oligomers inhibit myelin sheet formation.[114] This novel oligomer-dependent mechanism of pathogenesis could induce irreversible damage to myelinating neurons. Alzheimer's disease-associated perturbations of myelin organization and function could potentiate myelin defects caused by age-related decreases in GalCer/sulfatide expression. A clear-cut correlation between galactosphingolipid expression and myelin pathology was evidenced in multiple system atrophy (MSA), a neurodegenerative disease characterized by the accumulation of α-synuclein protein in the cytoplasm of oligodendrocytes.[115] Myelin instability is an early event in the pathogenesis of this disease.[116] Don et al. have recently studied the amount of galactosphingolipids in white matter from the brain of patients with MSA.[117] Compared with normal brains, the levels of both GalCer and sulfatide were significantly decreased (i.e., 40–69%) in affected white matter regions, whereas in unaffected white matter, no significant difference was detected. Overall, these data clearly indicated that dysregulation of myelin lipids in MSA pathogenesis may trigger myelin instability. The neurodegenerative alterations associated with MSA were recapitulated in a transgenic mouse model characterized by an accumulation of human alpha-synuclein in oligodendrocytes promoter.[118] Because α-synuclein has been shown to bind to both GalCer and sulfatide,[119] it is possible that a direct interaction between α-synuclein and myelin membranes could contribute to myelin pathology in this neurodegenerative disease. Moreover, both infectious (PrP^{sc}) and cellular (PrP^{c}) prion proteins have a high affinity for GalCer.[120,121] Amylin, an amyloid protein associated with type 2 diabetes, also binds to galactolipids.[122] Taken together, these data suggested that the capacity to bind to myelin sheath could be a general feature of amyloid proteins. From a structural point of view, this consequence is in fact expected from the systematic presence of a common sphingolipid-binding domain in amyloid proteins.[123] In any case, this point warrants further investigation because it could lead to the development of innovative therapeutic strategies aimed at blocking the deleterious association of amyloid proteins with myelin galactolipids.

4.5 IMPACT OF NUTRITION ON BRAIN LIPID CONTENT

"Young man says 'you are what you eat'—eat well. Old man says 'you are what you wear'—wear well." (from Genesis, "Dancing with the Moonlit Knight"). Faced with this intriguing manifestation of the conflict between the young and the older generation, one is tempted to side with the young. Our look may well suffer the consequences of that choice, but our brain will thank us. It is now well-documented that antioxidant and flavonoid-rich natural products (fruits, vegetables, and nuts) have potent neuroprotective effects.[124] Nutritional strategies are becoming important in the prevention of neurodegenerative diseases, including Alzheimer's and Parkinson's.[125–127] These strategies are chiefly based on dietary intake of polyphenol compounds that are known to attenuate the dramatic consequences of oxidative stress.[128] Alternatively, some natural compounds may have potent antiaggregative properties against various amyloid proteins, which is the case for the natural food pigment curcumin (Fig. 4.9), shown to

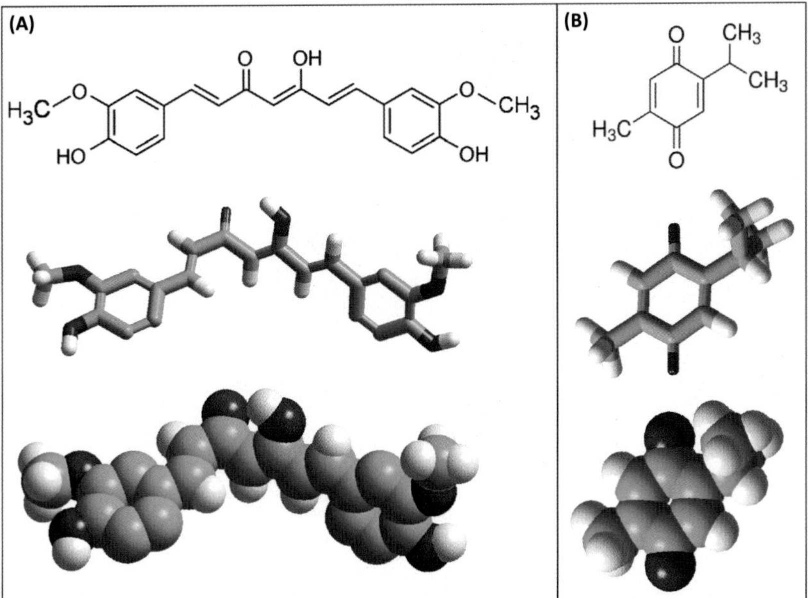

FIGURE 4.9 **Chemical structures of curcumin (A) and thymoquinone (B)**. Both compounds have potent neuroprotective properties and possible anti-Alzheimer and anti-Parkinson effects.

prevent the aggregation of Alzheimer's β-amyloid peptide.[129] Consistently, curcumin reduced oxidative damage, inflammation, and cognitive deficits in rats receiving brain infusions of toxic Alzheimer's β-amyloid peptide.[130] Curcumin also reduced the neurotoxic effects of α-synuclein in a cell model of Parkinson's disease.[131] Moreover, curcumin treatment could rescue aging-related loss of hippocampal synapse input specificity of long-term potentiation in mice, suggesting beneficial effects on memory.[132] *Nigella sativa* seeds contain high amounts of thymoquinone (Fig. 4.9), an antioxidant compound with interesting anti-Alzheimer and anti-Parkinson's properties.[133,134] The neuroprotective effects of various natural compounds found in food have been recently reviewed by Essa et al.[124]

More closely related to the main theme of this book is the link between dietary fatty acids, brain aging, and neurodegenerative diseases.[104] A growing line of evidence suggests that diets enriched in ω6 fatty acids may promote Alzheimer's disease pathogenesis. As a matter of fact, Western diets are typically high in ω6 fatty acids, especially linoleic acid (C18:2ω6), and comparatively low in ω3 fatty acids, such as α-linoleic acid (C18:3ω3). Because linoleic acid is the precursor of arachidonic acid (C20:4ω6), the preferential intake of ω6 versus ω3 fatty acids leads to a preponderance of arachidonic acid in brain glycerophospholipids (see Chapter 1 for further molecular details). Arachidonic acid is converted to prostaglandin, thereby creating a proinflammatory environment that favors neuronal pathogenesis in Alzheimer's disease.[135]

Cole et al.[104] listed no fewer than 14 neuroprotective or anti-Alzheimer's disease effects of docosahexaenoic acid (DHA, C22:6ω3) (Fig. 4.10). In preclinical models, dietary DHA has

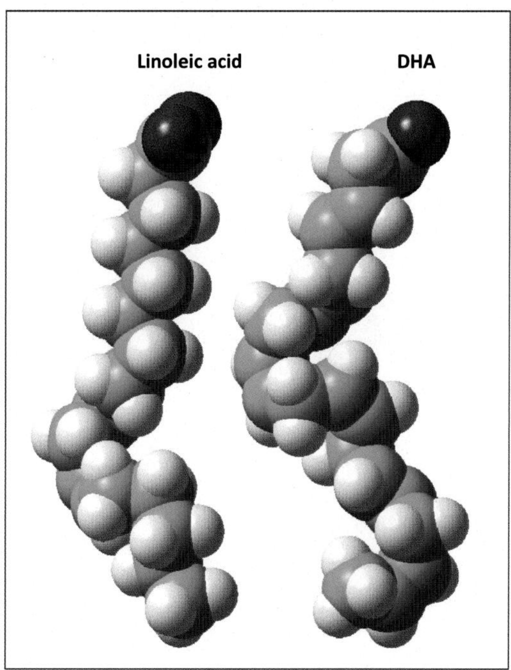

FIGURE 4.10 **Chemical structures of linoleic acid and DHA.** It is striking to see how close the structures of both fatty acids are. In fact, one is good for health (DHA), whereas the other one (linoleic acid) should be limited as much as possible in our diet.

been shown to promote neurogenesis and improved cognition.[136] Most importantly, DHA has been shown to exert remarkable protective effects against amyloid Alzheimer's β-oligomer-induced impairment of synaptic function in primary neurons.[137] Clinical trials in humans have confirmed that dietary supplementation of DHA could improve cognitive dysfunction.[138] In the absence of efficient treatments for neurodegenerative diseases, it is wise to incorporate in our diet some of the most potent antiaging and antineurodegenerative compounds naturally present in fish oils, fruits, vegetables, nuts, and herbs.[124] Definitely, the young man of the song mentioned at the beginning of this paragraph was wiser than the old man.

4.6 KEY EXPERIMENT: THE GM1/GM3 BALANCE AND ALZHEIMER'S DISEASE

As previously discussed, transgenic mice represent an interesting approach in studying the role of gangliosides in brain functions and their impact on neurodegenerative diseases. Oikawa et al.[139] studied the profile of amyloid deposition in the brain of transgenic mice coexpressing a mutant amyloid precursor protein (APP) and a disrupted GM2 synthase gene. In this animal model of Alzheimer's disease, amyloid deposition occurred in brain tissues expressing GM3 instead of GM1. Instead of the classical GM1-associated amyloid deposits found in the brain parenchyma of APP-mutated mice, the additional $GM1^-/GM3^+$ phenotype revealed a distinct type of amyloid pathology. Indeed, $GM1^-/GM3^+$ mice showed increased levels of amyloid plaques in the blood vessels surrounding brain parenchyma. In this respect, these data support the notion that the expression of brain gangliosides is a critical determinant for the type of amyloid pathology (senile plaque or amyloid angiopathy) in Alzheimer's disease. They also raise the intriguing possibility that the GM1/GM3 balance might control the regional localization of Alzheimer's β-amyloid peptide aggregation in brain tissues. In any case, these studies underscore the prominent role of gangliosides in brain functions and dysfunctions.

References

1. Sud M, Fahy E, Cotter D, et al. LMSD: LIPID MAPS structure database. *Nucleic Acids Res*. 2007;35:D527–D532.
2. Tajima Y, Ishikawa M, Maekawa K, et al. Lipidomic analysis of brain tissues and plasma in a mouse model expressing mutated human amyloid precursor protein/tau for Alzheimer's disease. *Lipids Health Dis*. 2013;12:68.
3. Woods AS, Jackson SN. Brain tissue lipidomics: direct probing using matrix-assisted laser desorption/ionization mass spectrometry. *AAPS J*. 2006;8(2):E391–395.
4. Adibhatla RM, Hatcher JF, Dempsey RJ. Lipids and lipidomics in brain injury and diseases. *AAPS J*. 2006;8(2):E314–321.
5. Abdullah L, Evans JE, Ferguson S, et al. Lipidomic analyses identify injury-specific phospholipid changes 3 mo after traumatic brain injury. *FASEB J*. 2014;28(12):5311–5321.
6. Rappley I, Myers DS, Milne SB, et al. Lipidomic profiling in mouse brain reveals differences between ages and genders, with smaller changes associated with alpha-synuclein genotype. *J Neurochem*. 2009;111(1):15–25.
7. Fantini J, Garmy N, Mahfoud RS, Yahi N. Lipid rafts: structure, function and role in HIV, Alzheimer's and prion diseases. *Exp Rev Mol Med*. 2002;4(27):1–22.
8. Schnaar RL, Gerardy-Schahn R, Hildebrandt H. Sialic acids in the brain: gangliosides and polysialic acid in nervous system development, stability, disease, and regeneration. *Physiol Rev*. 2014;94(2):461–518.
9. Fantini J, Yahi N, Garmy N. Cholesterol accelerates the binding of Alzheimer's beta-amyloid peptide to ganglioside GM1 through a universal hydrogen-bond-dependent sterol tuning of glycolipid conformation. *Front Physiol*. 2013;4:120.

10. Yahi N, Aulas A, Fantini J. How cholesterol constrains glycolipid conformation for optimal recognition of Alzheimer's beta amyloid peptide (Abeta1-40). *PloS One*. 2010;5(2):e9079.
11. Krengel U, Bousquet PA. Molecular recognition of gangliosides and their potential for cancer immunotherapies. *Front Immunol*. 2014;5:325.
12. Svennerholm L, Bostrom K, Jungbjer B. Changes in weight and compositions of major membrane components of human brain during the span of adult human life of Swedes. *Acta Neuropathol*. 1997;94(4):345–352.
13. Svennerholm L, Boström K, Jungbjer B, Olsson L. Membrane lipids of adult human brain: lipid composition of frontal and temporal lobe in subjects of age 20 to 100 years. *J Neurochem*. 1994;63(5):1802–1811.
14. Martin MG, Perga S, Trovo L, et al. Cholesterol loss enhances TrkB signaling in hippocampal neurons aging *in vitro*. *Mol Biol Cell*. 2008;19(5):2101–2112.
15. Soderberg M, Edlund C, Kristensson K, Dallner G. Lipid compositions of different regions of the human brain during aging. *J Neurochem*. 1990;54(2):415–423.
16. Thelen KM, Falkai P, Bayer TA, Lutjohann D. Cholesterol synthesis rate in human hippocampus declines with aging. *Neurosci Lett*. 2006;403(1–2):15–19.
17. Lutjohann D, Breuer O, Ahlborg G, et al. Cholesterol homeostasis in human brain: evidence for an age-dependent flux of 24S-hydroxycholesterol from the brain into the circulation. *Proc Natl Acad Sci USA*. 1996;93(18): 9799–9804.
18. Igbavboa U, Avdulov NA, Schroeder F, Wood WG. Increasing age alters transbilayer fluidity and cholesterol asymmetry in synaptic plasma membranes of mice. *J Neurochem*. 1996;66(4):1717–1725.
19. Linetti A, Fratangeli A, Taverna E, et al. Cholesterol reduction impairs exocytosis of synaptic vesicles. *J Cell Sci*. 2010;123(Pt 4):595–605.
20. Chen Z, Rand RP. The influence of cholesterol on phospholipid membrane curvature and bending elasticity. *Biophys J*. 1997;73(1):267–276.
21. Thiele C, Hannah MJ, Fahrenholz F, Huttner WB. Cholesterol binds to synaptophysin and is required for biogenesis of synaptic vesicles. *Nat Cell Biol*. 2000;2(1):42–49.
22. Tong J, Borbat PP, Freed JH, Shin YK. A scissors mechanism for stimulation of SNARE-mediated lipid mixing by cholesterol. *Proc Natl Acad Sci USA*. 2009;106(13):5141–5146.
23. Martin M, Dotti CG, Ledesma MD. Brain cholesterol in normal and pathological aging. *Biochim Biophys Acta*. 2010;1801(8):934–944.
24. Wood WG, Igbavboa U, Muller WE, Eckert GP. Cholesterol asymmetry in synaptic plasma membranes. *J Neurochem*. 2011;116(5):684–689.
25. Fantini J, Barrantes FJ. Sphingolipid/cholesterol regulation of neurotransmitter receptor conformation and function. *Biochim Biophys Acta*. 2009;1788(11):2345–2361.
26. Soderberg M, Edlund C, Alafuzoff I, Kristensson K, Dallner G. Lipid composition in different regions of the brain in Alzheimer's disease/senile dementia of Alzheimer's type. *J Neurochem*. 1992;59(5):1646–1653.
27. Svennerholm L, Gottfries CG. Membrane lipids, selectively diminished in Alzheimer brains, suggest synapse loss as a primary event in early-onset form (type I) and demyelination in late-onset form (type II). *J Neurochem*. 1994;62(3):1039–1047.
28. Gibson Wood W, Eckert GP, Igbavboa U, Muller WE. Amyloid beta-protein interactions with membranes and cholesterol: causes or casualties of Alzheimer's disease. *Biochim Biophys Acta*. 2003;1610(2):281–290.
29. Di Scala C, Troadec JD, Lelievre C, Garmy N, Fantini J, Chahinian H. Mechanism of cholesterol-assisted oligomeric channel formation by a short Alzheimer beta-amyloid peptide. *J Neurochem*. 2013.
30. Lal R, Lin H, Quist AP. Amyloid beta ion channel: 3D structure and relevance to amyloid channel paradigm. *Biochim Biophys Acta*. 2007;1768(8):1966–1975.
31. Butterfield SM, Lashuel HA. Amyloidogenic protein-membrane interactions: mechanistic insight from model systems. *Angew Chem Int Ed Engl*. 2010;49(33):5628–5654.
32. Shafrir Y, Durell S, Arispe N, Guy HR. Models of membrane-bound Alzheimer's Abeta peptide assemblies. *Proteins*. 2010;78(16):3473–3487.
33. Jang H, Arce FT, Capone R, Ramachandran S, Lal R, Nussinov R. Misfolded amyloid ion channels present mobile beta-sheet subunits in contrast to conventional ion channels. *Biophys J*. 2009;97(11):3029–3037.
34. Malchiodi-Albedi F, Paradisi S, Matteucci A, Frank C, Diociaiuti M. Amyloid oligomer neurotoxicity, calcium dysregulation, and lipid rafts. *Int JAlzheimers Dis*. 2011;906964.
35. Jang H, Arce FT, Ramachandran S, et al. Truncated beta-amyloid peptide channels provide an alternative mechanism for Alzheimer's disease and Down syndrome. *Proc Natl Acad Sci USA*. 2010;107(14):6538–6543.

36. Jang H, Arce FT, Ramachandran S, Capone R, Lal R, Nussinov R. beta-Barrel topology of Alzheimer's beta-amyloid ion channels. *J Mol Biol*. 2010;404(5):917–934.
37. Jang H, Connelly L, Arce FT, et al. Alzheimer's disease: which type of amyloid-preventing drug agents to employ? *Phys Chem Chem Phys*. 2013;15(23):8868–8877.
38. Arispe N, Diaz J, Durell SR, Shafrir Y, Guy HR. Polyhistidine peptide inhibitor of the Abeta calcium channel potently blocks the Abeta-induced calcium response in cells. Theoretical modeling suggests a cooperative binding process. *Biochemistry*. 2010;49(36):7847–7853.
39. Di Scala C, Chahinian H, Yahi N, Garmy N, Fantini J. Interaction of Alzheimer's beta-amyloid peptides with cholesterol: mechanistic insights into amyloid pore formation. *Biochemistry*. 2014;53(28):4489–4502.
40. Di Scala C, Yahi N, Lelievre C, Garmy N, Chahinian H, Fantini J. Biochemical identification of a linear cholesterol-binding domain within Alzheimer's beta amyloid peptide. *ACS Chem Neurosci*. 2013;4(3):509–517.
41. Molander-Melin M, Blennow K, Bogdanovic N, Dellheden B, Mansson JE, Fredman P. Structural membrane alterations in Alzheimer brains found to be associated with regional disease development; increased density of gangliosides GM1 and GM2 and loss of cholesterol in detergent-resistant membrane domains. *J Neurochem*. 2005;92(1):171–182.
42. Brown DA, Rose JK. Sorting of GPI-anchored proteins to glycolipid-enriched membrane subdomains during transport to the apical cell surface. *Cell*. 1992;68(3):533–544.
43. Lagerholm BC, Weinreb GE, Jacobson K, Thompson NL. Detecting microdomains in intact cell membranes. *Annu Rev Phys Chem*. 2005;56:309–336.
44. Munro S. Lipid rafts: elusive or illusive? *Cell*. 2003;115(4):377–388.
45. Fabelo N, Martin V, Marin R, et al. Evidence for premature lipid raft aging in APP/PS1 double-transgenic mice, a model of familial Alzheimer disease. *J Neuropathol Exp Neurol*. 2012;71(10):868–881.
46. Valenza M, Cattaneo E. Emerging roles for cholesterol in Huntington's disease. *Trends Neurosci*. 2011;34(9):474–486.
47. Leoni V, Caccia C. Study of cholesterol metabolism in Huntington's disease. *Biochem Biophys Res Comm*. 2014;446(3):697–701.
48. Valenza M, Leoni V, Karasinska JM, et al. Cholesterol defect is marked across multiple rodent models of Huntington's disease and is manifest in astrocytes. *J Neurosci*. 2010;30(32):10844–10850.
49. del Toro D, Xifro X, Pol A, et al. Altered cholesterol homeostasis contributes to enhanced excitotoxicity in Huntington's disease. *J Neurochem*. 2010;115(1):153–167.
50. Valenza M, Cattaneo E. Cholesterol dysfunction in neurodegenerative diseases: is Huntington's disease in the list? *Prog Neurobiol*. 2006;80(4):165–176.
51. Shah J, Atienza JM, Rawlings AV, Shipley GG. Physical properties of ceramides: effect of fatty acid hydroxylation. *J Lipid Res*. 1995;36(9):1945–1955.
52. Mullen TD, Hannun YA, Obeid LM. Ceramide synthases at the centre of sphingolipid metabolism and biology. *Biochem J*. 2012;441(3):789–802.
53. Johnston DS, Chapman D. The properties of brain galactocerebroside monolayers. *Biochim Biophys Acta*. 1988;937(1):10–22.
54. Yu RK, Nakatani Y, Yanagisawa M. The role of glycosphingolipid metabolism in the developing brain. *J Lipid Res*. 2009;50(suppl):S440–445.
55. Baumann N, Pham-Dinh D. Biology of oligodendrocyte and myelin in the mammalian central nervous system. *Physiol Rev*. 2001;81(2):871–927.
56. De Haas CG, Lopes-Cardozo M. Hydroxy- and non-hydroxy-galactolipids in developing rat CNS. *Int J Dev Neurosci*. 1995;13(5):447–454.
57. Nyholm PG, Pascher I, Sundell S. The effect of hydrogen bonds on the conformation of glycosphingolipids. Methylated and unmethylated cerebroside studied by X-ray single crystal analysis and model calculations. *Chem Phys Lipids*. 1990;52(1):1–10.
58. Barrier L, Ingrand S, Fauconneau B, Page G. Gender-dependent accumulation of ceramides in the cerebral cortex of the APP(SL)/PS1Ki mouse model of Alzheimer's disease. *Neurobiol Aging*. 2010;31(11):1843–1853.
59. Casas C, Sergeant N, Itier JM, et al. Massive CA1/2 neuronal loss with intraneuronal and N-terminal truncated Abeta42 accumulation in a novel Alzheimer transgenic model. *Am J Pathol*. 2004;165(4):1289–1300.
60. He X, Huang Y, Li B, Gong CX, Schuchman EH. Deregulation of sphingolipid metabolism in Alzheimer's disease. *Neurobiol Aging*. 2010;31(3):398–408.
61. van Echten-Deckert G, Walter J. Sphingolipids: critical players in Alzheimer's disease. *Prog Lipid Res*. 2012;51(4):378–393.

62. Schmidt R, Kienbacher E, Benke T, et al. Sex differences in Alzheimer's disease. *Neuropsychiatr*. 2008;22(1): 1–15.
63. Rubakhin SS, Hatcher NG, Monroe EB, Heien ML, Sweedler JV. Mass spectrometric imaging of the nervous system. *Curr Pharm Des*. 2007;13(32):3325–3334.
64. Sugiura Y, Shimma S, Konishi Y, Yamada MK, Setou M. Imaging mass spectrometry technology and application on ganglioside study; visualization of age-dependent accumulation of C20-ganglioside molecular species in the mouse hippocampus. *PloS One*. 2008;3(9):e3232.
65. Sonnino S, Chigorno V. Ganglioside molecular species containing C18- and C20-sphingosine in mammalian nervous tissues and neuronal cell cultures. *Biochim Biophys Acta*. 2000;1469(2):63–77.
66. Abbott SK, Li H, Muñoz SS, et al. Altered ceramide acyl chain length and ceramide synthase gene expression in Parkinson's disease. *Mov Disord*. 2014;29(4):518–526.
67. Wang G, Dinkins M, He Q, et al. Astrocytes secrete exosomes enriched with proapoptotic ceramide and prostate apoptosis response 4 (PAR-4): potential mechanism of apoptosis induction in Alzheimer disease (AD). *J Biol Chem*. 2012;287(25):21384–21395.
68. Fan M, Sidhu R, Fujiwara H, et al. Identification of Niemann-Pick C1 disease biomarkers through sphingolipid profiling. *J Lipid Res*. 2013;54(10):2800–2814.
69. Vitner EB, Futerman AH. Neuronal forms of Gaucher disease. *Handb Exp Pharmacol*. 2013;216:405–419.
70. Mencarelli C, Martinez-Martinez P. Ceramide function in the brain: when a slight tilt is enough. *Cell Mol Life Sci*. 2013;70(2):181–203.
71. Liu Q, Zhang J. Lipid metabolism in Alzheimer's disease. *Neurosci Bull*. 2014;30(2):331–345.
72. MacLachlan J, Wotherspoon AT, Ansell RO, Brooks CJ. Cholesterol oxidase: sources, physical properties and analytical applications. *J Steroid Biochem Mol Biol*. 2000;72(5):169–195.
73. Asou H, Hirano S, Uyemura K. Ganglioside composition of astrocytes. *Cell Struct Funct*. 1989;14(5):561–568.
74. Schwarz A, Futerman AH. The localization of gangliosides in neurons of the central nervous system: the use of anti-ganglioside antibodies. *Biochim Biophys Acta*. 1996;1286(3):247–267.
75. Kotani M, Kawashima I, Ozawa H, et al. Immunohistochemical localization of minor gangliosides in the rat central nervous system. *Glycobiology*. 1994;4(6):855–865.
76. Kotani M, Kawashima I, Ozawa H, Terashima T, Tai T. Differential distribution of major gangliosides in rat central nervous system detected by specific monoclonal antibodies. *Glycobiology*. 1993;3(2):137–146.
77. Bogoch S. Demonstration of serum precipitin to brain ganglioside. *Nature*. 1960;185:392–393.
78. Vajn K, Viljetic B, Degmecic IV, Schnaar RL, Heffer M. Differential distribution of major brain gangliosides in the adult mouse central nervous system. *PloS One*. 2013;8(9):e75720.
79. De Baecque C, Johnson AB, Naiki M, Schwarting G, Marcus DM. Ganglioside localization in cerebellar cortex: an immunoperoxidase study with antibody to GM1 ganglioside. *Brain Res*. 1976;114(1):117–122.
80. Hakomori SI, Teather C, Andrews H. Organizational difference of cell surface "hematoside" in normal and virally transformed cells. *Biochem Biophys Res Comm*. 1968;33(4):563–568.
81. Lingwood D, Binnington B, Rog T, et al. Cholesterol modulates glycolipid conformation and receptor activity. *Nat Chem Biol*. 2011;7(5):260–262.
82. Galban-Horcajo F, Halstead SK, McGonigal R, Willison HJ. The application of glycosphingolipid arrays to autoantibody detection in neuroimmunological disorders. *Curr Opin Chem Biol*. 2014;18:78–86.
83. Kracun I, Rosner H, Drnovsek V, Heffer-Lauc M, Cosovic C, Lauc G. Human brain gangliosides in development, aging and disease. *Int J Dev Biol*. 1991;35(3):289–295.
84. Tettamanti G, Bonali F, Marchesini S, Zambotti V. A new procedure for the extraction, purification and fractionation of brain gangliosides. *Biochim Biophys Acta*. 1973;296(1):160–170.
85. Svennerholm L, Bostrom K, Fredman P, Mansson JE, Rosengren B, Rynmark BM. Human brain gangliosides: developmental changes from early fetal stage to advanced age. *Biochim Biophys Acta*. 1989;1005(2): 109–117.
86. Takamiya K, Yamamoto A, Furukawa K, et al. Mice with disrupted GM2/GD2 synthase gene lack complex gangliosides but exhibit only subtle defects in their nervous system. *Proc Natl Acad Sci USA*. 1996;93(20): 10662–10667.
87. Sturgill ER, Aoki K, Lopez PH, et al. Biosynthesis of the major brain gangliosides GD1a and GT1b. *Glycobiology*. 2012;22(10):1289–1301.
88. Yamashita T, Wu YP, Sandhoff R, et al. Interruption of ganglioside synthesis produces central nervous system degeneration and altered axon-glial interactions. *Proc Natl Acad Sci USA*. 2005;102(8):2725–2730.

89. Xu YH, Barnes S, Sun Y, Grabowski GA. Multi-system disorders of glycosphingolipid and ganglioside metabolism. *J Lipid Res*. 2010;51(7):1643–1675.
90. Eliyahu E, Park JH, Shtraizent N, He X, Schuchman EH. Acid ceramidase is a novel factor required for early embryo survival. *FASEB J*. 2007;21(7):1403–1409.
91. Fenderson BA, Andrews PW, Nudelman E, Clausen H, Hakomori S. Glycolipid core structure switching from globo- to lacto- and ganglio-series during retinoic acid-induced differentiation of TERA-2-derived human embryonal carcinoma cells. *Dev Biol*. 1987;122(1):21–34.
92. Sbaschnig-Agler M, Dreyfus H, Norton WT, et al. Gangliosides of cultured astroglia. *Brain Res*. 1988;461(1): 98–106.
93. Ledeen RW, Diebler MF, Wu G, Lu ZH, Varoqui H. Ganglioside composition of subcellular fractions, including pre- and postsynaptic membranes, from Torpedo electric organ. *Neurochem Res*. 1993;18(11):1151–1155.
94. Yamamoto N, Matsubara T, Sato T, Yanagisawa K. Age-dependent high-density clustering of GM1 ganglioside at presynaptic neuritic terminals promotes amyloid beta-protein fibrillogenesis. *Biochim Biophys Acta*. 2008;1778(12):2717–2726.
95. Suzuki K, Katzman R, Korey SR. Chemical studies on Alzheimer's disease. *J Neuropathol Exp Neurol*. 1965;24: 211–224.
96. Maglione V, Marchi P, Di Pardo A, et al. Impaired ganglioside metabolism in Huntington's disease and neuroprotective role of GM1. *J Neurosci*. 2010;30(11):4072–4080.
97. Wu G, Lu ZH, Kulkarni N, Ledeen RW. Deficiency of ganglioside GM1 correlates with Parkinson's disease in mice and humans. *J Neurosci Res*. 2012;90(10):1997–2008.
98. Ando S, Toyoda Y, Nagai Y, Ikuta F. Alterations in brain gangliosides and other lipids of patients with Creutzfeldt–Jakob disease and subacute sclerosing panencephalitis (SSPE). *Jpn J Exp Med*. 1984;54(6):229–234.
99. Sanghera N, Correia BE, Correia JR, et al. Deciphering the molecular details for the binding of the prion protein to main ganglioside GM1 of neuronal membranes. *Chem Biol*. 2011;18(11):1422–1431.
100. Tamai Y, Ohtani Y, Miura S, et al. Creutzfeldt–Jakob disease--alteration in ganglioside sphingosine in the brain of a patient. *Neurosci Lett*. 1979;11(1):81–86.
101. Yu RK, Tsai YT, Ariga T. Functional roles of gangliosides in neurodevelopment: an overview of recent advances. *Neurochem Res*. 2012;37(6):1230–1244.
102. Svennerholm L, Bostrom K, Helander CG, Jungbjer B. Membrane lipids in the aging human brain. *J Neurochem*. 1991;56(6):2051–2059.
103. Pannese E. Morphological changes in nerve cells during normal aging. *Brain Struct Funct*. 2011;216(2):85–89.
104. Cole GM, Ma QL, Frautschy SA. Dietary fatty acids and the aging brain. *Nutr Rev*. 2010;68(Suppl 2):S102–111.
105. Noble M. The possible role of myelin destruction as a precipitating event in Alzheimer's disease. *Neurobiol Aging*. 2004;25(1):25–31.
106. Ngamukote S, Yanagisawa M, Ariga T, Ando S, Yu RK. Developmental changes of glycosphingolipids and expression of glycogenes in mouse brains. *J Neurochem*. 2007;103(6):2327–2341.
107. Boggs JM, Gao W, Zhao J, Park HJ, Liu Y, Basu A. Participation of galactosylceramide and sulfatide in glycosynapses between oligodendrocyte or myelin membranes. *FEBS Lett*. 2010;584(9):1771–1778.
108. Marcus J, Dupree JL, Popko B. Myelin-associated glycoprotein and myelin galactolipids stabilize developing axo-glial interactions. *J Cell Biol*. 2002;156(3):567–577.
109. Marcus J, Honigbaum S, Shroff S, Honke K, Rosenbluth J, Dupree JL. Sulfatide is essential for the maintenance of CNS myelin and axon structure. *Glia*. 2006;53(4):372–381.
110. Roberts DD. Sulfatide-binding proteins. *Chem Phys Lipids*. 1986;42(1–3):173–183.
111. Boggs JM, Menikh A, Rangaraj G. Trans interactions between galactosylceramide and cerebroside sulfate across apposed bilayers. *Biophys J*. 2000;78(2):874–885.
112. Ozgen H, Schrimpf W, Hendrix J, et al. The lateral membrane organization and dynamics of myelin proteins PLP and MBP are dictated by distinct galactolipids and the extracellular matrix. *PloS One*. 2014;9(7):e101834.
113. Schmued LC, Raymick J, Paule MG, Dumas M, Sarkar S. Characterization of myelin pathology in the hippocampal complex of a transgenic mouse model of Alzheimer's disease. *Curr Alzheimer Res*. 2013;10(1):30–37.
114. Horiuchi M, Maezawa I, Itoh A, et al. Amyloid beta1-42 oligomer inhibits myelin sheet formation *in vitro*. *Neurobiol Aging*. 2012;33(3):499–509.
115. Papp MI, Kahn JE, Lantos PL. Glial cytoplasmic inclusions in the CNS of patients with multiple system atrophy (striatonigral degeneration, olivopontocerebellar atrophy and Shy-Drager syndrome). *J Neurol Sci*. 1989; 94(1–3):79–100.

116. Ferguson B, Matyszak MK, Esiri MM, Perry VH. Axonal damage in acute multiple sclerosis lesions. *Brain*. 1997;120(Pt 3):393–399.
117. Don AS, Hsiao JH, Bleasel JM, Couttas TA, Halliday GM, Kim W. Altered lipid levels provide evidence for myelin dysfunction in multiple system atrophy. *Acta Neuropathol Comm*. 2014;2(1):150.
118. Shults CW, Rockenstein E, Crews L, et al. Neurological and neurodegenerative alterations in a transgenic mouse model expressing human alpha-synuclein under oligodendrocyte promoter: implications for multiple system atrophy. *J Neurosci*. 2005;25(46):10689–10699.
119. Fantini J, Yahi N. Molecular basis for the glycosphingolipid-binding specificity of alpha-synuclein: key role of tyrosine 39 in membrane insertion. *J Mol Biol*. 2011;408(4):654–669.
120. Klein TR, Kirsch D, Kaufmann R, Riesner D. Prion rods contain small amounts of two host sphingolipids as revealed by thin-layer chromatography and mass spectrometry. *Biol Chem*. 1998;379(6):655–666.
121. Mahfoud R, Garmy N, Maresca M, Yahi N, Puigserver A, Fantini J. Identification of a common sphingolipid-binding domain in Alzheimer, prion, and HIV-1 proteins. *J Biol Chem*. 2002;277(13):11292–11296.
122. Levy M, Garmy N, Gazit E, Fantini J. The minimal amyloid-forming fragment of the islet amyloid polypeptide is a glycolipid-binding domain. *FEBS J*. 2006;273(24):5724–5735.
123. Fantini J, Yahi N. Molecular insights into amyloid regulation by membrane cholesterol and sphingolipids: common mechanisms in neurodegenerative diseases. *Exp Rev Mol Med*. 2010;12:e27.
124. Essa MM, Vijayan RK, Castellano-Gonzalez G, Memon MA, Braidy N, Guillemin GJ. Neuroprotective effect of natural products against Alzheimer's disease. *Neurochem Res*. 2012;37(9):1829–1842.
125. Mi W, van Wijk N, Cansev M, Sijben JW, Kamphuis PJ. Nutritional approaches in the risk reduction and management of Alzheimer's disease. *Nutrition*. 2013;29(9):1080–1089.
126. Seidl SE, Santiago JA, Bilyk H, Potashkin JA. The emerging role of nutrition in Parkinson's disease. *Front Aging Neurosci*. 2014;6:36.
127. Hu N, Yu JT, Tan L, Wang YL, Sun L, Tan L. Nutrition and the risk of Alzheimer's disease. *Biomed Res Int*. 2013;2013:524820.
128. Bhullar KS, Rupasinghe HP. Polyphenols: multipotent therapeutic agents in neurodegenerative diseases. *Oxid Med Cell Longev*. 2013;2013:891748.
129. Ono K, Hasegawa K, Naiki H, Yamada M. Curcumin has potent anti-amyloidogenic effects for Alzheimer's beta-amyloid fibrils *in vitro*. *J Neurosci Res*. 2004;75(6):742–750.
130. Cole GM, Teter B, Frautschy SA. Neuroprotective effects of curcumin. *Adv Exp Med Biol*. 2007;595:197–212.
131. Wang MS, Boddapati S, Emadi S, Sierks MR. Curcumin reduces alpha-synuclein induced cytotoxicity in Parkinson's disease cell model. *BMC Neurosci*. 2010;11:57.
132. Cheng YF, Guo L, Xie YS, et al. Curcumin rescues aging-related loss of hippocampal synapse input specificity of long-term potentiation in mice. *Neurochem Res*. 2013;38(1):98–107.
133. Alhebshi AH, Gotoh M, Suzuki I. Thymoquinone protects cultured rat primary neurons against amyloid beta-induced neurotoxicity. *Biochem Biophys Res Comm*. 2013;433(4):362–367.
134. Alhebshi AH, Odawara A, Gotoh M, Suzuki I. Thymoquinone protects cultured hippocampal and human induced pluripotent stem cells-derived neurons against alpha-synuclein-induced synapse damage. *Neurosci Lett*. 2013.
135. Cudaback E, Jorstad NL, Yang Y, Montine TJ, Keene CD. Therapeutic implications of the prostaglandin pathway in Alzheimer's disease. *Biochem Pharmacol*. 2014;88(4):565–572.
136. Innis SM. Dietary (n-3) fatty acids and brain development. *J Nutr*. 2007;137(4):855–859.
137. Ma QL, Yang F, Rosario ER, et al. Beta-amyloid oligomers induce phosphorylation of tau and inactivation of insulin receptor substrate via c-Jun N-terminal kinase signaling: suppression by omega-3 fatty acids and curcumin. *J Neurosci*. 2009;29(28):9078–9089.
138. Kotani S, Sakaguchi E, Warashina S, et al. Dietary supplementation of arachidonic and docosahexaenoic acids improves cognitive dysfunction. *Neurosci Res*. 2006;56(2):159–164.
139. Oikawa N, Yamaguchi H, Ogino K, et al. Gangliosides determine the amyloid pathology of Alzheimer's disease. *Neuroreport*. 2009;20(12):1043–1046.

CHAPTER

5

A Molecular View of the Synapse

Jacques Fantini, Nouara Yahi

OUTLINE

5.1 THE SYNAPSE: A TRIPARTITE ENTITY?

The discovery that nerve cells are "independent elements which are never anastomosed" comes from the mastering use of the Golgi silver technique by Nobel Prize award winner Santiago Ramón y Cajal at the end of the nineteenth century.[1] However, the term *synapse* was coined by another Nobel laureate, Charles Scott Sherrington. Sherrington hypothesized that the lack of continuity between two neurons implied that the nervous impulse had to "*change in nature as it passes from one cell to the other.*" Etymologically, *synapsis* refers to a process of "contact." (Sherrington thought that *junction* was not appropriate and, motivated by the fear of committing a barbarism, he consulted Arthur Verrall, a Cambridge classicist, who helped him to choose the right word.)[2] In fact, the vision of Sherrington was more functional than structural, which explains why a Latin or Greek root for "junction" was not acceptable.

Anyway, the notion that nerve cells are individual and not a continuum is one of the most fundamental discoveries in biology. Electron microscopy studies of the neuromuscular junction revealed important anatomical features of the synapse: (1) the local thickenings of the pre- and

Brain Lipids in Synaptic Function and Neurological Disease. http://dx.doi.org/10.1016/B978-0-12-800111-0.00005-9

postsynaptic membranes (which reflects an intense activity), (2) the presence of numerous small vesicles at the nerve end near the presynaptic membrane, and (3) an extracellular space between the two membranes.[3] For the first time, the discontinuity between neuron membranes was evidenced at high microscopic resolution, providing definitive evidence supporting Cajal's thesis of the neuron doctrine. Most importantly, the functional orientation of the synapse was given by the exclusive localization of the vesicles on the presynaptic side. Correspondingly, a synapse can be described as an anatomical differentiation of two neuronal membranes separated by an extracellular space and functionally connected by a chemical transmission of information. This transfer of information is ensured by molecular messengers called neurotransmitters that are stored in the presynaptic vesicles, released in the extracellular space, and collected by specific receptors located on the postsynaptic membrane. This concept has gradually evolved with the discoveries of retrograde synapses (neurotransmission from the postsynaptic to the presynaptic neuron as it is the case for instance for endocannabinoids)[4] and with the inclusion of a third partner, the astrocyte.[5]

Historically, glia have been chiefly considered as accessory cells helping neurons to accomplish their noble tasks. Therefore, the concept of the tripartite synapse including the contribution of glial cells to the classical the pre- and postsynaptic neurons duet has been no less than a revolution in the neurosciences.[5] The concept emerged in 1990s when it was shown that glutamate could induce intracellular Ca^{2+} oscillations in astrocytes.[6] This discovery exploded the barrier between neurons and astrocytes, the latter being considered (until these findings) nonexcitable cells. Consequently, the old paradigm of brain function resulting exclusively from neuronal activity has been amended to acknowledge the contribution of glial cells.[5] It turned out that astrocytes, like neurons, can release several neurotransmitters, including glutamate, in response to an increases of Ca^{2+} concentration.[5,7] This property, referred to as *gliotransmission*, indicated a reciprocal communication between neurons and astrocytes. Hence, astrocytes are now considered as integral components of synapses.

5.2 ROLE OF GANGLIOSIDES IN GLUTAMATE CLEARANCE

Another critical aspect of astrocyte function is the clearance of glutamate released by the presynaptic neuron upon glutamatergic transmission. Indeed, excess glutamate leads to cell death via excitotoxicity.[8] Thus, the rapid removal of glutamate from the synaptic cleft is required for the survival of neurons.[9] Astrocytes fulfill this vital function via the expression on their cell surface of high-affinity glutamate transporters (excitatory amino acid transporters, EAATs).[10] These transporters mediate the astrocytic uptake of glutamate accompanied by the cotransport of 3 Na^+ and 1 H^+, and the countertransport of 1 K^+. The Na^+ ions are then expulsed from the cell via the Na^+/K^+-ATPase, thereby coupling the facilitated diffusion of glutamate (a passive transport) to an active (ATP-dependent) transport system that is in fact the driving force of glutamate transport (Fig. 5.1).

This "secondary active" transport ensures a high velocity of glutamate transport within the astrocyte independently of the respective extracellular and intracellular concentrations of the neurotransmitter. This point is an important because it is generally believed that an active transport allows the transport of a molecule against a concentration gradient. Given the high concentrations of glutamate in the extracellular space bathing a glutamatergic synapse,

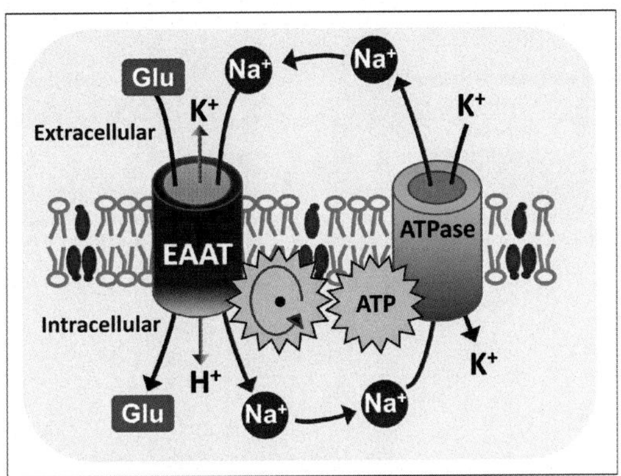

FIGURE 5.1 **Glutamate transport into astrocytes.** The transport of glutamate across the plasma membrane of astrocytes is mediated by a high-affinity transporter (EAAT). The passage of one glutamate is coupled to the co-transport of three Na^+ and one H^+ ions, and the counter transport of one K^+. This passive transport is functionally coupled to the Na^+/K^+-ATPase (the sodium/potassium pump), which exports three Na^+ and transports two K^+ ions inside the cell. Hence, the driving force of glutamate entry into the astrocyte is the active export of Na^+ ions by the ATP-dependent Na^+/K^+ pump.

it could be considered that glutamate enters the astrocyte via a passive mechanism controlled by its concentration gradient. Although a passive transport is advantageous for the cells (no ATP required), it suffers from a serious drawback: the concentration gradient regularly decreases as the transport goes on, until the transport is stopped when extracellular and intracellular concentrations are equilibrated. Clearly, if the astrocyte relied on passive transport to clear glutamate, the neuron would be killed within the minutes following the release of glutamate in the synaptic cleft. To avoid this fatal issue, the astrocyte has to maintain a maximal velocity of glutamate uptake independently of the extracellular and intracellular concentrations of the neurotransmitter. The price to pay is an ATP molecule (for the Na^+/K^+-ATPase) per glutamate transported.

A schematic rendition of the tripartite synapse, emphasizing the role of astrocytes in the management of excess glutamate and gliotransmitter release is described in Fig. 5.2.

In addition to glutamate, astrocytes have been shown to secrete several neuroactive molecules including γ-aminobutyric acid (GABA), ATP, adenosine, prostaglandins, and tumor necrosis factor α (TNF-α). All these compounds exert a fine control on synaptic transmission.[5] For instance, it has been shown that astrocytic activation evokes slow inward currents (SICs) through activation of *N*-methyl-aspartate (NMDA) receptors.[11,12] It should be noted that the astrocytic effects on neuronal physiology occur in the second time scale, whereas neuronal transmission is a matter of milliseconds, allowing astrocytes to efficiently control the strength of synaptic transmission. Astrocyte effects on long-term synaptic plasticity have also been reported, especially in the hypothalamus where ATP released by astrocytes enhanced the dynamic range for long-term potentiation and mediated activity-dependent, heterosynaptic depression.[13]

The plasma membrane of neurons and glial cells is equipped with a series of neurotransmitter receptors and transporters that coordinate the functions of the tripartite synapse. However, an important particularity in the membrane composition of these cells has been overlooked. Both cell types express high levels of gangliosides, but with a distinct specificity. Neurons

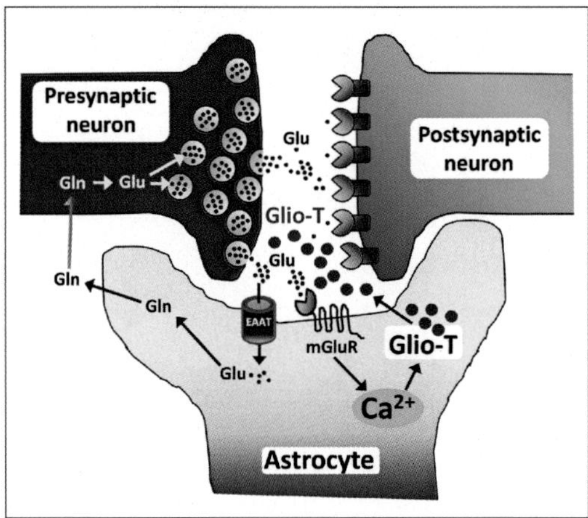

FIGURE 5.2 **The tripartite synapse.** The tripartite synapse model couples an astrocyte to the classic presynaptic/postsynaptic neurons duet. The astrocyte has two essential functions: (1) clearance of glutamate (Glu) in excess from the synapse, which is mediated by EAAT proteins (see Fig. 5.1); (2) the astrocyte also expresses metabotropic receptors for glutamate (mGluR). The activation of these receptors by glutamate induces an increase of intracellular Ca^{2+} that triggers the release of series of gliotransmitters (Glio-T) to regulate synaptic functions. (This model has been recently challenged though, as explained in the last paragraph of this chapter.) Note that the astrocyte transforms glutamate into glutamine (Gln). The recapture of glutamine by the presynaptic neuron allows the regeneration of glutamate at minimal energetic cost.

synthesize all four major brain gangliosides (GM1, GD1a, GD1b, and GT1b).[14] In contrast, astrocytes express essentially one ganglioside, GM3. This differential expression creates a significant dissymmetry in the tripartite synapse (Fig. 5.3). The presence of di- and trisialylated gangliosides (GD1a/GD1b and GT1b) in the pre- and postsynaptic membranes[14,15] confers a high density of negative charges (one per sialic acid residue). The astrocytic membrane lacks

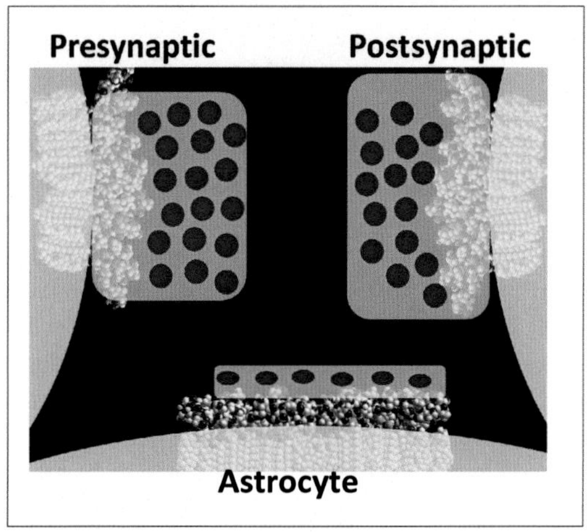

FIGURE 5.3 **Gangliosides generate a gradient of negative charges in the tripartite synapse.** Neurons and astrocytes have markedly distinct ganglioside composition. Neurons are enriched in polysialylated gangliosides, whereas astrocytes express chiefly a monosialyl ganglioside with a small polar head group, GM3. Therefore, the surface of the pre- and postsynaptic neurons display a higher density of negative charges (red disks) compared to that of astrocytes.

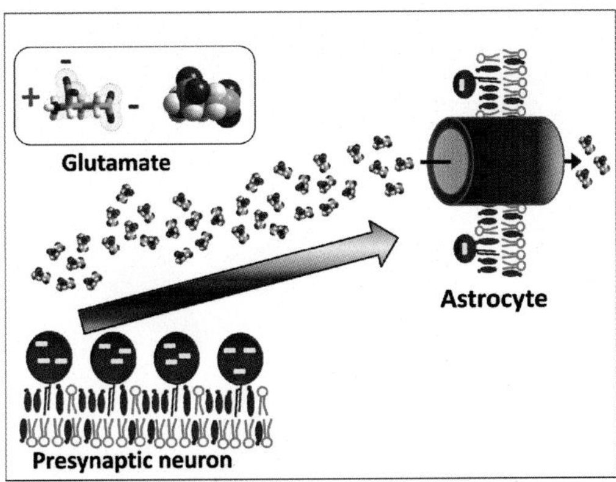

FIGURE 5.4 **Glutamate fluxes in the tripartite synapse.** At physiological pH, glutamate has two negative charges and one positive electrical charge (net charge −1). The chemical structure of glutamate is shown in the inset (tube and sphere representations). The high density of negative charges on the presynaptic neuronal membrane chases glutamate, which travels through the synaptic cleft with two possible destinations: receptors on the postsynaptic neuron (synaptic transmission) or transporters on the astrocyte surface (clearance).

these di- and trisialylated gangliosides. Its unique ganglioside, GM3, is a monosialylated glycolipid with a small head group (only three sugars, including the sialic acid per molecule). Thus, the surface of astrocytes is globally less charged than that of neurons.

Moreover, the negative charge of GM3 is close to the cell surface, whereas the di- and trisialylated gangliosides of neuronal membranes have their negative charges farther from the surface. This unique feature of the tripartite synapse is particularly important for electrically charged neurotransmitters (Fig. 5.3. The case of glutamate is highly significant. At the physiological pH, glutamate has a net electric charge of −1 (resulting from the combination of one positive and two negative charges). The high density of negative charges on neuronal membranes may exert a repulsive effect on glutamate. Therefore, glutamate has only two possibilities: (1) interact with a specific receptor on the postsynaptic membrane, which supposes that the glutamate-binding site is sufficiently far from membrane gangliosides; or (2) escape from the synaptic cleft to reach a less negatively charged area (the astrocytic membrane) (Fig. 5.4). In other words, the dissymmetric expression of gangliosides in the tripartite synapse is perfectly adjusted to create a flux of glutamate from neurons to astrocytes. Glutamate transporters in astrocytic membranes complete the equipment, allowing a rapid clearance of glutamate shortly after its release from the presynaptic neuron. In this scenario, the proteins (receptors and transporters) still play the first role, but the assistance of gangliosides, far from being anecdotic, deserves a better recognition.

5.3 NEUROTRANSMITTERS AND THEIR RECEPTORS: WHAT PHYSICOCHEMICAL PROPERTIES REVEAL

The schematic vision of a neurotransmitter that interacts with the extracellular domain of its receptor on the postsynaptic membrane is misleading. Indeed, it presupposes that the neurotransmitter is a water-soluble compound that is released from synaptic vesicles as free monomers (Fig. 5.5). This is not always the case.

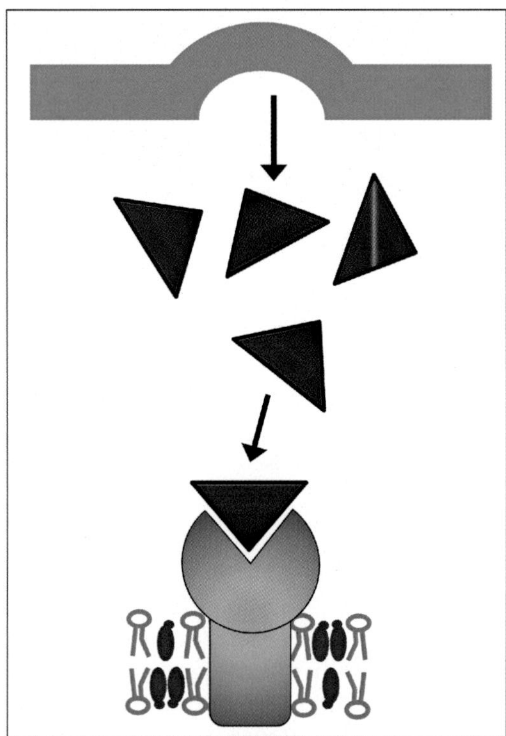

FIGURE 5.5 **A simple model of synaptic transmission.** The presynaptic neuronal membrane (green) releases the neurotransmitter (triangles) in the synaptic cleft. The neurotransmitter travels through the synapse and eventually binds to a receptor (blue) expressed by the postsynaptic neuronal membrane. If the neurotransmitter is soluble in water, it will bind to a surface-accessible region of its receptor. The postsynaptic membrane is represented with a lipid bilayer (cholesterol in red, phospholipids in green).

Let us consider three neurotransmitters: glutamate, serotonin, and anandamide. Glutamate is highly soluble in water, serotonin less, and anandamide very little. As already discussed, glutamate is negatively charged. When it comes close to synaptic membranes (either pre- or postsynaptic), it is pushed by the negative charges of the di- and trisialylated gangliosides of these membranes (GD1a, GD1b, and GT1b). Therefore, one could make the reasonable prediction that the binding site of glutamate in these receptors is far from the membrane (i.e., out of reach of repulsive gangliosides). Several types of glutamate receptors have been characterized. Three of them are ligand-gated ion channels, referred to as ionotropic receptors (Fig. 5.6).

Ionotropic glutamate receptors mediate the vast majority of excitatory neurotransmission in the brain.[16] These receptors are named after the agonists that activate them: NMDA, AMPA (α-amino-3-hydroxyl-5-methyl-4-isoxazolepropionate), and kainic acid (a marine compound). All three ionotropic receptors are integral membrane proteins composed of four large subunits (> 900 residues) that form a central ion channel pore. Glutamate binds to a pocket located in the large extracellular N-terminal domain, which is several nanometers from the membrane.[17] At this distance, the negative charges of membrane gangliosides are no longer repulsive (Fig. 5.7).

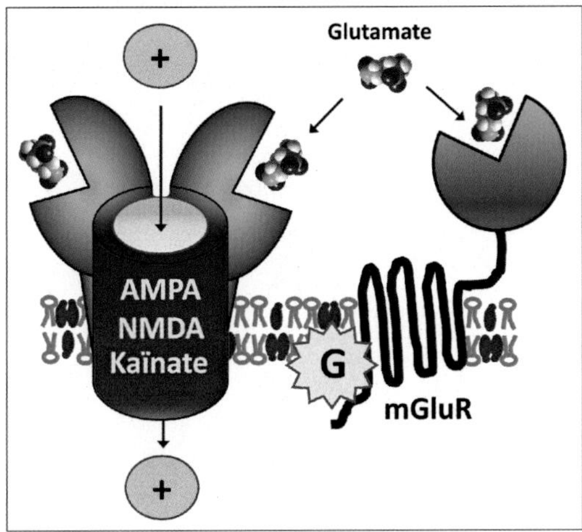

FIGURE 5.6 **Ionotropic and metabotropic receptors for glutamate.** Ionotropic receptors (on the left) are large transmembrane proteins that form a nonselective cationic channel (the cation is symbolized by an orange disk with a + sign). Three types of ionotropic glutamate receptors have been characterized and named after the agonists that activate them: AMPA, NMDA, and kainate. Metabotropic receptors for glutamate (mGluR, right panel) are G-protein coupled receptors (GPCRs) with seven transmembrane domains.

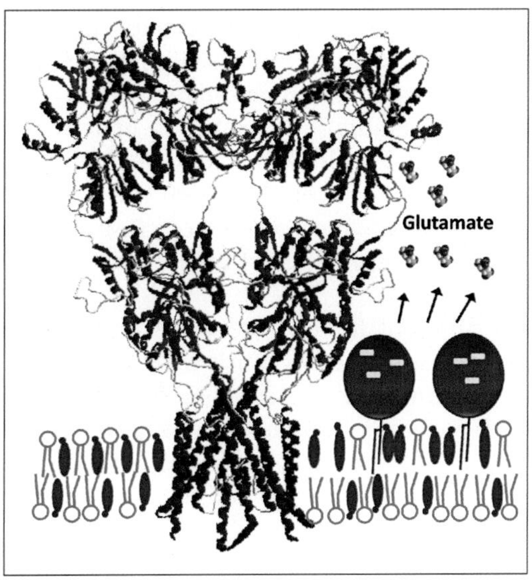

FIGURE 5.7 **How glutamate interacts with an ionotropic receptor.** The polysialylated gangliosides (red/purple disks with three negative charges) of the synaptic membrane exert a repulsive effect on glutamate whose binding site in the AMPA subtype ionotropic receptor is far from the membrane surface. The structure of the ionotropic receptor was obtained from PDB entry # 3KG2.[18]

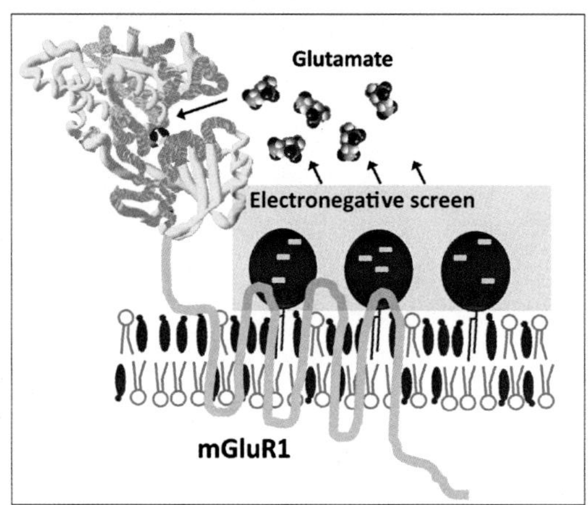

FIGURE 5.8 **How glutamate interacts with a metabotropic receptor.** Metabotropic glutamate receptors have an unusually large N-terminal domain, as illustrated for mGluR1 (PDB entry # 1EWK).[21] The polysialylated gangliosides (red/purple disks with three negative charges) of the synaptic membrane form an electronegative screen that prevents glutamate to reach the membrane surface. Correspondingly, the glutamate-binding site of mGluR1 is above this screen.

Apart from these ionotropic receptors, glutamate also binds to metabotropic receptors that belong to the family of G-protein-coupled receptors (GPCRs) with seven transmembrane domains. Usually, these receptors are flush with the membrane, yet metabotropic glutamate receptors display a large N-terminal extracellular domain (with ~ 600 amino acids) containing the ligand-binding site.[19] Such a large N-terminal region is totally unusual in the GPCRs family. Interestingly, when expressed as a recombinant protein, the extracellular region of the metabotropic glutamate receptor (subtype 1) retained a ligand-binding characteristic similar to that of the full-length receptor.[20] This indicates that the ligand-binding function of metabotropic glutamate receptors is totally uncoupled from the transmembrane domains.[21] Once again, this unique property of glutamate receptors is consistent with our prediction that the glutamate-binding site has to be out of reach of the repulsive influence of neuronal gangliosides (Fig. 5.8).

The situation is totally different for our second example, serotonin (5-hydroxytryptamine, 5-HT). Serotonin is an aromatic compound with a nitrogen-containing cycle (indole) bearing an OH group and an ethylamine chain (Fig. 5.9A). At physiological pH, the amine group is positively charged. Together with the OH group, this property tends to improve the solubility of serotonin in water. However, the aromatic structure counterbalances this effect by conferring π-stacking-associated self-aggregative properties to the molecule.[22] In synaptic vesicles, the concentration of serotonin is high (>0.1 mol L^{-1}) and the molecule forms oligomeric aggregates that dissociate upon release of the vesicle content in the synaptic cleft.[23] Like glutamate, serotonin can interact with either ionotropic or metabotropic receptors. Nevertheless, metabotropic receptors are far more abundant in this case: indeed, with the exception of 5-HT3, a ligand-gated ion channel, all serotonin receptors are G-protein coupled receptors.

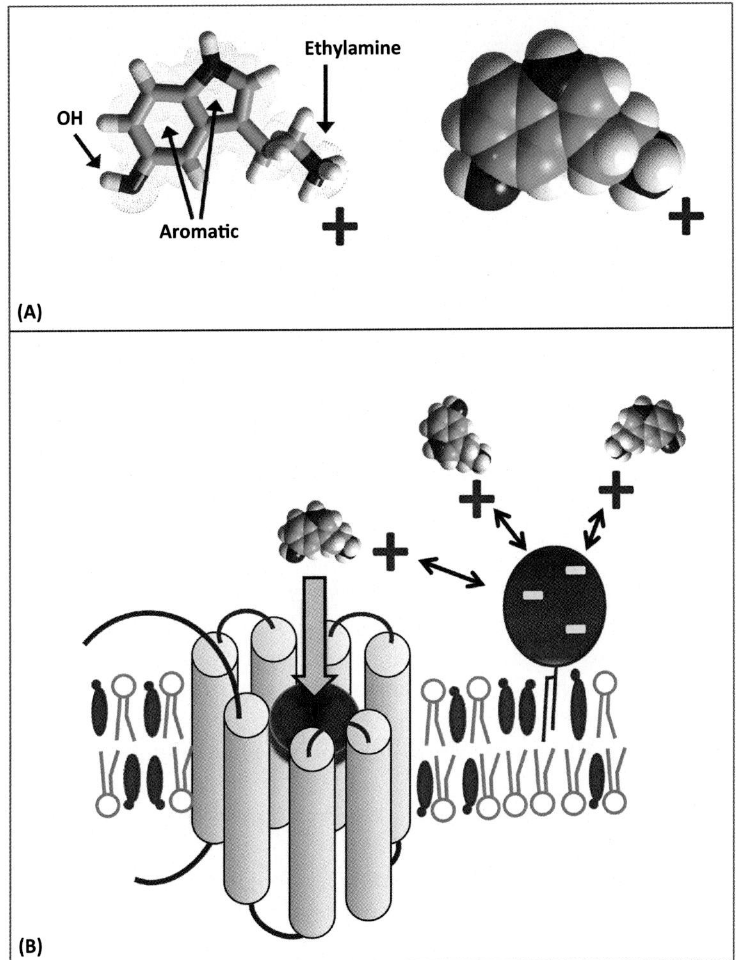

FIGURE 5.9 **Serotonin and its metabotropic receptors.** (A) Serotonin (5-hydroxytryptamine, 5-HT) is derived from L-tryptophan. It is an aromatic molecule (indole group) with an ethylamine chain (positively charged at pH 7) and an OH bound to the ring. These substitutions increase the solubility of serotonin, which is otherwise limited by its aromaticity. (B) The positive charge of serotonin is attracted by the negative charges of gangliosides on the postsynaptic membrane. Correspondingly, the serotonin-binding site of serotonin receptors (all but one are metabotropic receptors) is dipped into the membrane.

Membrane receptors for serotonin form a family with seven classes (5-HT1 to 5HT-7).[24] Can the chemical structure of serotonin help us to predict the localization of its binding site in these receptors? In contrast with glutamate, serotonin bears a positive charge at physiological pH. Hence, this neurotransmitter will be attracted by neuronal gangliosides.[25] On this basis, we can anticipate that the serotonin-binding site is probably close to the membrane, which is indeed the case (Fig. 5.9B). In all GPCR receptors for monoamine neurotransmitters (including the adrenergic, dopaminergic, and muscarinic receptors), the binding site is within the helical transmembrane domains.[26,27] To reach this intramembrane zone, serotonin relies

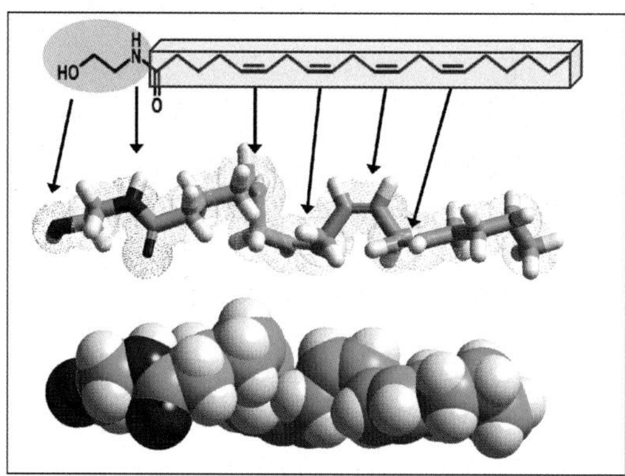

FIGURE 5.10 **Chemical structure of anandamide.** Three models of anandamide (arachidonoyl ethanolamide, AEA) are shown. In the first one (on the top) the region inherited from arachidonic acid (C20:4ω6) is framed. Note the four-double bonds in the *cis* (Z) configuration (see Chapter 1 for a description of unsaturated fatty acids). The ethanolamine part is shaded in rose. The two other models show an energy-minimized structure of anandamide in tube and sphere rendition. The arrows indicate the correspondence of chemical groups between the models.

on both its positive charge (which is attracted by cell surface gangliosides) and its aromatic structure (which fits with the apolar environment of the membrane). In the binding pocket, the charged amine group of serotonin binds to the carboxylate group of a conserved residue of aspartate (Asp-135 in the third transmembrane domain for human 5-HT2), and its aromatic ring interacts with a tryptophan (Trp-337 in the sixth transmembrane domain). Moreover, the OH group of a serine (Ser-139 in the third transmembrane domain) forms a hydrogen bond with the OH group of serotonin.[28]

Our last example concerns anandamide, an endocannabinoid neurotransmitter derived from arachidonic acid[29] (Fig. 5.10).

Due to its lipidic nature, anandamide is barely soluble in water. In this case, it is impossible for the neurotransmitter to bind to a solvent-exposed area of its receptor, as schematized for classical neurotransmitter/receptor couples (Fig. 5.5). Instead, the ligand-binding domain of anandamide receptors (e.g., the cannabinoid receptor CB1 in the brain) is made from a cluster of transmembrane helices. To reach this intramembrane binding site, anandamide has to penetrate the lipid bilayer *before* binding to the transmembrane embedded helices of the receptor. The molecular details of this intriguing mechanism are described in Section 5.2.

Overall, one can see that the chemical structure of neurotransmitters together with their solubility properties result in strikingly specific receptor-binding strategies. In particular, the net electric charge of the neurotransmitter determines whether its binding site is located in the extracellular part of the receptor (glutamate) or inside the membrane (serotonin, anandamide). In this latter case, the binding site may still be exposed to the water environment (serotonin) or totally dipped in the membrane (anandamide), according to the level of hydrophobicity of the neurotransmitter. The gangliosides expressed in synaptic membranes may either attract the neurotransmitter (serotonin) or repulse it (glutamate). In this respect, brain gangliosides behave as auxiliary factors helping the neurotransmitter find its way to the synapse, thereby speeding up the receptor-binding step of neural transmission.

5.4 A DUAL RECEPTOR MODEL FOR SEROTONIN

As described in Fig. 5.9, serotonin is a cationic/aromatic neurotransmitter. The combination of an aromatic ring with a positive electric charge forms a small structural motif that is remarkably complementary to the ganglioside head group.[22] Indeed, the cationic group may electrostatically bind to the carboxylate group of sialic acid, and the aromatic ring may stack onto a neutral sugar ring.[22] These interactions may confer a moderate affinity of serotonin for brain gangliosides. Hence, it is not surprising that gangliosides were the first receptors to be characterized for serotonin.[30] Binding studies in which serotonin was incubated with planar bilayers allowed for the determination that di- and trisialylated gangliosides (GD1a and GT1b) were more active than monosialyl species (GM1).[31] Molecular dynamics simulations help to figure out how serotonin may bind to GT1b (Fig. 5.11).

The driving force of the binding is the establishment of an electrostatic bond between the carboxylate group of a sialic acid of GT1b and the cationic group of serotonin. This primary interaction occurs at the tip of the ganglioside head group. The same sialic acid has the methyl of its N-acetyl group oriented toward the aromatic ring of serotonin, consistent with the establishment of a CH–π interaction.[32] Overall, this combination of electrostatic and CH–π bonds accounts for a weak and reversible binding of serotonin to GT1b, which fits with the low affinity of the serotonin–ganglioside complex.[33] It has been hypothesized that the transient binding of serotonin to the gangliosides of the postsynaptic membrane take place immediately after the release of the neurotransmitter in the synaptic cleft.[22] The underlying idea is that low-affinity binding of a neurotransmitter for gangliosides is an efficient way to concentrate neurotransmitters onto the postsynaptic membrane. In the specific case of serotonin, which displays self-aggregating properties in water,[23] this process would also be a simple way to prevent aggregation, ensuring the efficient delivery of serotonin monomers to its postsynaptic receptors. This scenario is illustrated by molecular modeling simulations (Fig. 5.12), supporting the notion that gangliosides could improve the bioavailability of serotonin on the postsynaptic membrane. In the synaptic vesicle, the concentration of serotonin can be as high as 600 mM.[23]

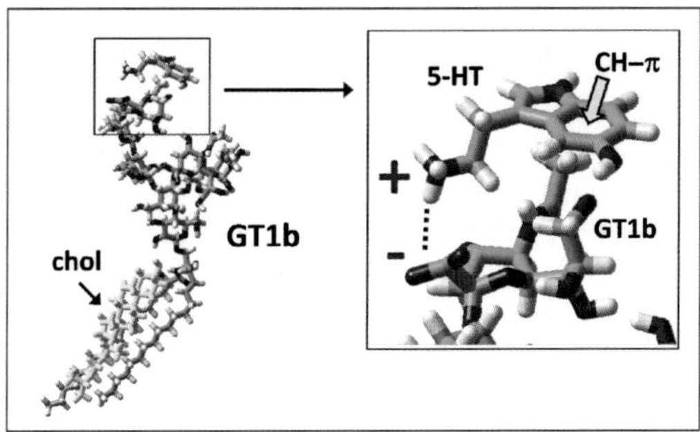

FIGURE 5.11 **Low-affinity binding of serotonin to ganglioside GT1b.** A GT1b–cholesterol complex (cholesterol in yellow) has been merged with serotonin (5-HT). Serotonin interacted with the tip of the ganglioside. The complex was stabilized with an electrostatic bridge and a CH–π interaction as indicated. This accounts for a low energy of interaction, in agreement with the literature.

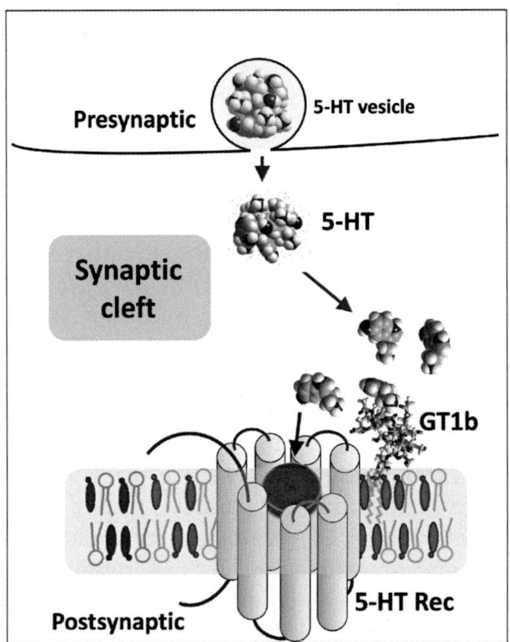

FIGURE 5.12 **How gangliosides help serotonin to reach the postsynaptic membrane.** In presynaptic vesicles, serotonin is highly concentrated and mostly aggregated. Upon release of the vesicles in the synaptic cleft, serotonin molecules start to disaggregate. They are attracted by the negative charges of polysialylated gangliosides (e.g., GT1b) on the postsynaptic membrane. Then serotonin binds to an intramembrane site on its metabotropic receptors (5-HT Rec). This dual receptor model involving both low-affinity (ganglioside) and high-affinity (protein) binding sites/receptors is also used by toxins and viruses to gain entry into brain cells (see Chapters 12 and 13).

Under these conditions, serotonin is mostly aggregated, and thus biologically inactive. Upon release in the synaptic cleft, some aggregates are spontaneously dislocated by dilution in the aqueous phase. Recalcitrant oligomers may be attracted by the gangliosides present in the postsynaptic membrane. Because these gangliosides can only bind monomeric serotonin (Fig. 5.11), they function as dislocating machinery for neurotransmitter aggregates. A critical aspect of this system is that the machine is prompt to release the neurotransmitter because the affinity of serotonin for the ganglioside is low. Therefore, serotonin monomers will easily reach specific protein receptors (chiefly GGPCRs) in the postsynaptic membrane. As emphasized in a recent review,[22] this concept of a dual ganglioside/receptor system for neurotransmitters in the postsynaptic membrane is similar to the double receptor model developed by Montecucco[34] for describing the binding of tetanus and botulinum toxins to neuronal membranes. The idea is that the combination of low-affinity binding sites (gangliosides) with classical high-affinity receptors (GPCRs) significantly increases the avidity of the ligand to the neuron surface. As the result of a purely thermodynamic process, this double ganglioside/receptor system is not associated with any particular molecular mechanism, which provides numerous possibilities for the ganglioside contribution to the whole process, including for instance the disaggregation step described for serotonin. Incidentally, one should note that neurotoxins simply use a routine system that can be viewed as a kind of greasing of synaptic transmission. If someone asks you, "What is the function of brain gangliosides?" you could answer, "to grease the synapse."

5.5 A DUAL RECEPTOR MODEL FOR ANANDAMIDE

Let us now treat the complex case of anandamide. From a chemical point of view, anandamide is a derivative of the polyunsaturated fatty acid, arachidonic acid (Fig. 5.10). The modification consists in grafting an ethanolamine ($HO–CH_2–CH_2–NH_2$) to the acyl chain of the C20:4ω6 fatty acid. The resulting molecule is an amide compound, which, in marked contrast with the fatty acid from which it derives, does not contain any ionizable group. At physiological concentrations, the carboxylate group of arachidonic acid has a $pK_a = 7$, so that it is present in the brain as an equimolar mixture of neutral and negatively charged forms. It is not the case for anandamide, which is a neutral compound at all pH values and at all concentrations. This is one of the main consequences of the chemical modification of C20:4ω6 into anandamide. The second change is an increased volume of the polar head group of anandamide versus arachidonic acid, together with the possibility of forming more hydrogen bonds (both the NH and OH groups of anandamide have donor and acceptor atoms). Overall, these modifications confer a slight increase of solubility in water for anandamide, and this gain is critical for its neurotransmitter activity. Nevertheless, one should keep in mind that anandamide remains a lipid. As a lipid neurotransmitter, anandamide will exert its neurological effects in a membrane environment. Once released from a neuron membrane, anandamide needs a vehicle to migrate to another neuron. This vehicle may be a transport protein, a lipid vesicle, or an exosome. Anandamide is usually produced by a postsynaptic neuron, released in the extracellular space, and transported to the presynaptic neuron in a typical retrograde synaptic transmission.[4] An illustration of the absolute requirement of a transport system for anandamide is given in Fig. 5.13.

When anandamide is mixed with water, it does not create a clear solution. In fact, most of the neurotransmitter molecules aggregate into a precipitate that is visible under the optical microscope (Fig. 5.13A). The addition of a small amount of the lipid-binding protein serum albumin induces an instantaneous disaggregation of anandamide (Fig. 5.13A). This impressive effect of serum albumin can be quantified in real time by measuring the surface pressure of the water/anandamide system (Fig. 5.13A). When a drop of anandamide is injected in water at a concentration of 150 nM, few molecules are truly dissolved in water. These rare molecules behave as tensioactive compounds and reach the air–water interface, inducing a slight increase of surface pressure immediately after injection. Then, some of the aggregates spontaneously dissociate and at their turn reach the interface, as evidenced by the progressive increase of surface pressure between 100 and 450 s postinjection. At this time, an equimolar amount of serum albumin is injected in the aqueous phase. As shown in Fig. 5.13A, serum albumin dissociates the anandamide aggregates, thereby generating monomers at a high rate. These monomers massively reach the interface, evoking a dramatic increase of the surface pressure (Fig. 5.13A). A control experiment in which serum albumin was added first demonstrated that the protein was not tensioactive by itself, so that the brutal increase in surface pressure observed upon albumin addition into the water/anandamide system was actually due to the migration of anandamide monomers at the interface (Fig. 5.13B). Finally,

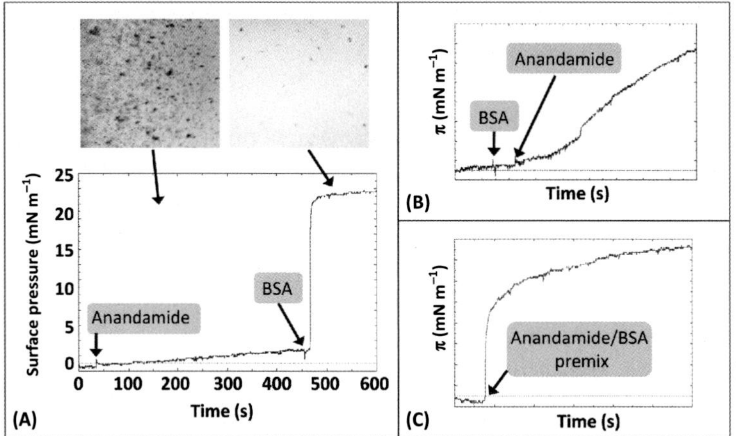

FIGURE 5.13 **Effect of serum albumin on anandamide solubility.** (A) The photomicrographs show how anandamide precipitates (left micrograph) disappear (right micrograph) in presence of bovine serum albumin (BSA). This striking effect of BSA can be quantified by measuring the surface pressure of a water droplet in which anandamide (150 nM) is injected. At this concentration, most anandamide is precipitated and only rare monomers can spontaneously reach the interface, as indicated by the slight increase of the surface pressure π during the first 450 s of the experiment. The injection of BSA (150 nM, hence in 1:1 molar ratio with anandamide) induced the dissociation of anandamide aggregates (as shown by the micrograph), allowing anandamide monomers to massively reach the interface. Correspondingly, a dramatic increase of π was recorded immediately after BSA injection. (B) BSA by itself did not affect π. However, when anandamide was added after BSA, the protein induced the disaggregation of the neurotransmitter and its delivery at the interface, resulting in a large increase of π. (C) When anandamide was premixed with BSA and then injected in the water droplet, it was already dissociated into monomers that massively reached the interface.

when anandamide and serum albumin were preincubated for 15 min before the injection of the mixture in water, the monomers were ready to go to the interface. Correspondingly, the surface pressure increased immediately upon injection (Fig. 5.13C). Taken together, these data demonstrate that anandamide requires a transport protein to move into aqueous phases as bioactive monomers. In the brain, this function could be fulfilled by various lipid-binding proteins including α-synuclein, which has been shown to interact with arachidonic acid,[35] or by lipid-based transport systems such as exosomes.[36–41] Interestingly, the involvement of α-synuclein in anandamide transport is suggested by surface pressure measurements (Fig. 5.14).

At the end of its journey through the extracellular space, anandamide eventually reaches the target neuronal membrane containing its receptor (e.g., the GPCR protein CB1). Due to its lipidic nature, anandamide will first penetrate the plasma membrane before interacting with this receptor. The insertion of anandamide in the membrane bilayer is a cholesterol-dependent process.[42] Anandamide interacts specifically with cholesterol in the plasma membrane of brain cells. Following this interaction, anandamide is pulled inside the membrane through a flip-flop mechanism (Fig. 5.15).

The initial interaction with cholesterol is mediated by the formation of a hydrogen bond linking the OH groups of both anandamide and cholesterol. Then, anandamide enters the membrane through a flip-flop mechanism. At this stage, its interaction with cholesterol is stabilized by both van der Waals forces and a new hydrogen bond between the OH group of

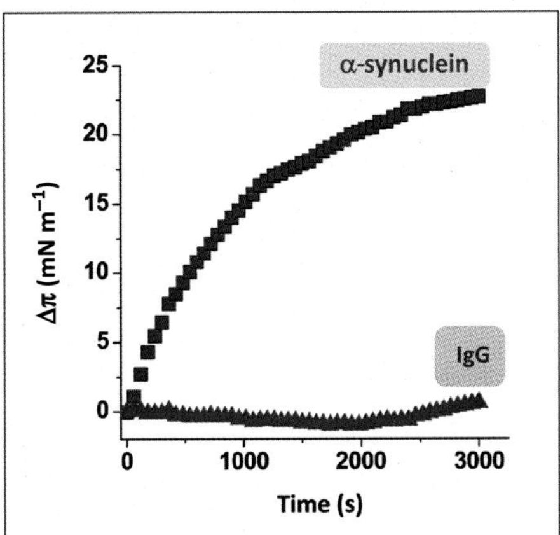

FIGURE 5.14 **Effect of α-synuclein on anandamide solubility.** The Parkinson's disease–associated protein α-synuclein is a lipid-binding protein that interacts intracellularly with anionic lipids (on synaptic vesicles) and extracellularly with ganglioside GM3 (on raft membranes). It has also been shown to interact with arachidonic acid. This prompted us to test whether α-synuclein could, like serum albumin, bind to anandamide, and induce its dissociation into tensioactive monomers. Thus, we used the same protocol as in Fig. 5.13C, and premixed anandamide with either α-synuclein (blue squares) or mouse IgG used as control (red triangles). The data indicated that α-synuclein (but not control IgG) could induce the dissociation of anandamide aggregates and allow the delivery of anandamide monomers to the interface (as shown by the marked increase in surface pressure Δπ). These data raised the intriguing possibility that one physiological function of α-synuclein could be to carry on anandamide (and perhaps other endocannabinoids) in the brain.

cholesterol and the NH of anandamide. The cholesterol–anandamide complex is sufficiently rigid to move laterally in the bidimensional plane of the plasma membrane until it reaches a CB1 receptor. Because the affinity of anandamide for CB1 is significantly higher than its affinity for cholesterol, anandamide leaves cholesterol whereas it remains attached to CB1 (Fig. 5.16). Once it releases in the aqueous phase, anandamide successively adopts three types of conformations: a condensed structure in water, an extended structure when bound to cholesterol, and a S shape when bound to the sixth transmembrane domain of CB1.[43] This conformational plasticity of anandamide results from the free rotation around the CH_2 groups intercalated between the double bonds of the carbon chain. These methylene groups act as hinges (or rotules) that allow the chain to adapt its shape to the environment. In this process, cholesterol has the same function as gangliosides for serotonin: to catch the neurotransmitter in the synaptic membrane and to transport it to its receptor through the bidimensional plane of the bilayer.

It is another example of the use of a dual lipid/receptor system. In Chapter 12, we will show that HIV-1 use both gangliosides and cholesterol in conjunction with the fusion coreceptor CCR5 to gain entry into brain cells. In this case, the dual system becomes a trio, yet the overall mechanism remains the same: lipid-assisted delivery of extracellular compounds to high-affinity receptors.

The interaction of anandamide with plasma membrane cholesterol may also be linked to the mechanism of anandamide transport through a biological membrane. This issue has been greatly debated.[44,45] As a lipid, anandamide is probably able to diffuse passively through the

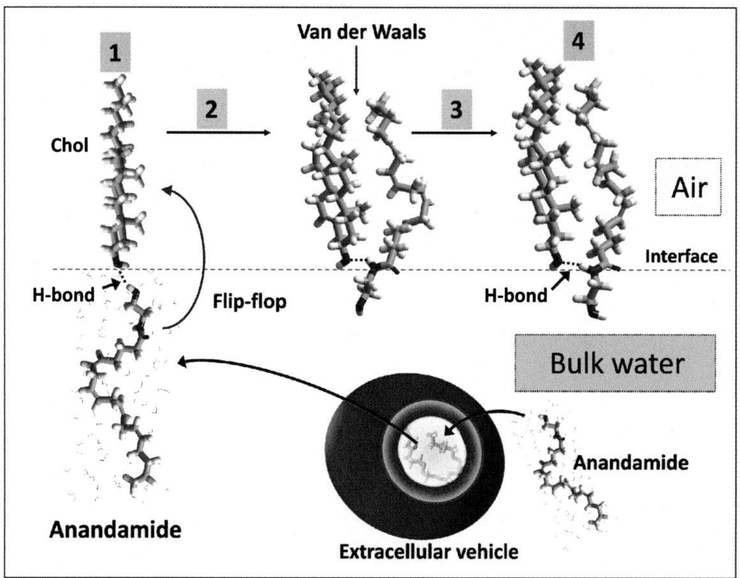

FIGURE 5.15 **Molecular mechanism of anandamide insertion in a cholesterol-containing membrane.** In the extracellular space, anandamide needs a vehicle (either a transport protein such as α-synuclein as shown in Fig. 5.14 or a lipid phase, e.g., exosomes). In the vicinity of the plasma membrane, anandamide is released from its carrier, allowing its terminal OH group to bind to cholesterol via a hydrogen bond (**step 1**). Then anandamide is attracted inside the membrane via a typical flip-flop process (**step 2**), allowing the formation of numerous van der Waals contacts between the respective hydrophobic parts of cholesterol and anandamide. Further conformational adjustments (**step 3**) facilitate the interaction between the methyl groups of the rough β face of cholesterol and the cavities determined by the successive double bonds of anandamide. This interaction improves the geometry of the hydrogen bond between the OH of cholesterol and the NH of anandamide (**step 4**). This process can be followed by measuring in real time the surface pressure of water in presence of anandamide (Figs 5.13 and 5.14). The air–water interface is indicated by a dotted blue line.

membrane, indicating no specific need for a membrane transport system. However, due to its limited solubility in aqueous phases, anandamide does require a carrier-assisted transport through the cytoplasm.

The group of D. Deutsch identified fatty-acid-binding proteins (FABPs) as intracellular carriers that transport AEA from the plasma membrane.[46,47] In particular, FABP5 delivers anandamide to fatty acid amide hydrolase (FAAH), which hydrolyzes anandamide into arachidonic acid and ethanolamine.[45] This mechanism maintains the transmembrane gradient of anandamide. The transport of anandamide through planar lipid bilayers data was strictly dependent on the presence of cholesterol, which, above a threshold molar value of 20%, stimulated the passage of anandamide.[42] Overall, these findings indicated that both the insertion and transmembrane transport of anandamide are cholesterol-dependent mechanisms. At the molecular level, the mechanisms allowing cholesterol and FABP5 to bind anandamide are remarkably similar. In both cases, the polar head group of anandamide forms a hydrogen bond with an OH group provided by its molecular partner: the OH group of cholesterol[42] or the OH group of a tyrosine residue (Tyr-131) in FABP5.[47] The apolar part of anandamide

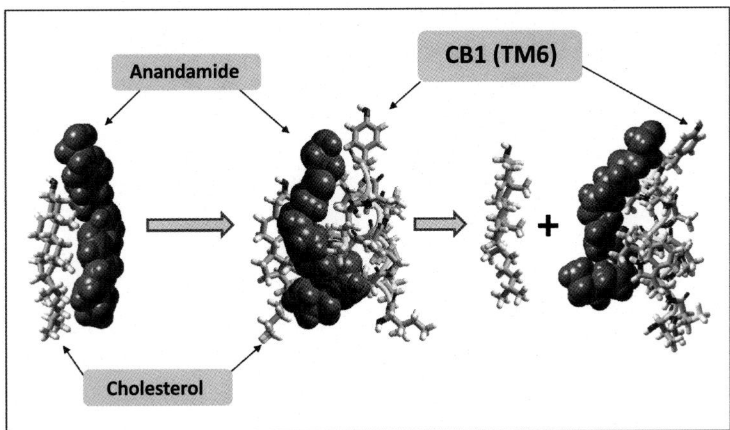

FIGURE 5.16 **A dual receptor model for anandamide.** Anandamide penetrates the membrane of the presynaptic neuron (retrograde transmission) via a primary interaction with cholesterol (see Fig. 5.15). Then the cholesterol–anandamide complex migrates in the membrane until it reaches a high-affinity receptor for anandamide (e.g., CB1, a GPCR protein with seven transmembrane domains). A trimolecular cholesterol–CB1–anandamide complex is formed transiently. In this complex, anandamide interacts with the transmembrane domains (e.g., TM6) of the CB1 receptor. Then cholesterol leaves the complex and anandamide remains bound to its receptor.

interacts with its partners through a series of van der Waals interaction involving either hydrophobic residues of FABP5[47] or the rough apolar β-face of cholesterol.[42] However, it should be noted that anandamide bound to cholesterol adopts an extended membrane-consistent conformation (Fig. 5.15, whereas the anandamide-binding pocket of FABP5 requires the folding of anandamide into a condensed hairpin structure.[47] This "folding" further illustrates the high plasticity of anandamide, a property conferred by the presence of four double bonds in the *cis* configuration in the lipophilic chain. In this respect, one can easily understand why the most important endocannabinoid neurotransmitters of anandamide and 2-arachidonoylglycerol (2-AG) are derived from arachidonic acid. Indeed, in the FABP5 binding pocket, 2-AG adopts a hairpin structure quite similar to that of anandamide.[47] On the other hand, molecular dynamics simulations indicated that cholesterol could bind to 2-AG and constrain its flexible arachidonoyl chain to adopt a more extended membrane-compatible conformation[48] (Fig. 5.17).

Thus, the high conformational plasticity of anandamide also applies to 2-AG, allowing both endocannabinoids to adapt their structures to various water and membrane environments.

5.6 CONTROL OF SYNAPTIC FUNCTIONS BY GANGLIOSIDES

Since the discovery of gangliosides by Klenk in the first half of the twentieth century,[49] considerable effort has been made to understand their functional role in the central nervous system.[50] In 1961, McIlwain reported that the addition of gangliosides could restore the lost electrical excitability of cortical slices kept at 0°C for 5 h.[51] Similarly, the administration of gangliosides could prevent the deleterious effects of ethanol on nerve function.[52] In cell cultures, exogenously added gangliosides could induce axonal sprouting and neurite outgrowth.[53–56]

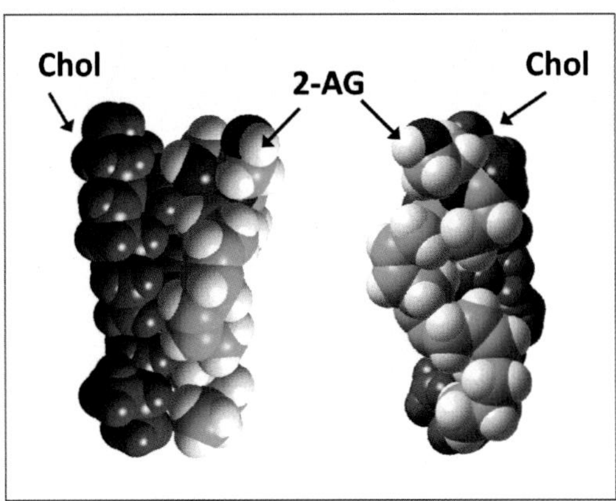

FIGURE 5.17 **Interaction of 2-AG with cholesterol.** Cholesterol is in purple and 2-arachidonoylglycerol (2-AG) in atom colors. Two views of the complex are shown. Note that the extended "double S" conformation of 2-AG is determined by the size of cholesterol and is compatible with a membrane insertion process.

Moreover, gangliosides have been shown to regenerate damaged nervous tissues in animal models.[57,58] Gangliosides have also been shown to improve synaptic transmission in the hippocampus.[59,60] The importance of gangliosides is further demonstrated by the deleterious effects of experimental neuraminidase treatments on neural functions.[61] Overall, brain gangliosides, especially GM1, have potent neurotrophic and neuroprotective properties.[62–64] This remarkable capacity has paved the way for clinical trial of gangliosides in human neurodegenerative diseases, especially for Parkinson's disease.[65] Indeed, an intriguing aspect of Parkinson's disease is its relationship with lowered GM1 expression in both animal models and in humans.[66] Correspondingly, GM1 treatment could stimulate the regeneration of functional dopaminergic neurons in the striatum.[67]

Gangliosides could play a role in synaptic transmission in several ways: attractive/repulsive effects on neurotransmitters (discussed earlier), direct interaction with neurotransmitter receptors (see Chapter 7), and regulation of neurotransmitter release from the presynaptic membrane.[50] We will discuss here the role of gangliosides in the presynaptic membrane. Both the pre- and postsynaptic membranes contain polysialylated gangliosides, which may interact with positively charged neurotransmitters (serotonin, dopamine, and acetylcholine) through electrostatic bonds. As already discussed, this characteristic may favor the diffusion of these neurotransmitters onto the postsynaptic membrane. However, the same reasoning applied to the presynaptic membrane, which also expresses high-levels of negatively charged gangliosides, would lead to contradictory conclusions. Indeed, because neurotransmitters are released in the vicinity of the presynaptic membrane, they should be stuck to the negative charges of gangliosides in these membranes. This issue has been brilliantly discussed in the past. In the brain cortex, ~ 65% of neuraminidase and ~ 40% of sialyltransferase activities map at nerve endings and are associated with synaptosomes.[68] The colocalization of these enzymes is consistent with a sialylation/desialylation cycle of gangliosides in presynaptic membranes[50] (Fig. 5.18).

Ganglioside desialylation would reduce the density of negative charges on the presynaptic membranes, thereby preventing the sticking of cationic neurotransmitters, which would in

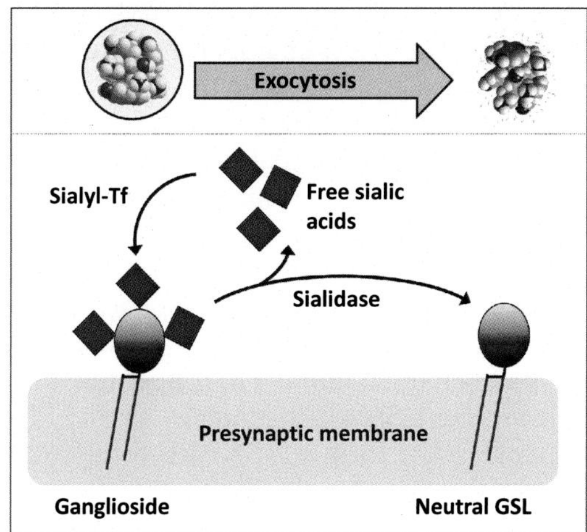

FIGURE 5.18 **Desialylation/sialylation cycle in the presynaptic membrane.** Upon release from synaptic vesicles, cationic neurotransmitters (monoamines) would be glued by the negatively charged polysialylated gangliosides of the presynaptic membrane. To avoid this problem, it has been suggested that those gangliosides may be transiently desialylated by neuraminidase (sialidase), and then resialylated by sialytransferase (sialyl-Tf). The concentration of both enzymes in synaptosomes is consistent with such a scenario.

turn facilitate the diffusion of these neurotransmitters onto the postsynaptic membranes. If we consider the energetic cost of this type of substrate cycle, we have to conclude that except during neurotransmitter release events, gangliosides are crucial for the presynaptic neuronal membrane (otherwise this membrane would contain neutral glycosphingolipids instead of gangliosides). If we are looking for a specific property of gangliosides that neutral glycolipid lack, the carboxylate group of sialic acid should immediately arouse our interest.[69] The binding of Ca^{2+} to these negatively charged groups may have a significant impact on synaptic transmission, especially during neurotransmitter exocytosis.[50] For instance, gangliosides have been shown to stimulate the release of acetylcholine through the potentiating Ca^{2+} influx in brain synaptosomes.[70] Conversely, the presence of Ca^{2+} appeared to potentiate the ganglioside-mediated release of dopamine.[71] Overall, these data indicated that gangliosides play a major role in neuronal functions. In this respect, the interesting parallel between the central role of Ca^{2+} on neuronal activity and the Ca^{2+} binding capacity of gangliosides is probably more than a coincidence. Finally, it has been recently demonstrated that neurite outgrowth is controlled by microRNAs.[72–74] Given that gangliosides also stimulate this neurogenesis function, it would be interesting to look for an effect of gangliosides on brain microRNA synthesis and function.

5.7 CONTROL OF SYNAPTIC FUNCTIONS BY CHOLESTEROL

Cholesterol is deeply involved in the mechanisms controlling the exocytosis of synaptic vesicles. Indeed, reducing cholesterol levels in cultured hippocampal neurons impaired vesicle exocytosis.[75] This effect was reversed by cholesterol reloading. These data are in line with the notion that cholesterol plays a critical role in synaptic vesicle exocytosis. Among the various hypotheses that could explain the involvement of cholesterol in this fundamental process, one could distinguish specific cholesterol effects on the proteins of the exocytosis machinery[76] or a modulation of physical membrane properties such as membrane curvature.[77] Synaptic vesicles contain high amounts of cholesterol that may induce an adequate curvature of their membrane. In support of this notion, a link among membrane curvature, cholesterol content, and the structure of SNARE proteins has been recently demonstrated.[78] The key point is that the release of neurotransmitters at the synapse requires the fusion of synaptic vesicles with the plasma membrane of the presynaptic neuron. Each of these membranes expresses specific fusion proteins, referred to as SNAREs (soluble *N*-ethylmaleimide-sensitive factor attachment protein receptors), which must interact to trigger the fusion process: v-SNARE synaptobrevin for the synaptic vesicle, t-SNAREs syntaxin 1A and SNAP-25 for the presynaptic membrane. These proteins interact to form the SNARE complex, which consists of a helical bundle that bridges both membranes, thereby facilitating fusion.[79,80] Using a fluorescence-based lipid-mixing assay, Tong et al.[78] investigated the effect of cholesterol on SNARE-mediated membrane fusion. They showed that lipid mixing was promoted when cholesterol is on the v-SNARE-carrying vesicles but not as much when it is on t-SNARE-carrying vesicles.[78] In fact, cholesterol induced a conformational change in the transmembrane domains of v-SNARE, converting its inactive open scissors-like structure into a closed scissors-like shape. This conformational change forced the vesicle bilayer to adopt a positive curvature compatible with the fusion process (Fig. 5.19).

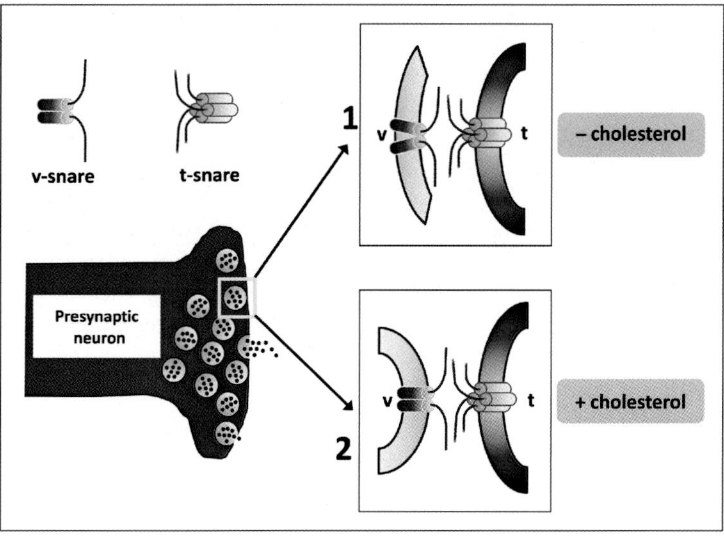

FIGURE 5.19 **Role of cholesterol in the exocytosis of synaptic vesicles.** Cholesterol constrains the conformation of v-SNARE proteins in such a way that their transmembrane domains adopt a parallel ("closed scissors") orientation. This particular conformation is required for the fusion process because it induces a fusion-compatible curvature of the synaptic vesicle. In this case, the interaction between v-SNARE and t-SNARE proteins triggers the fusion mechanism (lower inset, **2**). Without cholesterol, v-SNARE remains in an "open scissors" conformation (upper inset, **1**), and the curvature of the vesicle is not appropriate for fusion.[78]

Cholesterol has also been shown to interact with synaptophysin,[81] a protein involved in the formation and trafficking of synaptic vesicles.[82] Correspondingly, cholesterol depletion perturbed the biogenesis of synaptic-like microvesicles (SLMVs) from the plasma membrane of cultured cells. By determining the functional curvature of synaptic vesicles and deciding whether they can fuse with plasma membrane, cholesterol can control the whole process of synaptic transmission. This capability may explain why cholesterol is involved in a broad range of brain functions, including synaptic plasticity and degeneration.[83,84] In addition to cholesterol, sphingolipids, and phosphoinositides have also been involved in the regulation of the synaptic vesicle cycle.[85–88] It illustrates how important brain lipids are for synaptic transmission.[89]

Finally, cholesterol can specifically interact with a broad range of neurotransmitter receptors and regulate their activity.[22] The effects of cholesterol on neurotransmitter receptors, including ionotropic[90] and metabotropic[91–93] receptors are treated in Chapter 7.

5.8 KEY EXPERIMENTS: DEBUNKING MYTHS IN NEUROSCIENCES

Science evolves continuously. In the past 10 years, several important discoveries have shaken our certainties, contradicting some of our most solid theories. In neuroscience textbooks, it is taken for granted that the human brain contains much more glial cells than neurons (a ratio of 10 glial cells per neuron is often argued). However, as emphasized by Azevedo et al.,[94] "Despite the widespread quotes that the human brain contains 100 billion neurons and ten times more glial cells, the absolute number of neurons and glial cells in the human brain remains unknown." By using the isotropic fractionators method,[95] these authors determined that an adult human brain contains on average 86.1 ± 8.1 billion neurons and 84.6 ± 9.8 billion glial cells.[94] Therefore, the neuron-to-glial cell ratio of the human brain is not 1:10 as generally assumed, but 1:1. When applied to the adult rat brain, this new method found 200 million neurons for 130 million glial cells.[95] In fact, what makes the human brain so remarkable in its cognitive abilities could be simply its extremely large number of neurons.[96] Otherwise, as emphasized by S. Herculano-Houzel, "The human brain is a scaled-up primate brain in its cellular composition and metabolic cost."[96]

At the beginning of this chapter we described the tripartite synapse implicating two neurons and an astrocyte. The concept is that the astrocyte in contact with the pre- and postsynaptic neurons can modulate neuronal activity via a Ca^{2+}-dependent release of gliotransmitters.[5] This model has been developed on the wrong assumption that astrocytes exhibit metabotropic glutamate receptor 5 (mGluR5)-dependent increases of intracellular Ca^{2+} in response to glutamate.[6] In fact, these findings were based on studies of young rodents and extrapolated to adult animals. Unfortunately, the astrocytic expression of mGluR5 is developmentally regulated and totally undetectable in mouse and human adult astrocytes.[97] The only metabotropic glutamate receptor expressed in these cells is mGluR3, whose activation does not induce any Ca^{2+} increase.[97] These unexpected data do not call into question the existence of the tripartite synapse, yet they strongly challenge the concept of neuroglial signaling in the adult brain. In any case, this study suggested that the tripartite synapse concept may rely on signal transduction mechanisms totally distinct from those that operate during development. Most importantly, this reassessment warns against systematic extrapolations of experimental results obtained with a single experimental model.

The third and last principle that we want to discuss in this chapter concerns our mental representation of the synapse. In most cases, if a teacher asks a student to draw a synapse, the student will represent the transfer of neurotransmitters from presynaptic vesicles to the postsynaptic membrane through the synaptic cleft. "The most common type of electrical synapse," according to Bennett and Zukin,[98] is largely overlooked. These authors referred to the electrical synapse involving gap junctions.[99] These synapses led to renewed interest at the turning of the third millennium when their involvement in neuronal synchronization was demonstrated by several groups.[100–102] Curiously, the existence of electrical synapses was known for decades.[99] However, they remained of scant interest for most neuroscientists until it became evident that these synapses were more frequent than anticipated. Indeed, it appeared that many neuron types known to communicate chemically could also be electrically coupled via gap junctions.[99] Gap junctions allow an ultrafast neuronal communication through axoaxonal coupling[102] but are also used as a communication pathway between glial cells (astrocyte–astrocyte and astrocyte–oligodendrocyte).[103,104] Gap junction channels have a large internal diameter (~1.2 nm) that allows the passage of ions, small metabolites, and signaling molecules.[98] The proteins that make gap junctions are called connexins.[105] In adult neurons, the principal connexin is Cx36.[106] Like other (yet not all) members of the connexin family, the neuronal Cx36 connexin is targeted to lipid rafts.[107] Both cholesterol and sphingolipids have been shown to regulate the functional properties of gap junctions.[108,109] Thus, it is likely that these lipids may control the electrical synapse just like they control the chemical synapse.[110]

References

1. Bennett MR. The early history of the synapse: from Plato to Sherrington. *Brain Res Bull.* 1999;50(2):95–118.
2. Tansey EM. Not committing barbarisms: Sherrington and the synapse, 1897. *Brain Res Bull.* 1997;44(3):211–212.
3. Shepherd GM, Erulkar SD. Centenary of the synapse: from Sherrington to the molecular biology of the synapse and beyond. *Trends Neurosci.* 1997;20(9):385–392.
4. Alger BE. Endocannabinoids: getting the message across. *Proc Natl Acad Sci USA.* 2004;101(23):8512–8513.
5. Perea G, Navarrete M, Araque A. Tripartite synapses: astrocytes process and control synaptic information. *Trends Neurosci.* 2009;32(8):421–431.
6. Cornell-Bell AH, Finkbeiner SM, Cooper MS, Smith SJ. Glutamate induces calcium waves in cultured astrocytes: long-range glial signaling. *Science.* 1990;247(4941):470–473.
7. Parpura V, Basarsky TA, Liu F, Jeftinija K, Jeftinija S, Haydon PG. Glutamate-mediated astrocyte-neuron signalling. *Nature.* 1994;369(6483):744–747.
8. Westbrook GL. Glutamate receptors and excitotoxicity. *Res Publ Assoc Res Nerv Ment Dis.* 1993;71:35–50.
9. Anderson CM, Swanson RA. Astrocyte glutamate transport: review of properties, regulation, and physiological functions. *Glia.* 2000;32(1):1–14.
10. Kanai Y, Hediger MA. The glutamate/neutral amino acid transporter family SLC1: molecular, physiological and pharmacological aspects. *Pflugers Arch.* 2004;447(5):469–479.
11. Shigetomi E, Bowser DN, Sofroniew MV, Khakh BS. Two forms of astrocyte calcium excitability have distinct effects on NMDA receptor-mediated slow inward currents in pyramidal neurons. *J Neurosci.* 2008;28(26):6659–6663.
12. Perea G, Araque A. Astrocytes potentiate transmitter release at single hippocampal synapses. *Science.* 2007;317(5841):1083–1086.
13. Pascual O, Casper KB, Kubera C, et al. Astrocytic purinergic signaling coordinates synaptic networks. *Science.* 2005;310(5745):113–116.
14. van Echten-Deckert G, Walter J. Sphingolipids: critical players in Alzheimer's disease. *Prog Lipid Res.* 2012;51(4):378–393.

15. Ledeen RW, Diebler MF, Wu G, Lu ZH, Varoqui H. Ganglioside composition of subcellular fractions, including pre- and postsynaptic membranes, from Torpedo electric organ. *Neurochem Res*. 1993;18(11):1151–1155.
16. Traynelis SF, Wollmuth LP, McBain CJ, et al. Glutamate receptor ion channels: structure, regulation, and function. *Pharmacol Rev*. 2010;62(3):405–496.
17. Armstrong N, Jasti J, Beich-Frandsen M, Gouaux E. Measurement of conformational changes accompanying desensitization in an ionotropic glutamate receptor. *Cell*. 2006;127(1):85–97.
18. Sobolevsky AI, Rosconi MP, Gouaux E. X-ray structure, symmetry and mechanism of an AMPA-subtype glutamate receptor. *Nature*. 2009;462(7274):745–756.
19. Jingami H, Nakanishi S, Morikawa K. Structure of the metabotropic glutamate receptor. *Curr Opin Neurobiol*. 2003;13(3):271–278.
20. Okamoto T, Sekiyama N, Otsu M, et al. Expression and purification of the extracellular ligand binding region of metabotropic glutamate receptor subtype 1. *J Biol Chem*. 1998;273(21):13089–13096.
21. Kunishima N, Shimada Y, Tsuji Y, et al. Structural basis of glutamate recognition by a dimeric metabotropic glutamate receptor. *Nature*. 2000;407(6807):971–977.
22. Fantini J, Barrantes FJ. Sphingolipid/cholesterol regulation of neurotransmitter receptor conformation and function. *Biochim Biophys Acta*. 2009;1788(11):2345–2361.
23. Nag S, Balaji J, Madhu PK, Maiti S. Intermolecular association provides specific optical and NMR signatures for serotonin at intravesicular concentrations. *Biophys J*. 2008;94(10):4145–4153.
24. Hannon J, Hoyer D. Molecular biology of 5-HT receptors. *Behav Brain Res*. 2008;195(1):198–213.
25. Krishnan KS, Balaram P. A nuclear magnetic resonance study of the interaction of serotonin with gangliosides. *FEBS Lett*. 1976;63(2):313–315.
26. Kobilka B. Adrenergic receptors as models for G protein-coupled receptors. *Annu Rev Neurosci*. 1992;15:87–114.
27. Strader CD, Fong TM, Tota MR, Underwood D, Dixon RA. Structure and function of G protein-coupled receptors. *Annu Rev Biochem*. 1994;63:101–132.
28. Manivet P, Schneider B, Smith JC, et al. The serotonin binding site of human and murine 5-HT2B receptors: molecular modeling and site-directed mutagenesis. *J Biol Chem*. 2002;277(19):17170–17178.
29. Di Marzo V, Bifulco M, De Petrocellis L. The endocannabinoid system and its therapeutic exploitation. *Nat Rev Drug Discov*. 2004;3(9):771–784.
30. Woolley DW, Gommi BW. Serotonin receptors. VII. Activities of various pure gangliosides as the receptors. *Proc Natl Acad Sci USA*. 1965;53(5):959–963.
31. Matinyan NS, Melikyan GB, Arakelyan VB, Kocharov SL, Prokazova NV, Avakian TM. Interaction of ganglioside-containing planar bilayers with serotonin and inorganic cations. *Biochim Biophys Acta*. 1989;984(3):313–318.
32. Nishio M, Umezawa Y, Fantini J, Weiss MS, Chakrabarti P. CH–π hydrogen bonds in biological macromolecules. *Phys Chem Chem Phys*. 2014;16(25):12648–12683.
33. Ochoa EL, Bangham AD. *N*-Acetylneuraminic acid molecules as possible serotonin binding sites. *J Neurochem*. 1976;26(6):1193–1198.
34. Montecucco C. How do tetanus and botulinum toxins bind to neuronal membranes? *Trends Biochem Sci*. 1986;11(8):314–317.
35. Darios F, Ruiperez V, Lopez I, Villanueva J, Gutierrez LM, Davletov B. Alpha-synuclein sequesters arachidonic acid to modulate SNARE-mediated exocytosis. *EMBO Rep*. 2010;11(7):528–533.
36. Bellingham SA, Guo BB, Coleman BM, Hill AF. Exosomes: vehicles for the transfer of toxic proteins associated with neurodegenerative diseases? *Front Physiol*. 2012;3:124.
37. Chivet M, Hemming F, Pernet-Gallay K, Fraboulet S, Sadoul R. Emerging role of neuronal exosomes in the central nervous system. *Front Physiol*. 2012;3:145.
38. Chivet M, Javalet C, Hemming F, et al. Exosomes as a novel way of interneuronal communication. *Biochem Soc Trans*. 2013;41(1):241–244.
39. Danzer KM, Kranich LR, Ruf WP, et al. Exosomal cell-to-cell transmission of alpha synuclein oligomers. *Mol Neurodegener*. 2012;7:42.
40. Fruhbeis C, Frohlich D, Kuo WP, et al. Neurotransmitter-triggered transfer of exosomes mediates oligodendrocyte-neuron communication. *PLoS Biol*. 2013;11(7):e1001604.
41. Fruhbeis C, Frohlich D, Kuo WP, Kramer-Albers EM. Extracellular vesicles as mediators of neuron-glia communication. *Front Cell Neurosci*. 2013;7:182.
42. Di Pasquale E, Chahinian H, Sanchez P, Fantini J. The insertion and transport of anandamide in synthetic lipid membranes are both cholesterol-dependent. *PloS One*. 2009;4(3):4989.

43. Latek D, Kolinski M, Ghoshdastider U, et al. Modeling of ligand binding to G protein coupled receptors: cannabinoid CB1, CB2 and adrenergic beta 2 AR. *J Mol Model*. 2011;17(9):2353–2366.
44. Glaser ST, Kaczocha M, Deutsch DG. Anandamide transport: a critical review. *Life Sci*. 2005;77(14):1584–1604.
45. Glaser ST, Abumrad NA, Fatade F, Kaczocha M, Studholme KM, Deutsch DG. Evidence against the presence of an anandamide transporter. *Proc Natl Acad Sci USA*. 2003;100(7):4269–4274.
46. Kaczocha M, Glaser ST, Deutsch DG. Identification of intracellular carriers for the endocannabinoid anandamide. *Proc Natl Acad Sci USA*. 2009;106(15):6375–6380.
47. Sanson B, Wang T, Sun J, et al. Crystallographic study of FABP5 as an intracellular endocannabinoid transporter. *Acta Crystallogr D Biol Crystallogr*. 2014;70(Pt 2):290–298.
48. Tian X, Guo J, Yao F, Yang DP, Makriyannis A. The conformation, location, and dynamic properties of the endocannabinoid ligand anandamide in a membrane bilayer. *J Biol Chem*. 2005;280(33):29788–29795.
49. Klenk E. On cerebrosides gangliosides. *Prog Chem Fats Lipids*. 1969;10:411–431.
50. Thomas PD, Brewer GJ. Gangliosides and synaptic transmission. *Biochim Biophys Acta*. 1990;1031(3):277–289.
51. McIlwain H. Characterization of naturally occurring materials which restore excitability to isolated cerebral tissues. *Biochem J*. 1961;78:24–32.
52. Klemm WR, Boyles R, Mathew J, Cherian L. Gangliosides, or sialic acid, antagonize ethanol intoxication. *Life Sci*. 1988;43(22):1837–1843.
53. Ferrari G, Fabris M, Gorio A. Gangliosides enhance neurite outgrowth in PC12 cells. *Brain Res*. 1983;284 (2–3):215–221.
54. Tsuji S, Yamashita T, Tanaka M, Nagai Y. Synthetic sialyl compounds as well as natural gangliosides induce neuritogenesis in a mouse neuroblastoma cell line (Neuro2a). *J Neurochem*. 1988;50(2):414–423.
55. Ichikawa N, Iwabuchi K, Kurihara H, et al. Binding of laminin-1 to monosialoganglioside GM1 in lipid rafts is crucial for neurite outgrowth. *J Cell Sci*. 2009;122(Pt 2):289–299.
56. Mendez-Otero R, Santiago MF. Functional role of a specific ganglioside in neuronal migration and neurite outgrowth. *Braz J Med Biol Res*. 2003;36(8):1003–1013.
57. Kojima H, Gorio A, Janigro D, Jonsson G. GM1 ganglioside enhances regrowth of noradrenaline nerve terminals in rat cerebral cortex lesioned by the neurotoxin 6-hydroxydopamine. *Neuroscience*. 1984;13(4):1011–1022.
58. Mahadik SP, Vilim F, Korenovsky A, Karpiak SE. GM1 ganglioside protects nucleus basalis from excitotoxin damage: reduced cortical cholinergic losses and animal mortality. *J Neurosci Res*. 1988;20(4):479–483.
59. Ramirez OA, Gomez RA, Carrer HF. Gangliosides improve synaptic transmission in dentate gyrus of hippocampal rat slices. *Brain Res*. 1990;506(2):291–293.
60. She JQ, Wang M, Zhu DM, Sun LG, Ruan DY. Effect of ganglioside on synaptic plasticity of hippocampus in lead-exposed rats *in vivo*. *Brain Res*. 2005;1060(1–2):162–169.
61. Wieraszko A, Seifert W. Evidence for a functional role of gangliosides in synaptic transmission: studies on rat brain striatal slices. *Neurosci Lett*. 1984;52(1–2):123–128.
62. Allende ML, Proia RL. Lubricating cell signaling pathways with gangliosides. *Curr Opin Struct Biol*. 2002;12(5):587–592.
63. Hakomori S, Igarashi Y. Gangliosides and glycosphingolipids as modulators of cell growth, adhesion, and transmembrane signaling. *Adv Lipid Res*. 1993;25:147–162.
64. Schnaar RL. Brain gangliosides in axon-myelin stability and axon regeneration. *FEBS Lett*. 2010;584(9): 1741–1747.
65. Schneider JS, Sendek S, Daskalakis C, Cambi F. GM1 ganglioside in Parkinson's disease: results of a five-year open study. *J Neurol Sci*. 2010;292(1–2):45–51.
66. Wu G, Lu ZH, Kulkarni N, Ledeen RW. Deficiency of ganglioside GM1 correlates with Parkinson's disease in mice and humans. *J Neurosci Res*. 2012;90(10):1997–2008.
67. Toffano G, Savoini G, Moroni F, Lombardi G, Calza L, Agnati LF. GM1 ganglioside stimulates the regeneration of dopaminergic neurons in the central nervous system. *Brain Res*. 1983;261(1):163–166.
68. Tettamanti G, Preti A, Cestaro B, et al. Gangliosides, neuraminidase and sialyltransferase at the nerve endings. *Adv Exp Med Biol*. 1980;125:263–281.
69. Rahmann H. Brain gangliosides and memory formation. *Behav Brain Res*. 1995;66(1–2):105–116.
70. Tanaka Y, Waki H, Kon K, Ando S. Gangliosides enhance KCl^- induced Ca^{2+} influx and acetylcholine release in brain synaptosomes. *Neuroreport*. 1997;8(9–10):2203–2207.
71. Cumar FA, Maggio B, Caputto R. Neurotransmitter movements in nerve endings. Influence of substances that modify the interfacial potential. *Biochim Biophys Acta*. 1980;597(1):174–182.

72. Agostini M, Tucci P, Steinert JR, et al. MicroRNA-34a regulates neurite outgrowth, spinal morphology, and function. *Proc Natl Acad Sci USA*. 2011;108(52):21099–21104.
73. Wakabayashi T, Hidaka R, Fujimaki S, Asashima M, Kuwabara T. MicroRNAs and epigenetics in adult neurogenesis. *Adv Genet*. 2014;86:27–44.
74. Hong J, Zhang H, Kawase-Koga Y, Sun T. MicroRNA function is required for neurite outgrowth of mature neurons in the mouse postnatal cerebral cortex. *Front Cell Neurosci*. 2013;7:151.
75. Linetti A, Fratangeli A, Taverna E, et al. Cholesterol reduction impairs exocytosis of synaptic vesicles. *J Cell Sci*. 2010;123(Pt 4):595–605.
76. Barrantes FJ. Cholesterol effects on nicotinic acetylcholine receptor. *J Neurochem*. 2007;103(suppl 1):72–80.
77. Takamori S, Holt M, Stenius K, et al. Molecular anatomy of a trafficking organelle. *Cell*. 2006;127(4):831–846.
78. Tong J, Borbat PP, Freed JH, Shin YK. A scissors mechanism for stimulation of SNARE-mediated lipid mixing by cholesterol. *Proc Natl Acad Sci USA*. 2009;106(13):5141–5146.
79. Poirier MA, Xiao W, Macosko JC, Chan C, Shin YK, Bennett MK. The synaptic SNARE complex is a parallel four-stranded helical bundle. *Nat Struct Biol*. 1998;5(9):765–769.
80. Rizo J, Rosenmund C. Synaptic vesicle fusion. *Nat Struct Mol Biol*. 2008;15(7):665–674.
81. Thiele C, Hannah MJ, Fahrenholz F, Huttner WB. Cholesterol binds to synaptophysin and is required for biogenesis of synaptic vesicles. *Nat Cell Biol*. 2000;2(1):42–49.
82. Shin OH. Exocytosis and synaptic vesicle function. *Compre Physiol*. 2014;4(1):149–175.
83. Frank C, Rufini S, Tancredi V, Forcina R, Grossi D, D'Arcangelo G. Cholesterol depletion inhibits synaptic transmission and synaptic plasticity in rat hippocampus. *Exp Neurol*. 2008;212(2):407–414.
84. Koudinov AR, Koudinova NV. Essential role for cholesterol in synaptic plasticity and neuronal degeneration. *FASEB J*. 2001;15(10):1858–1860.
85. Puchkov D, Haucke V. Greasing the synaptic vesicle cycle by membrane lipids. *Trends Cell Biol*. 2013;23(10): 493–503.
86. Darios F, Wasser C, Shakirzyanova A, et al. Sphingosine facilitates SNARE complex assembly and activates synaptic vesicle exocytosis. *Neuron*. 2009;62(5):683–694.
87. Hoeferlin LA, Wijesinghe DS, Chalfant CE. The role of ceramide-1-phosphate in biological functions. *Handb Exp Pharmacol*. 2013;(215):153–166.
88. Honigmann A, van den Bogaart G, Iraheta E, et al. Phosphatidylinositol 4,5-bisphosphate clusters act as molecular beacons for vesicle recruitment. *Nat Struct Mol Biol*. 2013;20(6):679–686.
89. Davletov B, Montecucco C. Lipid function at synapses. *Curr Opin Neurobiol*. 2010;20(5):543–549.
90. Baier CJ, Fantini J, Barrantes FJ. Disclosure of cholesterol recognition motifs in transmembrane domains of the human nicotinic acetylcholine receptor. *Sci Rep*. 2011;1:69.
91. Hanson MA, Cherezov V, Griffith MT, et al. A specific cholesterol binding site is established by the 2.8 A structure of the human beta2-adrenergic receptor. *Structure*. 2008;16(6):897–905.
92. Chattopadhyay A, Paila YD. Lipid-protein interactions, regulation, and dysfunction of brain cholesterol. *Biochem Biophys Res Comm*. 2007;354(3):627–633.
93. Paila YD, Chattopadhyay A. Membrane cholesterol in the function and organization of G-protein coupled receptors. *Subcell Biochem*. 2010;51:439–466.
94. Azevedo FA, Carvalho LR, Grinberg LT, et al. Equal numbers of neuronal and nonneuronal cells make the human brain an isometrically scaled-up primate brain. *J Comp Neurol*. 2009;513(5):532–541.
95. Herculano-Houzel S, Lent R. Isotropic fractionator: a simple, rapid method for the quantification of total cell, and neuron numbers in the brain. *J Neurosci*. 2005;25(10):2518–2521.
96. Herculano-Houzel S. The remarkable, yet not extraordinary, human brain as a scaled-up primate brain and its associated cost. *Proc Natl Acad Sci USA*. 2012;109(suppl 1):10661–10668.
97. Sun W, McConnell E, Pare JF, et al. Glutamate-dependent neuroglial calcium signaling differs between young and adult brain. *Science*. 2013;339(6116):197–200.
98. Bennett MV, Zukin RS. Electrical coupling and neuronal synchronization in the mammalian brain. *Neuron*. 2004;41(4):495–511.
99. Bennett MV. Seeing is relieving: electrical synapses between visualized neurons. *Nat Neurosci*. 2000;3(1):7–9.
100. Galarreta M, Hestrin S. A network of fast-spiking cells in the neocortex connected by electrical synapses. *Nature*. 1999;402(6757):72–75.
101. Gibson JR, Beierlein M, Connors BW. Two networks of electrically coupled inhibitory neurons in neocortex. *Nature*. 1999;402(6757):75–79.

102. Schmitz D, Schuchmann S, Fisahn A, et al. Axo-axonal coupling. a novel mechanism for ultrafast neuronal communication. *Neuron*. 2001;31(5):831–840.
103. Nagy JI, Li X, Rempel J, et al. Connexin26 in adult rodent central nervous system: demonstration at astrocytic gap junctions and colocalization with connexin30 and connexin43. *J Comp Neurol*. 2001;441(4):302–323.
104. Nagy JI, Ionescu AV, Lynn BD, Rash JE. Coupling of astrocyte connexins Cx26, Cx30, Cx43 to oligodendrocyte Cx29, Cx32, Cx47: Implications from normal and connexin32 knockout mice. *Glia*. 2003;44(3):205–218.
105. Willecke K, Eiberger J, Degen J, et al. Structural and functional diversity of connexin genes in the mouse and human genome. *Biol Chem*. 2002;383(5):725–737.
106. Belluardo N, Mudo G, Trovato-Salinaro A, et al. Expression of connexin36 in the adult and developing rat brain. *Brain Res*. 2000;865(1):121–138.
107. Schubert AL, Schubert W, Spray DC, Lisanti MP. Connexin family members target to lipid raft domains and interact with caveolin-1. *Biochemistry*. 2002;41(18):5754–5764.
108. Defamie N, Mesnil M. The modulation of gap-junctional intercellular communication by lipid rafts. *Biochim Biophys Acta*. 2012;1818(8):1866–1869.
109. Hung A, Yarovsky I. Gap junction hemichannel interactions with zwitterionic lipid, anionic lipid, and cholesterol: molecular simulation studies. *Biochemistry*. 2011;50(9):1492–1504.
110. Goritz C, Mauch DH, Pfrieger FW. Multiple mechanisms mediate cholesterol-induced synaptogenesis in a CNS neuron. *Mol Cell Neurosci*. 2005;29(2):190–201.

CHAPTER

6

Protein–Lipid Interactions in the Brain

Jacques Fantini, Nouara Yahi

OUTLINE

6.1 GENERAL ASPECTS OF PROTEIN–LIPID INTERACTIONS

Molecular interactions between lipids and proteins do not follow the rules that describe the binding of two water-soluble compounds in an aqueous environment.[1] For years, the binding of lipids to proteins has been chiefly considered as a manifestation of the hydrophobic effect that placed the apolar partner, the lipid, in a hydrophobic pocket of a protein (Fig. 6.1A). In the beginning, membrane proteins were not concerned by these hydrophobic interactions because the proteins were thought to spread on both sides of the membrane, interacting with the charged head groups of phosphatidylcholine through electrostatic interactions[2] (Fig. 6.1B). It is

Brain Lipids in Synaptic Function and Neurological Disease. http://dx.doi.org/10.1016/B978-0-12-800111-0.00006-0

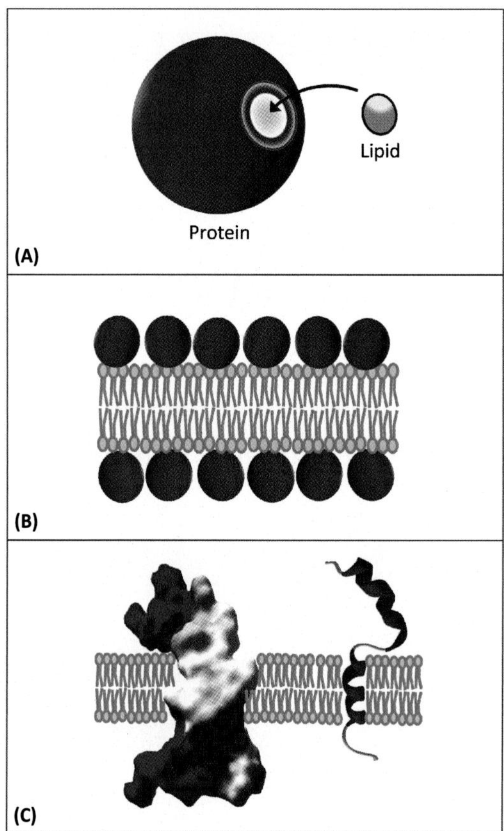

FIGURE 6.1 **Evolution of the notion of protein–lipid interactions during the past 60 years.** (A) Binding of a lipid molecule in a hydrophobic pocket of the protein. (B) Early model of the plasma membrane with proteins adsorbed on both sides and interacting with the head groups of phospholipids through electrostatic bonds. (C) Topology of an integral membrane protein inserted in the membrane bilayer (surface model on the left, ribbon model on the right). In this case, the protein may interact with membrane at three levels (extracellular, transmembrane, and intracellular domains).

only after the publication of the Singer–Nicolson model[3] that molecular mechanisms accounting for protein–lipid interactions within the apolar zone of the membrane were required.[4] In addition, the postulated existence of proteins that cross the membrane, referred to as integral membrane proteins, suggested that both the cytoplasmic and extracellular domains might also interact with the surface-exposed head groups of lipids in both leaflets of the bilayer (Fig. 6.1C).

Therefore, protein–lipid interactions are mediated by a broad range of molecular mechanisms, including van der Waals interactions, hydrogen bonds, electrostatic bridges, in fact all the molecular interactions at work in biology. The problem is further complicated by the fact that the membrane lipid that binds to the protein is also involved in various types of lipid–lipid interactions. Finally, we have to distinguish the lipids that are loosely attached to the protein, and thus might be exchanged for other membrane lipids (referred to as "annular" lipids), from lipids that are permanently interacting with the protein ("nonannular" lipids).[5]

6.2 ANNULAR VERSUS NONANNULAR LIPIDS

In a biological membrane, lipid molecules are generally considered to act as solvent for integral proteins, just like water surrounds soluble compounds in solution. In this lipid-solvent model, the lipids are loosely attached to the protein and thus can be easily displaced by other membrane lipids (Fig. 6.2). These "solvent" lipids have been referred to as annular lipids, because they form an annular shell of lipids around the protein.[6] Protein–lipid interactions of this type are relatively nonspecific.[7] On the other hand, some proteins interact with a high specificity with selected lipid molecules, such as cholesterol[5,8,9] or sphingolipids.[10,11] In this case, these lipids are referred to as "nonannular." Nonannular lipids may act as cofactors that control the biological activity of the protein through conformational effects.

Finally, membrane lipids that do not physically interact with a protein are referred to as bulk lipids, usually phosphatidylcholine.[7] These bulk glycerophospholipids can take the place of annular lipids via a rapid exchange mechanism (estimated to 10^{-7} s at 37°C).[12] The rate of exchange of nonannular lipids with bulk lipids is presumed to be significantly slower, given the high affinity of the protein for its nonannular lipid cofactors.[7] In this chapter, we will review the different mechanisms involved in the specific interaction of membrane proteins with nonannular lipids.

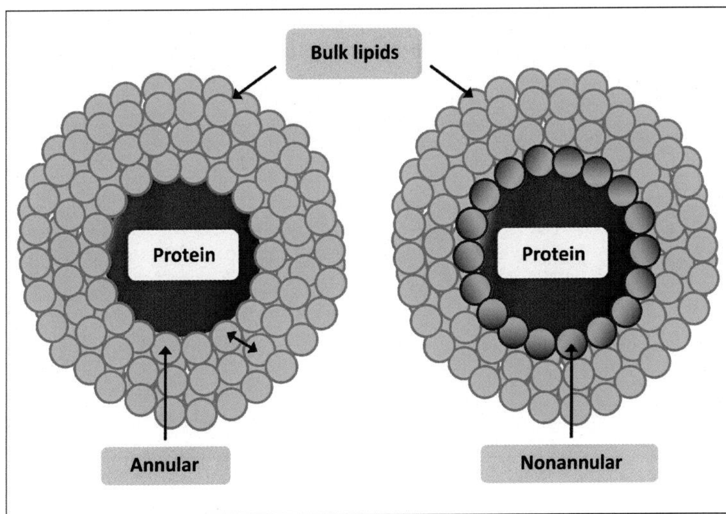

FIGURE 6.2 **Annular vs. nonannular lipids.** Annular lipids (left panel) form an annular shell of lipids that are loosely attached to the protein. These lipids act as "solvent" in the bidimensional plane of the membrane and are easily exchangeable with surrounding bulk lipids (green disks). Nonannular lipids (blue disks) interact more strongly with membrane proteins and are not exchangeable with bulk lipids. Cholesterol and sphingolipids are typical nonannular lipids, whereas phosphatidylcholine is a representative annular lipid.

6.3 INTERACTIONS BETWEEN MEMBRANE LIPIDS AND CYTOPLASMIC DOMAINS

The asymmetric distribution of lipids in each leaflet of the plasma membrane is often balanced by a complementary asymmetry of amino acids in the protein sequence. In this respect, the case of anionic phospholipids is particularly significant. Phosphatidylserine (PS) is exclusively located in the inner (cytoplasmic) leaflet of the plasma membrane. This phospholipid is formed by the condensation of serine onto the phosphatidic acid backbone. At pH 7, the head group of PS displays three electric charges: the positive charge of choline, and two negative charges, one for the phosphate group, the second one for the serine carboxylate. Overall, the net electrical charge of PS is –1. The simplest way for a protein to interact with a negatively charged lipid whose head group is flush with the membrane is to provide complementary positive charges (Fig. 6.3).

Correspondingly, the cytoplasmic domains of transmembrane proteins are often enriched in basic amino acids (lysine and/or arginine). This is the case for prominin, a protein with 5 TM domains concentrated in cellular protrusions[13] and interacting with both cholesterol and gangliosides in lipid raft domains.[14,15] Immediately after the fifth TM domain, the cytoplasmic region of human prominin starts with a cationic motif (**K**L**K**YY**RR**) that may electrostatically bind to anionic phospholipid such as PS (Fig. 6.3). This is the simplest mechanism of molecular interaction that can be described for a protein–lipid couple. This type of electrostatic bond also explains the interaction of the microtubule-associated tau protein with neural plasma membranes.[16] *In vitro* studies have revealed that the K19 fragment of tau, which contains the microtubule binding domain, binds to PS liposomes with a high affinity.[17] In comparison, the affinity for phosphatidylcholine was about 250 times lower.[17] This difference seems logical because this region of tau is enriched in lysine residues that can interact with the negative charge of PS through electrostatic interactions. It should be noted that the association of

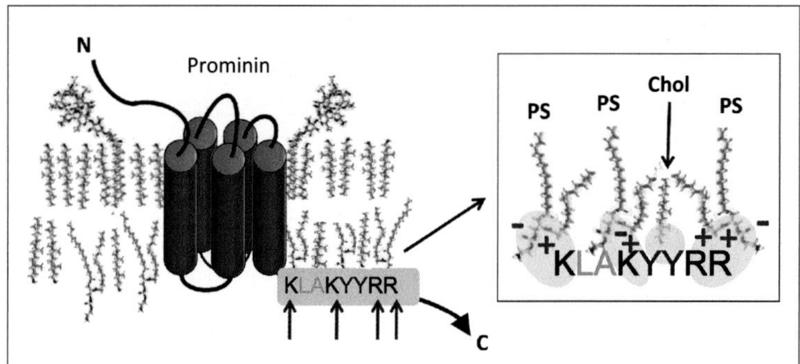

FIGURE 6.3 **How membrane proteins interact with anionic lipids of the inner membrane leaflet.** Cytoplasmic domains of proteins, especially when these proteins are located within lipid raft domains, are enriched in cationic amino acids (lysine and/or arginine). This is the case for prominin, whose juxtamembrane C-terminal region contains two lysine and two arginine residues in the first eight cytoplasmic amino acids (left panel). The positive charge of these basic residues may interact with the negative charge of anionic phospholipids such as phosphatidylserine (PS). Note that in the case of prominin, the OH group of tyrosine residues may also interact with cholesterol through hydrogen bonds (the right panel).

tau with neural membranes may occur in both physiological and pathological conditions. Indeed, tau is enriched in axons and growth cones where its simultaneous interaction with the plasma membrane and microtubules may be important for neuritic development.[16] In Alzheimer's disease, neural membranes may act as a concentration platform[18] that are thought to facilitate the formation of neurotoxic tau oligomers.[17,19]

Although it can be intuitively assumed that basic residues are prone to interact with the anionic head group of PS, it has been reported that the nature of the acyl chains anchoring the phospholipid in the membrane might also be important for the binding. The case of α-synuclein, the protein associated with Parkinson's disease, is highly informative. This protein interacts with negatively charged phospholipids, including phosphatidic acid and PS.[20] These interactions are mediated by lysine residues, present on membrane-induced α-helices of α-synuclein, which bind to the negatively charged phospholipid head groups.[21–23] Nevertheless, the interaction of α-synuclein to synthetic PS molecules appeared to be highly dependent on acyl chain composition.[24] Indeed, α-synuclein did not bind to synthetic PS with both acyl chains identical. It turned out that the interaction with α-synuclein requires a combination of PS with oleic (C18:1ω9) and polyunsaturated (e.g., arachidonic acid C20:4ω6). The requirement of both C18:1ω9 and C20:4ω6 for binding to PS suggested a role for phase transition between raft and nonraft domains.[24] In all cases, these data indicated that the electrostatic interaction between cationic amino acids of α-synuclein and the negative head group of PS is not as simple as anticipated. A fine tuning of membrane fluidity conferred by a specific acyl composition of PS plays an important role in the binding of the protein at the periphery of lipid raft domains.

6.4 INTERACTIONS BETWEEN MEMBRANE LIPIDS AND TRANSMEMBRANE DOMAINS

6.4.1 Back to First Principles

Integral membrane proteins have at least one transmembrane domain that crosses the lipid bilayer. These transmembrane (TM) domains are naturally enriched in apolar amino acids that allow a smooth insertion in the apolar phase of the lipid bilayer. Basically, a TM domain consists in a cluster of ~25 apolar amino acid residues with a α-helical structure. Classifying the amino acids according to their hydropathy had allowed Kyte and Doolittle to propose a hydropathy/hydrophobicity scale[25] that has been widely used as an algorithm for the prediction of membrane protein topology.[26] However, the rapid progress of bioinformatics approaches has rapidly supplanted this early approach by machine learning methods that extract statistical sequence preferences from databases of experimentally mapped topologies[27] and from endless alignments of homologous sequences.[28] That the best predictive methods relied on sequence statistics rather than physicochemical principles as the underlying basis for the prediction has been lucidly highlighted by Bernsel et al.[27] These authors proposed a return to basic principles for developing new algorithms[27] that take into account an experimental scale of position-specific amino acid contributions to the free energy of membrane insertion.[29] Their simplified approach was able to compete in terms of efficiency with the best statistics-based topology predictors. Most importantly, these data demonstrated that the prediction of membrane–protein topology and structure directly from first principles is an attainable goal. Our own contributions to the definition of cholesterol- and sphingolipid-binding domains have the same general objectives: study a particular protein–lipid binding process, understand the basic principles of this interaction, and derive general rules that can be applied predictively to other lipid–protein duets.

6.4.2 Cholesterol-Binding Domains

The length of cholesterol is ~ 20Å. Given that in an ideal α-helix, each amino acid residue advances the helix by ~ 1.5 Å,[30] a α-helix stretch of ~ 13 amino acid residues (20/1.5) has approximately the same size as cholesterol. The mean length of a TM domain crossing the plasma membrane of a vertebrate cell is 24.4 residues, with the most frequent length between 22 and 26 residues.[31] Therefore, the same TM domain can theoretically interact with two cholesterol molecules, one in the exofacial leaflet and the second one in the inner leaflet of the plasma membrane. Surprisingly, although TM domains are naturally enriched in apolar amino acids, the abundance of individual hydrophobic residues changes along the length of the TM domains, thereby creating a striking asymmetric distribution.[31] For instance, valine and glycine are preferentially located in exoplasmic positions, whereas leucine is more abundant in the inner leaflet. Similarly, aromatic residues are overrepresented in the inner leaflet of the plasma membrane, which contains more voluminous amino acids than the outer leaflet. Thus, the mechanisms controlling the interaction of TM domain with plasma membrane cholesterol might be adapted to this double chemical–geometric asymmetry.

Two algorithms are currently used to predict cholesterol-binding domains in TM domains. The CRAC domain (cholesterol recognition/interaction amino acid consensus sequence) has been defined after studying the binding of cholesterol to the peripheral-type benzodiazepine receptor (now referred to as translocator protein TSPO),[32,33] a mitochondrial membrane protein involved in the regulation of cholesterol transport into mitochondria.[34] Because the job of TSPO is not just to bind, but rather to bind-and-release cholesterol, it is intriguing that this particular protein has become the prototype of cholesterol-binding mechanisms of integral membrane proteins.[5,8,9] Moreover, mitochondria are considered cholesterol-poor organelles.[35] Nevertheless, CRAC domains have been found in a broad range of cholesterol-binding proteins.[5,8,9,32,33] The CRAC algorithm is, from the N-terminal to C-terminal direction, (L/V)–X_{1-5}–(Y)–X_{1-5}–(K/R).

This means that a CRAC domain consists in (1) a branched apolar residue (either leucine or valine), (2) a segment containing 1–5 of any residues, (3) a central tyrosine, (4) then again a segment containing 1–5 of any residues, and (5) a cationic residue (either lysine or arginine). The looseness of the CRAC definition for a motif that mediates binding to a unique lipid is intriguing, which has raised skepticism about its predictive value.[36] Nevertheless, the physical interaction between cholesterol and CRAC domains has been confirmed by various experimental approaches.[37] Moreover single mutations in the CRAC domain have been found to markedly affect cholesterol recognition, which demonstrates the robustness of the CRAC algorithm.[33,37]

The structure of the CRAC domain gives some clues about its mechanism of interaction with cholesterol (Fig. 6.4). The C-terminal basic residue (lysine or arginine) is naturally inclined to interact with the OH group of cholesterol. On the other hand, the N-terminal residue

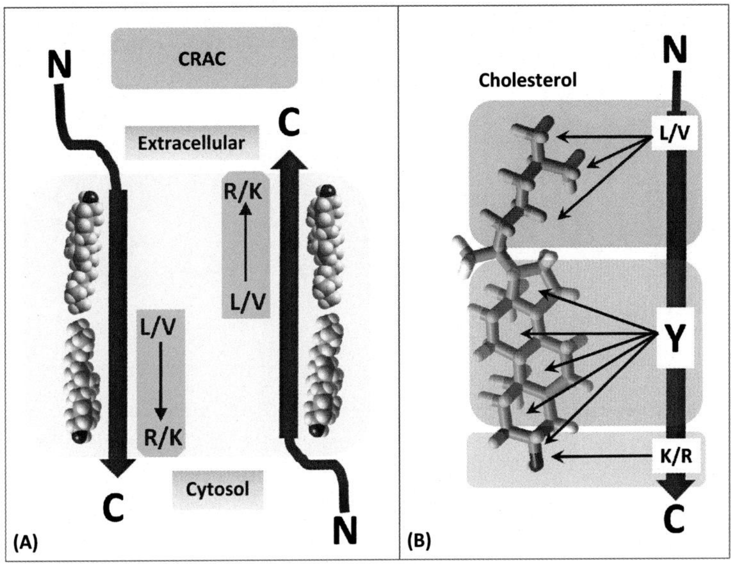

FIGURE 6.4 **Topology of membrane proteins and interaction with the CRAC domain.** (A) The CRAC motif has its basic amino acid (Lys or Arg) at its C-terminus. This basic residue is supposed to contact the polar OH group of cholesterol. Hence, for a type I membrane protein that crosses the bilayer in the [N-ter/extracellular → C-ter/cytoplasm] direction, it is adequately oriented for interacting with cholesterol in the cytosolic leaflet (left panel). In the case of a type II membrane protein, which crosses the membrane in the opposite direction [N-ter/cytoplasm → C-ter/extracellular], the CRAC domain will interact with cholesterol in the outer leaflet (right panel). (B) The CRAC–cholesterol interaction may involve van der Waals forces (Leu/Val with the isooctyl chain), CH–π stacking (Tyr with one of the four rings of sterane), and/or hydrogen bonding (Lys/Arg with the sterol OH group). The linkers joining the central tyrosine to each end may vary in length (1–5 of any amino acids), so that they can be compared to cursors. These cursors may finely tune the CRAC motif for an optimal interaction with cholesterol.

(valine or leucine) fits perfectly with the isooctyl chain of the sterol. These interactions give a unique possible topology of the CRAC/cholesterol complex. In this topology, both cholesterol and the CRAC domain are considered as amphipathic molecules that stick together through a parallel (head-to-head/tail-to-tail) orientation (Fig. 6.4A). The central aromatic residue will stack onto one of the rings of the sterane backbone, depending on the length of the linker (X_{1-5}) between this residue and the cationic group (several possibilities are explored in Fig. 6.4B).

If the TM domain crosses the bilayer in the N-terminal to C-terminal direction, the CRAC domain will probably interact with a cholesterol molecule located in the cytoplasmic leaflet of the membrane. This is the case for single-pass type I membrane proteins and for TM domains 1, 3, 5, and 7 of G-proteins coupled receptor (GPCR) proteins with seven TM domains. A representative example of this type of interaction is given by the fifth TM domain of the human type 3 somatostatin receptor.[5] An interaction of the CRAC domain with cholesterol in the exofacial leaflet is possible if the TM domain crosses the bilayer in the C-terminal to N-terminal direction, as is the case for single-pass type II membrane proteins and for TM domains 2, 4, and 6 of GPCR proteins.

An important caveat in the predictive value of the CRAC algorithm is that in some cases, it may correctly identify a cholesterol-binding domain, but outside the membrane bilayer.

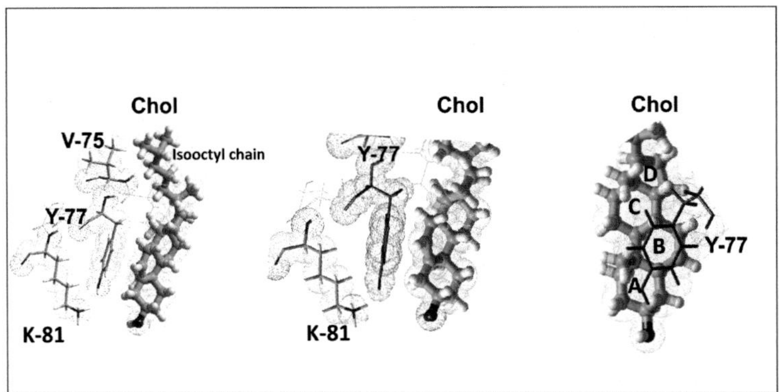

FIGURE 6.5 **Molecular mechanisms of a typical cholesterol–CRAC interaction.** These models show a detailed analysis of the interaction between cholesterol and the CRAC domain of human delta-type opioid receptor (74-**IVRYTKMK**-81). Three distinct views of the complex are shown, with residues Val-75, Tyr-77, and Lys-81 enlightened. The NH_3^+ of Lys-81 and the OH groups of Tyr-77 and cholesterol are rejected in a polar area where they form a network of energetically favored interactions (including hydrogen bonds). The aromatic side chain of Tyr-77 stacks onto the B ring of sterane through typical CH–π stacking interactions. The isooctyl chain of cholesterol interacts with the aliphatic side chain of Val-75. The four rings of sterane (named A, B, C, and D) are indicated for cholesterol in the right panel. *(Reproduced from Fantini and Barrantes[5] with permission.)*

This case has been documented for human delta-type opioid receptor, a GPCR protein.[5] This receptor contains a CRAC motif (74-**IVRYTKMK**-81) just upstream the second TM domain. Indeed, the second TM domain of this receptor encompasses residues 85–102. The interaction between cholesterol and this particular CRAC domain is representative of both the robustness and the weakness of the CRAC algorithm. The fit between cholesterol and the **IVRYTKMK** motif is impressive (Fig. 6.5). The energy of interaction of the complex is high (– 49 kJ mol^{-1}) and the tyrosine residue accounts for ~ 50% of the interaction (–26 kJ mol^{-1}). Specifically, the aromatic ring of Tyr-77 binds to the B ring of sterane through a quite perfect CH–π stacking interaction.[38] Finally, the branched aliphatic chain of Val-75 interacts with the isooctyl chain of cholesterol (Fig. 6.5).

Although the third amino acid that defines the CRAC domain (Lys-79 or Lys-81 in this case) is not directly involved in cholesterol binding, it stabilizes the orientation of Tyr-77, resulting in a remarkable fit of cholesterol for this CRAC domain. This interaction illustrates how well the algorithm performs. Nevertheless, the whole motif is rather polar than apolar because the nondefined X residues of this CRAC domain are chiefly arginine and lysine. Consequently, this domain is not located inside the membrane, but in the cytoplasm.[5] Thus, despite the fact that this domain can theoretically bind cholesterol with high affinity, as correctly predicted by the CRAC algorithm, its location outside the membrane precludes any functional interaction with membrane cholesterol.

More recently, an inverted version of the CRAC algorithm, referred to as "CARC" has been formulated by Baier et al.[39]: (K/R)–X_{1-5}–(Y/F/W)–X_{1-5}–(L/V). Although it looks similar to CRAC, the CARC domain displays specific remarkable features: (1) the central residue has to be aromatic, but contrarily to CRAC (which has a specific requirement for tyrosine), this aromatic residue can be either Tyr or Phe[5]; (2) in the CARC domain, the basic amino acid

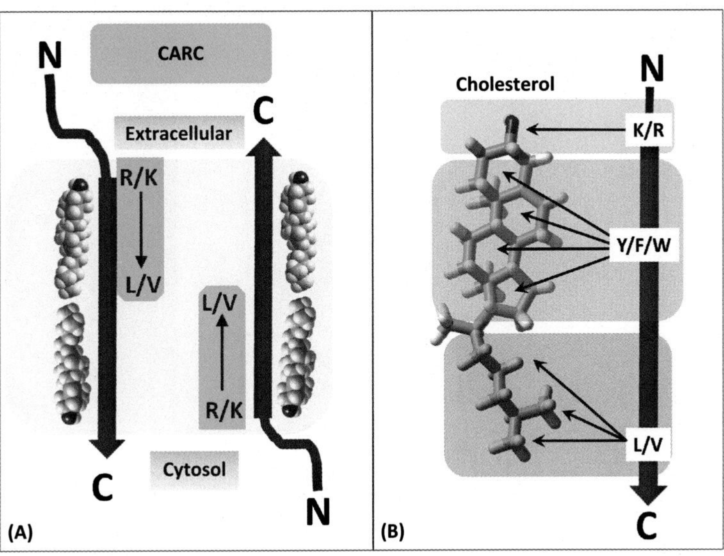

FIGURE 6.6 **Topology of membrane proteins and interaction with the CARC domain.** (A) In total contrast with CRAC (Fig. 6.4), the CARC motif has its basic amino acid (Lys or Arg) at its N-terminus. Thus, a CARC motif in a type I membrane protein can bind to cholesterol in the exofacial leaflet of the plasma membrane (left panel). In the case of a type II membrane protein (right panel), the CARC motif will bind to cholesterol in the cytoplasmic leaflet. (B) As for CRAC, the CARC–cholesterol interaction may involve van der Waals forces (Leu/Val with the isooctyl chain), CH–π stacking (Tyr/Phe/Trp with one of the four rings of sterane), and/or hydrogen bonding (Lys/Arg with the sterol OH group). The linkers joining the central aromatic residue of CARC to each end may also vary in length (1–5 of any amino acids). As for CRAC, these cursors may finely tune the CARC motif for an optimal interaction with cholesterol.

(Lys or Arg) is transferred on the N-terminal side. Therefore, the motif is adequately oriented for an interaction of type I membrane protein with cholesterol in the exofacial leaflet (Fig. 6.6).

Two main reasons explain this orientation. On one hand, arginine and lysine residues are more frequently found in the N-terminus than in the C-terminus of a TM domain.[40] On the other hand, after the first basic residue, the amino acids that are mandatory for defining the CARC motif (Y/F/W, L/V) have apolar side chains. Overall, these structural requirements are fully consistent with a TM domain, and thus the CARC algorithm is particularly reliable for the prediction of cholesterol-binding sites that are actually located within TM domains.[5] The fifth TM domain of the human type 3 somatostatin receptor has a typical CARC domain (203-RAGFIIYTAAL-213) ensuring a tight association with cholesterol in the outer leaflet of the plasma membrane (Fig. 6.7A). In this case, the binding of cholesterol involved the first basic residue (Arg-203), both aromatic residues (Phe-206 and Tyr-209), and the terminal aliphatic residue Leu-213, for a total energy of interaction of –54 kJ mol^{-1}. Moreover, as already discussed, this particular TM domain has also a CRAC domain facing the lipids of cytoplasmic leaflet. Docking studies indicated no particular steric hindrance forbids the simultaneous interaction of this domain with two cholesterol molecules, one in each leaflet of the membrane bilayer[5] (Fig. 6.7B). Both cholesterol molecules are in a typical tail-to-tail configuration, a topology that has been reproducibly found in artificial membrane bilayers.[41]

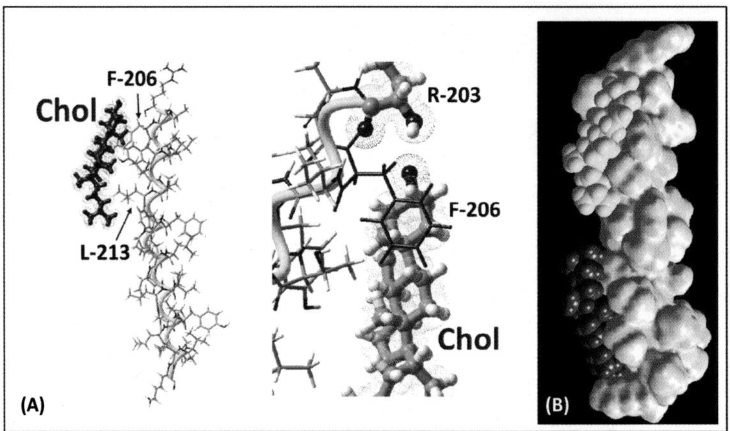

FIGURE 6.7 **Molecular mechanisms of a typical cholesterol–CARC interaction.** (A) The CARC domain of the human type 3 somatostatin receptor (203-**R**AG**F**IIYTAA**L**-213) is located in the exofacial leaflet of the fifth TM domain. Two distinct views of the complex are shown, one with the whole TM5 domain (left panel), the other with residues Arg-203 and Phe-206 enlightened. Note the CH–π stacking interaction of the phenyl ring of Phe-206 onto the A ring of sterane. The large aliphatic chain of Leu-213 interacts with the isooctyl group of cholesterol. (B) Whole view of the fifth TM domain of the receptor interacting with two cholesterol molecules, one in the exofacial leaflet (cholesterol in yellow bound to CARC) and the second on the inner leaflet (cholesterol in purple bound to CRAC). *(Reproduced from Fantini and Barrantes[5] with permission.)*

In conclusion, the CARC domain has been detected in a wide range of type I and type III membrane proteins, including ion channels and GPCR proteins.[39] Its unique biochemical features are particularly consistent with the definition of a consensus cholesterol-binding domain ensuring the specific interaction of TM domains with membrane cholesterol. Its definition is recapitulated in a simple algorithm that reflects basic principles of protein–lipid interactions in a membrane environment.

Apart from CRAC and CARC motifs that define a contiguous cholesterol-binding motif within a single TM domain, a three-dimensional cholesterol-binding site has been described in the human β-2 adrenergic receptor.[42] This motif, referred to as the *cholesterol consensus motif* (CCM), includes a series of amino acid residues belonging to adjacent TM domains: (W/Y)–(I/V/L)–(K/R) on one helix, and (F/Y/R) on the second helix.[43] Interestingly, the residues that constitute the CCM are also those defining both the CARC and CRAC motifs. Therefore, the same basic principles apply for cholesterol/CRAC, cholesterol/CARC, and cholesterol/CCM interactions. Finally, two novel cholesterol-binding sites were recently identified in inwardly rectifying K^+ channels Kir2.1, both in the inner leaflet of the membrane.[44] These cholesterol-binding domains involve in each case two adjacent helices of the channel. The interaction with cholesterol is chiefly mediated by bulky apolar residues such as isoleucine, leucine, valine, and methionine. Aromatic residues, required by both CRAC and CARC algorithms, are absent from these new cholesterol-binding motifs. Similarly, the cholesterol-binding domains of Alzheimer's β-amyloid peptide[45] and Parkinson's disease associated α-synuclein[46] are also devoid of aromatic residues. As for Kir2.1, the interaction of these amyloid proteins with cholesterol is mediated by large apolar residues such as isoleucine, valine, and methionine. Docking studies indicated that cholesterol contacts the α-helix of these domains in a tilted orientation that resembles that of virus fusion peptides.[47] This tilted orientation facilitates the oligomerization of amyloid proteins into a Ca^{2+}-permeable pore.[48,49] Tilted peptides with a similar oblique topology were initially discovered in viral fusion proteins.[50] These peptides are relatively short (10–20 residues), have a α-helix structure, and their net hydrophobicity increases from one end of the helix to the other.[51] This hydrophobic asymmetry induces a characteristic tilt when the peptides are located at an apolar/polar interface, such as the lipid–water interface of the plasma membrane.[52] This tilted orientation destabilizes the membrane in which the peptide is inserted, thereby facilitating membrane fusion (for virus) or membrane insertion followed by oligomeric pore formation (for amyloid proteins). The unique gradient of hydrophobicity along the α-helix of a tilted peptide is a structural requirement that can be achieved by numerous combinations of amino acid sequences. Hence, tilted peptides have no particular consensus motif. Nevertheless, it appeared that tilted peptides constitute a specific class of cholesterol-binding domains,[5,46] which can in some cases overlap with the other cholesterol-binding motifs defined by CRAC, CARC, or CCM algorithms. In conclusion, many types of cholesterol-binding sites can be either linear[32,33,39,45,46] or three-dimensional.[42,53] These motifs may vary in size and amino acid composition. Nevertheless, the molecular mechanisms controlling the physical association of cholesterol with these different motifs are remarkably convergent. In all cases, the OH group of cholesterol interacts with a polar group of the protein, and the apolar part of cholesterol (sterane rings, methyl groups, and terminal isooctyl chain) interacts with bulky apolar residues of the cholesterol-binding domain. These hydrophobic interactions can be mediated by branched aliphatic residues (Ile, Val, Leu) and/or aromatic residues (Phe, Tyr, Trp). These basic principles were illustrated in Fig. 6.7 and may apply to most if not all protein–cholesterol interactions that take place in biological membranes.[5]

6.4.3 Interaction of TM Domains with Sphingolipids

Although the interaction of TM domains with cholesterol has been well documented, much less is known about sphingolipid binding to TM domains. A recent study revealed that the TM domain of the COPI machinery protein p24 displays a striking high affinity for a single sphingomyelin species, *N*-stearoyl sphingomyelin (i.e., sphingomyelin with a C18:0 acyl chain).[54] This interaction seemed to depend on both the polar head group and the ceramide backbone of sphingomyelin, and on a specific sequence on the TM domain, VxxTLxxIY. A signature motif derived from this sequence was subsequently found in a set of mammalian plasma membrane proteins. In the model proposed by Contreras et al.,[54] the polar head group of sphingomyelin binds to the tyrosine side chain of the motif (Tyr-21 in the p24 protein studied), and the acyl chain interacts with the valine, threonine, and leucine residues. This set of amino acid residues, especially the presence of both aromatic and large branched residues, is strikingly similar to the composition of a typical cholesterol-binding motif. To check this point, we used molecular dynamics simulation to evaluate the capability of the sphingomyelin-binding domain of p24 to interact with cholesterol. The results of this modeling exercise are shown in Fig. 6.8A. The TM domain of this protein has 21 amino acid residues. The sphingomyelin signature motif lies at the C-terminus of this TM domain, between amino acids Val-13 and Tyr-21: 13-VAMTLGQIY-21. We modeled the α-helix encompassing residues 12–21 and

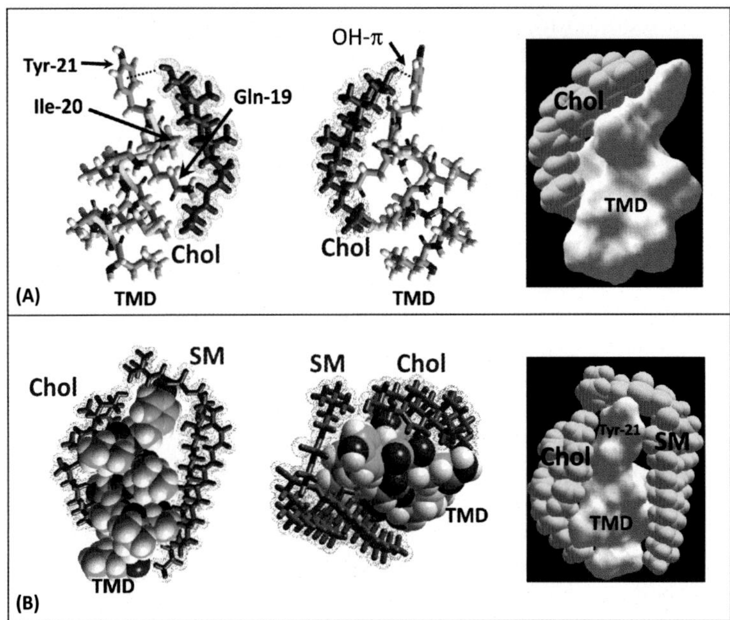

FIGURE 6.8 **The sphingomyelin-binding motif for transmembrane domains.** (A) Docking of cholesterol onto the sphingolipid-binding motif of the COPI machinery protein p24 located in the transmembrane domain (TMD). Note the curved shape of cholesterol (colored in purple). A typical OH–π bond is visible in the left and middle illustrations. At the right is a surface rendering of the TMD complex with cholesterol (in yellow). (B) Molecular dynamics simulations of a trimolecular complex involving cholesterol (chol), the TMD of p24, and sphingomyelin (SM).

merged the energy-minimized helix with cholesterol. Molecular dynamics simulations in the ns range of time indicated that the sphingomyelin-binding motif also displayed a remarkable fit for cholesterol.

As shown in Fig. 6.8B, cholesterol adopted a curved structure that allowed its insertion between the side chains of Gln-19 and Ile-20 residues. The polar OH group of cholesterol was at 3 Å of the aromatic ring of Tyr-21, consistent with the establishment of a weak hydrogen (OH–π) bond.[55] Overall, the interaction of cholesterol with the TM domain of the COPI machinery protein p24 involved both van der Waals interactions and an OH–π bond, for a total energy of interaction of –52 kJ mol^{-1}. Therefore, the affinity of this domain for cholesterol is equivalent to that of CARC or tilted domains.[5] Interestingly, when sphingomyelin was merged with this TM-cholesterol complex, it readily bound to both cholesterol and the TM domain (Fig. 6.8), raising the energy of interaction to –88 kJ mol^{-1}. Because in natural plasma membranes sphingomyelin is always associated with cholesterol, these data support the notion that the sphingomyelin-binding domain of the p24 transmembrane protein may still bind to sphingomyelin in presence of high cholesterol concentrations. Moreover, the topology of sphingomyelin in the trimolecular p24:cholesterol:sphingomyelin complex is not drastically different from the topology of the sphingomyelin:p24 complex described by Contreras et al.[54] In particular, the polar head group of sphingomyelin interacted with tyrosine, in either the absence or presence of cholesterol. In the cholesterol:p24 complex, the contribution of Tyr-21 to the total energy of interaction was estimated to be -12 kJ mol^{-1}. In the trimolecular complex, Tyr-21 interacted with both cholesterol and sphingomyelin, and its energy of interaction was -30 kJ mol^{-1} (i.e., -12 kJ mol^{-1} for cholesterol and -18 kJ mol^{-1} for sphingomyelin).

The finding that both the apolar and polar head groups of sphingomyelin are involved in the interaction with the TM domain of the p24 protein may explain why this protein selectively interacts with sphingomyelin and not with phosphatidylcholine, although both membrane lipids share the same phosphorylcholine head group.[54] Beyond the particular case of COPI machinery protein p24, these data have allowed the development of a simple algorithm for the detection of sphingomyelin-binding domains from amino acid sequences. Nevertheless, it should be noted that in the p24 protein, which has served as the basis for this algorithm, the consensus motif (VxxTLxxIY) is oriented from the N-terminus toward the C-terminus. Therefore, this may lead to the misleading identification of a sphingomyelin-binding domain in the cytoplasmic leaflet of the plasma membrane, which does not contain sphingomyelin (Fig. 6.9). This could be the case, for instance, for type I membrane proteins, which cross the plasma membrane in the N-terminus toward C-terminus. To ensure a correct detection of sphingomyelin-binding domains in the exofacial leaflet, where sphingolipids are exclusively located, the consensus motif should be reversed (i.e., YIxxLTxxV from the N-terminus toward the C-terminus).

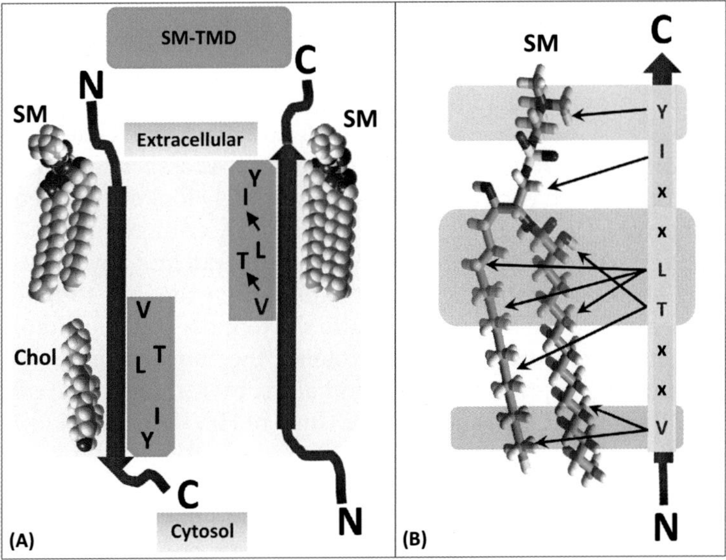

FIGURE 6.9 **Orientation of the sphingomyelin-binding motif in transmembrane domains.** (A) The tyrosine residue of the transmembrane sphingomyelin-binding motif interacts with the polar head group of sphingomyelin. Because sphingomyelin is exclusively located in the exofacial leaflet of the plasma membrane, the orientation of the motif is an important issue that should be considered for prediction of a functional sphingomyelin-binding motif in a transmembrane domain of an integral protein. Indeed, for a type I membrane protein (left panel), the motif is in the inner leaflet, which is devoid of sphingolipids. In this case, the motif may be used for cholesterol binding. In contrast, the motif is correctly oriented for type II membrane proteins (right panel). (B) The amino acid residues that define the sphingomyelin-binding motif may interact with distinct regions of sphingomyelin, resulting in a high specificity of interaction.

Like the reversion of the CRAC motif, which had allowed the identification of previously overlooked cholesterol-binding domains, the reversion of the sphingomyelin-binding domain could lead to the finding of new sphingomyelin-binding sites in TM domains. In any case, defining a sphingomyelin domain from sequence data represents a major breakthrough in our knowledge of protein–lipid interactions in biological membranes.

6.5 INTERACTIONS BETWEEN MEMBRANE LIPIDS AND EXTRACELLULAR DOMAINS

6.5.1 The Sphingolipid-Binding Domain (SBD)

Ten years before this major discovery, our group reported that unrelated viral and amyloid proteins shared a common structural fold that allowed them to bind to sphingolipids with high affinity.[56] We called this domain the sphingolipid-binding domain (SBD). The idea behind this research came from a course taught to master degree students. Consulting the literature on prions, we came across an article describing the association of prion rods with two specific membrane lipids, galactosylceramide (GalCer), and sphingomyelin.[57] This article instantly brought us back 10 years, to the time when we and others searched for alternative HIV-1 receptors that could explain the infection of CD4-negative cells. In 1991, Harouse et al. reported that the virus could use a glycosphingolipid abundantly expressed in the brain, GalCer, to gain entry into neural cells.[58] On our side, we were trying to identify the receptor used by HIV-1 to infect CD4-negative epithelial intestinal cells,[59] and in remarkable convergence with the discovery of the Gonzalez-Scarano team, it turned out that GalCer was also the intestinal HIV-1 receptor.[60] So, in 2001, learning that GalCer was associated with infectious prion rods was both fascinating and exciting, and we immediately looked for a sequence homology between the V3 loop of HIV-1 gp120, known to be the GalCer-binding domain of HIV-1[61] and the prion protein PrP. To our great disappointment, we did not find any homology between the V3 loop and PrP. At this time, one of us (Nouara Yahi), proposed to look for a structural homology and by browsing the Protein Databank she found a suitable bioinformatics tool for finding structural alignments between two proteins, the combinatorial extension (CE) program.[62] By comparing the peptide chain (carbon alpha by carbon alpha) of the extracellular domain of PrP (PDB entry # 1QLX) and of the V3 loop of HIV-1 gp120 (PDB entry # 1CE4), we found a significant structural alignment (Fig. 6.10A).

In fact, both the V3 loop and the 179–211 segment of PrP adopted a surprisingly similar loop-shaped structure.[56] Without the CE program, it would have been impossible to find this structural fold, because of the 33 amino acids aligned, only one position is identical (Thr-199 in PrP, and Thr-22 in the V3 loop). However, we noticed the presence, at the tip of the motif, of aromatic residues (Phe-20/Tyr-21for V3, Phe-198 for PrP). In addition, both motifs had turn-inducing amino acids (Gly and/or Pro) and an overrepresentation of charged residues (Asp, Glu, Lys, and Arg). Synthetic peptides derived from the V3 loop-like domain of PrP interacted with GalCer and sphingomyelin in a dose-dependent and specific manner, thereby fully confirming the data obtained *in silico*. We also showed that the Alzheimer's β-amyloid peptide could adopt a similar V3-like structure that accounted for its affinity for GalCer and sphingomyelin,[56] and probably other raft lipids.[63] Having found a common fold conferring

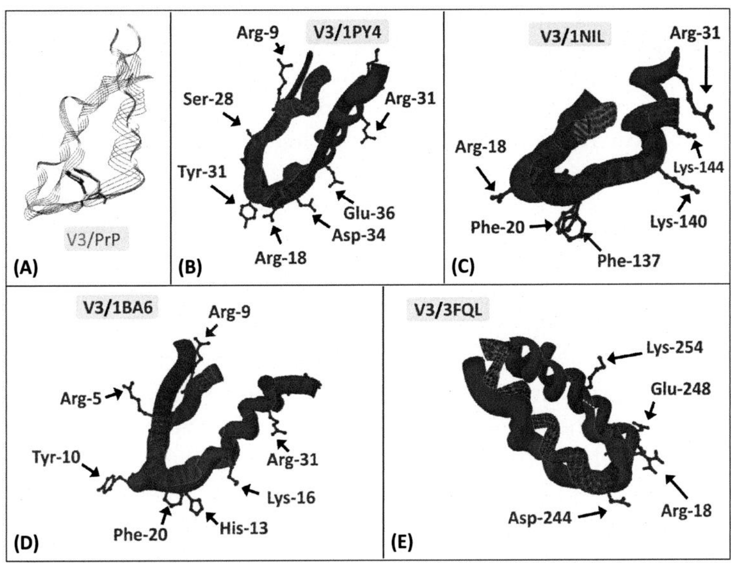

FIGURE 6.10 **The universal sphingolipid-binding domain (SBD).** (A) Superposition of the V3 loop of HIV-1 with the SBD of the PrP protein. (B–E) The four following SBDs were superposed with the V3 loop: β2-microglobulin (PDB entry # 1PY4); the receptor-binding domain of *Pseudomonas aeruginosa* pili (PDB entry # 1NIL); Alzheimer's β-amyloid peptide (PDB entry # 1BA6); and hepatitis C virus polymerase (PDB entry # 3CQF). (The V3/PrP model as shown in subpart (**A**) is reproduced from Mahfoud et al.[56] with permission.)

sphingolipid-binding properties for unrelated viral and amyloid proteins, we called this fold the sphingolipid-binding domain (SBD). Later on, the SBD was found in other amyloid proteins (amylin,[64] α-synuclein[65]) and a wide variety of proteins, including membrane receptors,[66,67] differentiation markers,[15] lipid-binding proteins,[68,69] bacterial toxins,[11] bacterial adhesins,[70] and viral proteins.[71] Because it has also been detected in insect proteins,[72] the SBD covers of a large part of the phylogenetic tree of living organisms.

Following its initial characterization on the basis of structural alignments with the V3 loop of HIV-1 gp120, a sequence-based detection algorithm has been developed.[70] Among the most notable achievements of this algorithm, one could mention the characterization of the SBD identified in a putative adhesion of *Helicobacter pylori*.[70] A short peptide (15 amino acids) derived from this SBD has been synthesized and tested for glycosphingolipid recognition. This synthetic SBD recapitulated the glycosphingolipid binding properties of the whole bacterium, especially the selective recognition of LacCer and the lack of interaction with Gb_3. More recently, we have been able to synthetize a universal ganglioside-binding peptide combining the properties of Alzheimer's β-amyloid peptide (binding to GM1) and α-synuclein (GM3 recognition) and extending the recognition pattern to all major brain gangliosides.[73]

Single mutations in synthetic SBDs at positions predicted to be important for sphingolipid recognition systematically abrogated or strongly decreased sphingolipid binding. For instance, in the case of *Helicobacter pylori*, the replacement of Phe-147 with Ala strongly affected LacCer recognition, whereas a conservative substitution with Trp did not.[70] Similarly, the substitution of Tyr-39 by Ala (but not by Phe) in the SBD of α-synuclein abrogated GM3 binding.[55] Overall, these data established the prediction algorithm of the SBD from sequence data as a robust and reliable strategy. Typical examples of SBD–sphingolipid interactions are illustrated in Fig. 6.10. Other examples can be found throughout this book, especially in Chapters 8–13). An important feature of the SBD is that it is defined as a symmetric motif centered on an aromatic residue and displaying charged residues (either acidic or basic) at both ends (Fig. 6.11). Hence, it is not necessary in this case to consider a reversed version of the motif, as it has been done for instance in the case of the CRAC algorithm.[39]

Another important feature is that although the initial definition of the SBD concerned both sphingomyelin and glycosphingolipids, the SBDs that have been characterized thereafter are essentially glycosphingolipids. In fact, the recognition of sphingomyelin requires molecular interactions with both the polar and apolar part of the lipid.[54] Therefore, sphingomyelin is essentially recognized by TM domains of host proteins, which are more difficult to study experimentally than water-soluble SBD peptides. Moreover, these SBD peptides have access only to the head group of the sphingolipids, which are more voluminous in the case of glycosphingolipids than for sphingomyelin. For all these reasons, it is more difficult to characterize a sphingomyelin-specific SBD than a SBD able to interact with neutral glycosphingolipids or gangliosides. Nevertheless, several cases of SBD-sphingomyelin interactions have been demonstrated,[56,71,74] so that the SBD can actually be considered as a generic domain that may interact with any sphingolipid, according to its biochemical structure.[11,75,76] Finally, we showed that the unique amino acid sequence of each SBD, in particular at "minor" positions that are not considered by the universal algorithm, confers the specific pattern of sphingolipid recognition.[65] According to this basic principle, minor variations in the SBD sequence may change the sphingolipid-binding specificity of the SBD. This guideline has allowed to us to create, through a rational and predictive strategy, a universal ganglioside-binding SBD peptide.[73]

6.5.2 Recognition of Membrane Cholesterol by Extracellular Proteins

In lipid rafts, extracellular proteins (and extracellular domains of membrane proteins) do not have access to cholesterol because its polar OH group is totally covered by the head groups of sphingolipids with which cholesterol interacts.[46] The situation is different outside lipid rafts where cholesterol is associated with phosphatidylcholine.[77,78] In this latter case, the OH group of cholesterol is not masked by phospholipids. Bacterial toxins such as cytolysins interact with the plasma membrane of host cells through a loop domain displaying at its tip two conserved amino acids, a leucine, and a threonine.[79–81] It is believed that this couple of amino acids acts as a sensor to detect plasma membrane cholesterol. This mechanism can be used by any protein interacting with the accessible head group of cholesterol in liquid disordered Ld phases of the plasma membrane. It is important to note that in the phosphatidylcholine-rich areas of the membrane, such OH groups are rather rare, because (1) the only constitutive lipid that displays a free accessible OH group in the liquid-disordered phase is cholesterol; (2) cholesterol is enriched in lipids rafts, which means that it is present in lower amounts outside lipid rafts. However, in the Ld phase, cholesterol is immediately recognized by external ligands that must find a OH group that is flush with the surface of the membrane. These ligands can be proteins (e.g., bacterial cytolysins[82]) or lipid neurotransmitters (e.g., endocannabinoids[83]). This initial interaction with cholesterol requires significant reinforcement, because at the membrane surface cholesterol has no more to offer than its OH group. Thus, this

Protein	Sphingolipid-binding domain (SBD)	Sphingolipid
PrP	KQHTVTTTTKGENFTETDVKMMER	GalCer SM
β-Amyloid peptide	RHDSGYEVHHQK	GM1 GalCer SM
α-Synuclein	KEGVLYVGSKTK	GM3 Gb_3 GalCer
Amylin	NFGAILSS	LacCer
Lipase BSDL	NATYEVYTEPWAQDSSQTETRKKTMVDLETDIL	GalCer GlcCer
GalCer transport protein CNL3	RWTVFKGLLWYIVPLVVVYFAEYFINQGL	GalCer
Insulin B chain	QHLCGDHLVEALYLVCGERGFFYTPK	sulfatide
Prominin	KAWNYELPATNYETQDSHK	GM1GD3
5-HT1A receptor	LNKWTLGQVTC	GM1
Potassium channel KCNB1	KSVLQFQNVRR	SM
TNF receptor Fas/CD95	KCRCKPNFFCNSTVCEH	LacCer Gb_3
TNF receptor Fas/CD95	KCKCKPDFYCDSPGCEH	LacCer Gb_3
Notch ligand Delta-like 1	RFPFGFTWPGTFSLIIEALH	LacCer GM1
Lysenin	KQRWAINKSLPLR .......KREDKWILEVVK	SM
Adhesin HpaA (*H. pylori*)	KKSEPGLLFSTGLDK	LacCer
Sphingomyelinase (*L. ivanovii*)	KSPQWSVKSWFKTYTYQDF	SM
Shiga-like toxin SLT1B	KELFTNRWNLQ	Gb_3
Serrate	VLPFTFRWTK	Neutral insect GSLs
HIV-1 gp120 V3 loop	RKSIRIQRGPGRAFVTIGKIGNMR	GalCer sulfatide Gb_3 GM3 SM
Hepatitis C virus polymerase	NDIRVEESIYQCCDLAPEARQAIR	SM
Synthetic peptide	DFRRLPGAFWQLRQP	GM1

FIGURE 6.11 **Sphingolipid-binding domains (SBDs) in viral, bacterial, worm, insect, and mammalian proteins.** The motif involved in sphingolipid binding is framed in yellow, basic residues are in blue, and aromatic amino acids are shown in red. SM, sphingomyelin. SBDs were detected in a broad range of species including human[15,56,64–69,106,107] (blue frame), mouse[66,108] (pink frame), worm[109] (gray frame), bacteria[11,70,74] (green frame), viruses[71,110] (yellow frame), and Drosophila[72] (purple frame). Finally a synthetic peptide displaying a high affinity for ganglioside GM1was selected with a phage display library[111] (brown frame). Most importantly, this nonnatural peptide contained the typical amino acid combination found in natural SBDs (central aromatic and lateral basic residues).

interaction is generally followed by an insertion process that ensures further molecular interactions involving the whole cholesterol molecule and its lipid or protein ligands. The process of cholesterol-dependent membrane insertion is described in Chapter 12 for bacterial toxins and in Chapter 5 for endocannabinoids.

6.6 CHAPERONE EFFECTS

An intriguing aspect of protein–lipid interactions is the "chaperone effect" exerted by the lipid on protein conformation. What do we mean exactly by *chaperone*? A biologist would probably associate the concept with proteins rather than with lipids. However, as natives of France, the authors of this book have a special linguistic relationship with this word. Etymologically, it comes from the Old French word *chaperon*, literally "hood." In French, the word *chaperon* has given the verb *chaperonner*, meaning "to cover with a hood." Later on this verb came to have the figurative sense "to protect." Under the influence of the verb sense, the French noun *chaperon* came to mean "escort." It is with this sense that the word was introduced into the English language in the eighteenth century. A chaperone is a person, commonly an older woman, who accompanies a young unmarried woman in public to protect her (i.e., to prevent her from acting inappropriately). In the scientific world, the term *molecular chaperone* was used for the first time by Laskey et al. in 1978 as a metaphor to describe the properties of nucleoplasmin,[84] a protein that appeared to assist the assembly of nucleosome cores from histones and DNA. As pointed out by Ellis,[85] nucleoplasmin is not a component of the nucleosome nor does it possess any of the steric information required for nucleosome assembly. Nucleoplasmin has several acidic side chains that interact with the basic residues of histones, thereby reducing their high positive charge density through transient electrostatic interactions and preventing the nonspecific aggregation of histones with DNA. In presence of nucleoplasmin, the self-assembly properties of histones with DNA predominate over the incorrect electrostatic attraction generated by their high densities of opposite charges. Because nucleoplasmin acts both transiently and negatively, it can be envisioned as a molecular mimic of the traditional role of human chaperones in preventing incorrect interactions between people. One key function of chaperones is to prevent both newly synthesized polypeptide chains and assembled subunits from aggregating into nonfunctional structures through incorrect interactions between surface-exposed apolar residues.[86] A number of distinct proteins belonging to more than 20 sequence-unrelated families (including chaperonins, which represent one the most important families of molecular chaperones) have been shown to display typical chaperone activity.[85] Nevertheless, according to the generalized view of Ellis, virtually any type of molecule able to prevent the incorrect folding of a macromolecule and promote its biologically active conformation should also be considered as a molecular chaperone.[85] Membrane lipids are one of them.[87] Indeed, the lack of phosphatidylethanolamine (PE) in a mutant of *Escherichia coli* resulted in the misfolding of the lactose permease protein.[88,89] An antibody directed against a conformational epitope in the native, biologically active form of lactose permease was then used as a probe for detecting the correct folding of the protein.[90] This allowed the study of the impact of membrane lipids on lactose permease folding. In absence of PE, lactose permease was incorrectly folded and did not express this epitope. However, this wrong conformation of the protein could be corrected through a lipid-dependent renaturation process. This process was lipid-specific, because both PE and phosphatidylserine (PS), but not other glycerophospholipids, could restore the correct shape of the protein. Interestingly, PE appeared to induce some fine conformational adjustments of lactose permease that occurred after its membrane insertion.[91] By analogy with protein chaperone, Bogdanov and Dowhan coined the term *lipochaperone* for those lipids capable of assisting the folding of membrane proteins.[87] Like protein chaperones, lipochaperones are supposed to detach from the protein once it has acquired its correct folding. However, although the release of a protein chaperone is easy in the 3D volume of an aqueous phase, it might be more difficult in a membrane bilayer where lipids are densely packed. Moreover, some lipids may have a high affinity for membrane proteins and remain firmly attached throughout the life of the protein. In particular, nonannular lipids are often required for the proper function of the protein and are generally considered as cofactors.[6,7] In most cases these lipid cofactors also exert a fine conformational tuning of the protein.[7] However, nonannular lipids are usually nonexchangeable with bulk membrane lipids.[6] Thus, we propose to extend the definition of lipochaperones to nonannular lipids that control the activity of membrane proteins through fine conformational tuning of their structure, which is particularly relevant to the complex relationships between the membrane surface of brain cells and amyloid proteins.

When disordered amyloid proteins interact with the plasma membrane surface, membrane lipids induce significant α-helix structuration of these proteins.[63] Because amyloid proteins have a marked preference for raft lipids (sphingolipids) rather than lipids of the liquid-disordered phase (glycerophospholipids), this α-helix structuration is a consequence of protein binding to raft sphingolipids. It is likely that the high order that reigns over lipid rafts is an important parameter of this conformational, chaperone-like effect exerted by sphingolipids on protein conformation. To illustrate this point, we have schematized the interaction of a disordered amyloid protein with a ganglioside-rich lipid raft (Fig. 6.12).

This protein could be for instance α-synuclein. Starting from an extended conformation, the protein will progressively acquire a helical structure through a series of coordinated interactions between its charged basic amino acid residues (chiefly lysine) that are concentrated in its N-terminal domain.[65] By providing a surface decorated by a high density of negative charges (the carboxylate groups of sialic acid residues), the raft will constrain the protein to adopt its typical membrane-associated α-helix structure.[63] A similar α-helical folding has been described for the microtubule-associated tau protein, which interacts with the cytoplasmic leaflet of the plasma membrane during neuronal differentiation.[16] Upon binding to membrane vesicles containing anionic phospholipids such as PS (a phospholipid-enriched in the inner leaflet of lipid raft membrane domains), tau rapidly adopts a α-helical structure.[17,18] In Alzheimer's disease, the association of tau with neural membranes may facilitate the formation of neurotoxic oligomers, illustrating how versatile the balance between chaperone and antichaperone effects of membrane lipids can be. A particularly strong chaperone effect occurs during the biosynthetic pathway of the cellular PrP protein. In the early secretory pathway, this protein interacts with protective sphingolipids that stabilize its nonpathological α-helical structure.[92] The finding of raft sphingolipids (GalCer and sphingomyelin) in infectious prion rods[57] attests of the strength of these protein–sphingolipid interactions,[63] even if in this case, these residual lipids are no longer able to maintain the protein in its physiological α-helix conformation.

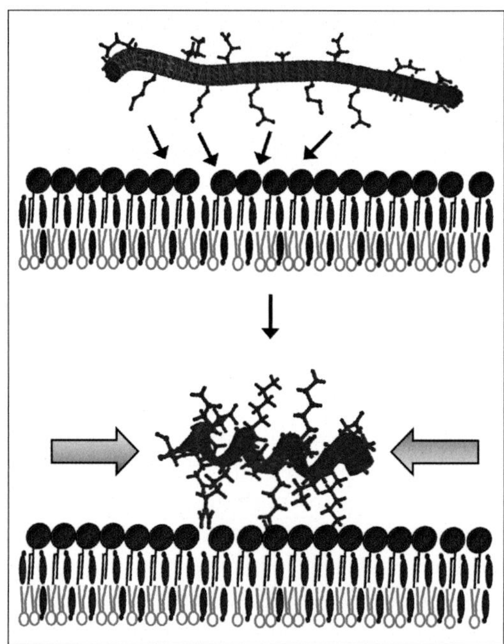

FIGURE 6.12 **Interaction of a disordered amyloid protein with a lipid raft domain.** In the upper panel, the disordered protein has an extended conformation. The charged residues (e.g., basic residues) may bind to the head groups of raft lipids (e.g., gangliosides) through electrostatic interactions. Because raft lipids are densely packed, the electrical charges are regularly arranged so that the protein chain has to condense its structure to ensure an optimal interaction with the raft surface. This induces the folding of the protein into a α-helix (lower panel), which is a hallmark of amyloid–membrane interactions.

The inhibitory effect of sphingolipids on the oligomerization and aggregation of amyloid proteins has been demonstrated for amylin, a pancreatic protein associated with type 2 diabetes. The amyloid core motif of this protein spontaneously self-aggregates in aqueous solution. However, when injected underneath a monolayer of LacCer, this core motif strongly interacts with the glycosphingolipids and no longer aggregates.[64] Beyond amyloid proteins, the chaperone effect of lipids on proteins is a widespread mechanism that efficiently controls the folding of most membrane proteins, including ionotropic[43] and metabotropic receptors of neurotransmitters.[42,93] The downside is that under pathological conditions, the same lipids (especially cholesterol and sphingolipids) may also exert antichaperone effects leading to the formation of neurotoxic oligomers such as amyloid pores.[94,95]

6.7 CONCLUSIONS

At the end of this survey of protein–lipid interactions, we can draw a schematic picture of a typical membrane protein (e.g., a class I membrane protein) whose TM domain is inserted in a lipid raft microdomain (Fig. 6.13).

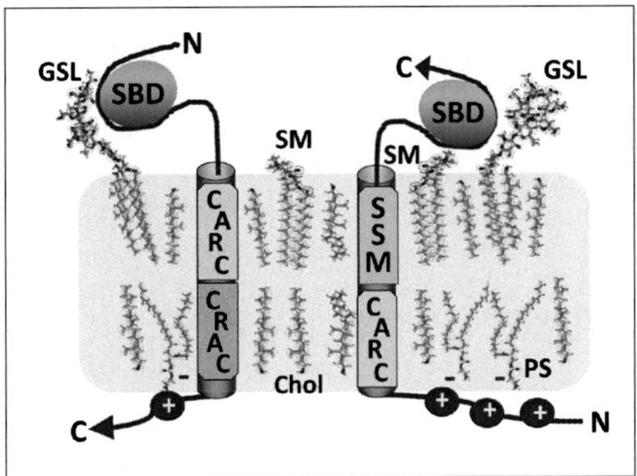

FIGURE 6.13 **Different types of protein–lipid interactions in membrane proteins.** A type I membrane protein is shown in the left panel, and a type II protein in the right panel. In both cases, the SBD located in the extracellular domain may interact with the head groups of glycosphingolipids (GSL). The cytoplasmic domain, often enriched in basic amino acids, may engage specific interactions with anionic phospholipids such as phosphatidylserine (PS). Cholesterol-binding domains (CRAC, CARC, and/or tilted) in the transmembrane domain may allow the protein to interact with cholesterol in each leaflet of the membrane bilayer, according to the orientation of the motif (see Figs 6.4 and 6.6). A sphingomyelin-sensor motif (SSM) similar to that of the p24 protein may be responsible for a specific interaction with sphingomyelin (see Fig. 6.8).

The different zones of interaction (extracellular, TM, and cytoplasmic) defined for an integral membrane protein may also concern extracellular proteins. Indeed, an extracellular protein will first bind to the membrane surface of the raft through mechanisms also used by the extracellular domain of a membrane protein. The insertion of the extracellular protein is rendered possible by the same mechanisms that control the interaction of a TM domain of an integral protein with the apolar phase of the raft. Finally, the insertion of an extracellular protein can result in the cytoplasmic exposure of the SBD. In this case, the SBD must manage the high concentration of acidic charges borne by anionic phospholipids, just as the cytoplasmic domains of integral membrane proteins do. In this respect, the SBDs of Alzheimer's β-amyloid peptides and α-synuclein, which display symmetric cationic residues,[65,73] are particularly concerned by these interactions.

Another important issue to discuss is the specificity of the lipid composition of brain membranes. At first glance, what differentiates brain membranes from other membranes is the high content in glycosphingolipids, especially gangliosides.[96] This unique composition affects both the physicochemical (fluidity, curvature) and recognition properties of brain membranes. Moreover, each brain cell type has its specific lipid raft equipment, which reflects a highly selective expression of glycosphingolipids. Even the pre- and postsynaptic neuronal membranes have distinct ganglioside contents.[97] These particularities of brain membranes have a profound biological significance whose most obvious aspects we have just begun to decipher. We should keep in mind that the brain is above all a lipid machine, ~ 50% of its organic matter being lipids.[98] The overexpression of glycosphingolipids in brain tissues is paralleled by the presence of a functional SBD in a broad range of brain proteins, including neurotransmitter receptors, membrane channels, and disease-associated amyloid proteins. The high content of cholesterol in brain membranes parallels the presence of various types of cholesterol-binding domains in brain proteins. Thus, brain proteins live in this lipid world and constantly interact with brain lipids through a handful of strikingly similar molecular mechanisms of protein–lipid interactions. These interactions control, either directly or indirectly, most of the physiological functions of the brain. The lipid composition of brain cells gradually changes with aging, rendering these functions more and more chaotic. Lipid–protein interactions are also involved in the pathogenesis of neurological disorders and brain infections, as discussed in Chapters 8–13.

6.8 KEY EXPERIMENT: THE LANGMUIR MONOLAYER AS A UNIVERSAL TOOL FOR THE STUDY OF LIPID–PROTEIN INTERACTIONS[48]

Of the several methods for studying lipid–protein interactions, the Langmuir monolayer system is both the most simple and the most powerful.[99–101] Basically, the setup consists of a lipid monolayer reconstituted at the air–water interface.[102] The lipid head groups are oriented toward the aqueous subphase, whereas their apolar tails are in the air. This lipid organization has been unraveled by Irving Langmuir (Nobel Prize 1932) through his famous works on fatty acids.[103] The term *Langmuir monolayer* was coined in his honor to describe a one-molecule-thick layer of an insoluble organic material, usually an amphiphilic compound, spread onto an aqueous subphase. It should not be confused with a cellular monolayer, which refers to a single layer of cells in culture or in an epithelial tissue. The presence of amphiphiles at the air–water interface decreases the surface tension of water, a force that is generated by the greater attraction of water molecules at the interface than in the bulk solvent (Fig. 6.14). In the bulk

FIGURE 6.14 **Surface tension of water: causes and effects.** In the bulk of the liquid, each water molecule is surrounded by other water molecules in all spatial directions. In contrast, water molecules at the air–water interface have no companions above them, so they interact more tightly with their lateral neighbors (left panel). This phenomenon explains why the surface of water is more solid than the bulk of the liquid. Insects take advantage of this striking property to "walk" on the water surface. The surface tension of water also explains why dewdrops adopt a minimal spheric shape on a blade of grass (black arrows, in the photographs).

of the liquid, each water molecule is pulled equally in every direction by neighboring water molecules. In contrast, the water molecules at the surface do not have companion molecules above them, so they interact more strongly. These stronger interactions force the water surface to contract to the minimal area.

Surface tension is responsible for the shape of liquid droplets, which can be illustrated by dewdrops on a blade of grass (Fig. 6.14). When amphiphile molecules are spread on the water surface, the last layer of water molecules recovers molecular partners above them so that the surface tension is relaxed (Fig. 6.15A). The amphiphiles exert a pressure on the water surface that is proportional to their effect on surface tension. A simple relationship can thus link the surface tension and the surface pressure: $\pi = \gamma_0 - \gamma$, where π is the surface pressure

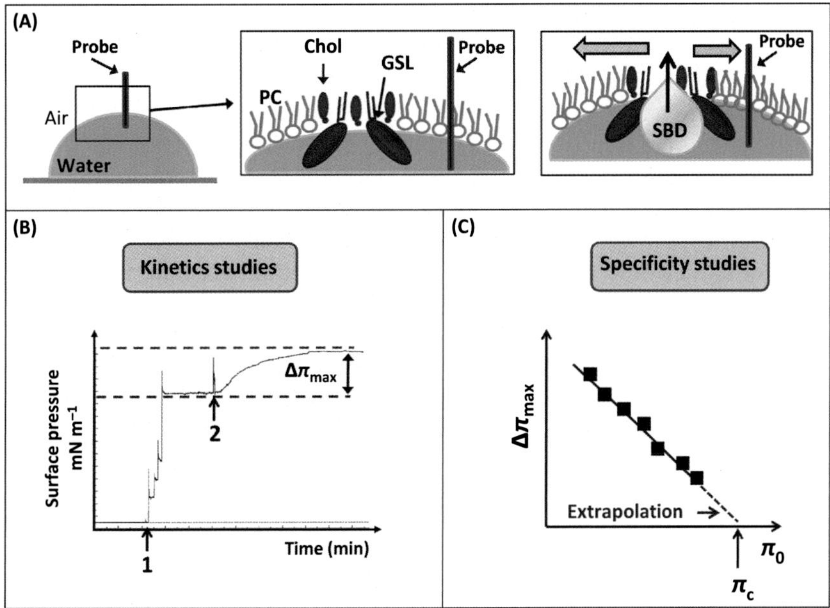

FIGURE 6.15 **Langmuir monolayers for studies of protein–lipid interactions.** (A) A microtensiometer can be used to measure the surface tension of pure water (γ_0) with a platinum probe. If a drop of lipid (or a lipid mixture) is deposited on the surface of water, it will readily spread until forming a monomolecular film (yet multilayers will form if the lipid concentration is too high). The head groups of lipids will interact with water molecules, thus decreasing the surface tension γ. This system is ideal for studying lipid–protein interactions. If a protein is added in the aqueous phase underneath the monolayer, it may interact with the lipid head groups. For instance, a SBD will interact with a lipid monolayer containing glycosphingolipids (GSL). This active lipid may dissolve in a matrix of phosphatidylcholine (PC) and/or cholesterol (chol). The SBD–GSL interaction will increase the surface pressure of the monolayer π (with $\pi = \gamma_0 - \gamma$). (B) Surface pressure measurements in real time follow the formation of the monolayer. (The surface pressure increases upon successive additions of lipids, until the desired initial pressure π_0 is reached, step 1.) The insertion of the protein in the lipid monolayer (step 2) evokes a progressive increase of the surface pressure, until a maximal value π_{max} is reached at the equilibrium. The binding process is thus fully characterized by $\Delta\pi_{max} = \pi_{max} - \pi_0$. (C) Specificity curves $\Delta\pi_{max} = f(\pi_0)$ are required to check that protein insertion gradually decreases as the density of lipids in the monolayers (reflected by π_0) increases. The extrapolated value of π_0 at $\Delta\pi_{max} = 0$ corresponds to the critical pressure of insertion π_c, which should be above the threshold value of 30 mN m^{-1} (i.e., the reference surface pressure value of a plasma membrane).

induced by the amphiphiles, γ_0 the surface tension of pure water (78.2 mN m^{-1}), and γ the surface tension measured in the presence of amphiphiles. The effect exerted by amphiphiles on surface tension is proportional to the number of amphiphile molecules spread on the water surface. Thus, the surface pressure gradually increases with the packing of amphiphiles at the air–water interface.

One of the main advantages of Langmuir monolayers over other techniques when studying the mechanisms of protein–lipid interactions is that the lipid content of the monolayer can be fully controlled. A lipid monolayer is prepared at the air–water interface by deposing a drop of lipid (either a pure lipid preparation of a lipid mixture, e.g., cholesterol with sphingomyelin) at the surface of a water droplet (ultrapure water devoid of any surfactant contaminant is required). This deposition induces an immediate increase in surface pressure followed by a rapid stabilization (Fig. 6.15B). After evaporation of the solvent (usually within 2 min), the protein is injected in the aqueous phase (which can be either pure water or a buffer); the kinetics of protein insertion within the monolayer are followed by real-time measurements of surface pressure. The initial surface pressure at which the protein is injected is noted π_0. If the protein simply adsorbs on the polar heads of the lipids, it will not modify the surface pressure. However, if the protein penetrates between the polar heads of the lipids, the lateral packing of the molecules in the monolayer (lipids + proteins) increases, and so the surface pressure also increases. At the equilibrium, a maximal surface pressure is reached. The difference between this maximum and π_0 is referred to as $\Delta\pi_{max}$. Specificity studies consist in determining $\Delta\pi_{max}$ at different values of π_0 (Fig. 6.15C). Experimental data are determined with a series of monolayers prepared at different values of π_0 and probed with the protein at a unique concentration (usually in the nM–μM range). The extrapolated value of π_0 for $\Delta\pi_{max} = 0$ is referred to as the critical pressure of insertion π_c. This parameter reflects the avidity of the protein for the monolayer and should ideally be >30 mN m^{-1}, which is the reference value measured for a natural plasma membrane.[104] Several excellent reviews on the use of Langmuir monolayers are available for studying a broad range of lipid–protein interactions.[99–101,105]

References

1. Israelachvili J, Wennerstrom H. Role of hydration and water structure in biological and colloidal interactions. *Nature*. 1996;379(6562):219–225.
2. Stoeckenius W. Structure of the plasma membrane. An electron-microscope study. *Circulation*. 1962;26: 1066–1069.
3. Singer SJ, Nicolson GL. The fluid mosaic model of the structure of cell membranes. *Science*. 1972;175(4023): 720–731.
4. Goni FM. The basic structure and dynamics of cell membranes: an update of the Singer–Nicolson model. *Biochim Biophys Acta*. 2014;183:1467–1476.
5. Fantini J, Barrantes FJ. How cholesterol interacts with membrane proteins: an exploration of cholesterol-binding sites including CRAC, CARC, and tilted domains. *Front Physiol*. 2013;4:31.
6. Lee AG. Lipid-protein interactions in biological membranes: a structural perspective. *Biochim Biophys Acta*. 2003;1612(1):1–40.
7. Lee AG. How lipids affect the activities of integral membrane proteins. *Biochim Biophys Acta*. 2004;1666(1–2): 62–87.
8. Epand RM. Proteins and cholesterol-rich domains. *Biochim Biophys Acta*. 2008;1778(7–8):1576–1582.
9. Epand RM, Thomas A, Brasseur R, Epand RF. Cholesterol interaction with proteins that partition into membrane domains: an overview. *Subcell Biochem*. 2010;51:253–278.
10. Snook CF, Jones JA, Hannun YA. Sphingolipid-binding proteins. *Biochim Biophys Acta*. 2006;1761(8):927–946.

11. Fantini J. How sphingolipids bind and shape proteins: molecular basis of lipid-protein interactions in lipid shells, rafts and related biomembrane domains. *Cell Mol Life Sci*. 2003;60(6):1027–1032.
12. East JM, Melville D, Lee AG. Exchange rates and numbers of annular lipids for the calcium and magnesium ion dependent adenosinetriphosphatase. *Biochemistry*. 1985;24(11):2615–2623.
13. Corbeil D, Marzesco AM, Wilsch-Brauninger M, Huttner WB. The intriguing links between prominin-1 (CD133), cholesterol-based membrane microdomains, remodeling of apical plasma membrane protrusions, extracellular membrane particles, and (neuro)epithelial cell differentiation. *FEBS Lett*. 2010;584(9):1659–1664.
14. Giebel B, Corbeil D, Beckmann J, et al. Segregation of lipid raft markers including CD133 in polarized human hematopoietic stem and progenitor cells. *Blood*. 2004;104(8):2332–2338.
15. Taieb N, Maresca M, Guo XJ, Garmy N, Fantini J, Yahi N. The first extracellular domain of the tumour stem cell marker CD133 contains an antigenic ganglioside-binding motif. *Cancer Lett*. 2009;278(2):164–173.
16. Brandt R, Leger J, Lee G. Interaction of tau with the neural plasma membrane mediated by tau's amino-terminal projection domain. *J Cell Biol*. 1995;131(5):1327–1340.
17. Kunze G, Barre P, Scheidt HA, Thomas L, Eliezer D, Huster D. Binding of the three-repeat domain of tau to phospholipid membranes induces an aggregated-like state of the protein. *Biochim Biophys Acta*. 2012;1818(9): 2302–2313.
18. Georgieva ER, Xiao S, Borbat PP, Freed JH, Eliezer D. Tau binds to lipid membrane surfaces via short amphipathic helices located in its microtubule-binding repeats. *Biophys J*. 2014;107(6):1441–1452.
19. Lasagna-Reeves CA, Castillo-Carranza DL, Guerrero-Muoz MJ, Jackson GR, Kayed R. Preparation and characterization of neurotoxic tau oligomers. *Biochemistry*. 2010;49(47):10039–10041.
20. Perrin RJ, Woods WS, Clayton DF, George JM. Interaction of human alpha-Synuclein and Parkinson's disease variants with phospholipids. Structural analysis using site-directed mutagenesis. *J Biol Chem*. 2000;275(44): 34393–34398.
21. Davidson WS, Jonas A, Clayton DF, George JM. Stabilization of alpha-synuclein secondary structure upon binding to synthetic membranes. *J Biol Chem*. 1998;273(16):9443–9449.
22. Eliezer D, Kutluay E, Bussell Jr R, Browne G. Conformational properties of alpha-synuclein in its free and lipid-associated states. *J Mol Biol*. 2001;307(4):1061–1073.
23. Chandra S, Chen X, Rizo J, Jahn R, Sudhof TC. A broken alpha -helix in folded alpha -synuclein. *J Biol Chem*. 2003;278(17):15313–15318.
24. Kubo S, Nemani VM, Chalkley RJ, et al. A combinatorial code for the interaction of alpha-synuclein with membranes. *J Biol Chem*. 2005;280(36):31664–31672.
25. Kyte J, Doolittle RF. A simple method for displaying the hydropathic character of a protein. *J Mol Biol*. 1982;157(1):105–132.
26. Elofsson A, von Heijne G. Membrane protein structure: prediction versus reality. *Annu Rev Biochem*. 2007;76:125–140.
27. Bernsel A, Viklund H, Falk J, Lindahl E, von Heijne G, Elofsson A. Prediction of membrane-protein topology from first principles. *Proceedings of the National Academy of Sciences of the United States of America*. 2008;105(20): 7177–7181.
28. Viklund H, Elofsson A. Best alpha-helical transmembrane protein topology predictions are achieved using hidden Markov models and evolutionary information. *Protein Sci*. 2004;13(7):1908–1917.
29. Hessa T, Meindl-Beinker NM, Bernsel A, et al. Molecular code for transmembrane-helix recognition by the Sec61 translocon. *Nature*. 2007;450(7172):1026–1030.
30. Pauling L, Corey RB, Branson HR. The structure of proteins; two hydrogen-bonded helical configurations of the polypeptide chain. *Proceedings of the National Academy of Sciences of the United States of America*. 1951;37(4):205–211.
31. Sharpe HJ, Stevens TJ, Munro S. A comprehensive comparison of transmembrane domains reveals organelle-specific properties. *Cell*. 2010;142(1):158–169.
32. Li H, Papadopoulos V. Peripheral-type benzodiazepine receptor function in cholesterol transport. Identification of a putative cholesterol recognition/interaction amino acid sequence and consensus pattern. *Endocrinology*. 1998;139(12):4991–4997.
33. Jamin N, Neumann JM, Ostuni MA, et al. Characterization of the cholesterol recognition amino acid consensus sequence of the peripheral-type benzodiazepine receptor. *Mol Endocrinol*. 2005;19(3):588–594.
34. Papadopoulos V, Baraldi M, Guilarte TR, et al. Translocator protein (18kDa): new nomenclature for the peripheral-type benzodiazepine receptor based on its structure and molecular function. *Trends Pharmacol Sci*. 2006;27(8):402–409.

35. Garcia-Ruiz C, Mari M, Colell A, et al. Mitochondrial cholesterol in health and disease. *Histol Histopathol.* 2009;24(1):117–132.
36. Palmer M. Cholesterol and the activity of bacterial toxins. *FEMS Microbiol Lett.* 2004;238(2):281–289.
37. Epand RF, Thomas A, Brasseur R, Vishwanathan SA, Hunter E, Epand RM. Juxtamembrane protein segments that contribute to recruitment of cholesterol into domains. *Biochemistry.* 2006;45(19):6105–6114.
38. Nishio M, Umezawa Y, Fantini J, Weiss MS, Chakrabarti P. CH–π hydrogen bonds in biological macromolecules. *Phys Chem Chem Phys.* 2014.
39. Baier CJ, Fantini J, Barrantes FJ. Disclosure of cholesterol recognition motifs in transmembrane domains of the human nicotinic acetylcholine receptor. *Sci Rep.* 2011;1:69.
40. Strandberg E, Killian JA. Snorkeling of lysine side chains in transmembrane helices: how easy can it get? *FEBS Lett.* 2003;544(1–3):69–73.
41. Harris JS, Epps DE, Davio SR, Kezdy FJ. Evidence for transbilayer, tail-to-tail cholesterol dimers in dipalmitoylglycerophosphocholine liposomes. *Biochemistry.* 1995;34(11):3851–3857.
42. Hanson MA, Cherezov V, Griffith MT, et al. A specific cholesterol binding site is established by the 2.8 A structure of the human beta2-adrenergic receptor. *Structure.* 2008;16(6):897–905.
43. Levitan I, Singh DK, Rosenhouse-Dantsker A. Cholesterol binding to ion channels. *Front Physiol.* 2014;:5.
44. Rosenhouse-Dantsker A, Noskov S, Durdagi S, Logothetis DE, Levitan I. Identification of novel cholesterol-binding regions in Kir2 channels. *J Biol Chem.* 2013;288(43):31154–31164.
45. Di Scala C, Yahi N, Lelievre C, Garmy N, Chahinian H, Fantini J. Biochemical identification of a linear cholesterol-binding domain within Alzheimer's beta amyloid peptide. *ACS Chem Neurosci.* 2013;4(3):509–517.
46. Fantini J, Carlus D, Yahi N. The fusogenic tilted peptide (67-78) of alpha-synuclein is a cholesterol binding domain. *Biochim Biophys Acta.* 2011;1808(10):2343–2351.
47. Charloteaux B, Lorin A, Brasseur R, Lins L. The "Tilted Peptide Theory" links membrane insertion properties and fusogenicity of viral fusion peptides. *Protein Pept Lett.* 2009;16(7):718–725.
48. Di Scala C, Chahinian H, Yahi N, Garmy N, Fantini J. Interaction of Alzheimer's beta-amyloid peptides with cholesterol: mechanistic insights into amyloid pore formation. *Biochemistry.* 2014;53(28):4489–4502.
49. Di Scala C, Troadec JD, Lelievre C, Garmy N, Fantini J, Chahinian H. Mechanism of cholesterol-assisted oligomeric channel formation by a short Alzheimer beta-amyloid peptide. *J Neurochem.* 2013.
50. Brasseur R, Cornet B, Burny A, Vandenbranden M, Ruysschaert JM. Mode of insertion into a lipid membrane of the N-terminal HIV gp41 peptide segment. *AIDS Res Hum Retroviruses.* 1988;4(2):83–90.
51. Lins L, Brasseur R. Tilted peptides: a structural motif involved in protein membrane insertion? *J Pept Sci.* 2008;14(4):416–422.
52. Brasseur R. Tilted peptides: a motif for membrane destabilization (hypothesis). *Mol Membr Biol.* 2000;17(1): 31–40.
53. Garmy N, Taieb N, Yahi N, Fantini J. Interaction of cholesterol with sphingosine: physicochemical characterization and impact on intestinal absorption. *J Lipid Res.* 2005;46(1):36–45.
54. Contreras FX, Ernst AM, Haberkant P, et al. Molecular recognition of a single sphingolipid species by a protein's transmembrane domain. *Nature.* 2012;481(7382):525–529.
55. Fantini J, Yahi N. The driving force of alpha-synuclein insertion and amyloid channel formation in the plasma membrane of neural cells: key role of ganglioside- and cholesterol-binding domains. *Adv Exp Med Biol.* 2013;991:15–26.
56. Mahfoud R, Garmy N, Maresca M, Yahi N, Puigserver A, Fantini J. Identification of a common sphingolipid-binding domain in Alzheimer, prion, and HIV-1 proteins. *J Biol Chem.* 2002;277(13):11292–11296.
57. Klein TR, Kirsch D, Kaufmann R, Riesner D. Prion rods contain small amounts of two host sphingolipids as revealed by thin-layer chromatography and mass spectrometry. *Biol Chem.* 1998;379(6):655–666.
58. Harouse JM, Bhat S, Spitalnik SL, et al. Inhibition of entry of HIV-1 in neural cell lines by antibodies against galactosyl ceramide. *Science.* 1991;253(5017):320–323.
59. Fantini J, Yahi N, Baghdiguian S, Chermann JC. Human colon epithelial cells productively infected with human immunodeficiency virus show impaired differentiation and altered secretion. *J Virol.* 1992;66(1):580–585.
60. Yahi N, Baghdiguian S, Moreau H, Fantini J. Galactosyl ceramide (or a closely related molecule) is the receptor for human immunodeficiency virus type 1 on human colon epithelial HT29 cells. *J Virol.* 1992;66(8): 4848–4854.
61. Cook DG, Fantini J, Spitalnik SL, Gonzalez-Scarano F. Binding of human immunodeficiency virus type I (HIV-1) gp120 to galactosylceramide (GalCer): relationship to the V3 loop. *Virology.* 1994;201(2):206–214.

62. Shindyalov IN, Bourne PE. Protein structure alignment by incremental combinatorial extension (CE) of the optimal path. *Protein Eng*. 1998;11(9):739–747.
63. Fantini J, Yahi N. Molecular insights into amyloid regulation by membrane cholesterol and sphingolipids: common mechanisms in neurodegenerative diseases. *Exp Rev Mol Med*. 2010;12:e27.
64. Levy M, Garmy N, Gazit E, Fantini J. The minimal amyloid-forming fragment of the islet amyloid polypeptide is a glycolipid-binding domain. *FEBS J*. 2006;273(24):5724–5735.
65. Fantini J, Yahi N. Molecular basis for the glycosphingolipid-binding specificity of alpha-synuclein: key role of tyrosine 39 in membrane insertion. *J Mol Biol*. 2011;408(4):654–669.
66. Chakrabandhu K, Huault S, Garmy N, et al. The extracellular glycosphingolipid-binding motif of Fas defines its internalization route, mode and outcome of signals upon activation by ligand. *Cell Death Differ*. 2008;15(12): 1824–1837.
67. Chattopadhyay A, Paila YD, Shrivastava S, Tiwari S, Singh P, Fantini J. Sphingolipid-binding domain in the serotonin(1A) receptor. *Adv Exp Med Biol*. 2012;749:279–293.
68. Aubert-Jousset E, Garmy N, Sbarra V, Fantini J, Sadoulet MO, Lombardo D. The combinatorial extension method reveals a sphingolipid binding domain on pancreatic bile salt-dependent lipase: role in secretion. *Structure*. 2004;12(8):1437–1447.
69. Persaud-Sawin DA, McNamara 2nd JO, Rylova S, Vandongen A, Boustany RM. A galactosylceramide binding domain is involved in trafficking of CLN3 from Golgi to rafts via recycling endosomes. *Pediatr Res*. 2004;56(3):449–463.
70. Fantini J, Garmy N, Yahi N. Prediction of glycolipid-binding domains from the amino acid sequence of lipid raft-associated proteins: application to HpaA, a protein involved in the adhesion of *Helicobacter pylori* to gastrointestinal cells. *Biochemistry*. 2006;45(36):10957–10962.
71. Sakamoto H, Okamoto K, Aoki M, et al. Host sphingolipid biosynthesis as a target for hepatitis C virus therapy. *Nat Chem Biol*. 2005;1(6):333–337.
72. Hamel S, Fantini J, Schweisguth F. Notch ligand activity is modulated by glycosphingolipid membrane composition in Drosophila melanogaster. *J Cell Biol*. 2010;188(4):581–594.
73. Yahi N, Fantini J. Deciphering the glycolipid code of Alzheimer's and Parkinson's amyloid proteins allowed the creation of a universal ganglioside-binding Peptide. *PloS One*. 2014;9(8):e104751.
74. Openshaw AE, Race PR, Monzo HJ, Vazquez-Boland JA, Banfield MJ. Crystal structure of SmcL, a bacterial neutral sphingomyelinase C from *Listeria*. *J Biol Chem*. 2005;280(41):35011–35017.
75. Fantini J, Garmy N, Mahfoud R, Yahi N. Lipid rafts: structure, function and role in HIV, Alzheimer's and prion diseases. *Exp Rev Mol Med*. 2002;4(27):1–22.
76. Fantini J. Interaction of proteins with lipid rafts through glycolipid-binding domains: biochemical background and potential therapeutic applications. *Curr Med Chem*. 2007;14(27):2911–2917.
77. Mattjus P, Slotte JP. Does cholesterol discriminate between sphingomyelin and phosphatidylcholine in mixed monolayers containing both phospholipids? *Chem Phys Lipids*. 1996;81(1):69–80.
78. Ramstedt B, Slotte JP. Interaction of cholesterol with sphingomyelins and acyl-chain-matched phosphatidylcholines: a comparative study of the effect of the chain length. *Biophys J*. 1999;76(2):908–915.
79. Hamon MA, Ribet D, Stavru F, Cossart P. Listeriolysin O: the Swiss army knife of *Listeria*. *Trends Microbiol*. 2012;20(8):360–368.
80. Koster S, van Pee K, Hudel M, et al. Crystal structure of listeriolysin O reveals molecular details of oligomerization and pore formation. *Nat Comm*. 2014;5:3690.
81. Soltani CE, Hotze EM, Johnson AE, Tweten RK. Structural elements of the cholesterol-dependent cytolysins that are responsible for their cholesterol-sensitive membrane interactions. *Proceedings of the National Academy of Sciences of the United States of America*. 2007;104(51):20226–20231.
82. Tweten RK. Cholesterol-dependent cytolysins, a family of versatile pore-forming toxins. *Infect Immun*. 2005;73(10):6199–6209.
83. Di Pasquale E, Chahinian H, Sanchez P, Fantini J. The insertion and transport of anandamide in synthetic lipid membranes are both cholesterol-dependent. *PloS One*. 2009;4(3):e4989.
84. Laskey RA, Honda BM, Mills AD, Finch JT. Nucleosomes are assembled by an acidic protein which binds histones and transfers them to DNA. *Nature*. 1978;275(5679):416–420.
85. Ellis RJ. Do molecular chaperones have to be proteins? *Biochem Biophys Res Comm*. 1997;238(3):687–692.
86. Horwich AL, Low KB, Fenton WA, Hirshfield IN, Furtak K. Folding in vivo of bacterial cytoplasmic proteins: role of GroEL. *Cell*. 1993;74(5):909–917.

87. Bogdanov M, Dowhan W. Lipid-assisted protein folding. *J Biol Chem*. 1999;274(52):36827–36830.
88. Dowhan W. Molecular basis for membrane phospholipid diversity: why are there so many lipids? *Annu Rev Biochem*. 1997;66:199–232.
89. Dowhan W. Genetic analysis of lipid-protein interactions in *Escherichia coli* membranes. *Biochim Biophys Acta*. 1998;1376(3):455–466.
90. Sun J, Wu J, Carrasco N, Kaback HR. Identification of the epitope for monoclonal antibody 4B1 which uncouples lactose and proton translocation in the lactose permease of *Escherichia coli*. *Biochemistry*. 1996;35(3):990–998.
91. Bogdanov M, Dowhan W. Phospholipid-assisted protein folding: phosphatidylethanolamine is required at a late step of the conformational maturation of the polytopic membrane protein lactose permease. *EMBO J*. 1998;17(18):5255–5264.
92. Sarnataro D, Campana V, Paladino S, Stornaiuolo M, Nitsch L, Zurzolo C. PrP(C) association with lipid rafts in the early secretory pathway stabilizes its cellular conformation. *Mol Biol Cell*. 2004;15(9):4031–4042.
93. Paila YD, Ganguly S, Chattopadhyay A. Metabolic depletion of sphingolipids impairs ligand binding and signaling of human serotonin1A receptors. *Biochemistry*. 2010;49(11):2389–2397.
94. Lal R, Lin H, Quist AP. Amyloid beta ion channel: 3D structure and relevance to amyloid channel paradigm. *Biochim Biophys Acta*. 2007;1768(8):1966–1975.
95. Quist A, Doudevski I, Lin H, et al. Amyloid ion channels: a common structural link for protein-misfolding disease. *Proceedings of the National Academy of Sciences of the United States of America*. 2005;102(30):10427–10432.
96. van Echten-Deckert G, Walter J. Sphingolipids: critical players in Alzheimer's disease. *Prog Lipid Res*. 2012;51(4):378–393.
97. Ledeen RW, Diebler MF, Wu G, Lu ZH, Varoqui H. Ganglioside composition of subcellular fractions, including pre- and postsynaptic membranes, from Torpedo electric organ. *Neurochem Res*. 1993;18(11):1151–1155.
98. O'Brien JS, Sampson EL. Lipid composition of the normal human brain: gray matter, white matter, and myelin. *J Lipid Res*. 1965;6(4):537–544.
99. Leblanc RM. Molecular recognition at Langmuir monolayers. *Curr Opin Chem Biol*. 2006;10(6):529–536.
100. Thakur G, Pao C, Micic M, Johnson S, Leblanc RM. Surface chemistry of lipid raft and amyloid Abeta (1-40) Langmuir monolayer. *Colloids Surf B Biointerfaces*. 2011;87(2):369–377.
101. Maggio B. The surface behavior of glycosphingolipids in biomembranes: a new frontier of molecular ecology. *Prog Biophys Mol Biol*. 1994;62(1):55–117.
102. Hammache D, Pieroni G, Maresca M, Ivaldi S, Yahi N, Fantini J. Reconstitution of sphingolipid-cholesterol plasma membrane microdomains for studies of virus-glycolipid interactions. *Methods Enzymol*. 2000;312:495–506.
103. Langmuir I. The constitution and fundamental properties of solids and liquids. II. Liquids. *J Am Chem Soc*. 1917;39:1848–1906.
104. Seelig A. Local anesthetics and pressure: a comparison of dibucaine binding to lipid monolayers and bilayers. *Biochim Biophys Acta*. 1987;899(2):196–204.
105. Brockman H. Lipid monolayers: why use half a membrane to characterize protein-membrane interactions? *Curr Opin Struct Biol*. 1999;9(4):438–443.
106. Osterbye T, Jorgensen KH, Fredman P, et al. Sulfatide promotes the folding of proinsulin, preserves insulin crystals, and mediates its monomerization. *Glycobiology*. 2001;11(6):473–479.
107. Milescu M, Bosmans F, Lee S, Alabi AA, Kim JI, Swartz KJ. Interactions between lipids and voltage sensor paddles detected with tarantula toxins. *Nat Struct Mol Biol*. 2009;16(10):1080–1085.
108. Heuss SF, Tarantino N, Fantini J, et al. A glycosphingolipid binding domain controls trafficking and activity of the mammalian notch ligand delta-like 1. *PloS One*. 2013;8(9):e74392.
109. Kiyokawa E, Makino A, Ishii K, Otsuka N, Yamaji-Hasegawa A, Kobayashi T. Recognition of sphingomyelin by lysenin and lysenin-related proteins. *Biochemistry*. 2004;43(30):9766–9773.
110. Hammache D, Pieroni G, Yahi N, et al. Specific interaction of HIV-1 and HIV-2 surface envelope glycoproteins with monolayers of galactosylceramide and ganglioside GM3. *J Biol Chem*. 1998;273(14):7967–7971.
111. Matsubara T, Ishikawa D, Taki T, Okahata Y, Sato T. Selection of ganglioside GM1-binding peptides by using a phage library. *FEBS Lett*. 1999;456(2):253–256.

CHAPTER

7

Lipid Regulation of Receptor Function

Jacques Fantini, Nouara Yahi

OUTLINE

7.1 SPECIFIC LIPID REQUIREMENT OF MEMBRANE PROTEINS

The consideration of membrane lipids in the regulation of membrane protein function is a relatively recent issue that arose in the second half of the 1970s. For a while, it was thought that lipids acted essentially by altering membrane fluidity.[1,2] Controlled modulations of membrane fluidity were achieved by changing the cholesterol/phospholipid ratio of various model membrane systems, including reconstituted receptor-containing vesicles and synaptic membranes.[3,4] Beyond the biophysical state of the membrane, specific lipids appeared to be required for the functional activity of membrane receptors and channels. For instance, the activity of reconstituted Na^+/K^+-ATPase (the sodium/potassium pump) was tightly associated with the cholesterol content of lipid vesicles.[5] Similarly, cholesterol and phosphatidylethanolamine restored the activity of a delipidated preparation of the β-adrenergic receptor, whereas sphingomyelin and anionic glycerophospholipids were inactive.[6] Although these studies suggested that some membrane lipids could be involved in the regulation of receptor function, the molecular mechanisms associated with these effects were not immediately identified.

Brain Lipids in Synaptic Function and Neurological Disease. http://dx.doi.org/10.1016/B978-0-12-800111-0.00007-2

In 1978, pioneer studies at Changeux's laboratory showed that purified preparations of the nicotinic acetylcholine receptor displayed a higher affinity for cholesterol than for other membrane lipids.[7] This affinity was demonstrated by studying the association of the receptor with various lipid monolayers according to the Langmuir approach (see Chapter 6 for a discussion of the usefulness of this technique for studies of lipid–protein interactions). Several examples of modulation of receptor function by cholesterol were then published.[8] Progressively, a new concept emerged, taking into account specific interactions between the transmembrane domains of neurotransmitter receptors/transporters and cholesterol.[9–12]

The case of cholesterol is interesting for several reasons. Cholesterol is generally considered to decrease membrane fluidity. This notion is correct for cholesterol in the liquid-disordered (Ld) phase, where it intercalates between the acyl chains of glycerophospholipids and limits their mobility.[13] However, cholesterol interferes with the dense packing of sphingolipids, thereby slightly increasing the fluidity of the liquid-ordered (Lo) phase in lipid raft domains.[13] This dual effect of cholesterol on membrane fluidity was recognized several years before the formulation of the lipid raft concept.[3] A second effect of cholesterol is its ability to interact with sphingomyelin and glycosphingolipids and modulate the conformation of their large head groups.[13–15].Therefore, the whole structure of lipid rafts depends on the presence of cholesterol, which in turn may affect the function of the receptors localized in these membrane areas.[16] Experimental modulation of cholesterol homeostasis will thus induce broad effects on lipid raft structure and function, especially signal transduction. Finally, several neurotransmitter receptors display specific nonannular binding sites for cholesterol and sphingolipids.[16,17]

At this stage we would like to make a clear distinction between the segregation of neurotransmitter receptors in lipid domains of the plasma membrane, and the fact that sphingolipids and/or cholesterol could interact with these receptors, modify their conformation, and regulate their function. The point is that some confusion may result because both properties rely on the same biochemical mechanism (i.e., a direct interaction between the receptor and sphingolipid/cholesterol). First of all, even though cholesterol is actually concentrated in lipid domains, a biologically active pool of this lipid also exists outside these microdomains (i.e., in the Ld phase of the plasma membrane).[18,19] Thus, cholesterol has the possibility of regulating receptor function outside lipid domains. In addition, lipid rafts have a complex biochemical composition that reconstituted Lo domains of model membranes may only partially recapitulate.[20] In this respect, it is of interest to note that some of the gangliosides present in the outer leaflet of a cell membrane may have an extracellular origin. For instance, extracellular gangliosides shed from vicinal cells (or circulating as micelles) can be inserted into the membrane of neural cells and control growth, differentiation, and tumor progression.[21–25] In fact, it might be easier for these gangliosides to gain entry into the external leaflet in membrane areas by intercalating between lipids in the Ld phase, which are not as tightly packed as those in the Lo phase.[13] If this proved to be the case, then we would have to consider the possibility of some receptor–ganglioside interactions occurring outside lipid domains. Overall, it would be too simplistic to envision the complex interplay between neurotransmitter receptors and cholesterol or sphingolipids as solely a lipid domain-associated process. Instead we believe that this important level of regulation of receptor function is primarily driven by specific lipid–protein interactions that could theoretically take place anywhere in the membrane, and definitely not only within lipid domains.

One could argue that most (if not all) receptors coupled to a signal transduction pathway have to be associated with some form of lipid domain, if we define as such precisely those zones of the membrane in which neurotransmitter receptors and transducers are brought together.[26] In recent years, considerable effort has been made to characterize the biochemical composition of lipid domains. Today it would appear that these initial efforts were somewhat premature in light of parallel efforts being made to demonstrate the mere existence of domains and thus the reality of the "raft concept." In particular, the reliability of the techniques used to obtain detergent-resistant fractions, often assumed to represent the biochemical counterpart of raft domains, has been matter of intense debate.[27,28] Fortunately, nondisruptive methods such as fluorescence correlation spectroscopy have brought solid experimental support to the existence of raft domains.[29]

Numerous neurotransmitter receptors have been found in lipid domain,[26] including the receptors for acetylcholine,[30] glutamate,[31] neurokinin 1,[32] NMDA,[33] and serotonin.[34] However, the biological significance of the association of these receptors with lipid domains is not always perfectly understood.[35] One important reason for this is that domains are not stable entities: A continuous exchange of lipids and proteins with the bulk membrane results in geometry, structure, and boundaries that are in perpetual motion. The rigorous characterization of the molecular mechanisms underlying the physical association of receptors with individual domain lipids will be an important step toward understanding the role of lipid domains in neurotransmitter receptor function. With this mind, we will now give some representative examples of specific lipid–receptor interactions and discuss the impact of these interactions on receptor distribution, conformation, and function.

7.2 NICOTINIC ACETYLCHOLINE RECEPTOR

As strange as it may seem, acetylcholine was chemically synthesized 40 years before it was characterized as a naturally occurring bioactive compound.[36] Synthesized in 1867 and then discovered as a contaminant in a batch of ergot, acetylcholine started its neurotransmitter career extremely modestly. In an historical description of acetylcholine discovery, Tansey[36] emphasized that "By 1914 there was little evidence that acetylcholine was of any physiological value." The first direct evidence for the role acetylcholine in neurotransmission came from Henry Dale's laboratory in the mid-1930s. In 1936, Henry Dale and Otto Loewi shared the Nobel Prize for the discovery of chemical transmission. One of the most remarkable intuitions of Henry Dale was to coin the terms *cholinergic* and *adrenergic* to designate nerve fibers by the nature of the chemical that they used as transmitters.[36] Dale anticipated that "such a usage would assist clear thinking, without committing us to precise chemical identifications, which may be long in coming."[37,38] Indeed, the characterization of "adrenergic-like" and "acetylcholine like" neurotransmitters would take time.

As with most other neurotransmitter receptors, acetylcholine receptors are classified into ionotropic and metabotropic receptors. Ionotropic acetylcholine receptors are referred to as "nicotinic" because they are responsive to nicotine, an alkaloid compound present in cigarettes. Metabotropic acetylcholine receptors are referred to as "muscarinic" because they are sensitive to muscarine, a mushroom compound. Cholesterol has been reported to modulate the activity of both receptor types,[39–42] yet most studies have been focused on the interaction between membrane cholesterol and nicotinic receptors. Indeed, extensive experimental evidence indicates that cholesterol modulates the functional properties and membrane distribution of ionotropic acetylcholine receptors.[43,44] Photoaffinity labeling techniques have been widely used to identify cholesterol binding sites on the receptor.[45–47] The label was found in transmembrane segments, especially TM1, TM3, and TM4 of each of the five subunits of the receptor.[47] Given that the remaining transmembrane domain (TM2) is oriented toward the center of the channel, these data demonstrated that the cholesterol-binding domain fully overlaps the lipid–protein interface of the nicotinic acetylcholine receptor (nAchR).

The search for cholesterol-binding sites in the transmembrane domains of nAchR led to the discovery of a new cholesterol recognition motif referred to as *CARC* by F. Barrantes.[48] This new domain is similar to the previously described CRAC (cholesterol recognition/interaction amino acid consensus) sequence,[49] but in the opposite direction along the polypeptide chain. Thus, it corresponds to an inverted CRAC domain, and it was logically coined *CARC*.[48] (See Chapter 6 for a discussion of the respective structural and functional features of CRAC and CARC domains.) A modeling procedure derived from the fragment-based approach was used for each individual transmembrane segment (Fig. 7.1), after which molecular dynamics simulations were performed on the whole receptor-cholesterol complex.[48]

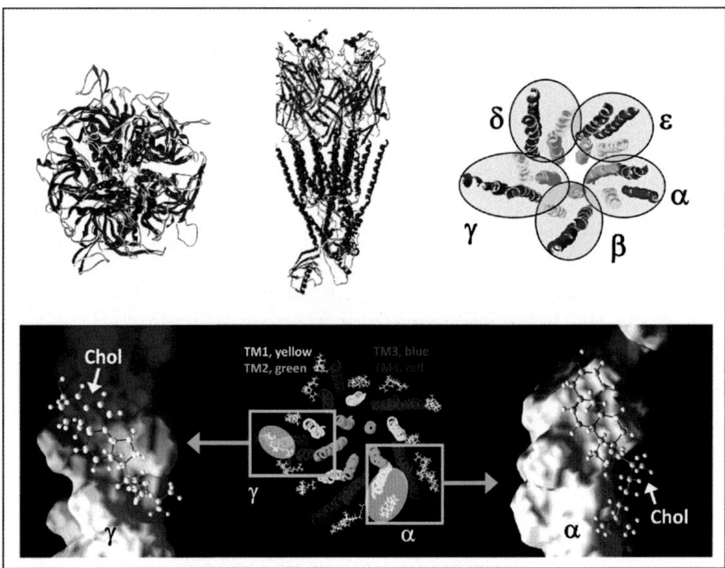

FIGURE 7.1 **Fragment-based approach to modeling cholesterol–nAchR interactions.** The nicotinic acetylcholine receptor (nAchR) is a complex quaternary protein formed by five subunits (α–ε). Each subunit displays four transmembrane domains. Top and lateral views of the secondary structure of nAchR are shown in the upper panels. The fragment-based approach consists of modeling separately each transmembrane domain merged with cholesterol to find an optimal fit. The docking of cholesterol on two transmembrane domains (αTM1 and γTM4) is illustrated in the lower panels. Once cholesterol has been docked on all individual transmembrane domains, the whole receptor is reconstituted and the model is corrected for any incompatibilities due to steric hindrance.

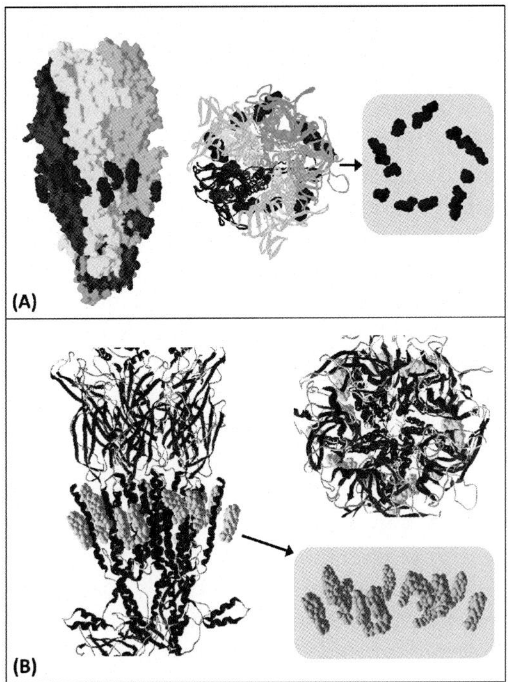

FIGURE 7.2 **Model of nAchR in complex with 15 cholesterol molecules.** (A) Lateral and top views of the receptor–cholesterol complex in surface rendering. Each subunit has its own color and cholesterol molecules are in deep blue. The 15 cholesterol molecules formed a ring surrounding the receptor as shown in the right panel. (B) Lateral and top views of the secondary structure of the receptor in complex with cholesterol (in yellow). All the cholesterol molecules are located in the outer leaflet of the plasma membrane.

Three cholesterol molecules could be docked on each AChR subunit, rendering a total of 15 cholesterol molecules per nAChR molecule. In full agreement with the results of photoaffinity labeling studies, Baier et al. found specific cholesterol-binding sites in transmembrane domains TM1, TM3, and TM4.[48] Interestingly, once bound to TM1, cholesterol could not interact with TM2 due to steric hindrance. Molecular models of nAchR in complex with 15 cholesterol molecules are shown in Fig. 7.2.

The discovery of the CARC motif has allowed identification of new cholesterol-binding sites in the transmembrane domains of a broad range of membrane proteins, including G protein–coupled receptors (GPCRs) with seven transmembrane domains.[17,48] Because both GPCRs and nAChR are thought to play a role in the pathology of Alzheimer's disease,[50–52] these data provide an interesting link between Alzheimer's disease, high cholesterol, and neurotransmitter receptors displaying CARC motifs in their transmembrane domains.[48] Moreover, nicotine has been reported to induce a marked increase in the aggregation and phosphorylation state of tau.[53] The relationships between cholesterol homeostasis and its impact on AchR function in the context of Alzheimer's disease have been recently discussed.[52]

7.3 CHOLESTEROL- AND GANGLIOSIDE-BINDING DOMAINS IN SEROTONIN RECEPTORS

Before describing the modalities of membrane–lipid binding to serotonin receptors, we will take the time to ask why one of the most important neurotransmitters has been called *serotonin*, a term that is not particularly suggestive of a neural function. As a matter of fact, the etymology of biochemical terms reveals lots of surprises. Most students learn that one of the most famous sugars is called "ribose" and that this sugar is responsible for the classification of nucleic acids into ribose and deoxyribose-containing nucleic acids, RNA and DNA. Yet who really knows today why ribose was called ribose? Several possible explanations have been proposed, among which an arbitrary shortening and rearrangement of *arabinose*, a sugar that gained its name because of its presence in gum arabic. May be this story is weird enough to be true. Then, what about serotonin? This neurotransmitter plays a major role in the history of neurosciences, as emphasized by Whitaker-Azmitia in a fascinating description of its discovery and characterization.[54] However, serotonin was not immediately recognized as a neurotransmitter. Indeed, the molecule was first isolated and characterized in chromaffin cells of the intestinal mucosa ("enterochromaffin" cells) and was thus logically called *enteramine*.[55] The structure of enteramine was resolved as 5-hydroxytryptamine (5-HT), which appeared to be identical to serotonin, a vasoconstrictor substance of bovine serum independently isolated by another group in 1948.[56,57] The pharmaceutical company that first synthesized the compound and made it available for research preferred the name serotonin, and so it remained.[54] Five years later, serotonin was found in mammalian brain,[58] and 5-HT started its brilliant neurotransmitter career during which it contributed to the recognition of neurochemistry as a scientific discipline.

Seven distinct families of serotonin receptors, referred to as 5-HT1 to 5-HT7, have been identified so far, and subpopulations have been described for several of these categories.[59] With the exception of the 5-HT3 receptor, a ligand-gated cation channel (thus a ionotropic receptor), all serotonin receptors are coupled to transducer G proteins. The 5-HT1 and 5-HT5A subtypes are coupled to Gi (decreasing cAMP levels), whereas the 5-HT4, 5-HT6, and 5-HT7 are coupled to Gs (increasing cAMP levels). The last subtype, 5-HT2, is coupled to Gq (increasing inositol triphosphate and diacylglycerol levels). Interestingly, gangliosides were initially thought to serve as receptors for serotonin on neuronal membranes, but it turned out that these anionic glycolipids were rather used as electrostatic attachment sites for cationic neurotransmitters on postsynaptic membranes.[16] Nevertheless, the localization of both ionotropic and metabotropic serotonin receptors in lipid rafts still suggests a functional role for glycosphingolipids in serotonergic transmission.[34,60–62] Moreover, the raft localization of the 5-HT1A receptor appeared to be involved in receptor-mediated signaling.[34] Consistently, metabolic depletion of sphingolipids altered both serotonin binding and signal transduction in transfected cells.[63] For these reasons we looked for a glycosphingolipid-binding domain (SBD) in the extracellular loops of the 5-HT1A receptor.[64] The second extracellular loop of 5-HT1A retained our attention. It is a short domain with the following amino acid sequence: 99-LNKWTLGQVTC-109. This motif contains the typical combination of basic, aromatic, and turn-inducing residues that are usually found in sphingolipid-binding domains[65,66]: Lys-101 (basic), Trp-102 (aromatic), and Gly-105 (turn-inducing amino acid). Physicochemical experiments with monolayers of gangliosides at the air–water interface showed that synthetic peptides derived from the second extracellular domain of 5-HT1A (LNKWTLGQVTC)

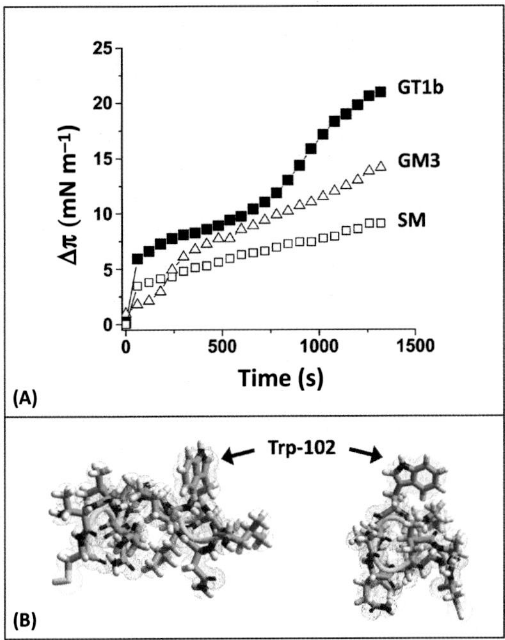

FIGURE 7.3 **The second extracellular domain of the 5-HT1A receptor binds to gangliosides.** (A) Surface pressure increases (**Δπ** expressed in mN m^{-1}) induced by the 5-HT1A peptide 99-LNKWTLGQVTC-109 injected underneath various sphingolipid monolayers. The strongest effect was obtained with ganglioside GT1b, whereas GM3 was less active and sphingomyelin (SM) was poorly recognized. (B) Molecular dynamics simulations of the 5-HT1A peptide in water reveal a curved structure with the aromatic side chain of Trp-102 exposed to the solvent.

actually interacted with GT1b and, although to a lesser extent, with GM3 (Fig. 7.3A). In contrast, the peptide interacted poorly with sphingomyelin. These data suggested that the SBD of the 5-HT1A receptor binds preferentially to gangliosides rather than to other sphingolipids. Nevertheless, analysis of the intrinsic tryptophan fluorescence of the SBD peptide upon incubation with GM1-containing membranes did not reveal any particular ganglioside-binding properties.[64] A possible explanation of these data could be that the 5-HT1A receptor has a higher affinity for GT1b than for GM1, yet this remains to be formally established.

Molecular dynamics simulations were conducted in order to study the conformation of the synthetic 5-HT1A peptide in water and its ganglioside-binding properties. We found that the peptide readily adopted a typical curved structure with the aromatic rings of Trp-102 exposed to the solvent molecules on the tip of the loop (Fig. 7.3B). This topology is consistent with a short-loop motif joining two transmembrane domains, which is the case for the second extracellular domain of the 5-HT1A receptor in its membrane environment. When merged with two GT1b molecules, a stable complex was formed (Fig. 7.4). A remarkable feature of the complex was the insertion of the aromatic rings of Trp-102 in the space between the head groups of both GT1b gangliosides(Fig. 7.4). This interaction was stabilized by a hydrogen bond linking the carboxylate group of a sialic acid in the sugar head group of GT1b with the nitrogen atom of the indole ring system of Trp-102. Overall, the energy of interaction

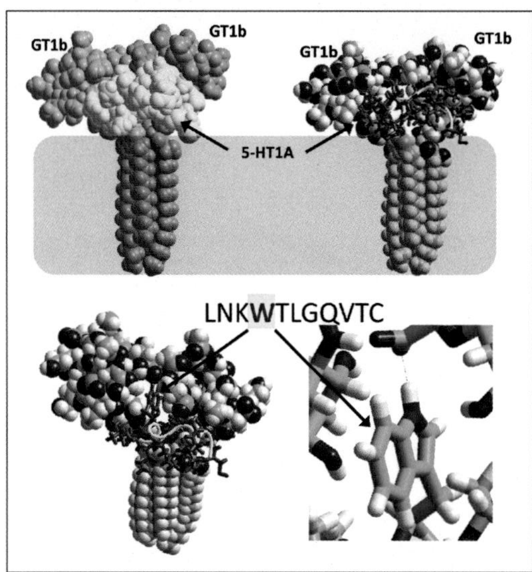

FIGURE 7.4 **Molecular dynamics simulations of 5-HT1A association with ganglioside GT1b.** The second extracellular domain of 5-HT1A has a remarkable fit for a chalice-shaped dimer of GT1b (upper panels). A remarkable feature of this interaction is the insertion of the aromatic side chain of Trp-102, which forms a hydrogen bond with the carboxylate group of a sialic acid in the head group of GT1b (lower panels).

of the complex was estimated to –150 kJ mol^{-1}, of which Trp-102 by itself accounted for –48 kJ mol^{-1} (i.e., one third of the whole energy of interaction). The importance of this tryptophan residue in the ganglioside-binding domain of the 5-HT1A receptor is emphasized by its status as a highly conserved residue among all human metabotropic serotonin receptors.[64]

Once bound to GT1b, the second extracellular domain of 5-HT1A is no longer accessible to external ligands because it is totally covered by the head groups of the gangliosides. Correspondingly, this domain is not involved in serotonin binding. However, it is closed to acidic amino acids Asp-82 in TM2 and Asp-116 in TM3 that are part of the three-dimensional serotonin-binding site[16] (Fig. 7.5). Interestingly, our molecular modeling studies also showed that the gangliosides bound to the second extracellular domain of 5-HT1A form a negatively charged funnel that can attract the cationic serotonin and direct it to its intramembrane binding site (Fig. 7.5). This attraction may not be an absolute requirement for serotonin binding, but it is obviously an excellent accessory mechanism for speeding up the binding process.

The region of the 5-HT1A receptor defined by the second and third transmembrane domains is also involved in cholesterol binding and in signal transduction.[16] Fantini and Barrantes proposed that this region might function as a regulatory cassette that concentrates several regulatory functions coordinated by serotonin binding.[16] It is highly significant that the same transmembrane domain (TM2) contains amino acid residues that are critical for interactions with both cholesterol (Tyr-73) and serotonin (Asp-82).[67–70] In addition, the second intracellular loop displays a triad motif (Asp-133/Arg-134/Tyr-135) which functions as a switch for protein G activation following serotonin binding.[69] The ganglioside-binding motif in the second extracellular loop completes the regulatory cassette (Fig. 7.5). It should

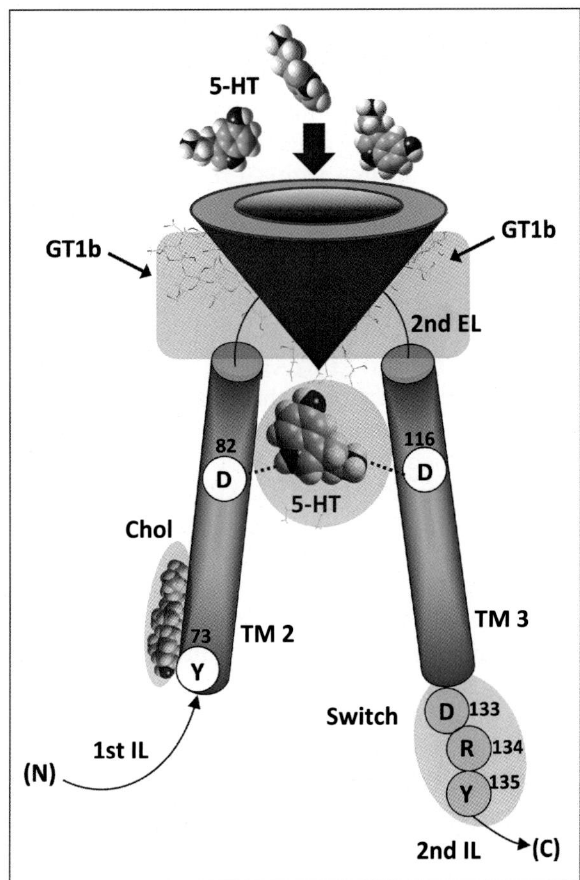

FIGURE 7.5 **Lipid regulation of the 5-HT1A receptor.** The regulatory unit includes: (1) the attractive electronegative funnel formed by GT1b gangliosides; (2) the second and third transmembrane helices of 5-HT1A (TM2 and TM3), connected by the second extracellular loop (2nd EL); (3) the serotonin (5-HT) binding pocket involving amino acid residues Asp-82 in TM2 and Asp-116 in TM3; the [Asp-133/Arg-134/Tyr-135] triad in the second intracellular loop (2nd IL), which functions as a switch for G protein activation; and (4) the cholesterol-binding motif involving residue Tyr-73 in TM2. The first intracellular loop (1st IL) and the orientation of the peptidic chain of the receptor (N- to C-terminus) are indicated.

be emphasized that this cassette is created by the condensing effect of cholesterol on the whole receptor. This typical effect of cholesterol, observed for several metabotropic receptors, implies that the cholesterol-binding site involves several amino acid residues located in distinct transmembrane domains.[17,68,71] In contrast, when cholesterol interacts with a single transmembrane domain at the periphery of the receptor, it may facilitate the recruitment of individually inactive receptors into a functional multimeric complex. The dimerization of G protein–coupled neurotransmitter receptors is a perfect illustration of such cholesterol-driven mechanisms.[17] A detailed discussion on the different aspects of cholesterol–receptor interactions has been recently published by Fantini and Barrantes.[17]

7.4 CHOLESTEROL- AND GalCer-BINDING DOMAINS IN SIGMA-1 RECEPTORS

Sigma-1 receptors are endoplasmic reticulum proteins that have been involved in the regulation of various brain functions including neurotransmitter release[72] and oligodendrocyte differentiation.[73] In rat hippocampal cultures, sigma-1 receptors have been colocalized with GalCer and cholesterol in detergent-resistant membranes.[73] This finding prompted Palmer et al. to search for a cholesterol-binding domain in the sigma-1 receptor protein.[74] They found that cholesterol binds to a purified preparation of recombinant sigma-1 receptor expressed in HEK cells. They found two linear consensus cholesterol-binding CRAC motifs[49] in the C-terminal region of the protein: 171-**V**E**Y**G**R**-175 and 199-LFYTLRSYAR-208 (the latter contains two overlapping CRAC motifs **LFYTLR** and **LRSYAR**. This region was not initially identified as a membrane-buried area,[75] but a reevaluation of the hydrophobicity of the amino acid sequence revealed a further possible membrane-embedded region encompassing amino acid residues 176–206. Docking studies confirmed that the CRAC motifs identified in this region could bind cholesterol.[74] Sigma-1 receptors also displayed a CARC motif[48] in the first transmembrane domain (7-**RRW**A**W**AA**LLL**-16). Molecular dynamics simulations indicated that this CARC motif forms a pocket lined by two arginine and two tryptophan residues (Fig. 7.6). Cholesterol could be docked onto this CARC motif (Fig. 7.6) with an energy of interaction of –50 kJ mol^{-1}. These data suggested that cholesterol may interact with the first transmembrane domain of the sigma-1 receptor. The position of the CARC motif in the transmembrane domain is consistent with an interaction with cholesterol in the inner leaflet of the membrane.

Then we looked for a sphingolipid-binding domain (SBD)[76] that could explain the association of sigma-1 receptor with GalCer-enriched microdomains during oligodendrocyte differentiation.[73] We found that the extracellular region of the receptor contains a SBD that could mediate GalCer binding: 31-GTQSFVFQR-39. Docking studies indicated that this SBD could interact with two molecules of GalCer in GalCer-enriched microdomains (Fig. 7.7). Because the SBD domain is close to the CARC motif, we could propose a topology for the sigma-1 receptor that is fully consistent with a dual interaction with cholesterol and GalCer (Fig. 7.7). Moreover, these interactions with GalCer glycolipids concern the N-terminal region

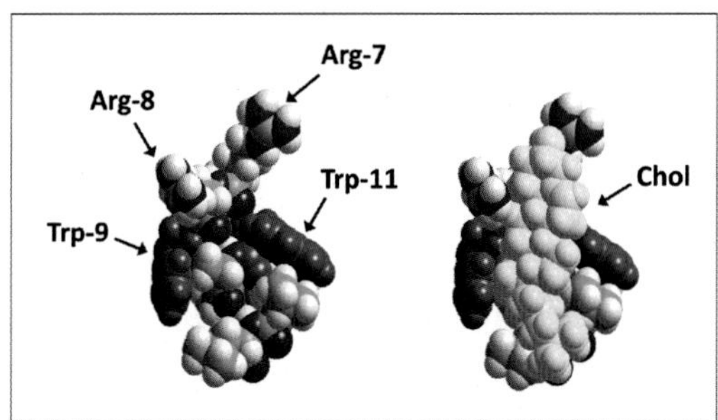

FIGURE 7.6 **Cholesterol binding to the CARC motif of the sigma-1-receptor.** A CARC motif was identified in the first transmembrane domain of the sigma-1 receptor: 7-RRWAWAALLL-16. Molecular dynamics simulations indicated that this region formed a pocket lined with two basic (Arg-7/Arg-8) and two aromatic (Trp-9/Trp-11) residues (left panel). Docking of cholesterol in this CARC domain revealed a high-affinity interaction (right panel, cholesterol in yellow).

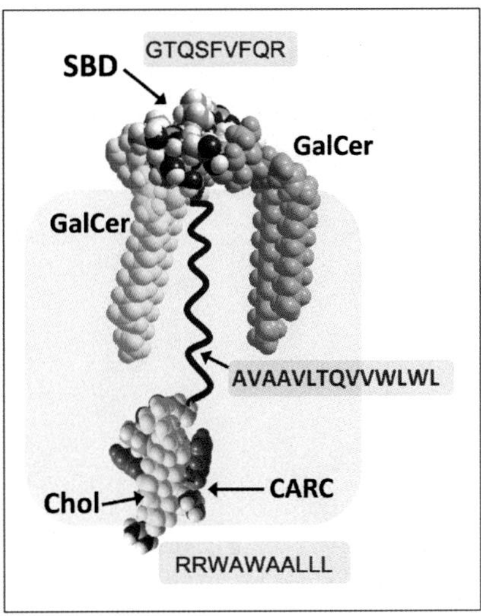

FIGURE 7.7 **GalCer and cholesterol binding to the sigma-1-receptor.** A sphingolipid-binding domain (SBD) was identified in the extracellular region of the sigma-1 receptor: 31-GTQSFVFQR-39. This domain binds to a pair of distant GalCer molecules located in the outer plasma membrane leaflet. The first transmembrane domain contains the CARC motif 7-RRWAWAALLL-16 that binds cholesterol in the inner leaflet (see Fig. 7.6). The CARC motif and the SBD are linked by the remaining residues of the transmembrane domain (17-AVAAVLTQVVWLWL-38). Thus, the membrane topology of the sigma-1 receptor is consistent with simultaneous interactions with GalCer and cholesterol in a lipid raft domain.

of the sigma-1 receptor. Thus, they may reinforce the association of the protein with cholesterol that is mediated by the C-terminal domain of the receptor. Overall, the sigma-1 receptor may interact simultaneously with two cholesterol and two GalCer molecules belonging to the same plasma membrane microdomain, as illustrated in Fig. 7.7.

7.5 GM1-BINDING DOMAIN IN HIGH-AFFINITY NGF RECEPTOR

Researchers have known for a long time that ganglioside GM1 enhances the activity of nerve growth factor (NGF) and stimulates neuronal sprouting.[77–80] Nevertheless, the molecular mechanisms associating GM1 with NGF function remained elusive until the finding of a functional interaction between GM1 and the high-affinity NGF receptor protein Trk (for tyrosine kinase receptor).[81] These data suggested that GM1 is an endogenous activator of NGF receptor function and that its action is mediated by direct binding to Trk.

Thus we looked for the presence of a GM1-binding domain in the extracellular domain of Trk. The SBD algorithm[82] was applied to the amino acid sequence of this region of the receptor. We found a motif located at the C-terminal ending of this extracellular domain (i.e., in the juxtamembrane region preceding the unique transmembrane domain of the receptor). This region is ideally located for mediating the binding to the sugar head group of ganglioside GM1. A complex between GM1 and this domain of Trk is illustrated in Fig. 7.8.

In this complex, the NH_3^+ group of Lys-410 faces the negatively charged carboxylate group of GM1. Overall the fit between GM1 and Trk is quite remarkable, the peptide forming a cavity

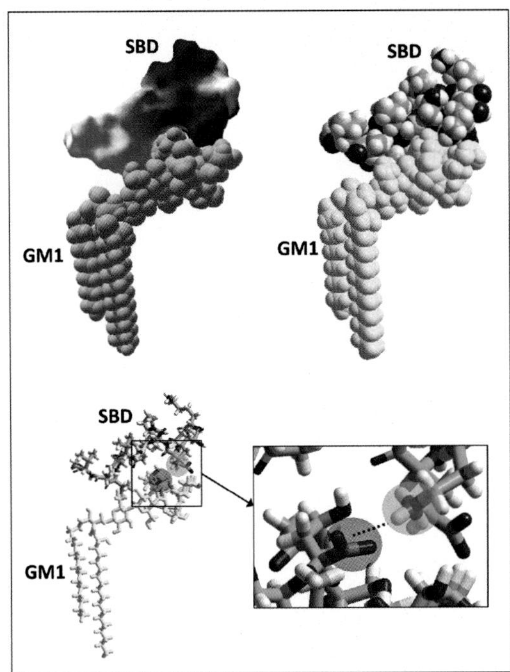

FIGURE 7.8 **Identification of a GM1-binding domain in the high-affinity NGF receptor.** A sphingolipid-binding domain (SBD) was identified at the C-terminus of the extracellular region of the high-affinity NGF receptor: 410-KKDETPFGVSVAVG-423. Molecular dynamics studies indicated that this region formed a cavity for the sialic acid of GM1 (upper panels). The complex is stabilized by an electrostatic bond between the carboxylate group of the sialic acid (rose disk) and the cationic group of Lys-410 (blue disk) (lower panels).

into which the sialic acid of GM1 could enter. The driving force of the binding process is the electrostatic bridge between Lys-410 and the carboxylate group of the sialic acid. The presence of a pair of lysine residues in the SBD of Trk may allow different geometries of the complex, so that exogenously added GM1 could also bind to the receptor through similar electrostatic bonds. This would explain the potentiating effects induced by GM1 on neurite outgrowth when the ganglioside is added in cell culture media. Finally one should note that in the GM1–Trk complex, the head group of the ganglioside is tilted with respect to the main axis of the ceramide anchor (Fig. 7.8). This particular geometry of glycosphingolipids is stabilized by cholesterol, as shown for GalCer, Gb_3, GM1, and GM3.[14,83,84] It has been shown that upon NGF binding, Trk translocates and concentrates in cholesterol-rich membrane microdomains.[85] The cholesterol-assisted interaction of GM1 with Trk may thus facilitate this translocation.

7.6 PHOSPHOINOSITIDE BINDING TO PURINERGIC RECEPTORS

In addition to its well-documented activity as a lipid precursor for G protein–coupled Ca^{2+}/protein kinase C signaling cascades,[86] phosphatidylinositol bisphosphate (PIP2) may also modulate synaptic functions via a direct interaction with neurotransmitter receptors.

A typical example of PIP2–receptor interactions is given by synaptic purinergic receptors. Purinergic signaling is crucial for neuron–glia communication. Extracellular ATP is released from neurons and glial cells through the exocytosis of vesicles containing ATP alone or ATP in combination with other neurotransmitters (glutamate, GABA, or acetylcholine).[87–89] Astrocytes express both ionotropic (P2X) and metabotropic (P2Y) receptors for extracellular ATP.[89] Metabotropic receptors P2Y receptors trigger Ca^{2+}-mediated signaling.[90] P2X receptors are ATP-gated channels permeable to Na^+, K^+, and Ca^{2+} cations.[91] In astrocytes, these receptors are formed by the functional association of P2X1 and P2X5 subunits.[92] Interestingly, the P2X1 subunit displays a lipid-binding domain that mediates an association with PIP2.[92] This domain, located in the C-terminal region of the protein, is enriched in basic amino acid residues that bind electrostatically to the negative charge of PIP2.[93] Replacing these basic amino acids by glutamine residues resulted in loss of PIP2 binding.[93] Using a lipid-binding assay, Ase et al. showed that the C-terminal peptide of P2X1 recognized several acidic lipids including phosphatidylserine and PIP2.[92] In contrast, the corresponding C-terminal domain of P2X5 did not interact with these lipids. This result is both intriguing and interesting because both subunits display a basic motif. The amino acid sequences of this motif in P2X1 and P2X5 proteins are shown in Fig. 7.9. A possible explanation for the lack of binding of P2X5 is the presence of three acidic residues (Glu-366, Asp-370, and Glu-374) that may counterbalance the attractive

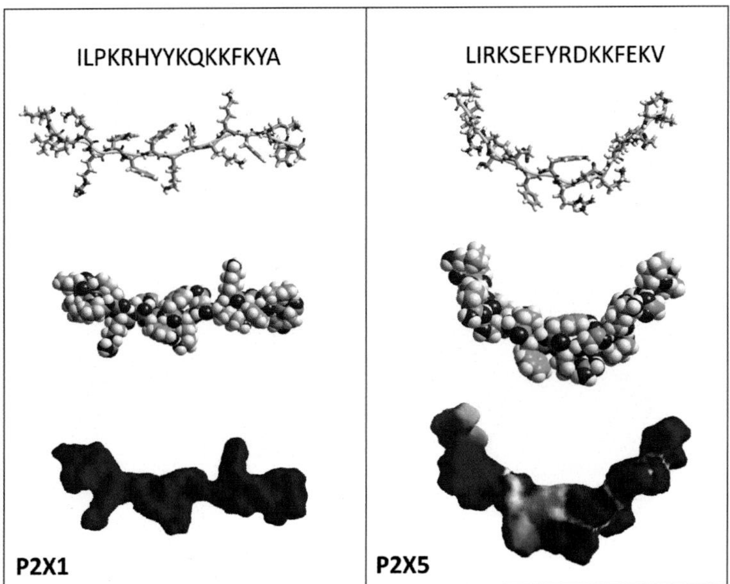

FIGURE 7.9 **Sequence, conformation, and electrostatic surfaces of the basic motifs in the P2X1 and P2X5 subunits of ionotropic purinergic receptor.** Note that in the case of P2X1 (left panels), the basic residues are oriented toward the solvent and thus are fully accessible for interacting with PIP2. In contrast, the presence of acidic residues in the P2X5 motif (right panels) induced a curved structure that decreased the accessibility of basic residues. The electrostatic surface of P2X1 is more cationic (blue surface) than that of P2X5, which also display negative clusters (red surface).

effect of basic residues on anionic lipids. On the other hand, the presence of acidic residues modified both the conformation and electrostatic surface of the motif (Fig. 7.9).

In all cases, this study revealed that the P2X1 subunit is sufficient to confer the sensitivity for PIP2 modulation to the heteromeric P2X1/5 channel.[92] By these means, ionotropic purinergic P2X channels contribute to the regulation of astroglial Ca^{2+} signaling.[94]

7.7 KEY EXPERIMENT: TRANSFECTION OF MEMBRANE RECEPTORS: WHAT ABOUT LIPIDS?

After these concrete examples of lipid–receptor interactions, the reader will not be surprised that the lipid composition of cells used for heterologous expression of membrane proteins is a critical issue. As summarized in Fig. 7.10, lipids exert a fine control of receptor function through direct lipid–protein interactions. Moreover, specific lipid–lipid interactions might be required to transform structural yet "inert" lipids into regulatory lipids able to modulate receptor conformation and function. For instance, cholesterol programs sphingolipids in such a way that they acquire a specific conformation, which in turn is recognized by the receptor.

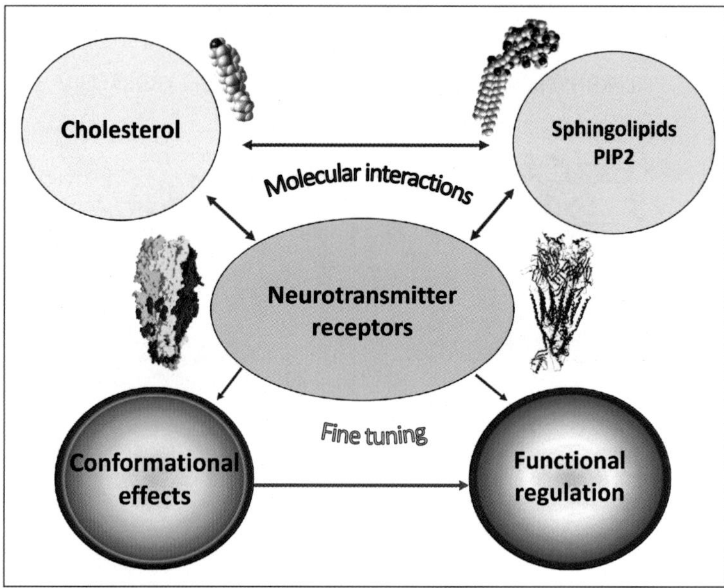

FIGURE 7.10 **Lipid regulation of neurotransmitter receptor function.** Neurotransmitter receptors interact specifically with nonannular lipids, essentially cholesterol, sphingolipids, or PIP2. These interactions affect the conformation of these receptors, so that these lipids may modulate the functions of these receptors through fine tuning of their active structure. Molecular interactions between lipids (e.g., sphingolipid/cholesterol) may be required to transform inert "structural" lipids into "regulatory" lipids independently of classical signal transduction mechanisms. The whole concept of lipid regulation of neurotransmitter function is based on molecular interactions including lipid–lipid and lipid–receptor associations.

Therefore, if the recipient cell does not express the specific lipids required by the receptor (which may concern the acyl chain content of sphingolipids), or an adequate cholesterol–sphingolipid balance, the transfection experiment may lead to an abundant expression of a totally inactive receptor. In 2003, Opekarova and Tanner published a list of more than 30 membrane proteins whose activity is specifically affected by lipids.[95] The list covered a broad range of proteins expressed by various bacteria, yeasts, insect, and mammalian cells. The problem is particularly acute when mammalian receptors or transporters are expressed in bacteria. For instance, the failure to express functional serotonin transporters in *E. coli* has been attributed to the lack of cholesterol in bacteria.[96] Moreover, the recovery of fully active neurotensin and adenosine receptors in transfected bacteria required the presence of cholesteryl hemisuccinate (a cholesterol derivative) during solubilization.[97,98] Paradoxical results have also been obtained for some proteins whose activity requires cholesterol[99] but can be functionally expressed in bacterial hosts.[100] In this case, one can exclude a direct interaction of cholesterol with the protein but rather consider a more general effect of the sterol on membrane properties.[95] As a matter of fact, we are just at the beginning of our comprehension of the complex molecular ballet that involves both lipid and protein actors in the plasma membrane of excitable cells.

References

1. el Battari A, Ah-Kye E, Muller JM, Sari H, Marvaldi J. Modification of HT 29 cell response to the vasoactive intestinal peptide (VIP) by membrane fluidization. *Biochimie*. 1985;67(12):1217–1223.
2. Lazar DF, Medzihradsky F. Altered microviscosity at brain membrane surface induces distinct and reversible inhibition of opioid receptor binding. *J Neurochem*. 1992;59(4):1233–1240.
3. Criado M, Eibl H, Barrantes FJ. Effects of lipids on acetylcholine receptor. Essential need of cholesterol for maintenance of agonist-induced state transitions in lipid vesicles. *Biochemistry*. 1982;21(15):3622–3629.
4. Maguire PA, Druse MJ. The influence of cholesterol on synaptic fluidity, dopamine D1 binding and dopamine-stimulated adenylate cyclase. *Brain Res Bull*. 1989;23(1-2):69–74.
5. Yeagle PL. Modulation of membrane function by cholesterol. *Biochimie*. 1991;73(10):1303–1310.
6. Kirilovsky J, Schramm M. Delipidation of a beta-adrenergic receptor preparation and reconstitution by specific lipids. *J Biol Chem*. 1983;258(11):6841–6849.
7. Popot JL, Demel RA, Sobel A, van Deenen LL, Changeux JP. Preferential affinity of acetylcholine receptor protein for certain lipids studied using monolayer cultures. *C R Acad Sci Hebd Seances Acad Sci D*. 1977;285(9): 1005–1008.
8. Papaphilis A, Deliconstantinos G. Modulation of serotonergic receptors by exogenous cholesterol in the dog synaptosomal plasma membrane. *Biochem Pharmacol*. 1980;29(24):3325–3327.
9. Albert AD, Young JE, Yeagle PL. Rhodopsin-cholesterol interactions in bovine rod outer segment disk membranes. *Biochim Biophys Acta*. 1996;1285(1):47–55.
10. Gimpl G, Fahrenholz F. Cholesterol as stabilizer of the oxytocin receptor. *Biochim Biophys Acta*. 2002;1564(2): 384–392.
11. Gimpl G, Burger K, Fahrenholz F. Cholesterol as modulator of receptor function. *Biochemistry*. 1997;36(36): 10959–10974.
12. Scanlon SM, Williams DC, Schloss P. Membrane cholesterol modulates serotonin transporter activity. *Biochemistry*. 2001;40(35):10507–10513.
13. Fantini J, Garmy N, Mahfoud R, Yahi N. Lipid rafts: structure, function and role in HIV, Alzheimer's and prion diseases. *Exp Rev Mol Med*. 2002;4(27):1–22.
14. Yahi N, Aulas A, Fantini J. How cholesterol constrains glycolipid conformation for optimal recognition of Alzheimer's beta amyloid peptide (Abeta1-40). *PloS One*. 2010;5(2):e9079.
15. Krengel U, Bousquet PA. Molecular recognition of gangliosides and their potential for cancer immunotherapies. *Front Immunol*. 2014;5:325.

16. Fantini J, Barrantes FJ. Sphingolipid/cholesterol regulation of neurotransmitter receptor conformation and function. *Biochim Biophys Acta*. 2009;1788(11):2345–2361.
17. Fantini J, Barrantes FJ. How cholesterol interacts with membrane proteins: an exploration of cholesterol-binding sites including CRAC, CARC, and tilted domains. *Front Physiol*. 2013;4:31.
18. Wustner D. Plasma membrane sterol distribution resembles the surface topography of living cells. *Mol Biol Cell*. 2007;18(1):211–228.
19. Shah WA, Peng H, Carbonetto S. Role of non-raft cholesterol in lymphocytic choriomeningitis virus infection via alpha-dystroglycan. *J Gen Virol*. 2006;87(Pt 3):673–678.
20. Gombos I, Steinbach G, Pomozi I, et al. Some new faces of membrane microdomains: a complex confocal fluorescence, differential polarization, and FCS imaging study on live immune cells. *Cytometry A*. 2008;73(3): 220–229.
21. Thorne RF, Mhaidat NM, Ralston KJ, Burns GF. Shed gangliosides provide detergent-independent evidence for type-3 glycosynapses. *Biochem Biophys Res Comm*. 2007;356(1):306–311.
22. Olshefski R, Ladisch S. Synthesis, shedding, and intercellular transfer of human medulloblastoma gangliosides: abrogation by a new inhibitor of glucosylceramide synthase. *J Neurochem*. 1998;70(2):467–472.
23. Li RX, Ladisch S. Shedding of human neuroblastoma gangliosides. *Biochim Biophys Acta*. 1991;1083(1):57–64.
24. Valentino L, Moss T, Olson E, Wang HJ, Elashoff R, Ladisch S. Shed tumor gangliosides and progression of human neuroblastoma. *Blood*. 1990;75(7):1564–1567.
25. Masserini M, Freire E. Kinetics of ganglioside transfer between liposomal and synaptosomal membranes. *Biochemistry*. 1987;26(1):237–242.
26. Allen JA, Halverson-Tamboli RA, Rasenick MM. Lipid raft microdomains and neurotransmitter signalling. *Nat Rev Neurosci*. 2007;8(2):128–140.
27. Edidin M. Lipids on the frontier: a century of cell-membrane bilayers. *Nature Rev Mol Cell Biol*. 2003;4(5): 414–418.
28. Cottingham K. Do you believe in lipid rafts? Biologists are turning to several analytical techniques to find out whether lipid rafts really exist? *Anal Chem*. 2004;76(21):403a–406a.
29. Lenne PF, Wawrezinieck L, Conchonaud F, et al. Dynamic molecular confinement in the plasma membrane by microdomains and the cytoskeleton meshwork. *EMBO J*. 2006;25(14):3245–3256.
30. Bruses JL, Chauvet N, Rutishauser U. Membrane lipid rafts are necessary for the maintenance of the (alpha)7 nicotinic acetylcholine receptor in somatic spines of ciliary neurons. *J Neurosci*. 2001;21(2):504–512.
31. Suzuki T, Ito J, Takagi H, Saitoh F, Nawa H, Shimizu H. Biochemical evidence for localization of AMPA-type glutamate receptor subunits in the dendritic raft. *Brain Res Mol Brain Res*. 2001;89(1-2):20–28.
32. Monastyrskaya K, Hostettler A, Buergi S, Draeger A. The NK1 receptor localizes to the plasma membrane microdomains, and its activation is dependent on lipid raft integrity. *J Biol Chem*. 2005;280(8):7135–7146.
33. Besshoh S, Bawa D, Teves L, Wallace MC, Gurd JW. Increased phosphorylation and redistribution of NMDA receptors between synaptic lipid rafts and post-synaptic densities following transient global ischemia in the rat brain. *J Neurochem*. 2005;93(1):186–194.
34. Renner U, Glebov K, Lang T, et al. Localization of the mouse 5-hydroxytryptamine(1A) receptor in lipid microdomains depends on its palmitoylation and is involved in receptor-mediated signaling. *Mol Pharmacol*. 2007;72(3):502–513.
35. Allen JA, Halverson-Tamboli RA, Rasenick MM. Lipid raft microdomains and neurotransmitter signalling. *Nat Rev Neurosci*. 2006;8(2):128–140.
36. Tansey EM. Henry Dale and the discovery of acetylcholine. *C R Biol*. 2006;329(5-6):419–425.
37. Dale H. Chemical transmission of the effects of nerve impulses. *Br Med J*. 1934;1(3827):835–841.
38. Dale HH. The chemical transmission of nerve impulses. *Science*. 1934;80(2081):450.
39. Colozo AT, Park PS, Sum CS, Pisterzi LF, Wells JW. Cholesterol as a determinant of cooperativity in the M2 muscarinic cholinergic receptor. *Biochem Pharmacol*. 2007;74(2):236–255.
40. Michal P, Rudajev V, El-Fakahany EE, Dolezal V. Membrane cholesterol content influences binding properties of muscarinic M2 receptors and differentially impacts activation of second messenger pathways. *Eur J Pharmacol*. 2009;606(1-3):50–60.
41. Barrantes FJ. Structural basis for lipid modulation of nicotinic acetylcholine receptor function. *Brain Res Brain Res Rev*. 2004;47(1-3):71–95.
42. Barrantes FJ. Cholesterol effects on nicotinic acetylcholine receptor. *J Neurochem*. 2007;103(Suppl 1):72–80.

43. Baier CJ, Gallegos CE, Levi V, Barrantes FJ. Cholesterol modulation of nicotinic acetylcholine receptor surface mobility. *Eur Biophys J*. 2010;39(2):213–227.
44. Barrantes FJ. Cholesterol effects on nicotinic acetylcholine receptor: cellular aspects. *Subcell Biochem*. 2010; 51:467–487.
45. Middlemas DS, Raftery MA. Identification of subunits of acetylcholine receptor that interact with a cholesterol photoaffinity probe. *Biochemistry*. 1987;26(5):1219–1223.
46. Corbin J, Wang HH, Blanton MP. Identifying the cholesterol binding domain in the nicotinic acetylcholine receptor with [125I]azido-cholesterol. *Biochim Biophys Acta*. 1998;1414(1-2):65–74.
47. Hamouda AK, Chiara DC, Sauls D, Cohen JB, Blanton MP. Cholesterol interacts with transmembrane alpha-helices M1, M3, and M4 of the Torpedo nicotinic acetylcholine receptor: photolabeling studies using [3H] Azicholesterol. *Biochemistry*. 2006;45(3):976–986.
48. Baier CJ, Fantini J, Barrantes FJ. Disclosure of cholesterol recognition motifs in transmembrane domains of the human nicotinic acetylcholine receptor. *Sci Rep*. 2011;1:69.
49. Jamin N, Neumann JM, Ostuni MA, et al. Characterization of the cholesterol recognition amino acid consensus sequence of the peripheral-type benzodiazepine receptor. *Mol Endocrinol*. 2005;19(3):588–594.
50. Maudsley S, Martin B, Luttrell LM. G protein-coupled receptor signaling complexity in neuronal tissue: implications for novel therapeutics. *Curr Alzheimer Res*. 2007;4(1):3–19.
51. Thathiah A, De Strooper B. The role of G protein-coupled receptors in the pathology of Alzheimer's disease. *Nat Rev Neurosci*. 2011;12(2):73–87.
52. Barrantes FJ, Borroni V, Valles S. Neuronal nicotinic acetylcholine receptor-cholesterol crosstalk in Alzheimer's disease. *FEBS Lett*. 2010;584(9):1856–1863.
53. Oddo S, Caccamo A, Green KN, et al. Chronic nicotine administration exacerbates tau pathology in a transgenic model of Alzheimer's disease. *Proc Natl Acad Sci USA*. 2005;102(8):3046–3051.
54. Whitaker-Azmitia PM. The discovery of serotonin and its role in neuroscience. *Neuropsychopharmacology*. 1999;21(2 Suppl):2s–8s.
55. Erspamer V, Asero B. Identification of enteramine, the specific hormone of the enterochromaffin cell system, as 5-hydroxytryptamine. *Nature*. 1952;169(4306):800–801.
56. Rapport MM, Green AA, Page IH. Partial purification of the vasoconstrictor in beef serum. *J Biol Chem*. 1948;174(2):735–741.
57. Rapport MM, Green AA, Page IH. Serum vasoconstrictor, serotonin; isolation and characterization. *J Biol Chem*. 1948;176(3):1243–1251.
58. Twarog BM, Page IH. Serotonin content of some mammalian tissues and urine and a method for its determination. *Am J Physiol*. 1953;175(1):157–161.
59. Nichols DE, Nichols CD. Serotonin receptors. *Chem Rev*. 2008;108(5):1614–1641.
60. Eisensamer B, Uhr M, Meyr S, et al. Antidepressants and antipsychotic drugs colocalize with 5-HT3 receptors in raft-like domains. *J Neurosci*. 2005;25(44):10198–10206.
61. Sjogren B, Svenningsson P. Depletion of the lipid raft constituents, sphingomyelin and ganglioside, decreases serotonin binding at human 5-HT7(a) receptors in HeLa cells. *Acta Physiol*. 2007;190(1):47–53.
62. Ponimaskin EG, Heine M, Joubert L, et al. The 5-hydroxytryptamine(4a) receptor is palmitoylated at two different sites, and acylation is critically involved in regulation of receptor constitutive activity. *J Biol Chem*. 2002;277(4):2534–2546.
63. Paila YD, Ganguly S, Chattopadhyay A. Metabolic depletion of sphingolipids impairs ligand binding and signaling of human serotonin1A receptors. *Biochemistry*. 2010;49(11):2389–2397.
64. Chattopadhyay A, Paila YD, Shrivastava S, Tiwari S, Singh P, Fantini J. Sphingolipid-binding domain in the serotonin(1A) receptor. *Adv Exp Med Biol*. 2012;749:279–293.
65. Fantini J. How sphingolipids bind and shape proteins: molecular basis of lipid-protein interactions in lipid shells, rafts and related biomembrane domains. *Cell Mol Life Sci*. 2003;60(6):1027–1032.
66. Fantini J, Yahi N. Molecular insights into amyloid regulation by membrane cholesterol and sphingolipids: common mechanisms in neurodegenerative diseases. *Exp Rev Mol Med*. 2010;12:e27.
67. Paila YD, Chattopadhyay A. The function of G-protein coupled receptors and membrane cholesterol: specific or general interaction? *Glycoconj J*. 2009;26(6):711–720.
68. Paila YD, Chattopadhyay A. Membrane cholesterol in the function and organization of G protein-coupled receptors. *Subcell Biochem*. 2010;51:439–466.

69. Pucadyil TJ, Kalipatnapu S, Chattopadhyay A. The serotonin1A receptor: a representative member of the serotonin receptor family. *Cell Mol Neurobiol*. 2005;25(3-4):553–580.
70. Lopez JJ, Lorch M. Location and orientation of serotonin receptor 1a agonists in model and complex lipid membranes. *J Biol Chem*. 2008;283(12):7813–7822.
71. Hanson MA, Cherezov V, Griffith MT, et al. A specific cholesterol binding site is established by the 2.8 A structure of the human beta2-adrenergic receptor. *Structure*. 2008;16(6):897–905.
72. Nuwayhid SJ, Werling LL. Steroids modulate *N*-methyl-D-aspartate-stimulated [3H] dopamine release from rat striatum via sigma receptors. *J Pharmacol Exp Ther*. 2003;306(3):934–940.
73. Hayashi T, Su TP. Sigma-1 receptors at galactosylceramide-enriched lipid microdomains regulate oligodendrocyte differentiation. *Proc Natl Acad Sci USA*. 2004;101(41):14949–14954.
74. Palmer CP, Mahen R, Schnell E, Djamgoz MB, Aydar E. Sigma-1 receptors bind cholesterol and remodel lipid rafts in breast cancer cell lines. *Cancer Res*. 2007;67(23):11166–11175.
75. Aydar E, Palmer CP, Klyachko VA, Jackson MB. The sigma receptor as a ligand-regulated auxiliary potassium channel subunit. *Neuron*. 2002;34(3):399–410.
76. Mahfoud R, Garmy N, Maresca M, Yahi N, Puigserver A, Fantini J. Identification of a common sphingolipid-binding domain in Alzheimer, prion, and HIV-1 proteins. *J Biol Chem*. 2002;277(13):11292–11296.
77. Doherty P, Dickson JG, Flanigan TP, Walsh FS. Ganglioside GM1 does not initiate, but enhances neurite regeneration of nerve growth factor-dependent sensory neurones. *J Neurochem*. 1985;44(4):1259–1265.
78. Katoh-Semba R, Skaper SD, Varon S. Interaction of GM1 ganglioside with PC12 pheochromocytoma cells: serum- and NGF-dependent effects on neuritic growth (and proliferation). *J Neurosci Res*. 1984;12(2-3):299–310.
79. Cuello AC, Garofalo L, Kenigsberg RL, Maysinger D. Gangliosides potentiate *in vivo* and *in vitro* effects of nerve growth factor on central cholinergic neurons. *Proc Natl Acad Sci USA*. 1989;86(6):2056–2060.
80. Cannella MS, Oderfeld-Nowak B, Gradkowska M, et al. Derivatives of ganglioside GM1 as neuronotrophic agents: comparison of *in vivo* and *in vitro* effects. *Brain Res*. 1990;513(2):286–294.
81. Mutoh T, Tokuda A, Miyadai T, Hamaguchi M, Fujiki N. Ganglioside GM1 binds to the Trk protein and regulates receptor function. *Proc Natl Acad Sci USA*. 1995;92(11):5087–5091.
82. Fantini J, Garmy N, Yahi N. Prediction of glycolipid-binding domains from the amino acid sequence of lipid raft-associated proteins: application to HpaA, a protein involved in the adhesion of *Helicobacter pylori* to gastrointestinal cells. *Biochemistry*. 2006;45(36):10957–10962.
83. Fantini J. Interaction of proteins with lipid rafts through glycolipid-binding domains: biochemical background and potential therapeutic applications. *Curr Med Chem*. 2007;14(27):2911–2917.
84. Fantini J, Yahi N, Garmy N. Cholesterol accelerates the binding of Alzheimer's beta-amyloid peptide to ganglioside GM1 through a universal hydrogen-bond-dependent sterol tuning of glycolipid conformation. *Front Physiol*. 2013;4:120.
85. Limpert AS, Karlo JC, Landreth GE. Nerve growth factor stimulates the concentration of TrkA within lipid rafts and extracellular signal-regulated kinase activation through c-Cbl-associated protein. *Mol Cell Biol*. 2007;27(16):5686–5698.
86. Berridge MJ, Irvine RF. Inositol trisphosphate, a novel second messenger in cellular signal transduction. *Nature*. 1984;312(5992):315–321.
87. North RA, Verkhratsky A. Purinergic transmission in the central nervous system. *Pflugers Arch*. 2006;452(5): 479–485.
88. Pankratov Y, Lalo U, Verkhratsky A, North RA. Quantal release of ATP in mouse cortex. *J Gen Physiol*. 2007;129(3):257–265.
89. Franke H, Verkhratsky A, Burnstock G, Illes P. Pathophysiology of astroglial purinergic signalling. *Purinergic Signal*. 2012;8(3):629–657.
90. Verkhratsky A, Kettenmann H. Calcium signalling in glial cells. *Trends Neurosci*. 1996;19(8):346–352.
91. Lalo U, Pankratov Y, Wichert SP, et al. P2X1 and P2X5 subunits form the functional P2X receptor in mouse cortical astrocytes. *J Neurosci*. 2008;28(21):5473–5480.
92. Ase AR, Bernier LP, Blais D, Pankratov Y, Seguela P. Modulation of heteromeric P2X1/5 receptors by phosphoinositides in astrocytes depends on the P2X1 subunit. *J Neurochem*. 2010;113(6):1676–1684.
93. Fujiwara Y, Kubo Y. Regulation of the desensitization and ion selectivity of ATP-gated P2X2 channels by phosphoinositides. *J Physiol*. 2006;576(Pt 1):135–149.
94. James G, Butt AM. P2X and P2Y purinoreceptors mediate ATP-evoked calcium signalling in optic nerve glia in situ. *Cell Calcium*. 2001;30(4):251–259.

95. Opekarova M, Tanner W. Specific lipid requirements of membrane proteins—a putative bottleneck in heterologous expression. *Biochim Biophys Acta*. 2003;1610(1):11–22.
96. Tate CG. Overexpression of mammalian integral membrane proteins for structural studies. *FEBS Lett*. 2001; 504(3):94–98.
97. Tucker J, Grisshammer R. Purification of a rat neurotensin receptor expressed in *Escherichia coli*. *Biochem J*. 1996;317(Pt 3):891–899.
98. Weiss HM, Grisshammer R. Purification and characterization of the human adenosine A(2a) receptor functionally expressed in *Escherichia coli*. *Eur J Biochem*. 2002;269(1):82–92.
99. Rothnie A, Theron D, Soceneantu L, et al. The importance of cholesterol in maintenance of P-glycoprotein activity and its membrane perturbing influence. *Eur Biophys J*. 2001;30(6):430–442.
100. Bibi E, Gros P, Kaback HR. Functional expression of mouse mdr1 in Escherichia coli. Proc Natl Acad Sci USA. 1993;90(19):9209–9213.

CHAPTER

8

Common Mechanisms in Neurodegenerative Diseases

Jacques Fantini, Nouara Yahi

OUTLINE

8.1 AMYLOIDOSIS: A BRIEF HISTORY

As striking as it may seems, the term *amyloid* is in fact a misnomer. It was coined by Virchow to describe cellular deposits that appeared to behave as starch-like, hence amyloid, structures are similar to the complex carbohydrates found in plants.[1] Indeed, these deposits, although found in various tissues, shared the common property to develop a blue stain after binding iodine. However, the so-called amyloid deposits are in fact protein aggregates that have nothing to do with plant carbohydrates. It turned out that iodine stains the accessory material (merely glycoconjugates of extracellular matrix molecules) that represents less than 2% of the deposits.[1] Nevertheless, the term *amyloid* is still universally used to describe various types of protein aggregates, including those found in brain tissues and associated with

Brain Lipids in Synaptic Function and Neurological Disease. http://dx.doi.org/10.1016/B978-0-12-800111-0.00008-4

several neurodegenerative diseases. The proteins found in the deposits are referred to as *amyloid proteins*, and the diseases caused by these proteins are called *amyloidoses*. Each amyloidose has its protein culprit: α-synuclein for Parkinson's disease,[2,3] β-amyloid peptide (Aβ) for Alzheimer's disease,[4,5] PrP for Creutzfeldt–Jakob and prion diseases,[6,7] huntingtin for Huntington's disease.[8] Biochemical and biophysical studies revealed that all these proteins have an exceptional plasticity. They can undergo major conformational changes affecting their secondary structure (α→β transition) and form various types of oligomers and aggregates with high neurotoxic potential. Understanding the molecular mechanisms underlying the conformational plasticity of amyloid proteins requires basic notions of protein structure that are summarized in the following section.

8.2 PROTEIN STRUCTURE

Proteins are polymeric biological macromolecules formed by the condensation of amino acids. The covalent bond between two successive amino acid residues in a protein is called the peptide bond. Hence, what makes each protein unique is the sequential order of the amino acid residues, which can be compared, for instance, to the succession of rosary beads (Fig. 8.1). Because the peptide bond links the carboxyl group of one amino acid to the amino group of the

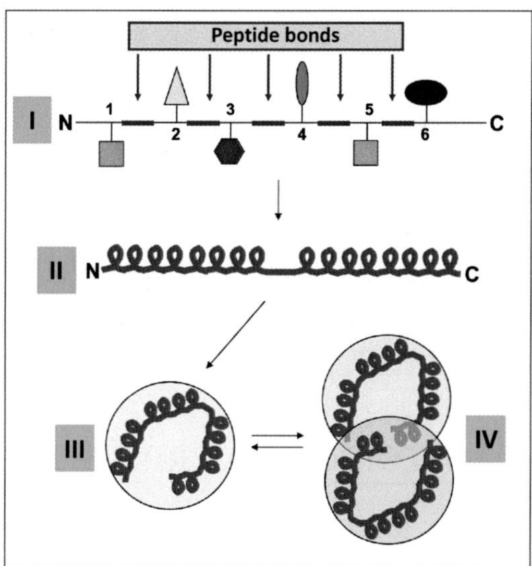

FIGURE 8.1 **The different levels of protein structure.** The primary structure (**I**) is the amino acid sequence. The sequence is numbered from the N-terminus to the C-terminus of the chain. Amino acid residues are the building blocks linked by a unique type of linkage, the peptide bond. Hence, what makes a protein unique is its amino acid sequence. The side chains of amino acid residues are represented as colored geometric symbols. The secondary structure (**II**) is formed by a local folding of the peptide chain into a highly regular segment (e.g., a helix). Two successive helices separated by a short linker are represented. To improve clarity, the amino acid side chains are not shown. The tertiary structure (**III**) corresponds to the spatial (3D) structure of the protein. Finally, the quaternary structure (**IV**) refers to multi-subunit proteins generally interacting through noncovalent bonds.

next one, the peptide chain is sequentially oriented. The amino group NH_3^+ of the first amino acid of the chain is not engaged in a peptide bond. Thus, this amino acid, numbered 1 in the sequence, defines the N-terminus of the peptide chain. At the other extremity of the chain, the carboxyl group of the last amino acid is also free and it logically defines the C-terminus.

Students should realize how practical this situation is to locate any amino acid residue of a protein, just by its number in the sequence. Proteins can be schematized by drawing a simple line, generally oriented with the N-terminus on the left and the C-terminus on the right. If we ask for amino acid 39 of α–synuclein, an amyloid protein with 140 amino acid residues, we can immediately spot it without any ambiguity. The amino acid sequence of a protein is referred to as the *primary structure*. It is the first level of protein structure, directly translated from the mRNA by the genetic code. Schematically it is an oriented wire with numerical markers. Such a wire that is folded on itself locally defines a *secondary structure*. The main secondary structure elements of a protein are helices (e.g., the α-helix) and strands (β structures). The wire has only one dimension, but the secondary structure is projected on a plane so that the succession of the secondary structure elements of a protein can be viewed as a bidimensional formation. The three-dimensional (3D) structure of the protein is the third level of protein structure (*tertiary structure*). Finally, some proteins gain their biological activity when associated in oligomeric assemblies consisting of several subunits, each of these subunits corresponding to a single peptide chain, which defines the *quaternary structure* of the protein. This is the case for both circulating proteins such as hemoglobin and membrane receptors, including ion channel receptors for neurotransmitters. In most cases, the interaction between protein subunits involves the side chains of specific amino acid residues and is not covalent.

8.3 PROTEIN FOLDING

Understanding how proteins fold into a 3D structure is not an easy task. If we know the primary structure of a protein, the first task is to decipher its secondary structure. Schematically, the two types of secondary structures are referred to as α and β. Molecular models of a typical α-helix and of a β-strand are shown in Fig. 8.2A and B. In both cases, the repetitive structure results from a fixed value of the angle formed by two successive amino acid residues of the chain. In fact, they are two torsion angles that control the process: ϕ and ψ in Fig. 8.2C. When these angles are repeated with the same value over a segment of the chain, a secondary structure is created. It is important to realize how different the α-helix and the β-strand are.

The α-helix is a condensed structure that requires a great deal of energy to be stabilized. The helical shape allows the atoms of the peptide bonds to be at a distance, favoring the formation of a hydrogen bond. Each i and $i + 4$ residues is linked by a hydrogen bond, so that the whole helix benefits from a large stabilizing energy (Fig. 8.2A). The side chains of the amino acid residues, which point outside the helix, are not involved in these forces of stabilization. Overall, one can consider the α-helix as an autostabilized structure. For this reason, the α-helix is not, by itself, particularly prone to oligomerization. Of course α-helix–α-helix interactions may occur, for instance when the transmembrane domains of G protein–coupled receptors form a compact structure surrounded by membrane lipids.[9,10] But in this case, the interactions between vicinal helices are chiefly mediated by the lateral chains of amino acids,[11,12] so they do not affect the autostabilizing potential of these

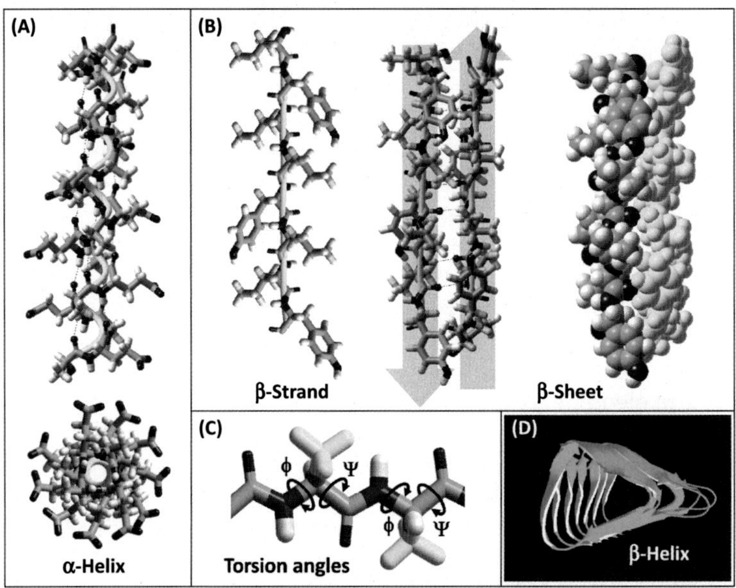

FIGURE 8.2 **Secondary structure of proteins: α-helix and β-strands.** (A) A typical α-helix. (B) A β-strand and a β-sheet. (C) Torsion angles in a peptide chain. (D) A β-helix. The typical α-helix (A) is autostabilized by hydrogen bonds formed by the C = O and NH groups of peptide bonds. Indeed, the C = O of an amino acid (referred to as *i*) forms a hydrogen bond with the NH of the amino acid four residues later (*i* + 4), which is possible because the amino acid residues in an α-helix adopt backbone dihedral angles (φ, ψ) around (–60°, –45°). These repeating values of φ and ψ torsion angles (C) are characteristic of the α-helix. The side chains of the amino acid residues are pointing outside of the α-helix (lower panel in cartoon in subpart A), like the branches of an evergreen tree. A typical β-strand is shown in subpart B (left panel). In this case, the values of the torsion angles of φ and ψ (–135°, 135°) give to the chain an extended conformation that does not allow the formation of hydrogen bonds between the atoms of the peptide bonds. Hence, a β-strand is not autostabilized and will look for other β-strands to form either β-sheets (in subpart B, middle and right side of the panel) or β-helices (D). In this respect, β-strands are typically aggregative-prone motifs. Note that in a β-sheet, both β-strands are at a short distance (subpart B, right panel with one β-strand in atom colors and its companion in yellow). In subpart C, the side chains are in yellow.

α-helices. It is a different story for β-strands. As shown in Fig. 8.2B, a β-strand is an extended structure, with torsion angles close to 180°. In this case, it is not possible to form any hydrogen bond between the atoms of the peptide bonds. Hence, in marked contrast with the α-helix, a β-strand is not autostabilized. For this reason, β-strands have to look for stabilizing partners. In a protein, two vicinal β-strands can achieve an interstabilization process by forming a series of hydrogen bonds between the atoms of their respective peptide bonds (Fig. 8.2B). The compact structure created by two β-strands is called a β-sheet. Although β-strands can be easily characterized by the repetition of ϕ and ψ angles (with values close to 180° for both), there are myriads of possibilities for creating β-sheets with either parallel or antiparallel orientation of β-strands, and with 2, 3, 4, or more β-strands. In some extreme cases, a superstructure called a β-barrel is formed by the regular assembly of β-strands arranged in an antiparallel fashion, as is the case for porins.[13] Some proteins that

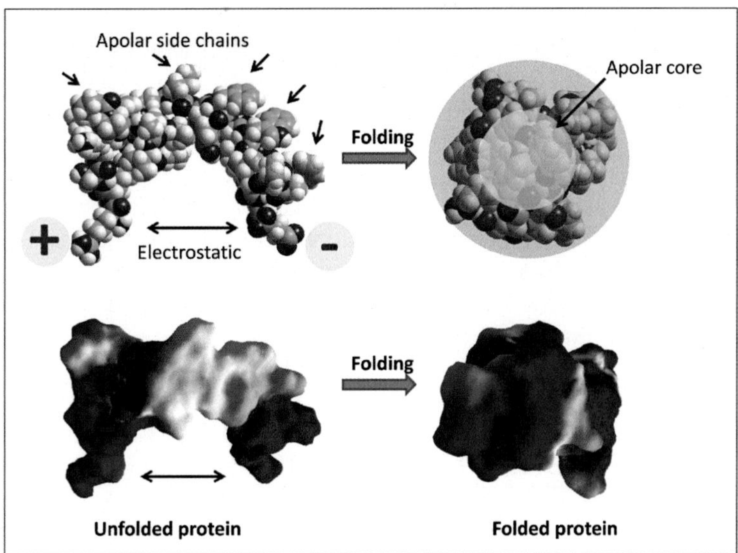

FIGURE 8.3 **Protein folding in water.** The hydrophobic effect constrains the apolar residues of a protein to escape water. These apolar residues converge to form an apolar core of the protein, leaving more polar residues at the surface where they can form hydrogen bonds with water molecules (upper panel). The enthalpic contribution to the folding reaction is illustrated in the lower panel (electrostatic interaction between the positive N-terminus in blue and the negative C-terminus in red). Apolar surfaces (white) disappear from the surface when the protein folds.

are highly resistant to extreme conditions of temperature and salts are folded into another type of superstructure, roughly a triangular prism, called a β-helix (Fig. 8.2D). In this case, the interaction between the constituting β-strands are reinforced by numerous interactions between the lateral chains of the amino acids, especially aromatic rings that form regular arrangements of π–π stacking pairs.[14]

A somewhat paradoxical feature is that the lateral chains of amino acids that belong to α-helices or β-strands are pointing outside the structure and do not participate in their stabilization. Nevertheless, each amino acid has its own propensity for belonging to an α-helix or to a β-strand. In 1976, Chou and Fasman published the P_α and P_β parameters of all amino acids, which correspond to the probability of finding a residue in given type of secondary structure.[15] The Chou and Fasman parameters were in fact derived from the study of approximately 50 proteins whose 3D structure was available at the time. More recent studies[16] confirmed the values of P_α and P_β initially published by Chou and Fasman, which indicate that being part of an α or β structure is a strong intrinsic property of each amino acid. For instance, alanine will often be found in an α-helix but rarely in β-strands; in contrast, tyrosine has a higher propensity to be part of a β-strand rather than an α-helix. The weak point is that most of the studied proteins are water-soluble proteins whose structuration is in part entropically driven by the hydrophobic effect (Fig. 8.3). Membrane-embedded proteins, which face an apolar environment and physically interact with lipids, are not likely to follow the same rules.

The folding of a protein in an aqueous milieu follows a clear-cut biochemical logic. Apolar residues isolate themselves from water by burrowing into the apolar phase that they autocreate. The apolar core of the protein groups most of the apolar amino acid residues into the center of a roughly spherical structure from which water molecules are totally excluded (Fig. 8.3).

The surface of this sphere bears polar residues that can form hydrogen bonds with water molecules, which results in a tightly bound solvation layer around the protein (see Chapter 2). In most cases, enthalpy-driven mechanisms also contribute to the folding process. In the example shown in Fig. 8.3, the N-terminus of the protein is positively charged, whereas the C-terminus is negative. Hence, the strong attraction between both ends favors the compaction of the protein into a globular shape.

8.4 INTRINSICALLY DISORDERED PROTEINS (IDPs): THE DARK SIDE OF THE PROTEOME

The entropic contribution of protein folding has a prerequisite: the protein should contain a sufficient percentage of apolar residues. If this is not the case, an apolar core will not form and the protein will not fold in water. Therefore, the protein has no well-defined 3D structure and will instead oscillate between a huge number of possible conformations. Such a situation is illustrated in Fig. 8.4.

Let us consider a protein with the following sequence: **A**EE**I**DED**A**SEDQEES. The percentage of apolar residues in this sequence is only 20% (3 of 15 residues, bold and underscored in the example sequence). It is not enough to constitute an apolar core, however. Thus this protein will not adopt any 3D structure in water and its global shape will fluctuate between a series of extended structures. These proteins are referred to as *intrinsically disordered proteins* (IDPs). They constitute an important part (perhaps up to 30%) of the proteome that has been called the *unfoldome* by Uversky.[17] IDPs perform a wide range of cellular functions related to signal transduction, proliferation, and differentiation[18] and may also play a major role in virus infection and pathogenesis.[19] It is precisely because they do not have a stable 3D structure that they can adapt their shape to several distinct molecular partners. Thus, these proteins have a high level of conformational plasticity that render them as actors of choice for various functions that rely on the establishment of reversible molecular interactions.[18] The conformational changes of IDPs should not be confounded with the "induced fit" mechanism, which refers to the mutual and generally slight conformational adjustment accomplished by both partners of a binding process during the course of the binding reaction.[20] In the induced fit mechanism, the binding partners have a well-defined 3D structure before binding and the conformational change occurring during the binding is minor. In the case of IDPs, the protein is totally disordered before binding, and the binding partner triggers the folding of the IDP, which can adopt distinct 3D structures according to the nature of the partner. Most neurological disease-associated amyloid proteins, including α-synuclein and Aβ peptides, are IDPs.[21] These IDP/amyloid proteins have a privileged partnership with membrane lipids, especially glycosphingolipids and cholesterol, which control their conformation in lipid raft domains of the plasma membrane of brain cells.[22]

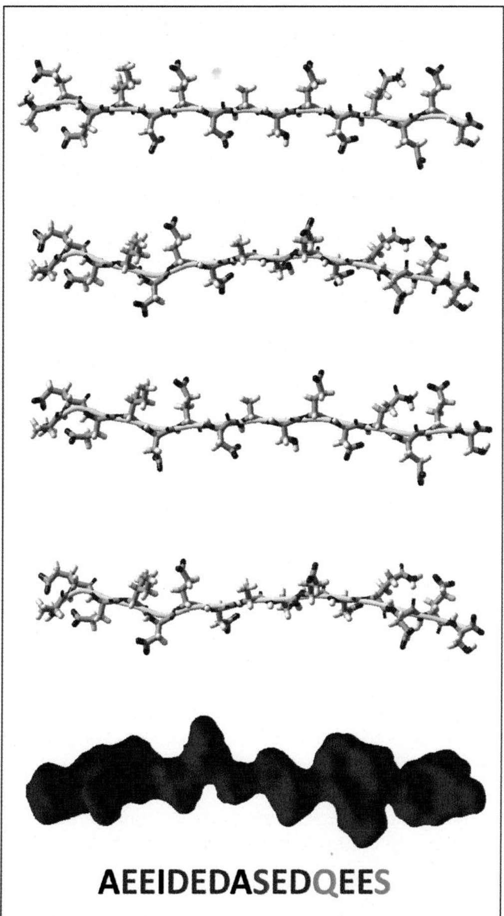

FIGURE 8.4 **Intrinsically disordered proteins (IDPs).** The balance between polar and apolar residues in a protein is a key parameter that controls the formation of a hydrophobic core in water. Proteins with too much polar versus apolar residues will not fold through an entropically driven mechanism and will instead adopt a myriad of conformations in water (4 possible conformers of the same protein are shown from the top of the cartoon). These proteins, referred to as "intrinsically disordered proteins" (IDPs), lack a precise 3D structure in water. IDPs have a high level of conformational plasticity and they can adapt their conformation to various molecular partners, which can be either a protein or a membrane. In this example, the protein is acidic (red surface in the lower model). The sequence of the protein is shown at the bottom of the illustration (apolar residues in black, acidic in red, and polar but not acidic in green).

8.5 LIPID RAFTS AS PLATFORMS FOR AMYLOID LANDING AND CONVERSIONS

Lipid rafts play both global and specific roles that contribute to control the conformation of amyloid proteins. The first effect is a reduction of dimensionality, a process that occurs for each soluble molecule transferred from the bulk extracellular water phase (a 3D space) to a

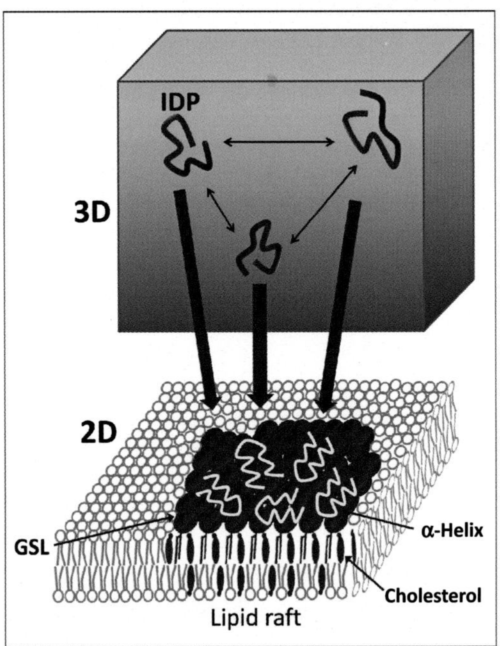

FIGURE 8.5 **Structuration of an IDP upon binding to membrane lipids.** When an IDP comes close to a membrane, specific interactions with selected lipids (such as gangliosides in lipid rafts) will favor the efficient adhesion of the protein to this membrane. Moreover, the repetitive head groups of raft lipids will induce α-helical folding of the protein through a typical chaperone-facilitated reaction. The reduction of dimensionality from 3D (extracellular milieu) to 2D (lipid raft surface) also favors the concentration of the protein on the membrane surface. The Parkinson's and Alzheimer's disease–associated proteins α-synuclein and β-amyloid peptides (Aβ) are typical examples of such IDPs that acquire a helical structure when bound to lipid rafts of neural cells.

membrane (a 2D surface).[23,24] In the 3D space, the mean distance between two amyloid proteins is important (Fig. 8.5).

However, when the proteins fall on the 2D surface of the membrane, this distance is significantly decreased. The effect is potentiated by the fact that the attachment of the amyloid protein on the membrane is not stochastic. Indeed, a common feature of amyloid proteins is that they have a marked preference for glycosphingolipids (e.g., gangliosides) that are concentrated in relatively small microdomains of the plasma membrane (lipid rafts). Thus, landing on a lipid raft will result in an important concentration of the amyloid protein, and this effect is potentiated by the reduction of dimensionality (Fig. 8.5). Moreover, the head groups of glycolipid clusters will exert a chaperone effect[22] that forces the amyloid protein to adopt a secondary structure, which, depending on the protein–lipid stoichiometry, can be either an α-helix[25] or a β-rich structure.[26]

8.6 AMYLOID PORES

The α-helical structuration of amyloid proteins is compatible with the insertion of the protein into the membrane. This process shares some analogy with bacterial toxins, such as colicin, which also acquire an α-helical structure upon binding to membrane lipids.[27]

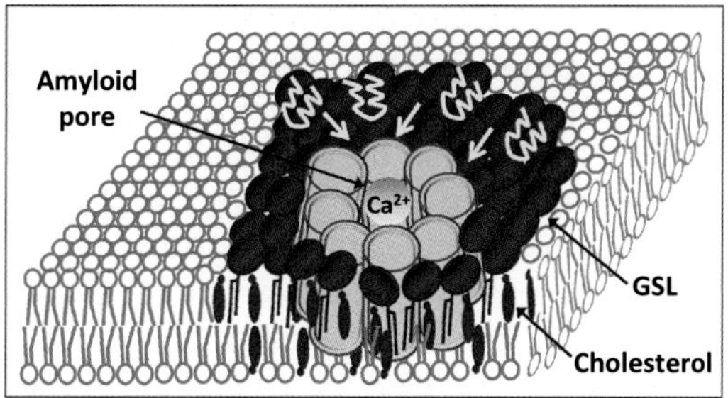

FIGURE 8.6 **Pore formation by amyloid proteins: a membrane-assisted process.** Amyloid proteins that have acquired an α-helical structure upon binding to lipid rafts can penetrate the membrane. The presence of cholesterol underneath the glycosphingolipids (GSL) in lipid raft areas stimulates both the insertion of the protein within the membrane and its oligomerization into Ca^{2+}-permeable pores.

The following step is the penetration of the newly formed α-helix within the membrane (Fig. 8.6).

In the case of amyloid/IDP proteins such as α-synuclein, this insertion step is fully controlled by cholesterol[28] (see Chapter 10). In this respect, the marked preference of amyloid proteins for glycosphingolipids ensures that the protein will find cholesterol during its insertion into the membrane.[22,28] All amyloid proteins have an α-helical cholesterol-binding domain that belongs to the part of the protein that penetrates the membrane.[28,29] The interaction between cholesterol and the protein is of sufficient energy to cause the cholesterol-binding domain, although both apolar and α-helical, to actually bind cholesterol and not other membrane lipids.[30] This distinction is important because upon binding to cholesterol, the amyloid protein adopts a typically tilted topology[28] particularly suited for the recruitment of vicinal amyloid/cholesterol complexes.[29,31,32] Overall, the tilted orientation of amyloid peptides in cholesterol-enriched plasma membrane domains favors the oligomerization of α-helical amyloid proteins into an annular pore (Fig. 8.6). Thus, one important mechanism of formation of annular oligomers is a coordinated process that sequentially involves the binding of the amyloid protein to raft lipids (e.g., gangliosides), the cholesterol-dependent insertion of the protein into the membrane, and the cholesterol-assisted oligomerization of annular pore. Such annular pores function as ion channels that can dramatically affect Ca^{2+} fluxes in neural cells.[33–35] A growing line of evidence suggests that these pore-forming oligomers are the most toxic assemblies of Aβ peptides in Alzheimer's disease.[5,36,37] The simultaneous presence of a glycosphingolipid-binding domain and a cholesterol-binding site in other amyloid proteins that are also able to form oligomeric channels[22,28] strongly suggest that this mechanism is operative in other neurodegenerative diseases.[37] Deciphering the molecular mechanisms controlling the formation of amyloid pores in lipid raft domains of neural membranes may lead to new therapeutic strategies for curing these diseases.

8.7 AMYLOID FIBRILS

Another hallmark of amyloid proteins is their capability to self-aggregate into insoluble fibrils with extensive β structures.[38–40] This capacity further increases the complexity of the problem we are facing, because it adds a new type of structure to proteins already characterized by high conformational versatility. The first issue to address is to understand why the same protein, starting from a totally disordered structure, can sometimes acquire an α and sometimes a β structure. As we will see, this matter is both environmental (membrane vs. water) and protein crowding (protein concentration in the extracellular space, and protein–lipid ratio in membranes). The α-helical structure is usually achieved when the protein is bound to membrane lipids, but changing the lipid–protein ratio can lead to β structuration instead.[29] The β-strand structure concerns merely the protein in a water environment. Let us consider the Aβ1−42 peptide. This peptide adopts α-helical structure when bound to detergent micelles[41] or to membrane lipids.[25] However, the same Aβ1−42 peptide also self-aggregates into various β-rich structures in water (Fig. 8.7).

On one hand, it forms annular β-oligomers with a central channel-like pore that can perforate the plasma membrane of brain cells and affect Ca^{2+} fluxes.[34] On the other hand, the peptide can form β-rich protofibrils and fibrils. Kallberg et al. studied the propensity of various amyloid proteins to form either α or β secondary structures.[16] They report the interesting observation that all amyloid proteins contain α/β discordant segments. These α/β discordant stretches correspond to a part of the protein that has been shown to adopt an α-helix structure in spite of being composed by amino acids that have a higher propensity

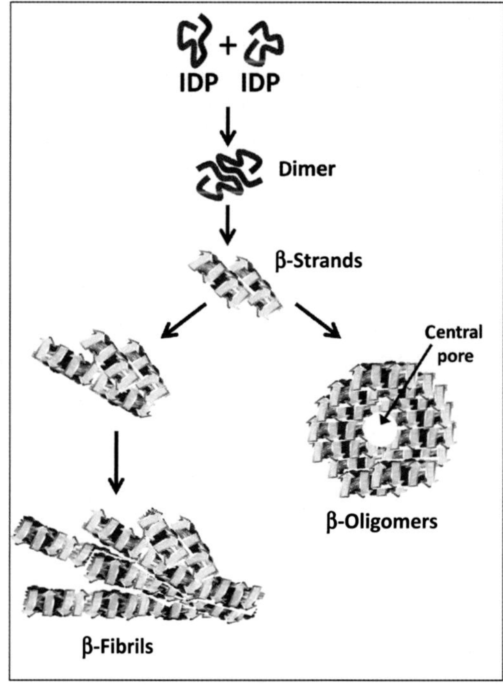

FIGURE 8.7 **Aggregation in water: β-oligomers and β-fibrils.** Aggregation of amyloid proteins in water starts from the structuration of a short core fragment into a β-strand. For instance, the fragment of Aβ 15–25 has a high propensity to form a β structure in water. Depending on their concentration, these peptides will further aggregate into β-rich annular oligomers or into β-fibrils.

to form a β-strand. In the case of Aβ, the α/β discordant part is the 15–25 segment.[16] Our interpretation of this paradoxical situation is that this segment will adopt an α-helical structure only when bound to membrane lipids that act as chaperone.[22] In absence of lipids (in bulk water), this segment will form a β-strand and not an α-helix. This short β-strand can then serve as a nucleation motif for the recruitment of other peptides, which will lead to the formation of larger β structures until we have protofibrils and fibrils (Fig. 8.7). In support of this notion, alanine substitutions K16A, L17A, and F20A, which reverted the α/β discordance of the 15–25 sequence, reduced amyloid fibril formation.[16] Similarly, α/β discordant stretches have been detected in other amyloid proteins, including the cellular prion protein PrP. Taken together, these data suggest that amyloid proteins are IDPs that, above a critical concentration in water, will fold into aggregative-prone β-strand-rich structures. Secondary structure predictions based on the known propensity of amino acid to form α or β structures in water help to understand why these proteins adopt a β structure in the extracellular milieu. When bound to membrane lipids, these proteins can be constrained to adopt an α-helical structure that is naturally prone to membrane insertion and oligomerization into Ca^{2+}-permeable annular pores. In fact the α/β discordance adequately reflects the different possible behaviors of the protein, basically an IDP, in water and membrane environments.

8.8 COMMON MOLECULAR MECHANISMS OF OLIGOMERIZATION AND AGGREGATION

Now we will discuss the molecular mechanisms controlling the self-aggregation of amyloid proteins. Two of these mechanisms are illustrated in Fig. 8.8. The first mechanism concerns amyloid proteins enriched in glutamine (e.g., the N-terminal domain of huntingtin[42] or synthetic polyglutamine that aggregates into classical β-sheet-rich amyloid-like structure).[43,44] The side chain of glutamine is ended by an amide group that bears both hydrogen bond donors and acceptor atoms. For this reason, glutamine is particularly suited for establishing a hydrogen bond network between two vicinal peptide chains. As shown in Fig. 8.8, both the amide group of glutamine and those of the peptide bonds contribute to the stabilization of the oligomer structure. The second mechanism of aggregation involves aromatic residues that can form π–π stacking interactions.[14] These noncovalent aromatic–aromatic bonds significantly contribute to the stabilization of the 3D structure of proteins, especially in the apolar core where these aromatic residues are concentrated. An illustration of these stacking interactions is given in Fig. 8.8.

In an elegant study of the amyloid aggregation of amylin, Azriel and Gazit[45] demonstrated that the substitution of the unique phenylalanine residue by an alanine abrogated the amyloid potential of the core peptide NFGAILSS. In contrast, alanine scanning of the whole peptide indicated that the other amino acid residues were not critical for amyloid aggregation. Moreover, rat amylin, which lacks this aromatic residue, does not form amyloid fibrils. As a matter of fact, aromatic residues are present in the minimal amyloid core-forming fibrils of most amyloid proteins.[45] This suggests that π–π stacking interactions play a major role in the formation of amyloid fibrils.[46] Whatever the molecular mechanisms involved in the oligomerization/aggregation of amyloid proteins, we would like to emphasize that

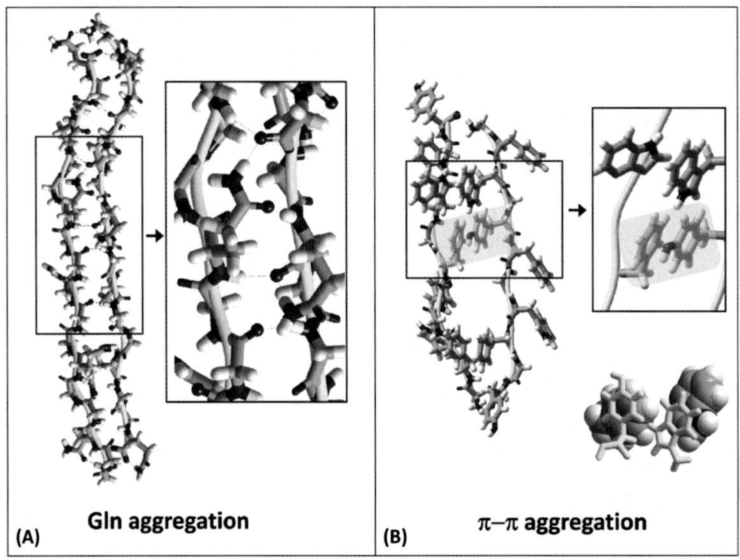

FIGURE 8.8 **Molecular mechanisms of amyloidogenesis.** (A) Glutamine-based aggregation driven by hydrogen bond formation. (B) Aromatic–aromatic-based aggregation driven by π–π stacking (note the mutual adjustment of aromatic rings as shown for Tyr–Phe and Trp–Phe pairs.

speaking of correct folding/misfolding for these proteins is inappropriate. At the molecular level, there is no correct or incorrect folding. The environment in which the protein lives is largely responsible for the 3D structure of an IDP. In turn, the 3D structure of the IDP determines, by itself, which types of molecular interactions the protein is able to share with its neighbors. Let us consider an IDP that has just become associated with a brain cell membrane in a lipid raft domain. It was disordered in the aqueous extracellular milieu; it is now α-helical. Which of both structures is the correct one? In fact neither, or both. The same is true for an amyloid protein that adopts a β-rich structure in water. The only thing that should be considered is whether a given conformation of the protein is associated with pathogenesis. In summary, it is difficult to decide which type of folding of an amyloid protein is correct, but it is definitely advisable to assess whether a given structure is toxic for brain cells. Thus, at first sight, interpreting neurodegenerative diseases as caused by protein "misfolding"[47–49] may be legitimate, but a thorough mechanistic analysis at the molecular level is encouraged.[22]

8.9 THERAPEUTIC STRATEGIES BASED ON LIPID RAFTS

This survey of the intrinsic and environmental factors that govern the oligomerization and aggregation properties of amyloid proteins indicates that lipid rafts exert an important control on these phenomena. Individual cases of lipid–amyloid interactions are discussed in the following chapters. However, at this stage, we should mention that raft lipids

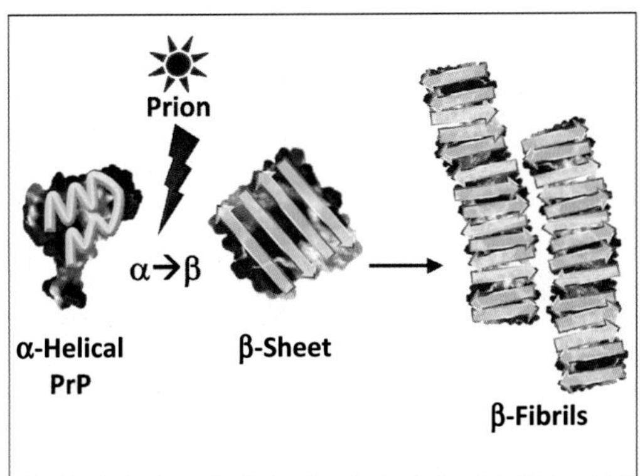

FIGURE 8.9 **Molecular plasticity in neurodegenerative diseases: α→β transition.** The replication of infectious prions is based on a conformational α→β change in the cellular protein PrP. This conformational change, which occurs in lipid raft domains, is induced by contact with the prion-associated infectious protein. The PrP protein is forced to adopt a β-sheet structure. Several β-structured PrP will then form larger β-fibrils.

may in fact play opposite roles in the formation of amyloid structures. In some instances, raft lipids may lock the protein into a nontoxic α-helical–rich structure that is probably the physiologically active form of the protein. This is the case for the cellular counterpart of the prion protein, a protein that is constitutively associated with lipid rafts through a glycosylphosphatidylinositol (GPI) anchor.[50] The PrP protein also interacts with sphingomyelin and glycosphingolipids[51] that stabilize its α-helical structure.[52] This good side of raft lipids prevent amyloid formation.[22] Unfortunately, the dark side of lipid rafts occurs when an α-helical is touched by an infectious prion, which is made of a β-rich version of PrP; then the "normal" protein is constrained to adopt the β-rich structure of the infectious protein (Fig. 8.9).

This process, referred to as α→β transition, requires that both the normal and infectious proteins are located in the same lipid raft.[53] At this stage, the protective lipid cocoon turns bad and the infection disseminates in the plasma membrane, which progressively accumulates newly formed and fully infectious β-fibrils of prions.[53] Several other manifestations of the dark side of lipid rafts occur in the context of brain diseases. As long as they remain without a defined structure, extracellular IDPs are not harmful. However, the α-helix structuration of these proteins, under the rigorous control of raft lipids, will most often induce membrane insertion and formation of Ca^{2+}-permeable amyloid pores. Finally, ganglioside GM1 has been shown to interact with Aβ peptides, leading to the formation of Aβ/GM1 complexes with high toxicity potential.[54] Given the multiple effects of lipid rafts on amyloid proteins (Fig. 8.10), it could be advantageous to design new drugs that can either counteract the prodisease activity of lipid rafts and/or reinforce their antidisease potential.

Deciphering the molecular mechanisms by which lipid rafts interact with amyloid proteins and control their oligomerization/aggregation into neurotoxic superstructures is a prerequisite for developing these new therapeutic strategies. The discovery of strikingly common mechanisms shared by distinct amyloid proteins raises some hope for the discovery of a common cure for all these diseases.

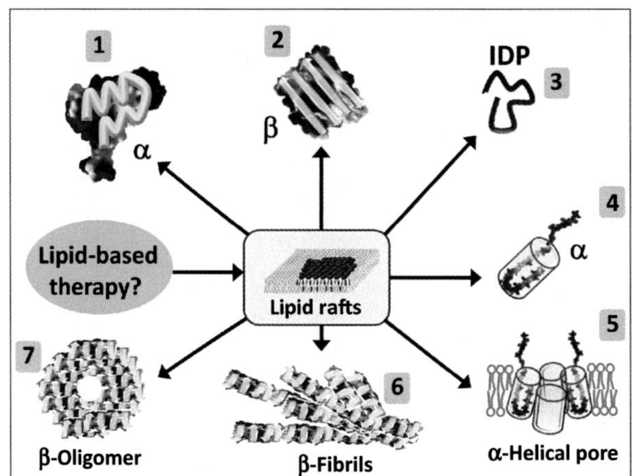

FIGURE 8.10 **Lipid rafts and neurodegenerative diseases.** By controlling the conformation of proteins involved in neurodegenerative diseases, lipid rafts play a key role in the pathogenesis of these diseases. In some cases, lipid rafts act as protective chaperones that lock the protein in a nonpathological form (**1**). The contact with an infectious prion will induce a β-sheet conformation and lipid rafts play an active role in this reaction (**2**). In other cases, raft lipids may transform a disordered protein (IDP) (**3**) into a highly pathological form (**4**) prone to form oligomeric channels in the plasma membrane of neural cells (**5**). Lipid rafts may also promote the formation of highly toxic β-fibrils (**6**) or pore-like β-oligomers (**7**). Hence, the development of lipid-based therapeutic approaches able to prevent the formation of neurotoxic oligomers and fibrils in lipid raft domains of brain membranes is crucial.

8.10 A KEY EXPERIMENT: COMMON STRUCTURE OF AMYLOID OLIGOMERS IMPLIES COMMON MECHANISM OF PATHOGENESIS[55,56]

One of the most spectacular findings on amyloid proteins was published in 2003 by the group of C. Glabe from the University of California at Irvine.[55] At the time of publication of this article, several data already suggested that soluble oligomers of amyloid proteins could be in fact more important pathologically than the fibrillar amyloid deposits. Because these oligomers had yet to be detected in the brain of patients suffering from neurodegenerative diseases, these authors decided to obtain an antibody directed against an oligomeric form of Aβ. Surprisingly, it turned out that this antibody recognized, in addition to Aβ oligomers, several other types of oligomeric aggregates formed by a series of sequence-unrelated amyloid proteins: α-synuclein, amylin, PrP, and polyglutamine. Moreover, this antibody did not react with low-molecular-weight species or with the fibrils formed by these proteins. The authors concluded that their "oligomer-specific antibody recognizes a unique common structural feature of the polypeptide backbone in the amyloid soluble oligomers that is independent of the amino acid side chains."[55] In other words, the oligomers display a common conformation-dependent structure that is not found in monomers or in fibrils.[57] The oligomer-specific antibody efficiently inhibited the toxicity induced by all soluble oligomers in neural cells, but it had no effect on the toxicity of Aβ fibrils. Overall, these data showed that amyloid oligomers have a common structure that may underlie a common pathogenic mechanism. This common mechanism is probably the permeabilization of membrane induced by these oligomers, either in the form of annular pores[58] or through less-specific membrane-permeabilizing effects.[59] In any case, the finding that a conformation-specific antibody could recognize and prevent the toxicity of the oligomers formed by distinct amyloid proteins reinforces our view that a common therapy, based on amyloid–membrane interactions (Fig. 8.10) is a reachable goal. As stated by Glabe, "the generic structures and common toxicity of amyloids raise the interesting possibility that a single 'silver bullet' drug might be an effective therapeutic for a wide variety of degenerative diseases."[56] Antioligomer antibodies that show promising effects in animal models of Alzheimer's disease[60] or a universal antiganglioside peptide active on all amyloid proteins[61] could already be serious candidates for such Graal compounds.

The biochemical nature of the common epitope shared by the oligomers formed by sequence-unrelated amyloid proteins has remained elusive. Conceptually, it is difficult to figure out how such a common motif can be generated during the folding and oligomerization of a series of proteins that lack sequence homology. Because in this case, antibody recognition is strikingly independent of the amino acid sequence, it has been speculated that the epitope could be a common peptide backbone motif, such as the array of hydrogen bond donors and acceptors at the edge of a β-sheet or a turn motif.[56] Coincidentally, we have shown that amyloid proteins precisely share a common sphingolipid-binding domain (SBD) that mediates the functional interactions of these proteins with cell surface glycosphingolipids, including gangliosides.[22,61] The SBD is a common structural hairpin motif found in numerous sequence-unrelated proteins that interact with lipid rafts.[62,63] Figure 8.11 shows the striking structural

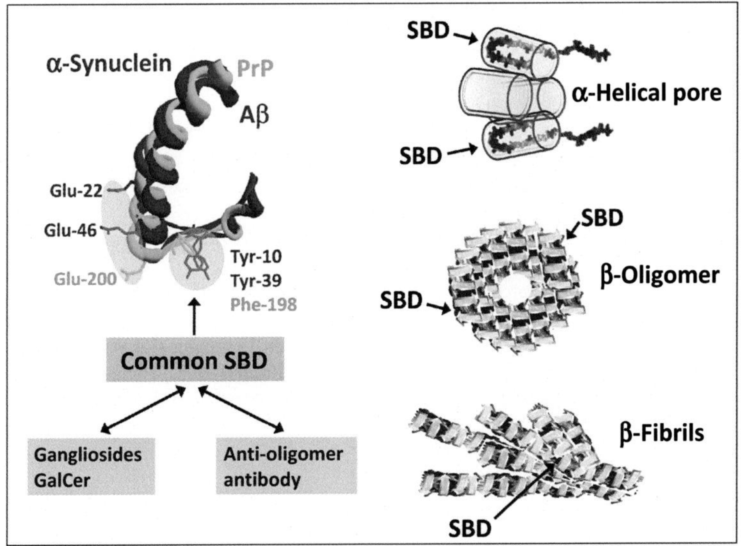

FIGURE 8.11 **Is the SBD the common epitope recognized by a universal antioligomer antibody?** In the left panel, one can see the superposition of the regions of α-synuclein (red), PrP (green), and Aβ (blue) that include the SBD of these proteins in their membrane-bound conformations. It appears that the SBD of these proteins is a common helix-turn-helix motif with a central aromatic residue. Interestingly, the SBD is close to an acidic Glu residue that is mutated into a basic lysine in inherited forms of Alzheimer's (Glu-22), Creutzfeldt–Jakob (Glu-200), and Parkinson's (Glu-46) diseases. All these mutations may affect the interaction of proteins with raft lipids. The SBD of Aβ, PrP, and α-synuclein is a common structural motif recognized by several glycolipids, including gangliosides and GalCer. Apart from the central aromatic residue, little or no sequence homology is present between the different SBDs, so that glycolipid recognition relies on a peptide backbone (hairpin or turn). This is also the case for the common antioligomer antibody. In the right panel, we can see that the SBD is fully accessible in both α- and β-oligomers, but it becomes progressively masked as β-fibrils grow. This typical feature is also a characteristic of the antioligomer antibody.

similarities between the SBDs of Aβ, PrP, and α-synuclein. In this respect, the SBD fulfils the criteria enunciated by Glabe[56] for the chemical structure of the common epitope expressed by different amyloid oligomers. Interestingly, we observed that in some cases, the same glycosphingolipid can be recognized by the SBD from distinct proteins.[64] This situation is remarkably parallel to the recognition of various amyloid oligomers by the same antibody.[55] If the SBD is accessible on the surface of amyloid oligomers (Fig. 8.11), then it could be recognized by the Glabe antibody during the immunization process and generate a universal antioligomer antibody. It should be emphasized that the SBD of amyloid proteins is structurally related to the V3 loop domain of HIV-1 gp120, which is the principal neutralization epitope of the virus.[65] Thus, it is likely that the SBDs of amyloid proteins are also immunogenic. This conclusion is supported by the occurrence of amino acid residues with high immunogenic potential (His, Lys, Ala, Leu, Asp, and Arg)[66] in these domains.[51,61] The possibility to generate broadly neutralizing antioligomer antibodies by immunizing with SBD peptides is currently under evaluation.

References

1. Buxbaum JN, Linke RP. A molecular history of the amyloidoses. *J Mol Biol*. 2012;421(2–3):142–159.
2. Goedert M. Alpha-synuclein and neurodegenerative diseases. *Nat Rev Neurosci*. 2001;2(7):492–501.
3. Lashuel HA, Overk CR, Oueslati A, Masliah E. The many faces of alpha-synuclein: from structure and toxicity to therapeutic target. *Nat Rev Neurosci*. 2013;14(1):38–48.
4. Querfurth HW, LaFerla FM. Alzheimer's disease. *N Engl J Med*. 2010;362:329–344.
5. Jang H, Connelly L, Arce FT, et al. Alzheimer's disease: which type of amyloid-preventing drug agents to employ? *Phys Chem Chem Phys*. 2013;15(23):8868–8877.
6. Prusiner SB. Novel proteinaceous infectious particles cause scrapie. *Science*. 1982;216(4542):136–144.
7. Prusiner SB. Prions. *Proc Natl Acad Sci USA*. 1998;95(23):13363–13383.
8. McGowan DP, van Roon-Mom W, Holloway H, et al. Amyloid-like inclusions in Huntington's disease. *Neuroscience*. 2000;100(4):677–680.
9. Fantini J, Barrantes FJ. Sphingolipid/cholesterol regulation of neurotransmitter receptor conformation and function. *Biochim Biophys Acta*. 2009;1788(11):2345–2361.
10. Fantini J, Barrantes FJ. How cholesterol interacts with membrane proteins: an exploration of cholesterol-binding sites including CRAC, CARC, and tilted domains. *Front Physiol*. 2013;4:31.
11. Hanson MA, Cherezov V, Griffith MT, et al. A specific cholesterol binding site is established by the 2.8 A structure of the human beta2-adrenergic receptor. *Structure*. 2008;16(6):897–905.
12. Sal-Man N, Gerber D, Bloch I, Shai Y. Specificity in transmembrane helix-helix interactions mediated by aromatic residues. *J Biol Chem*. 2007;282(27):19753–19761.
13. Fairman JW, Noinaj N, Buchanan SK. The structural biology of beta-barrel membrane proteins: a summary of recent reports. *Curr Opin Struct Biol*. 2011;21(4):523–531.
14. McGaughey GB, Gagne M, Rappe AK. pi-Stacking interactions. Alive and well in proteins. *J Biol Chem*. 1998;273(25):15458–15463.
15. Chou PY, Fasman GD. Empirical predictions of protein conformation. *Ann Rev Biochem*. 1978;47:251–276.
16. Kallberg Y, Gustafsson M, Persson B, Thyberg J, Johansson J. Prediction of amyloid fibril-forming proteins. *J Biol Chem*. 2001;276(16):12945–12950.
17. Uversky VN. The mysterious unfoldome: structureless, underappreciated, yet vital part of any given proteome. *J Biomed Biotechnol*. 2010;2010:568068.
18. Dunker AK, Uversky VN. Signal transduction via unstructured protein conduits. *Nat Chem Biol*. 2008;4(4):229–230.
19. Xue B, Mizianty MJ, Kurgan L, Uversky VN. Protein intrinsic disorder as a flexible armor and a weapon of HIV-1. *Cell Mol Life Sci*. 2012;69(8):1211–1259.
20. Zavodszky P, Hajdu I. Evolution of the concept of conformational dynamics of enzyme functions over half of a century: a personal view. *Biopolymers*. 2013;99(4):263–269.

21. Uversky VN, Oldfield CJ, Midic U, et al. Unfoldomics of human diseases: linking protein intrinsic disorder with diseases. *BMC Genomics*. 2009;10(suppl 1):S7.
22. Fantini J, Yahi N. Molecular insights into amyloid regulation by membrane cholesterol and sphingolipids: common mechanisms in neurodegenerative diseases. *Exp Rev Mol Med*. 2010;12:e27.
23. Adam G, Delbrück M, Reduction of dimensionality in biological diffusion processes. Rich A, Davidson N, eds. Structural Chemistry and Molecular Biology. San Francisco, USA: W.H. Freeman & Co; 1968:198–215.
24. Aisenbrey C, Borowik T, Bystrom R, et al. How is protein aggregation in amyloidogenic diseases modulated by biological membranes? *Eur Biophys J*. 2008;37(3):247–255.
25. McLaurin J, Franklin T, Fraser PE, Chakrabartty A. Structural transitions associated with the interaction of Alzheimer beta-amyloid peptides with gangliosides. *J Biol Chem*. 1998;273(8):4506–4515.
26. Choo-Smith LP, Surewicz WK. The interaction between Alzheimer amyloid beta(1-40) peptide and ganglioside GM1-containing membranes. *FEBS Lett*. 1997;402(2–3):95–98.
27. Stroud RM, Reiling K, Wiener M, Freymann D. Ion-channel-forming colicins. *Curr Opin Struct Biol*. 1998;8(4): 525–533.
28. Fantini J, Carlus D, Yahi N. The fusogenic tilted peptide (67-78) of alpha-synuclein is a cholesterol binding domain. *Biochim Biophys Acta*. 2011;1808(10):2343–2351.
29. Di Scala C, Chahinian H, Yahi N, Garmy N, Fantini J. Interaction of Alzheimer's beta-amyloid peptides with cholesterol: mechanistic insights into amyloid pore formation. *Biochemistry*. 2014;53(28):4489–4502.
30. Di Scala C, Yahi N, Lelievre C, Garmy N, Chahinian H, Fantini J. Biochemical identification of a linear cholesterol-binding domain within Alzheimer's beta amyloid peptide. *ACS Chem Neurosci*. 2013;4(3):509–517.
31. Di Scala C, Troadec JD, Lelievre C, Garmy N, Fantini J, Chahinian H. Mechanism of cholesterol-assisted oligomeric channel formation by a short Alzheimer beta-amyloid peptide. *J Neurochem*. 2013.
32. Fantini J, Di Scala C, Yahi N, et al. Bexarotene blocks calcium-permeable ion channels formed by neurotoxic Alzheimer's beta-amyloid peptides. *ACS Chem Neurosci*. 2014;5(3):216–224.
33. Arispe N, Diaz JC, Simakova O. Abeta ion channels. Prospects for treating Alzheimer's disease with Abeta channel blockers. *Biochim Biophys Acta*. 2007;1768(8):1952–1965.
34. Shafrir Y, Durell S, Arispe N, Guy HR. Models of membrane-bound Alzheimer's Abeta peptide assemblies. *Proteins*. 2010;78(16):3473–3487.
35. Jang H, Arce FT, Ramachandran S, et al. Truncated beta-amyloid peptide channels provide an alternative mechanism for Alzheimer's disease and Down syndrome. *Proc Natl Acad Sci USA*. 2010;107(14):6538–6543.
36. Lal R, Lin H, Quist AP. Amyloid beta ion channel: 3D structure and relevance to amyloid channel paradigm. *Biochim Biophys Acta*. 2007;1768(8):1966–1975.
37. Quist A, Doudevski I, Lin H, et al. Amyloid ion channels: a common structural link for protein-misfolding disease. *Proc Natl Acad Sci USA*. 2005;102(30):10427–10432.
38. Ghahghaei A. Review: structure of amyloid fibril in diseases. *J Biomed Sci Eng*. 2009;02(05):345–358.
39. Vilar M, Chou HT, Luhrs T, et al. The fold of alpha-synuclein fibrils. *Proc Natl Acad Sci USA*. 2008;105(25): 8637–8642.
40. Stohr J, Weinmann N, Wille H, et al. Mechanisms of prion protein assembly into amyloid. *Proc Natl Acad Sci USA*. 2008;105(7):2409–2414.
41. Coles M, Bicknell W, Watson AA, Fairlie DP, Craik DJ. Solution structure of amyloid beta-peptide(1-40) in a water-micelle environment. Is the membrane-spanning domain where we think it is? *Biochemistry*. 1998;37(31): 11064–11077.
42. Scherzinger E, Sittler A, Schweiger K, et al. Self-assembly of polyglutamine-containing huntingtin fragments into amyloid-like fibrils: implications for Huntington's disease pathology. *Proc Natl Acad Sci USA*. 1999;96(8): 4604–4609.
43. Chen S, Berthelier V, Hamilton JB, O'Nuallain B, Wetzel R. Amyloid-like features of polyglutamine aggregates and their assembly kinetics. *Biochemistry*. 2002;41(23):7391–7399.
44. Kar K, Hoop CL, Drombosky KW, et al. Beta-hairpin-mediated nucleation of polyglutamine amyloid formation. *J Mol Biol*. 2013;425(7):1183–1197.
45. Azriel R, Gazit E. Analysis of the minimal amyloid-forming fragment of the islet amyloid polypeptide. An experimental support for the key role of the phenylalanine residue in amyloid formation. *J Biol Chem*. 2001;276(36):34156–34161.
46. Gazit E. A possible role for pi-stacking in the self-assembly of amyloid fibrils. *FASEB J*. 2002;16(1):77–83.

47. Cheng B, Gong H, Xiao H, Petersen RB, Zheng L, Huang K. Inhibiting toxic aggregation of amyloidogenic proteins: a therapeutic strategy for protein misfolding diseases. *Biochim Biophys Acta*. 2013;1830(10):4860–4871.
48. Kraus A, Groveman BR, Caughey B. Prions and the potential transmissibility of protein misfolding diseases. *Ann Rev Microbiol*. 2013;67:543–564.
49. San Sebastian W, Samaranch L, Kells AP, Forsayeth J, Bankiewicz KS. Gene therapy for misfolding protein diseases of the central nervous system. *Neurotherapeutics*. 2013;10(3):498–510.
50. Stahl N, Baldwin MA, Hecker R, Pan KM, Burlingame AL, Prusiner SB. Glycosylinositol phospholipid anchors of the scrapie and cellular prion proteins contain sialic acid. *Biochemistry*. 1992;31(21):5043–5053.
51. Mahfoud R, Garmy N, Maresca M, Yahi N, Puigserver A, Fantini J. Identification of a common sphingolipid-binding domain in Alzheimer, prion, and HIV-1 proteins. *J Biol Chem*. 2002;277(13):11292–11296.
52. Sarnataro D, Campana V, Paladino S, Stornaiuolo M, Nitsch L, Zurzolo C. PrP(C) association with lipid rafts in the early secretory pathway stabilizes its cellular conformation. *Mol Biol Cell*. 2004;15(9):4031–4042.
53. Baron GS, Wehrly K, Dorward DW, Chesebro B, Caughey B. Conversion of raft associated prion protein to the protease-resistant state requires insertion of PrP-res (PrP(Sc)) into contiguous membranes. *EMBO J*. 2002;21(5):1031–1040.
54. Yanagisawa K, Odaka A, Suzuki N, Ihara Y. GM1 ganglioside-bound amyloid beta-protein (A beta): a possible form of preamyloid in Alzheimer's disease. *Nat Med*. 1995;1(10):1062–1066.
55. Kayed R, Head E, Thompson JL, et al. Common structure of soluble amyloid oligomers implies common mechanism of pathogenesis. *Science*. 2003;300(5618):486–489.
56. Glabe CG. Conformation-dependent antibodies target diseases of protein misfolding. *Trends Biochem Sci*. 2004;29(10):542–547.
57. Glabe CG. Structural classification of toxic amyloid oligomers. *J Biol Chem*. 2008;283(44):29639–29643.
58. Jang H, Zheng J, Lal R, Nussinov R. New structures help the modeling of toxic amyloid beta ion channels. *Trends Biochem Sci*. 2008;33(2):91–100.
59. Kayed R, Sokolov Y, Edmonds B, et al. Permeabilization of lipid bilayers is a common conformation-dependent activity of soluble amyloid oligomers in protein misfolding diseases. *J Biol Chem*. 2004;279(45):46363–46366.
60. Rasool S, Martinez-Coria H, Wu JW, LaFerla F, Glabe CG. Systemic vaccination with anti-oligomeric monoclonal antibodies improves cognitive function by reducing Abeta deposition and tau pathology in 3xTg-AD mice. *J Neurochem*. 2013;126(4):473–482.
61. Yahi N, Fantini J. Deciphering the glycolipid code of Alzheimer's and Parkinson's amyloid proteins allowed the creation of a universal ganglioside-binding peptide. *PloS One*. 2014;9(8):e104751.
62. Fantini J, Garmy N, Mahfoud R, Yahi N. Lipid rafts: structure, function and role in HIV, Alzheimer's and prion diseases. *Exp Rev Mol Med*. 2002;4(27):1–22.
63. Fantini J. How sphingolipids bind and shape proteins: molecular basis of lipid-protein interactions in lipid shells, rafts and related biomembrane domains. *Cell Mol Life Sci*. 2003;60(6):1027–1032.
64. Fantini J, Yahi N. Molecular basis for the glycosphingolipid-binding specificity of alpha-synuclein: key role of tyrosine 39 in membrane insertion. *J Mol Biol*. 2011;408(4):654–669.
65. Javaherian K, Langlois AJ, LaRosa GJ, et al. Broadly neutralizing antibodies elicited by the hypervariable neutralizing determinant of HIV-1. *Science*. 1990;250(4987):1590–1593.
66. Welling GW, Weijer WJ, van der Zee R, Welling-Wester S. Prediction of sequential antigenic regions in proteins. *FEBS Lett*. 1985;188(2):215–218.

CHAPTER

9

Creutzfeldt–Jakob Disease

Jacques Fantini, Nouara Yahi

OUTLINE

9.1 PRION DISEASES

In the origin, the search for the causative agent for scrapie, an infectious neurodegenerative disease in sheep, was just a routine effort to characterize a presumably classical microbe. Basically, what you need is a sick animal from which you take a sample containing the infectious agent. You inoculate this sample to a healthy animal, and you see if it gets sick in turn. Once this inoculation protocol is set up, you can work on the sample that is supposed to contain the infectious agent and try to identify this agent. The first characterization step concerns the size of the infectious agent, which has been for a long time the key feature used for distinguishing viruses from bacteria. If the infectious agent can pass through glazed porcelain, then it is referred to as an "ultrafiltrable" agent, that is, basically, the definition of a virus (a term etymologically derived from the Latin *virus*, meaning "poison"). By contrast, bacteria are "nonultrafiltrable" pathogens efficiently retained by glazed porcelain. Here we should mention that recently, several viruses that are in fact bigger than the smaller bacteria have been discovered,[1–4] but at the time the characterization of the scrapie agent was investigated (the mid-1960s), nobody was aware of these weird oversized viruses. The beginning

Brain Lipids in Synaptic Function and Neurological Disease. http://dx.doi.org/10.1016/B978-0-12-800111-0.00009-6

of the story was quite commonplace: the scrapie agent proved to be a member of the class of filter-passing viruses.[5] The following was most striking because the scrapie agent was much more resistant to ionization and ultraviolet radiation than any known virus. It also had a low molecular weight (ca. 2.10^5), which seemed incompatible with a virus.[6]

In the late 1960s, the central dogma of molecular biology, which states that biological information is stored in DNA and transferred into proteins through a unidirectional [DNA → RNA → protein] flux was more than a sacred tablet, it governed any thought in biology. Coined by Francis Crick at a time when neither messenger nor transfer RNAs had yet been characterized, this dogma helped molecular biology to emerge from biochemistry and led to resounding breakthroughs such as the deciphering of the genetic code and the discovery of the enzymatic mechanism of DNA replication. Francis Crick himself considered with skepticism the possibility that a protein could carry genetic information.[7] Nevertheless, a passionate debate ensued on the issue of whether the agent of scrapie could actually replicate without nucleic acid.[6] The main difficulty, then, was that the absolute requirement of nucleic acids for replication and genetic control was so firmly established that this possibility was considered totally unacceptable.[5,7] Nevertheless, some authors acknowledged the fact that the scrapie agent did not contain any nucleic acid and proposed alternative mechanisms for explaining its replication in animals. Proteins, polysaccharides, and even cellular membranes were incriminated,[5] but it is only a decade later, when Prusiner et al. obtained by sedimentation highly purified preparation of infectious particles,[8] that an in-depth investigation of the chemical nature of the scrapie agent could be achieved. In 1981, it was demonstrated that protease digestion resulted in the destruction of more than 99.9% of the infectivity of purified scrapie particles.[9] In line with these data, the demonstration that these particles contained a hydrophobic protein[9] strongly suggested that the scrapie agent was in no way conventional and definitely fell somewhere outside the dogma. The logical conclusion of these studies was that the transmissible biological information stored in this "unconventional" infectious agent was protein in nature. In 1982, Prusiner published in *Science* an iconoclast article in which he crossed the Rubicon, stating that "because the novel properties of the scrapie agent distinguish it from viruses, plasmids, and viroids, a new term 'prion' is proposed to denote a small proteinaceous infectious particle which is resistant to inactivation by most procedures that modify nucleic acids."[10]

In his 1998 review following his Nobel Prize, Prusiner mentions that he "never imagined the irate reaction of some scientists to the word *prion*." Perhaps the most striking feature of the prion story is that in fact, the "proteinaceous" infectious particle is made of a unique protein species of 27-30 kDa,[11] later designated PrP (for <u>pr</u>ion <u>p</u>rotein) by Prusiner.[12]

9.2 PrP: STRUCTURAL FEATURES, BIOLOGICAL FUNCTIONS, AND ROLE IN NEUROLOGICAL DISEASES

The gene coding for the PrP protein, located on chromosome 20, schematically consists of a short untranslated exon (nucleotides 1–362) and a larger exon (363–2740), which contains the totality of the open reading frame (373–1134) encoding the 253-amino acid PrP (GenBank entry #NM_000311.3). Mutations in this gene have been linked to several brain diseases, either inherited, sporadic, or infectious in origin, including Creutzfeldt–Jakob disease (CJD), Gerstmann–Straüssler–Sheinker syndrome (GSS), and fatal familial insomnia (FFI), which are thus considered prion diseases.[13] The structure of the human PrP is described in Fig. 9.1.

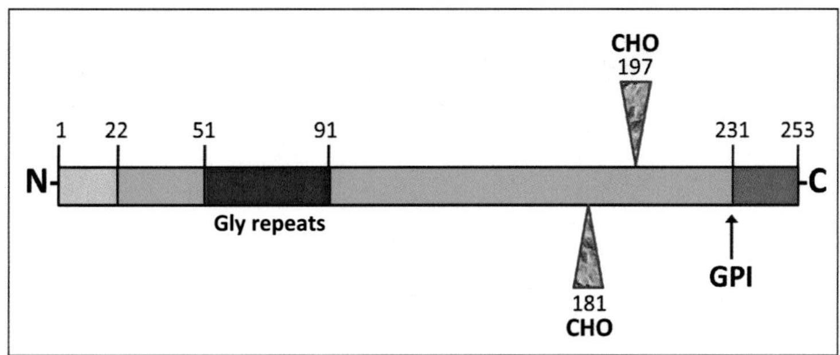

FIGURE 9.1 **The structure of human PrP.** The cellular PrP protein is a membrane glycoprotein (UNIPROT entry # P04156). The mRNA coding for PrP has 762 nucleotides (761 + STOP codon), resulting in the translation of a 253-amino acid protein. Residues 1–22 correspond to the signal sequence (yellow box), which is cleaved, leading to the 23–253 form of the protein. The C-terminal sequence (orange box, residues 231–253) is then removed to generate the mature 23–230 form of PrP. Posttranslational modifications also include *N*-glycosylation (CHO) at two sites (Asn-181 and Asn-197) and addition of a glycosylphosphatidylinositol (GPI) anchor at the C-terminus of the chain (GPI-anchor amidated Ser-230). Five successive Gly-rich repeats (red box, amino acid residues 51–59, 60–37, 68–75, 76–83, and 84–91), probably involved in binding of Cu^{2+} or Zn^{2+} (on Gly or His residues), are present in the sequence.

The GPI anchor is an efficient sorting signal for plasma membrane proteins that segregate within lipid rafts.[14] Indeed, GPI-anchored proteins generally contain two saturated acyl chains that fit better with the apolar chains of sphingolipids of the Lo phase of the membrane than with the glycerophospholipids of the Ld phase.[15] The 3D structure of the extracellular domain of recombinant human PrP (residues 23–230) has been determined by nuclear magnetic resonance spectroscopy.[16] It consists of three α-helices comprising the residues 144–154, 173–194, and 200–228 and a short antiparallel β-sheet (residues 128–131 and 161–164). On the basis of this structure, one can draw a schematic, yet realistic, topology of human PrP in a lipid raft domain (Fig. 9.2).

Now it is time to ask three important questions: (1) What is the physiological function of PrP? (2) Why is it associated with CJD? and (3) Why is this particular protein also found in infectious prions?

As a matter of fact, we have to acknowledge that little is known about the physiological function of PrP. The protein is expressed in most tissues and is especially abundant in brain. A number of functions have been proposed, including protection against apoptotic and oxidative stress, cellular uptake or binding of copper ions, transmembrane signaling, formation and maintenance of synapses, and adhesion to the extracellular matrix.[17] Despite the fact that most of these functions (if not all) are critical for life, PRNP-knockout mice develop normally and do not suffer from significant pathology.[18] Hence, whatever the exact physiological function of PrP, this function is not crucial and/or it can be easily fulfilled by other brain components. From this we can deduce that the link between PrP and CJD is probably not a loss of function of the protein, which markedly contrasts with classical inherited diseases caused by genetic mutations, such as for instance cystic fibrosis. Instead one can speculate that an alteration of PrP structure could result in the acquisition of new properties and that these new properties are responsible for the disease. Epidemiologic studies have established that the disease is essentially sporadic (one per million) whereas 15% are familial, autosomal dominant. Moreover, accidental transmission of CJD to humans appears to be iatrogenic – for

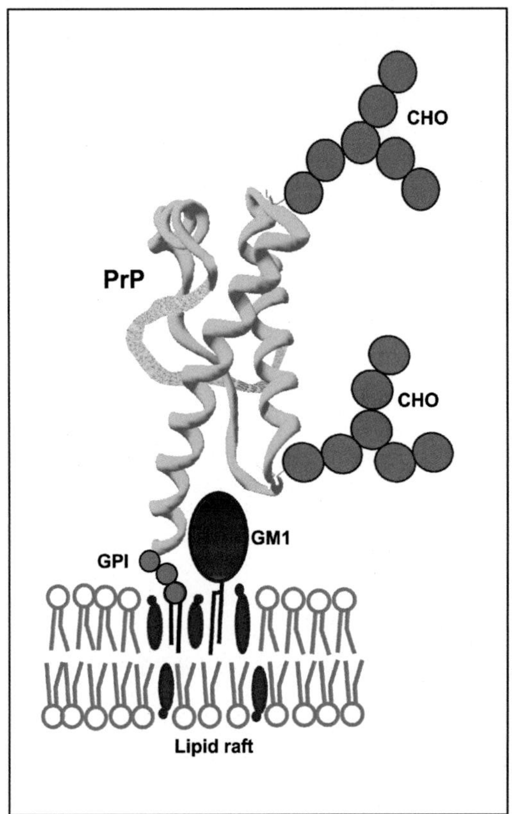

FIGURE 9.2 **Schematic structure of lipid raft-associated human PrP.** The protein is anchored in the extracellular leaflet of a lipid raft microdomain by a glycosylphosphatidylinositol (GPI) anchor at the C-terminus of the chain. The extracellular part of the protein (retrieved from PDB entry #1QLX) bears two glycosylation (CHO) sites. Note that the protein also interacts with the sugar head group of raft lipids such as gangliosides (e.g., GM1). Cholesterol (in red) does not physically interact with the PrP protein, but the sterol can indirectly affect the conformation of glycolipids (GPI and GM1) and thus contributes to stabilize the raft–PrP complex. Glycerophospholipids are in green.

example, children treated with contaminated human growth hormone, corneal transplantation, or electroencephalographic electrode implantation (information available through the UniProt entry# P04156).

The link between PrP and CJD has been established by the identification of single-point mutations in the PRNP gene, most of them being in the C-terminal part of the protein: Asp-178/Asn,[19–21] Val-180/Ile,[22] Glu-196/Lys,[23] Glu-200/Lys,[24] Asp-202/Asn,[25] Val-203/Ile,[23] Arg-208/His,[26,27] Val-210/Ile,[28] Glu-211/Gln,[23] Met-232/Arg.[22] Because CJD is associated with amyloid plaque formation, it is tempting to speculate that all these mutations trigger PrP-dependent amyloid fibrillation, providing a mechanistic link between mutation-induced structural alteration of PrP and amyloidogenesis potential. If this scenario is correct, an important issue that should be solved is why the wild-type PrP does not form amyloid fibers or plaques. This question, together with the reason explaining the presence of this cellular protein in infectious prion particles, will be the subject of a thorough inquiry in the second part of this chapter.

The finding that PrP was the unique protein species found in prion particles led to the logical assumption that the prion PrP was not identical to cellular PrP (otherwise everybody would die of CJD). These two isoforms of PrP are referred to as the nonpathological cellular PrP^c ("c" for cellular) and the infectious prion-associated PrP^{sc} ("sc" for scrapie). By analogy with the familial forms of CJD, which are caused by mutations in the PRNP gene, it was first assumed that PrP^{sc} was simply a mutant derivative of PrP^c. Although perfectly logical, this hypothesis proved to be incorrect. Strikingly, PrP^{sc} and PrP^c shared the same amino acid sequence. Moreover, analysis of posttranslational modifications (including N-glycosylation pattern) revealed that both proteins are chemically identical. This paradoxical situation had only two possible outcomes: (1) if we admit that PrP^c and PrP^{sc} are actually distinct species but have the same chemical composition, they could only differ by their 3D structure; in other words, they correspond to two distinct conformers of the same protein; and (2) there is no difference at all between PrP^c and PrP^{sc} and the infectivity of prions involves another compound that has been overlooked in Prusiner's studies.

That PrP^c and PrP^{sc} are not totally identical has been proven by their different sensitivity to proteolysis: upon mild proteinase K treatment, PrP^c is totally hydrolyzed in amino acids, whereas under the same conditions PrP^{sc} is partially resistant to the protease.[12] This finding is important for two reasons: first, it proves that PrP^c and PrP^{sc} have distinct properties that would be impossible to explain if they were totally identical; and second, it allows differentiation between PrP^c and PrP^{sc} with a simple assay based on a routine Western blot analysis of brain extracts that either are or are not incubated with proteinase K prior electrophoresis, transfer on nitrocellulose, and immunodetection with an anti-PrP antibody.[13,29,30] The principle of this assay is described in Fig. 9.3.

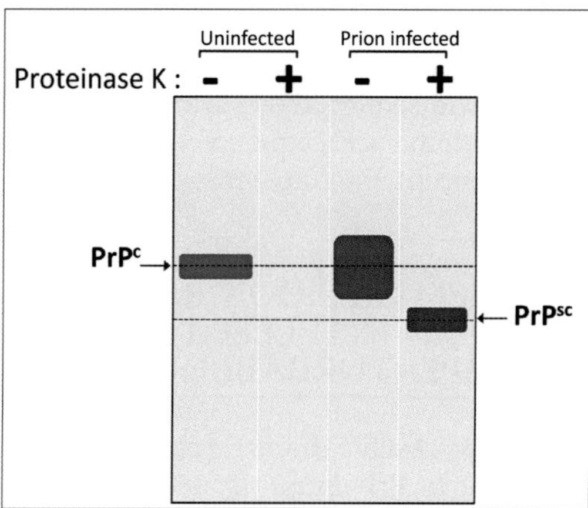

FIGURE 9.3 **Immunodetection of PrP^c and PrP^{sc} in brain extracts.** Western immunoblot of brain homogenates from uninfected and prion-infected animals (e.g., Syrian hamsters). Samples were either not digested (–) or digested (+) with proteinase K for 30 min at 37°C prior to polyacrylamide gel electrophoresis (PAGE). After electrotransfer, the blot is developed with a polyclonal rabbit anti-PrP antiserum that recognizes both PrP^c and PrP^{sc}. PrP^c (blue band) is completely hydrolyzed under these conditions, whereas PrP^{sc} (red band) is partially resistant to the protease treatment. In the third lane, the purple spot represents a mixture of PrP^c and PrP^{sc} that are codetected in brain extracts from infected animals.

Because neither PrP^{c} nor PrP^{sc} could be crystallized, it was not possible to establish the 3D structure of these proteins by X-ray crystallography. To circumvent this unfavorable situation, Prusiner used circular dichroïsm and Fourier transformed infrared (FTIR) spectroscopy, both techniques that allow determination of the respective percentages of the secondary structures (α, β, and turns) in a peptide or a protein. Applied to PrP^{c} and PrP^{sc} samples, these spectroscopic approaches definitely demonstrated that both isoforms have a distinct 3D structure. In a seminal article published in 1993,[31] the Prusiner's group established that PrP^{c} is chiefly an α-helical protein (43% of α-helix structure) with few β-structures (3%). In contrast, PrP^{sc} has less α-helix (30%) but has gained a high percentage of β-structures (43%).

For the first time in the history of biology, it was reported that the same protein could exist in two fundamentally distinct conformations: an α-helical version (PrP^{c}) and a β-strand variant (PrP^{sc}). That the same amino acid sequence (referred to as the primary structure) of a protein could generate two 3D structures with such different percentages of α-helix and β-strands was not less than a revolutionary finding that challenged the foundations of our knowledge of protein structure. In 1972, Christian Anfinsen has been the laureate of the Nobel Prize "for his work on ribonuclease, especially concerning the connection between the amino acid sequence and the biologically active conformation." His experiment is detailed in most biochemistry textbooks: starting from the biologically active "native" ribonuclease, a protein whose 3D structure requires the correct formation of four disulfide bridges involving eight cysteine residues, he totally denatured the protein by using a combination of β-mercaptoethanol (which breaks disulfide bridges and regenerates free cysteine residues) and urea (a chaotropic agent). Following these treatments, ribonuclease lost both its 3D structure and its enzymatic activity. After dialysis (which eliminates both β-mercaptoethanol and urea) and bubbling of oxygen (which allowed reformation of disulfide bridges), ribonuclease recovered its biological activity, which implies that the protein has again acquired its correct 3D structure. Anfinsen concluded that the information stored in the amino acid sequence led, upon protein folding, to only one 3D structure that corresponds to the biologically active form of the protein. The data obtained with PrP proved that this is not the case for all proteins. Beyond the classic rule [1 amino acid sequence → 1 tridimensional structure], the prion story taught us a new striking rule: [1 amino acid sequence → 2 possible tridimensional structures]. It was the beginning of a new concept, the conformational plasticity of proteins.

9.3 THE MECHANISM OR PRION REPLICATION: A GREAT INTUITION AND AN INTELLECTUAL JOURNEY OF AN IMPERTURBABLE LOGIC

Which mechanism could induce such a dramatic change of the 3D structure of a protein? How could the α → β conversion of PrP^{c} into PrP^{sc} account for the pathogenic properties of this protein? In 1967 (25 years before Prusiner invented the term *prion*), a remarkably inspired mathematician, J. S. Griffith, proposed a simple but prophetic model of a disease caused by the replication of an infectious protein.[32] One of his hypothesis was that the brain expressed a "normal" form of the protein, whereas the infectious particle contained an "abnormal" form of the same protein. The abnormal form of the protein was then supposed to bind to the normal brain protein and to convert it into a disease-causing protein. This simple mechanism

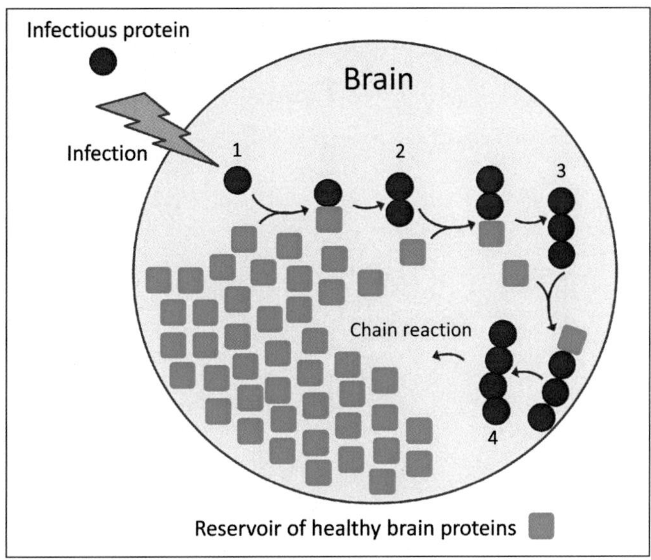

FIGURE 9.4 **Griffith's view of the replication of infectious proteins.** The infectious protein is symbolized by a red disk. The brain contains a full reservoir of healthy proteins (green squares) that are converted into infectious entities following contact with the infectious protein. The infectious protein has an abnormal form. Each contact of an infectious protein with a healthy one generates a new infectious protein so that an irreversible chain reaction gradually leads to the transformation of billions of proteins that acquire an abnormal form and cause disease.

describes in fact a chain reaction triggered by the initial contact between the abnormal/infectious protein and the normal protein (Fig. 9.4). The resulting formation of a complex of two abnormal proteins can, in turn, convert another normal protein to obtain three abnormal proteins, and so on until the brain contains billions of abnormal proteins that cause disease.

Replace "abnormal" protein by PrP^{sc} and "normal" protein by PrP^{c} and consider that PrP^{c} is mainly α-helical and PrP^{sc} rich in β-structures, and you have the mechanism of prion replication. It is remarkable that Griffith could propose this mechanism with so little information available at the time (the protein hypothesis was hardly debated) and that this mechanism proved to be absolutely correct. Nevertheless, it is fair to mention that its formal demonstration required several decades of intense research efforts, in which Prusiner played a prominent role.

This long quest was marked by milestones. As already discussed, the obtention of highly purified preparations of infectious scrapie particles was necessary to identify the PrP protein.[13] Circular dichroïsm studies of PrP^{c} and PrP^{sc} demonstrated the existence of two totally distinct conformations of PrP, one corresponding to the physiologically expressed brain protein and the other to the infectious protein.[31] However, the masterly demonstration of the mechanism of prion replication is that $PRNP^{0/0}$ mice, which do not express the PrP^{c} protein, failed to propagate prion infectivity.[33] Hence, without the brain reservoir of normal PrP^{c} proteins, infectious PrP^{sc} proteins are harmless and unable to cause any disease. If we link this information with the respective structures of PrP^{c} and PrP^{sc}, then we have a molecular mechanism accounting for the replication, by force, of prions invaders in the brain of healthy animals (Fig. 9.5).

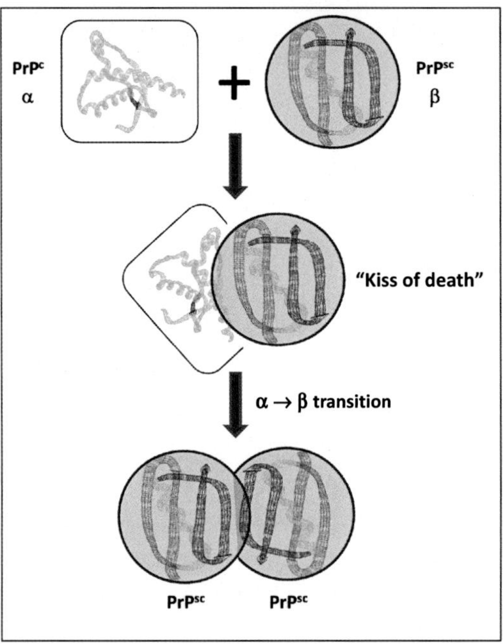

FIGURE 9.5 **Molecular mechanism of prion replication.** A healthy PrP^c protein, chiefly α-helical, binds to an infectious PrP^{sc} protein that has a high β-strand and low α-helix content. This physical contact, which has been compared to a "kiss of death," induces a major conformational change in PrP^c corresponding to the $\alpha \rightarrow \beta$ transition.

9.4 ROLE OF LIPID RAFTS IN THE CONFORMATIONAL PLASTICITY OF PrP

Because PrP^c is a membrane protein, anchored by a GPI glycolipid in the external leaflet of a lipid raft domain, it is likely that lipids have to play a role in the conversion of PrP^c into PrP^{sc}. Indeed, Baron et al. have established that the conversion reaction occurs in lipid rafts and that it requires the insertion of PrP^{sc} into the membrane containing PrP^c.[34] In other words, both PrP^c and PrP^{sc} have to be in the same contiguous membranes for triggering PrP^c–PrP^{sc} binding that leads to PrP^c conversion into PrP^{sc}. Therefore, the lipids surrounding PrP^c and PrP^{sc} might play a critical role in the conversion process.[35–37] What do we know of lipid-PrP interactions? First of all, two raft lipids, GalCer and sphingomyelin, are copurified with infectious prions.[38] This process indicates that either these two sphingolipids are initially associated with PrP^c and remained stuck to PrP^{sc} after the conversion reaction, or they become associated with PrP^{sc} during the reaction. In fact it is likely that sphingolipid association is an early step of the biosynthetic pathway of PrP^c that occurs in the endoplasmic reticulum when the protein is still in its immature form.[39] The consequence of this early interaction of PrP^c with raft lipids is a stabilization of its nonpathological α-helical structure. Conversely, sphingolipid depletion has been shown to increase PrP^{sc} formation in cultured cells.[40] Thus, the binding of PrP to model membranes appear to have a high impact on its conformation. In line with this notion, the interaction of PrP with negatively charged lipids such as phosphatidylglycerol promotes β-sheet formation.[41] In contrast, PrP binding to raft-like membranes containing both sphingomyelin and cholesterol induces α-helical structure.[41] The incorporation of ganglioside GM1 in raft-like membranes also stabilizes the native α-helical structure of PrP.[42] Finally, the prominent role of sphingolipids for maintaining the physiological α-helical conformation of PrP is consistent with the involvement of PrP^c in sphingolipid-associated signaling.[43] Overall these findings support the view that lipid membranes could play a role both in the physiological activity of native PrP^c and in the PrP^c-to-PrP^{sc} conversion.

The presence of GalCer and sphingomyelin in highly purified preparation of infectious prions suggested that PrP (either PrP^c or PrP^{sc} or both) contains a functional domain ensuring the specific association with these raft lipids. We noticed that the HIV-1 gp120 surface envelope glycoprotein can also bind to GalCer[44,45] and sphingomyelin,[46] and, logically, we looked for a sequence homology between PrP and HIV-1 gp120. Amino acid sequence alignments of both proteins did not reveal any homology. Despite these negative results, we decided to go ahead and to look for a common structural fold, using the combinatorial extension (CE) algorithm.[47] This computational method builds an alignment of two peptidic chains, independently of amino acid side chains. The only requirement is that the 3D structure of the proteins of interest has been characterized. Fortunately, both the V3 loop domain of HIV-1 gp120 (which mediates binding to GalCer) and the extracellular domain of α-helical PrP were available in the Protein Data Bank (PDB). The structural alignment of these two structures with the CE algorithm revealed a common hairpin loop, which, in the case of PrP, corresponded to a disulfide-linked motif encompassing amino acid residues 179–214.[48] The results of this study are shown in Fig. 9.6.

Moreover, a similar V3-like domain, now referred to as a universal sphingolipid-binding domain (SBD), was detected in Alzheimer's β-amyloid peptide[48] and later in several other cellular,[49–54] microbial,[35,55,56] and amyloid proteins.[57–62] The ability of all these SBDs to interact with sphingolipids was demonstrated by various techniques, including studies with sphingolipid monolayers at the air–water interface.[63,64] This method, derived from Langmuir's works on the biophysics of lipid monolayers,[65] has numerous advantages for studying lipid–protein interactions. Because a lipid monolayer corresponds to a half-membrane, it is possible to mimic either the inner (cytoplasmic) or the outer (exofacial) leaflets of the plasma membrane with a controlled molar ratio of lipids in mixed monolayers containing 2, 3, or more lipid species.[62,66] This point is particularly important, because in bilayer systems such as liposomes, the lipid distribution in each half-membrane leaflet is not known. Moreover, as discussed in Chapter 6, the Langmuir system works finely with biologically compatible concentrations of proteins.[48,67]

A synthetic peptide derived from the SBD of the PrP protein ($PrP_{185-204}$) has been synthesized and its interaction with GalCer and other sphingolipids was studied with the monolayer assay.[48] First of all, it was critical to determine that the synthetic peptide had no tensioactive property by itself, which would otherwise interfere with the interpretation of the data. This property is routinely checked by measuring the superficial tension of pure water (γ_o = 72.8 mN m^{-1}) in presence of the peptide. If the peptide is tensioactive, it will readily go to the air–water interface, inducing an immediate decrease of γ_o. This was not the case with the synthetic PrP SBD, indicating that this peptide is not tensioactive. A convenient way to express the results of a Langmuir assay is to use the surface pressure π instead of the surface tension. This approach provides a simple mathematical treatment of the data, because the surface

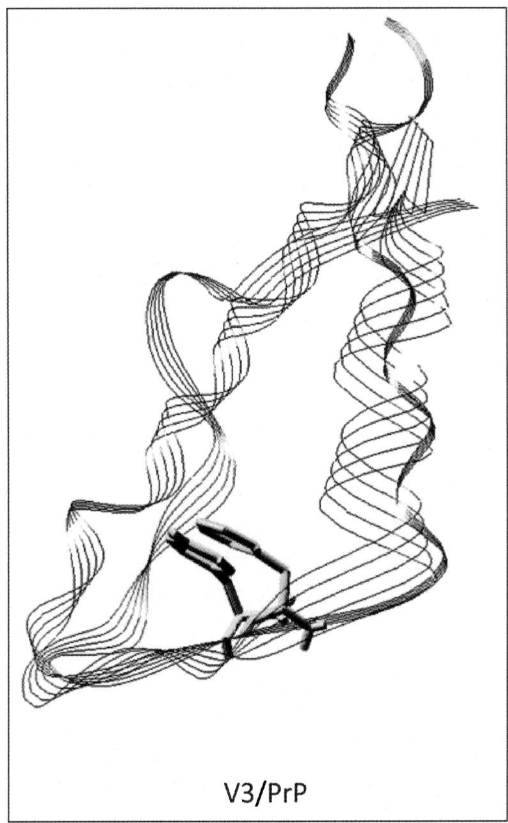

FIGURE 9.6 **The structural alignment between the V3 loop of HIV-1 gp120 and PrP.** The central aromatic residues of PrP (Phe-198 in green) and of the HIV-1 gp120 V3 loop (Tyr-21 in yellow) are indicated. *(From Mahfoud et al.[48] Used with permission.)*

pressure and the surface tension are linked by the relationship $\pi = \gamma_0 - \gamma$, where γ_0 is the surface tension of water and γ the surface tension measured in presence of a tensioactive compound. Thus, any increase of the surface pressure π of a lipid monolayer induced by the presence of a peptide in the subphase would mean that this peptide physically interacts with the lipids of this monolayer. When injected underneath a GalCer monolayer, the peptide elicited a dose-dependent increase of the surface pressure (Fig. 9.7A). The half-maximal effect was reached for a concentration of peptide of 50 nM.[48] Increasing the value of the initial surface pressure π_0 (i.e., deposing more and more GalCer molecules at the interface) before adding the SBD peptide in the subphase gradually decreased the level of interaction of the peptide with the GalCer monolayer (Fig. 9.7B). This phenomenon explains why the $\Delta\pi = f(\pi)$ curve has a negative slope. The critical pressure of insertion (π_c), which corresponds to the extrapolated value of π_0 at which no interaction can occur, because density of packed lipids is too high, was 45 mN m^{-1}.[48] This value is significantly higher than the 30 mN m^{-1} usually considered as the mean surface pressure of a cell membrane.[68] Thus, the interaction of the SBD of PrP with GalCer is likely to occur in densely packed raft-like membrane domains of live neural cells.

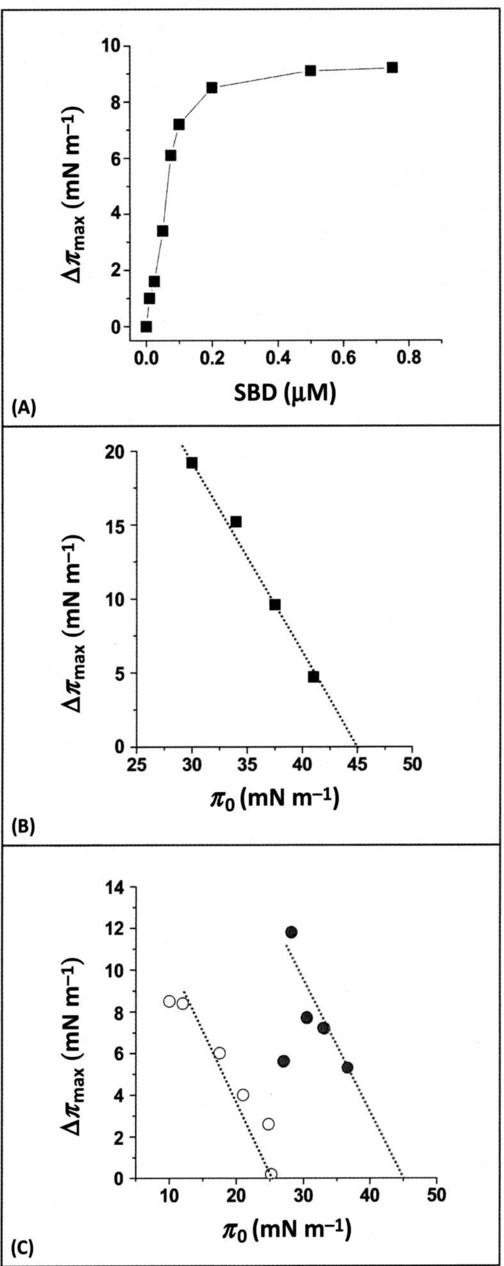

FIGURE 9.7 **Interaction of a synthetic peptide derived from the SBD of PrP with GalCer and sphingomyelin monolayers.** (A) Dose-dependent insertion of a synthetic peptide derived from the SBD of PrP (185-KQHTVTTTTKGENFTETDVKMMER-208) with GalCer monolayers. (B) Specificity of interaction of the synthetic SBD of PrP with GalCer monolayers prepared at increasing values of the initial surface pressure π_0. The intercept between the slope and the x-axis corresponds the critical pressure of insertion π_c (45 mN m^{-1}). (C) Interaction of the SBD of PrP with sphingomyelin monolayers. At initial pressures π_0 ranging from 10 to 27.5 mN m^{-1}, the maximal surface pressure increase induced by the synthetic SBD gradually decreased as the π_0 increased. A first critical pressure of insertion π_c could be determined (x-axis intercept, black dotted slope). At higher values of π_0, a second type of interaction occurred (red symbols) with a π_c of 47.5 mN m^{-1} (red dotted slope). *(From Mahfoud et al.[48] Used with permission.)*

The pattern of interaction of the SBD peptide of PrP with sphingomyelin was more complex than for GalCer. It appeared that the peptide interacted with sphingomyelin at both low and high surface pressures (Fig. 9.7C), with two distinct slopes allowing the determination of two π_c values of 25 and 40 mN m^{-1}, respectively.[48] Only the highest of these π_c values is compatible with a physiological interaction, the lowest one corresponding to a loose interaction with poorly packed assemblies of sphingomyelin that probably do not exist *in vivo*. Overall, these data indicate that the 185–204 fragment of PrP is a functional SBD able to recognize both GalCer and sphingomyelin in lipid microdomains of neural cells. A molecular dynamics simulation study of the interaction between the SBD of PrP and such raft-like membranes containing equimolar amounts of GalCer, sphingomyelin and cholesterol is shown in Fig. 9.8.

It is interesting to notice how these three types of lipids cooperate to provide a functional binding site on the surface of a lipid raft. Cholesterol, although dipped in the membrane and out of reach of PrP, plays a critical role in this process. Indeed, cholesterol constrains the

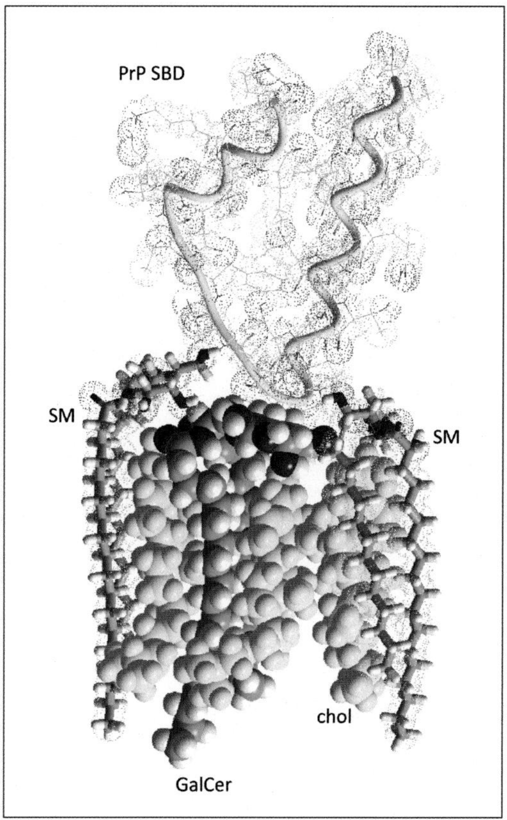

FIGURE 9.8 **Interaction of the SBD of PrP with raft-like membranes.** Three cholesterol molecules (yellow in sphere rendering) were merged with two sphingomyelin (in tube rendering) and one GalCer (in sphere rendering). This raft-like surface was then merged with the SBD of PrP (peptidic chain in yellow, dotted atoms). After geometry optimization with the Polak-Ribiere conjugate gradient method, molecular modeling simulations were performed with the CHARMM force field of HyperChem 8.

conformation of both sphingomyelin and GalCer so that the sphingolipids adapt the shape of their head groups for an optimal interaction with PrP. This powerful effect of cholesterol can be compared to a radio tuner that adjusts the frequency to obtain an optimal reception without parasites. For this reason, we have called this effect of cholesterol on sphingolipids a fine conformational "tuning."[59,66] In addition to GalCer and sphingomyelin, the SBD of PrP has a good fit for chalice-shaped dimers of gangliosides,[62] including GM1 and GM3 (Fig. 9.9). Each of these complexes has its specific topology. In particular, the SBD inserts more deeply in GM3 than in GM1 dimers, and this affects the global shape of the protein. These in silico data suggest that PrP can functionally interact with several brain sphingolipids, which may play a role in its conformational plasticity.

In complement with this SBD, a high-affinity binding site for GM1 has been recently characterized in the C-terminal domain of PrP on the basis of nuclear magnetic resonance studies.[69] This GM1-binding domain consists of a part of the third helix (chiefly residues 218–226 of helix C) and the loop between the second helix (helix B) and the short β-strand S2 (i.e., residues 164–173). The key residues involved in the physical interaction between the glycone part of GM1 and this specific region of PrP include Val-166, Asp-167, Tyr-218, Gln-219, Gln-223, Tyr-225, and Tyr-226. A series of molecular dynamics simulations of this PrP-GM1 interaction found that most of these amino acid residues were also involved in our model (Fig. 9.10).

The energy of interaction between the glycone part of GM1 and the C-terminal domain of PrP was estimated to -104 kJ mol^{-1}. In line with the study of Sanghera et al.,[69] we could determine that Tyr-225, Gln-172, Ser-222, and Tyr-226 contributed to 60% of this total energy of interaction. Overall these data suggest that PrP may contain two distinct sites involved in the recognition of sphingolipids: the SBD and the GM1-binding domain. An examination of the

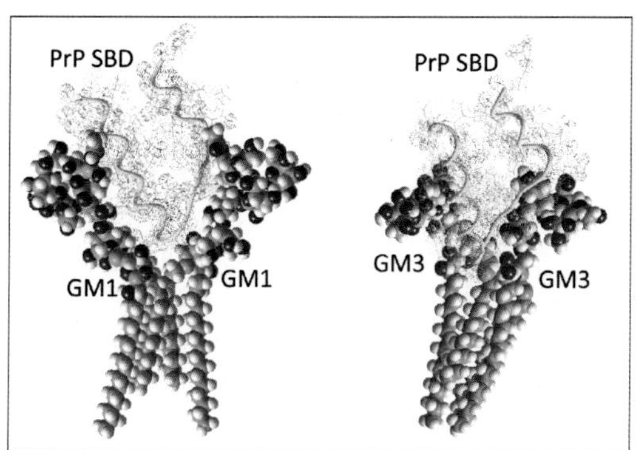

FIGURE 9.9 **Interaction of the SBD of PrP gangliosides GM1 and GM3.** The interaction of the SBD with each ganglioside dimer has a specific conformational effect. The insertion is significantly deeper with GM3 than with GM1. This could play a role in the conformational plasticity of PrP bound to various brain cell types.

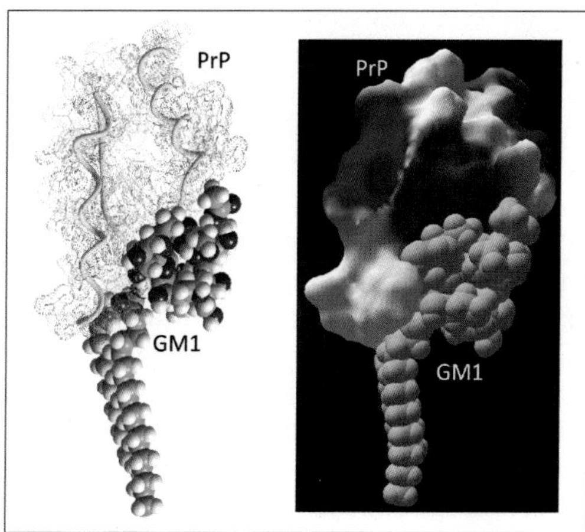

FIGURE 9.10 **Interaction of the GM1-binding domain of PrP with ganglioside GM1.** The extracellular domain of PrP (retrieved from PDB entry #1QLX) is represented with dotted atoms (left panel) or in surface electropotential colors (negative in red, positive in blue, apolar in white). Note the good complementarity of shape between GM1 and the C-terminal domain of PrP.

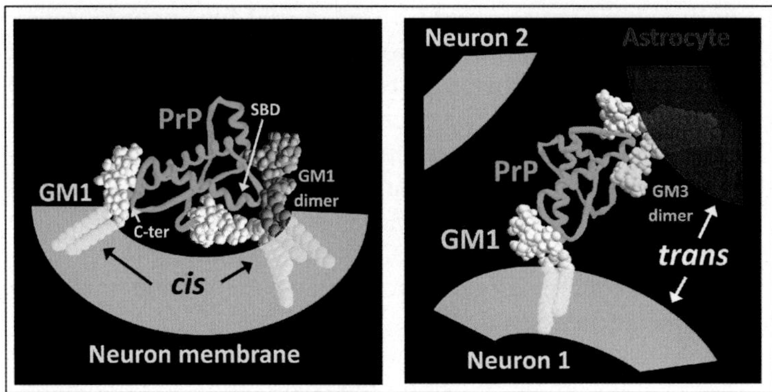

FIGURE 9.11 **Two possible topologies (*cis* or *trans*) of a sphingolipid-PrP complex.** An interaction in *cis* involving a dimer of GM1 bound to the SBD and a GM1 bound to the C-terminal domain of PrP is schematized in the left panel. In this case, all three GM1 molecules belong to the same cell (e.g., a neuron). In contrast, an interaction in *trans* involves two distinct cells in the right panel (e.g., a neuron for GM1 and an astrocyte for the GM3 dimer). This would be the case for a tripartite synapse involving two neurons and a glial cell (astrocyte).

topology of GPI-anchored PrP in its membrane environment shows that these sites have two distant locations in the 3D structure of the protein. Thus, both sites can simultaneously interact with target sphingolipids, either in *cis* (all sphingolipids in the same plasma membrane) or in *trans* (sphingolipids belonging to two distinct cell membranes). These two possible topologies of sphingolipid-bound PrP are shown in Fig. 9.11.

These two tolopogies do not have the same biological significance. If all sphingolipids bound to PrP are in the same membrane, the result would probably be an efficient locking of the α-helical structure of PrP. For instance, if we consider a single PrP protein bound to three GM1 gangliosides (one to the GM1-binding domain and a chalice-shaped dimer bound to the SBD) as shown in Fig. 9.11, the total energy of interaction reaches -242 kJ mol^{-1}. Thus, it would require more than 242 kJ mol^{-1} to displace the protein from these gangliosides. Because most amino acid residues involved in the interaction with these three GM1 molecules actually belong to α-helix structures, it is likely that the role of GM1 in this case is to stabilize the α-helical PrP in its physiological, nonpathological form. We have compared this effect of sphingolipids to a thermodynamical wedge.[35,36,58] Now, PrP can also bind to GM1 through its GM1-binding domain and to GM3 through its SBD. GM1 abundantly expressed in neurons and GM3 in astrocytes raises the interesting possibility that PrP could be a major actor of the tripartite synapse, bringing together neuron and glial membranes through the simultaneous recognition of neuron and astrocyte gangliosides. In this case, the total energy of interaction of a complex between PrP, GM1 (neuron[70]), and a GM3 dimer (astrocytes[71]) can be estimated to -292 kJ mol^{-1}. This high value illustrates how functional this type of glycosynapse can be. Moreover, this mechanism would attribute a precise role for PrP, whose physiological function remains elusive.[72] Interestingly, among the various postulated biological activities of PrP, a participation in cell–cell junctions has been recently considered.[73] Finally, the mechanistic role of PrP in the tripartite synapse would give a logical explanation of the necessity to maintain distinct gangliosides composition in neighboring brain cells belonging to distinct cell types.

The last issue that we would like to discuss now is why, when separated from its sphingolipid wedge, PrP has such a high propensity to adopt a β-strand–rich structure instead of α-helices, and why this structural change is harmful for the brain. This question leads to a discussion of the role of mutations in inherited prion disease and the mechanisms of the other forms of transmission. The third α-helix (helix C) of membrane-bound PrP consists of amino acid residues 200–227. Secondary structure predictions reveal that this segment has a high propensity to form an α-helix. Thus, the α-helical structuration of this part of PrP is spontaneous and does not require any chaperone effect of GM1. Nevertheless, apart from amino acid residues often found in α-helices (Ala, Gln, Glu, or Met), this GM1-binding domain also contains some residues that have a high propensity to belong to a β-strand, especially Tyr, which is the strongest β-forming amino acid.[74] It is interesting that both Sanghera et al.[69] and us identified Tyr-225 and Tyr-226 as critical for GM1 recognition. One could hypothesize that stabilizing these aromatic side chains in the α-helix through a strong interaction with the glycone part of GM1 further secures the nonpathological 3D structure of PrP in the lipid raft environment. Alternatively, Sanghera et al. proposed that this association could in fact induce a slight conformational rearrangement of PrP that could precede its major α → β transition.[69] In our model of the GM1–PrP complex (Fig. 9.10), this conformational change was of little impact on the 3D structure of PrP. Thus, we postulated that the interaction of GM1 with PrP is relevant to the normal function of this protein, a notion also considered by Sanghera et al.[69] Given the capacity of PrP to bind simultaneously to GM1 and to GM3 that can belong to two distinct cells, we propose that PrP plays a role in the recruitment and coalescence of both neuron and glial plasma membranes leading to the establishment of functional tripartite synapses. In absence of PrP (KO mice), other adhesion molecules would probably fulfill the same function in the glycosynapse, explaining why PrP-free animals can live without major health problems.[18]

In contrast with the GM1-binding domain, the SBD of PrP does not have a high propensity to form an α-helix. Indeed, 50% of the 185–204 fragment is predicted to form a β-strand, the other 50% being an unstructured coil. In the 3D structure of recombinant PrP (PDB entry # 1QLX), 11 of these 20 residues (55%) belong to an α-helix whereas none of them belong to a β-strand. Thus, the SBD of PrP is typically an α/β discordant stretch as defined by Kallberg,[74] (i.e., a fragment that is predicted to form a β-strand but is actually found in an α-helix in the 3D structure of the protein). Such α/β discordant segments are believed to play a key role in the α → β transition of amyloid proteins.[74] We have postulated that it is because they are constrained by sphingolipids to form an α-helix that these segments do not follow their natural inclination to fold into β-structures.[58] Correspondingly, it is only when these stabilizing sphingolipids are detached from the protein that the "natural" folding of the protein (i.e., the folding imposed by their amino acid sequence independently of any stabilizing wedge) can spontaneously occur. If so, we can estimate the energy that is required for the α→ β transition: slightly more than the 242 kJ mol^{-1}, which corresponds to the energy of interaction of PrP with three GM1 gangliosides. Providing this amount of energy (which represents the equivalent of ca. 11 hydrogen bonds in water) will allow the dissociation of the PrP–GM1 complex. Following this initial step, the amino acid residues that have a high propensity to form a β-strand will be suddenly surrounded by water molecules that, in contrast with "protecting" GM1 gangliosides, facilitate the natural (genetically programmed) folding of the protein into a β-strand-rich structure. This mechanism is schematically described in Fig. 9.12.

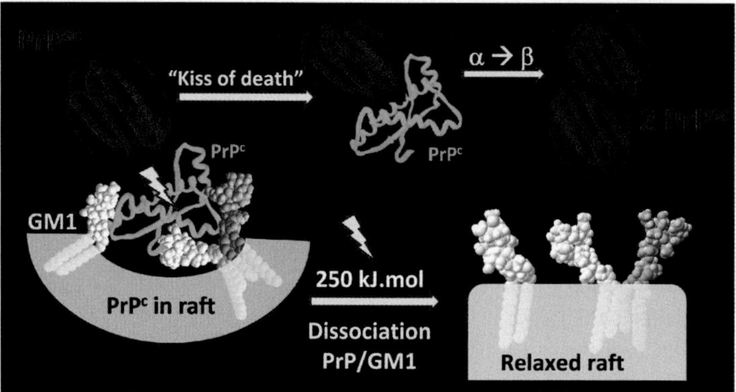

FIGURE 9.12 **How GM1 stabilizes the α-helical structure of PrP in a lipid raft domain.** To displace PrP^c bound to a GM1-enriched lipid raft (e.g., at the surface of a neuron), the infectious PrP^{sc} has to furnish an amount of energy of approximately 250 kJ mol^{-1} (i.e., a little bit more than the 242 kJ mol^{-1} of the interaction between PrP and three GM1 gangliosides). This energy exceeds the whole energy of interaction of PrP^c for three GM1 molecules (two GM1 bound to the SBD of PrP^c, the last GM1 to its C-terminal domain). This binding requires a close contact between PrP^{sc} and PrP^c, metaphorically referred to as the "kiss of death." Without GM1 that stabilizes its α-helical structure, PrP^c will spontaneously adopt a β-rich-structure that is favored by the water environment of the extracellular milieu. This $\alpha \rightarrow \beta$ transition transforms the "good" PrP^c into the "bad" PrP^{sc}. The conversion occurs within the lipid raft domain. During this process, the local curvature induced by the clustering of GM1 molecules is probably reversed to a more relaxed membrane topology. It is likely that several PrP^{sc} proteins remain attached to the raft before they leave the membrane to infect new brain cells.

Overall, this mechanism is consistent with data showing that the $\alpha \rightarrow \beta$ transition of PrP occurs in lipid rafts.[34] It also fits well with the observation that membrane lipids, including GM1, can stabilize the α-helical structure of PrP.[39,69] Although α-helical PrP may not be, by itself, the more stable structure of PrP, this mechanism is remarkably efficient, providing that the interaction with GM1 (or other sphingolipids such as GalCer and sphingomyelin) occurs as early as possible during the biosynthesis process of the PrP protein.[39] Once locked by raft lipids, the protein will remain both harmless and functional at a minimal energy cost, just like a wedge can immobilize a car on a sloping street.[36] Correspondingly, sphingolipid depletion will facilitate the $\alpha \rightarrow \beta$ transition.[40] In the normal life, it is not reasonable to fear a sudden blockade of sphingolipid renewing (although an acute intoxication with the mycotoxin fumonisin B1, a potent inhibitor of sphingolipid biosynthesis,[75] should be avoided). However, the breaking of the GM1–PrP complex can be achieved when a compound that has a higher affinity for PrP that GM1 meets PrP (Fig. 9.12). Among these potential ligands is the pathological form of PrP (i.e., the infectious prion protein PrP^{sc}). When PrP^{sc} binds to PrP^c, the interaction between PrP^c and GM1 is broken, so that the protection initially provided by GM1 molecules is no longer active. PrP^c becomes tightly attached to PrP^{sc} and the battle that follows this "kiss of death"[76] leads inevitably to Prp^{sc}'s victory. As discussed at the beginning of this chapter, prions replicate by force. The "force" consists in dissociating PrP^c from its protecting sphingolipid wedges (Fig. 9.12) and to provide compatible/complementary surfaces for aromatic acid side chains that will induce an oligomerization process of PrP^{sc} (Fig. 9.13). The aromatic side chains that were initially "buffered" by the sugars of protective

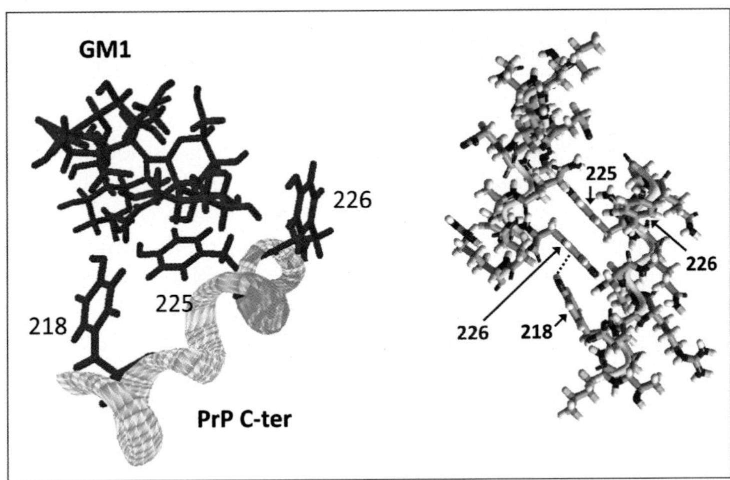

FIGURE 9.13 **PrP^{sc} replication through aromatic–aromatic interactions.** When PrP^{c} is embedded within lipid rafts, its aromatic residues that are close to the membrane (Tyr-218, Tyr-225, and Tyr-226) are neutralized by the sugar head group of glycosphingolipids such as GalCer or GM1 (left panel). These interactions contribute to stabilize the α-helical structure of the C-terminal part of PrP. However, when PrP–GM1 interactions are broken (as in Fig. 9.12), tyrosine residues of two vicinal PrP proteins form a network of stacking π–π and OH–π interactions (right panel). Molecular dynamics simulations indicate that the 215–228 fragment of PrP, which contain three tyrosine residues, spontaneously aggregate through the coordinated recruitment of tyrosine rings, for a total energy of interaction of –56.3 kJ mol^{-1}. Tyrosine by themselves contribute to 80% of this energy. Tyr-226 of one peptide interacts with both Tyr-218 (OH–π) and Tyr-225 (π–π stacking) of its partner. These energetic interactions contribute significantly to change the conformation of the protein into β-structures that are prone to oligomerization and aggregation.

sphingolipids (e.g., by CH–π stacking interactions[77]) become suddenly free to interact with other aromatic rings through stacking π–π interactions.[58] In this new sphingolipid-free environment, interpeptide aromatic–aromatic associations will induce a drastic conformational change from an initial α-helix to a β-strand (Fig. 9.13). This molecular mechanism can explain the various modes of oligomerization of PrP leading to small-sized oligomers as well as to amyloid fibers and plaques.

That sphingolipids can maintain an amyloid protein in an "inoffensive" conformation and thus constitutively inhibit its aggregation has been demonstrated in acellular conditions for the amylin–LacCer couple.[57] Mutations in the PrP gene that are associated with inherited CJD may also induce a loss of affinity of PrP^{c} for raft sphingolipids, as demonstrated for mutation E200K (Glu-200/Lys) that decreases PrP binding to sphingomyelin.[48] Thus, all these data are consistent with the notion that sphingolipids exert a constitutive protective effect on PrP^{c} by stabilizing the α-helical structure of the protein. Once dislodged from these thermodynamical wedges, which requires a high-energy input provided by a pathogenic prion protein (PrP^{sc}), the protein will adopt a distinct β-structure that is particularly prone to oligomerization/aggregation.[78] This process is how prions replicate and infect the whole brain.

9.5 CONCLUSION OF THE INVESTIGATION: WHO IS GUILTY, WHO IS INNOCENT?

From the preceding discussion, one can probably conclude that PrP is a "weak" protein. As a matter of fact, this protein would like to fold into β-strands rather than into α-helices, but sphingolipids (especially GM1) prevent it from doing so. In presence of infectious PrP^{sc}, the PrP^{c} protein eventually gets rid of these lipids that dictate which 3D structure it should adopt. For a brief period of time, the free PrP protein seems to control the situation, folding itself into its preferred β-structure. The drawback is that once converted into a β-structure by its mentor PrP^{sc}, it becomes an anonymous brick in infectious PrP^{sc} oligomers/aggregates. Because these PrP^{sc} assemblies are harmful, the protein is guilty, yet chiefly by weakness. The lipids (GM1 and some other sphingolipids) act as chaperones for the benefit of the organism. They give up the fight when PrP^{sc} enters the scene, because PrP^{sc} is stronger.

9.6 KEY EXPERIMENT: ADENINE IS A MINIMAL AROMATIC COMPOUND THAT SELF-AGGREGATES IN WATER THROUGH π–π STACKING INTERACTIONS[79]

Adenine, one of the four aromatic bases found in DNA and RNA, is poorly soluble in water. Indeed, even low-concentration adenine solutions that look perfectly limpid contain a series of small oligomers.[80] The driving force for the oligomerization of adenine is its aromatic structure, which is rejected by water. In addition to this entropic contribution, the noncovalent interactions between adenine molecules are further stabilized by aromatic π–π stacking, a type of interaction that plays a major role in protein structure.[81] In this respect, adenine oligomers in water behave just as aromatic bases in the double helix of DNA. When the concentration of adenine reaches approximately 80 mM in water, precipitation occurs spontaneously at room temperature. We took advantage of this unique feature to quantify the aggregation of adenine by spectrophotometric turbidity measurements.[79] Interestingly, we showed that galactose induced a dose-dependent inhibition of adenine aggregation. In fact, among a series of common sugars, galactose displayed the most potent antistacking activity, as assessed by the determination of their respective 50% inhibitory concentration (IC50): 100 mM for galactose, compared with 500 mM for glucose, 1,000 mM (1M) for ribose, and 1,200 mM (1.2M) for fructose. Molecular dynamics studies of sugar conformations in water indicated that galactose has a dissymmetric structure with two distinct faces, one apolar with C–H groups and the other one polar with O–H groups (Fig. 9.14).

When galactose meets adenine, the apolar side of the sugar will stack onto the aromatic rings of adenine, as shown in Fig. 9.14. This interaction is stabilized by a coordinated network of CH–π interactions with the acceptor π clouds of adenine.[79] The polar side of galactose is thus available to interact with water molecules through numerous hydrogen bonds. Under these conditions, galactose buffers the aromatic structure of adenine and efficiently prevents its oligomerization. The spontaneous oligomerization/aggregation of adenine in water, and its inhibition by galactose is, at a lower molecular scale, a perfect image of the chaperone effect of cell surface glycosphingolipids on the α-helical structure of PrP. Here, galactose plays the role of the glycolipid, and adenine represents the amyloidogenic protein whose healthy structure

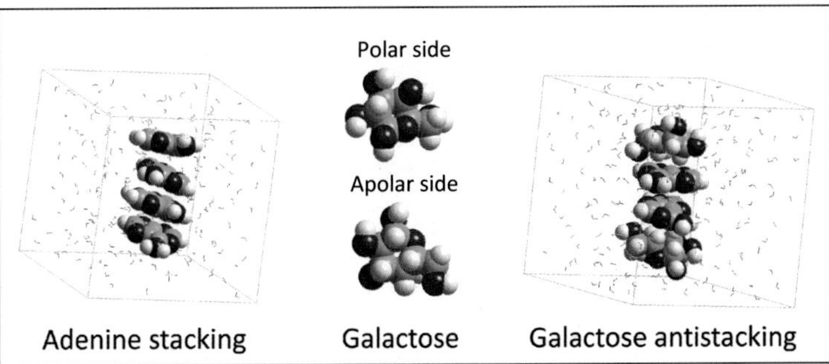

FIGURE 9.14 **Mechanisms of adenine aggregation in water and its inhibition by galactose.** Even at low concentrations, adenine does not exist as free monomers in water. For instance, vertical stacking including 2–6 adenine molecules (in sphere rendering) spontaneously occur in water (red sticks). These oligomers are stabilized by π–π interactions between aromatic rings (left panel). Galactose has an apolar side with the C–H groups and on the opposite, a polar side with the O–H groups (center panel). When galactose and adenine are mixed in water, the apolar side of galactose stacks onto the aromatic structure of adenine. The galactose–adenine complex is stabilized by CH–π stacking interactions. The other side of galactose, which bears the polar O–H groups, interacts with water through several hydrogen bonds. This limits the oligomerization process of adenine.

is locked by the glycolipid wedge. It is fascinating that a minimal experimental system only based on water, adenine, and galactose can mimic such a complex biological phenomenon associated with major illness. It is perhaps why galactose, in cell surface glycolipids, is often recognized by the aromatic residues of infectious proteins, such as HIV-1 gp120[35,45] or α-synuclein.[60]

References

1. Boyer M, Yutin N, Pagnier I, et al. Giant Marseillevirus highlights the role of amoebae as a melting pot in emergence of chimeric microorganisms. *Proc Natl Acad Sci USA*. 2009;106(51):21848–21853.
2. La Scola B, Audic S, Robert C, et al. A giant virus in amoebae. *Science*. 2003;299(5615):2033.
3. La Scola B, Desnues C, Pagnier I, et al. The virophage as a unique parasite of the giant mimivirus. *Nature*. 2008;455(7209):100–104.
4. Slimani M, Pagnier I, Raoult D, La Scola B. Amoebae as battlefields for bacteria, giant viruses, and virophages. *J Virol*. 2013;87(8):4783–4785.
5. Gibbons RA, Hunter GD. Nature of the scrapie agent. *Nature*. 1967;215(5105):1041–1043.
6. Alper T, Cramp WA, Haig DA, Clarke MC. Does the agent of scrapie replicate without nucleic acid? *Nature*. 1967;214(5090):764–766.
7. Crick FHC. The biochemistry of genetics. *Proc 6th Int Congr Biochem*. 1964;109–128.
8. Prusiner SB, Hadlow WJ, Eklund CM, Race RE. Sedimentation properties of the scrapie agent. *Proc Natl Acad Sci USA*. 1977;74(10):4656–4660.
9. Prusiner SB, McKinley MP, Groth DF, et al. Scrapie agent contains a hydrophobic protein. *Proc Natl Acad Sci USA*. 1981;78(11):6675–6679.
10. Prusiner SB. Novel proteinaceous infectious particles cause scrapie. *Science*. 1982;216(4542):136–144.
11. Bolton DC, McKinley MP, Prusiner SB. Identification of a protein that purifies with the scrapie prion. *Science*. 1982;218(4579):1309–1311.
12. McKinley MP, Bolton DC, Prusiner SB. A protease-resistant protein is a structural component of the scrapie prion. *Cell*. 1983;35(1):57–62.
13. Prusiner SB. Prions. *Proc Natl Acad Sci USA*. 1998;95(23):13363–13383.
14. Mayor S, Riezman H. Sorting GPI-anchored proteins. *Nat Rev Mol Cell Biol*. 2004;5(2):110–120.

15. Fantini J, Garmy N, Mahfoud R, Yahi N. Lipid rafts: structure, function and role in HIV, Alzheimer's and prion diseases. *Exp Rev Mol Med*. 2002;4(27):1–22.
16. Zahn R, Liu A, Luhrs T, et al. NMR solution structure of the human prion protein. *Proc Natl Acad Sci USA*. 2000;97(1):145–150.
17. Westergard L, Christensen HM, Harris DA. The cellular prion protein (PrP(C)): its physiological function and role in disease. *Biochim Biophys Acta*. 2007;1772(6):629–644.
18. Weissmann C, Flechsig E. PrP knockout and PrP transgenic mice in prion research. *Brit Med Bull*. 2003;66:43–60.
19. Goldfarb LG, Haltia M, Brown P, et al. New mutation in scrapie amyloid precursor gene (at codon 178) in Finnish Creutzfeldt–Jakob kindred. *Lancet*. 1991;337(8738):425.
20. Lee S, Antony L, Hartmann R, et al. Conformational diversity in prion protein variants influences intermolecular beta-sheet formation. *EMBO J*. 2010;29(1):251–262.
21. Medori R, Montagna P, Tritschler HJ, et al. Fatal familial insomnia: a second kindred with mutation of prion protein gene at codon 178. *Neurology*. 1992;42(3 Pt 1):669–670.
22. Kitamoto T, Ohta M, Doh-ura K, Hitoshi S, Terao Y, Tateishi J. Novel missense variants of prion protein in Creutzfeldt–Jakob disease or Gerstmann-Straüssler syndrome. *BiochemBiophys Res Comm*. 1993;191(2):709–714.
23. Peoc'h K, Manivet P, Beaudry P, et al. Identification of three novel mutations (E196K, V203I, E211Q) in the prion protein gene (PRNP) in inherited prion diseases with Creutzfeldt–Jakob disease phenotype. *Hum Mutat*. 2000;15(5):482.
24. Zhang Y, Swietnicki W, Zagorski MG, Surewicz WK, Sonnichsen FD. Solution structure of the E200K variant of human prion protein. Implications for the mechanism of pathogenesis in familial prion diseases. *J Biol Chem*. 2000;275(43):33650–33654.
25. Piccardo P, Dlouhy SR, Lievens PM, et al. Phenotypic variability of Gerstmann-Straüssler-Scheinker disease is associated with prion protein heterogeneity. *J Neuropathol Exp Neurol*. 1998;57(10):979–988.
26. Mastrianni JA, Iannicola C, Myers RM, DeArmond S, Prusiner SB. Mutation of the prion protein gene at codon 208 in familial Creutzfeldt–Jakob disease. *Neurology*. 1996;47(5):1305–1312.
27. Nitrini R, Rosemberg S, Passos-Bueno MR, et al. Familial spongiform encephalopathy associated with a novel prion protein gene mutation. *Ann Neurol*. 1997;42(2):138–146.
28. Pocchiari M, Salvatore M, Cutruzzola F, et al. A new point mutation of the prion protein gene in Creutzfeldt–Jakob disease. *Ann Neurol*. 1993;34(6):802–807.
29. Kaneko K, Zulianello L, Scott M, et al. Evidence for protein X binding to a discontinuous epitope on the cellular prion protein during scrapie prion propagation. *Proc Natl Acad Sci USA*. 1997;94(19):10069–10074.
30. Xiong LW, Raymond LD, Hayes SF, Raymond GJ, Caughey B. Conformational change, aggregation and fibril formation induced by detergent treatments of cellular prion protein. *J Neurochem*. 2001;79(3):669–678.
31. Pan KM, Baldwin M, Nguyen J, et al. Conversion of alpha-helices into beta-sheets features in the formation of the scrapie prion proteins. *Proc Natl Acad Sci USA*. 1993;90(23):10962–10966.
32. Griffith JS. Self-replication and scrapie. *Nature*. 1967;215(5105):1043–1044.
33. Prusiner SB, Groth D, Serban A, et al. Ablation of the prion protein (PrP) gene in mice prevents scrapie and facilitates production of anti-PrP antibodies. *Proc Natl Acad Sci USA*. 1993;90(22):10608–10612.
34. Baron GS, Wehrly K, Dorward DW, Chesebro B, Caughey B. Conversion of raft associated prion protein to the protease-resistant state requires insertion of PrP-res (PrP(Sc)) into contiguous membranes. *EMBO J*. 2002;21(5):1031–1040.
35. Fantini J. How sphingolipids bind and shape proteins: molecular basis of lipid-protein interactions in lipid shells, rafts and related biomembrane domains. *Cell Mol Life Sci*. 2003;60(6):1027–1032.
36. Fantini J. Interaction of proteins with lipid rafts through glycolipid-binding domains: biochemical background and potential therapeutic applications. *Curr Med Chem*. 2007;14(27):2911–2917.
37. Castilla J, Goni FM. Lipids, a missing link in prion propagation. *Chem Biol*. 2011;18(11):1345–1346.
38. Klein TR, Kirsch D, Kaufmann R, Riesner D. Prion rods contain small amounts of two host sphingolipids as revealed by thin-layer chromatography and mass spectrometry. *Biol Chem*. 1998;379(6):655–666.
39. Sarnataro D, Campana V, Paladino S, Stornaiuolo M, Nitsch L, Zurzolo C. PrP(C) association with lipid rafts in the early secretory pathway stabilizes its cellular conformation. *Mol Biol Cell*. 2004;15(9):4031–4042.
40. Naslavsky N, Shmeeda H, Friedlander G, et al. Sphingolipid depletion increases formation of the scrapie prion protein in neuroblastoma cells infected with prions. *J Biol Chem*. 1999;274(30):20763–20771.
41. Sanghera N, Pinheiro TJ. Binding of prion protein to lipid membranes and implications for prion conversion. *J Mol Biol*. 2002;315(5):1241–1256.

42. Re F, Sesana S, Barbiroli A, et al. Prion protein structure is affected by pH-dependent interaction with membranes: a study in a model system. *FEBS Lett*. 2008;582(2):215–220.
43. Schmalzbauer R, Eigenbrod S, Winoto-Morbach S, et al. Evidence for an association of prion protein and sphingolipid-mediated signaling. *J Neurochem*. 2008;106(3):1459–1470.
44. Harouse JM, Bhat S, Spitalnik SL, et al. Inhibition of entry of HIV-1 in neural cell lines by antibodies against galactosyl ceramide. *Science*. 1991;253(5017):320–323.
45. Yahi N, Baghdiguian S, Moreau H, Fantini J. Galactosyl ceramide (or a closely related molecule) is the receptor for human immunodeficiency virus type 1 on human colon epithelial HT29 cells. *J Virol*. 1992;66(8): 4848–4854.
46. Van Mau N, Misse D, Le Grimellec C, Divita G, Heitz F, Veas F. The SU glycoprotein 120 from HIV-1 penetrates into lipid monolayers mimicking plasma membranes. *J Membr Biol*. 2000;177(3):251–257.
47. Shindyalov IN, Bourne PE. Protein structure alignment by incremental combinatorial extension (CE) of the optimal path. *Protein Eng*. 1998;11(9):739–747.
48. Mahfoud R, Garmy N, Maresca M, Yahi N, Puigserver A, Fantini J. Identification of a common sphingolipid-binding domain in Alzheimer, prion, and HIV-1 proteins. *J Biol Chem*. 2002;277(13):11292–11296.
49. Aubert-Jousset E, Garmy N, Sbarra V, Fantini J, Sadoulet MO, Lombardo D. The combinatorial extension method reveals a sphingolipid binding domain on pancreatic bile salt-dependent lipase: role in secretion. *Structure*. 2004;12(8):1437–1447.
50. Chakrabandhu K, Huault S, Garmy N, et al. The extracellular glycosphingolipid-binding motif of Fas defines its internalization route, mode and outcome of signals upon activation by ligand. *Cell Death and Diff*. 2008;15(12):1824–1837.
51. Taieb N, Maresca M, Guo XJ, Garmy N, Fantini J, Yahi N. The first extracellular domain of the tumour stem cell marker CD133 contains an antigenic ganglioside-binding motif. *Cancer Lett*. 2009;278(2):164–173.
52. Hamel S, Fantini J, Schweisguth F. Notch ligand activity is modulated by glycosphingolipid membrane composition in Drosophila melanogaster. *J Cell Biol*. 2010;188(4):581–594.
53. Chattopadhyay A, Paila YD, Shrivastava S, Tiwari S, Singh P, Fantini J. Sphingolipid-binding domain in the serotonin(1A) receptor. *Adv Exp Med Biol*. 2012;749:279–293.
54. Heuss SF, Tarantino N, Fantini J, et al. A glycosphingolipid binding domain controls trafficking and activity of the mammalian notch ligand delta-like 1. *PloS One*. 2013;8(9):e74392.
55. Taieb N, Yahi N, Fantini J. Rafts and related glycosphingolipid-enriched microdomains in the intestinal epithelium: bacterial targets linked to nutrient absorption. *Adv Drug Deliv Rev*. 2004;56(6):779–794.
56. Fantini J, Garmy N, Yahi N. Prediction of glycolipid-binding domains from the amino acid sequence of lipid raft-associated proteins: application to HpaA, a protein involved in the adhesion of *Helicobacter pylori* to gastrointestinal cells. *Biochemistry*. 2006;45(36):10957–10962.
57. Levy M, Garmy N, Gazit E, Fantini J. The minimal amyloid-forming fragment of the islet amyloid polypeptide is a glycolipid-binding domain. *FEBS J*. 2006;273(24):5724–5735.
58. Fantini J, Yahi N. Molecular insights into amyloid regulation by membrane cholesterol and sphingolipids: common mechanisms in neurodegenerative diseases. *Exp Rev Mol Med*. 2010;12:e27.
59. Yahi N, Aulas A, Fantini J. How cholesterol constrains glycolipid conformation for optimal recognition of Alzheimer's beta amyloid peptide (Abeta1-40). *PloS One*. 2010;5(2):e9079.
60. Fantini J, Yahi N. Molecular basis for the glycosphingolipid-binding specificity of alpha-synuclein: key role of tyrosine 39 in membrane insertion. *J Mol Biol*. 2011;408(4):654–669.
61. Fantini J, Yahi N. The driving force of alpha-synuclein insertion and amyloid channel formation in the plasma membrane of neural cells: key role of ganglioside- and cholesterol-binding domains. *Adv Exp Med Biol*. 2013;991:15–26.
62. Yahi N, Fantini J. Deciphering the glycolipid code of Alzheimer's and Parkinson's amyloid proteins allowed the creation of a universal ganglioside-binding peptide. *PloS One*. 2014;9(8):e104751.
63. Di Scala C, Chahinian H, Yahi N, Garmy N, Fantini J. Interaction of Alzheimer's beta-amyloid peptides with cholesterol: mechanistic insights into amyloid pore formation. *Biochemistry*. 2014;53(28):4489–4502.
64. Hammache D, Pieroni G, Maresca M, Ivaldi S, Yahi N, Fantini J. Reconstitution of sphingolipid-cholesterol plasma membrane microdomains for studies of virus-glycolipid interactions. *Methods Enzymol*. 2000;312: 495–506.
65. Langmuir I. The constitution and fundamental properties of solids and liquids. II. Liquids. *J Am Chem Soc*. 1917;39:1848–1906.

66. Fantini J, Yahi N, Garmy N. Cholesterol accelerates the binding of Alzheimer's beta-amyloid peptide to ganglioside GM1 through a universal hydrogen-bond-dependent sterol tuning of glycolipid conformation. *Front Physiol*. 2013;4:120.
67. Lahdo R, Coillet-Matillon S, Chauvet JP, de La Fourniere-Bessueille L. The amyloid precursor protein interacts with neutral lipids. *Eur J Biochem*. 2002;269(8):2238–2246.
68. Seelig A. Local anesthetics and pressure: a comparison of dibucaine binding to lipid monolayers and bilayers. *Biochim Biophys Acta*. 1987;899(2):196–204.
69. Sanghera N, Correia BE, Correia JR, et al. Deciphering the molecular details for the binding of the prion protein to main ganglioside GM1 of neuronal membranes. *Chem Biol*. 2011;18(11):1422–1431.
70. Hansson HA, Holmgren J, Svennerholm L. Ultrastructural localization of cell membrane GM1 ganglioside by cholera toxin. *Proc Natl Acad Sci USA*. 1977;74(9):3782–3786.
71. Asou H, Hirano S, Uyemura K. Ganglioside composition of astrocytes. *Cell Struct Func*. 1989;14(5):561–568.
72. Nieznanski K. Interactions of prion protein with intracellular proteins: so many partners and no consequences? *Cell Mol Neurobiol*. 2010;30(5):653–666.
73. Petit CS, Besnier L, Morel E, Rousset M, Thenet S. Roles of the cellular prion protein in the regulation of cell-cell junctions and barrier function. *Tissue Barriers*. 2013;1(2):e24377.
74. Kallberg Y, Gustafsson M, Persson B, Thyberg J, Johansson J. Prediction of amyloid fibril-forming proteins. *J Biol Chem*. 2001;276(16):12945–12950.
75. Vesper H, Schmelz EM, Nikolova-Karakashian MN, Dillehay DL, Lynch DV, Merrill Jr AH. Sphingolipids in food and the emerging importance of sphingolipids to nutrition. *J Nutr*. 1999;129(7):1239–1250.
76. Caughey B. Interactions between prion protein isoforms: the kiss of death? *Trends Biochem Sci*. 2001;26(4): 235–242.
77. Nishio M, Umezawa Y, Fantini J, Weiss MS, Chakrabarti P. CH–π hydrogen bonds in biological macromolecules. *Phys Chem Chem Phys*. 2014;16(25):12648–12683.
78. Stohr J, Weinmann N, Wille H, et al. Mechanisms of prion protein assembly into amyloid. *Proc Natl Acad Sci USA*. 2008;105(7):2409–2414.
79. Maresca M, Derghal A, Caravagna C, Dudin S, Fantini J. Controlled aggregation of adenine by sugars: physicochemical studies, molecular modelling simulations of sugar-aromatic CH-pi stacking interactions, and biological significance. *Phys Chem Chem Phys*. 2008;10(19):2792–2800.
80. Martel P. Base crystallization and base stacking in water. *Eur J Biochem*. 1979;96(2):213–219.
81. McGaughey GB, Gagne M, Rappe AK. pi-stacking interactions. Alive and well in proteins. *J Biol Chem*. 1998;273(25):15458–15463.

CHAPTER

10

Parkinson's Disease

Jacques Fantini, Nouara Yahi

OUTLINE

10.1 PARKINSON'S DISEASE AND SYNUCLEOPATHIES

Parkinson's disease is a progressive neurodegenerative brain disorder that affects ~ 1% of people beyond 65 years of age, with a higher prevalence in men.[1] Indeed, the relative risk of developing Parkinson's disease is 1.5–2 times greater in men than women.[2,3] Clinically, Parkinson's disease is characterized by severe motor symptoms including uncontrollable resting tremor, postular imbalance, muscular rigidity, and bradykinesia.[4] Pathologically, Parkinson's disease is unique among other neurodegenerative disorders because it is characterized by the specific degeneration of dopaminergic neurons in the substantia nigra. This degeneration leads to dramatic

Brain Lipids in Synaptic Function and Neurological Disease. http://dx.doi.org/10.1016/B978-0-12-800111-0.00010-2

dopamine depletion in the striatum, to which these neurons project, with consequent disruption of the neuronal systems controlling motor functions.[5] Neurodegeneration in this case is associated with the presence of rounded eosinophilic inclusions, called Lewy bodies, in the cytoplasm of certain neuronal populations, especially in the substantia nigra.[5] However, Lewy bodies are in no way a specific feature of Parkinson's disease. In fact, these inclusions can be found in the normal aging brain and in the brains of patients with Alzheimer's disease and other neurodegenerative diseases such as dementia with Lewy bodies (DLB).[6] The mechanism of formation of Lewy bodies (or Lewy neurites when associated with abnormal neuritic profiles) and their pathogenic relevance is poorly understood. In 1997, a major breakthrough came with the identification of α-synuclein as the main biochemical component of Lewy bodies from patients with Parkinson's disease and DLB.[7] The same year, a mutation in the gene coding for α-synuclein was found to be associated with familial Parkinson's disease.[8] Since then, the involvement of α-synuclein in the pathogenesis of Parkinson's disease has been firmly established.[9] Incidentally, α-synuclein has been incriminated in several neurodegenerative diseases that are collectively referred to as "synucleopathies."[1,10] Apart from Parkinson's disease[8] and DLB,[11] α-synucleopathies also include Down,[12] Shy-Drager,[13] and Hallervorden-Spatz[14] syndromes. Lewy bodies are also detected in 30–40% of Alzheimer's disease cases,[15] so that these inclusions may be in fact a common hallmark of neurodegenerative disorders and that α-synuclein might be involved in the pathogenesis of most of these diseases.

10.2 α-SYNUCLEIN

α-synuclein is a 140-amino acid protein, which is encoded by a single gene (SNCA) located in human chromosome 4.[16] The SNCA gene spans ~112 kb and contains 6 exons.[17] The protein belongs to the synuclein family that also includes β-synuclein and γ-synuclein.[18,19] However, to date, despite a high sequence homology between all the members of the synuclein family, only α-synuclein has been associated with the pathogenesis of neurological disorders. The schematic structure of α-synuclein is shown in Fig. 10.1.

The protein is composed of three main domains. The N-terminal part (amino acids 1–60) contains a series of imperfect amphipathic repeats of 11 amino acids with a conserved motif KT-KEGV. These amphipathic motifs may allow α-helical structuration, especially when the protein interacts with cellular membranes.[9] The N-terminal region also contains a sphingolipid-binding domain (SBD) that controls α-synuclein binding to cell surface glycosphingolipids such as GalCer or GM3.[20] The central part of α-synuclein, encompassing residues 61–95, is a highly apolar region that contains the principal cholesterol-binding domain of the protein.[21] Interestingly, the non-Aβ component (NAC) of amyloid-enriched fractions from Alzheimer's disease patients proved to be a 35-amino acid fragment derived from the 61–95 region of α-synuclein.[18] Thus, α-synuclein is the precursor of the NAC peptide, explaining why the 61–95 region of α-synuclein is referred to as NAC. Biochemical analysis of the NAC peptide revealed that the N-terminal region (NAC1–18, corresponding to 61–78 in α-synuclein) aggregates to form amyloid fibrils, whereas the C-terminal region (NAC19–35, thus 79–95 in α-synuclein) remains soluble.[22] Overall this aggregation indicates that the cholesterol-binding domain, which is located in the N-terminal region of the NAC peptide, could play a prominent role in the formation of amyloid plaques and in the pathogenesis of α-synucleinopathies. Finally, the C-terminal region of α-synuclein is acidic

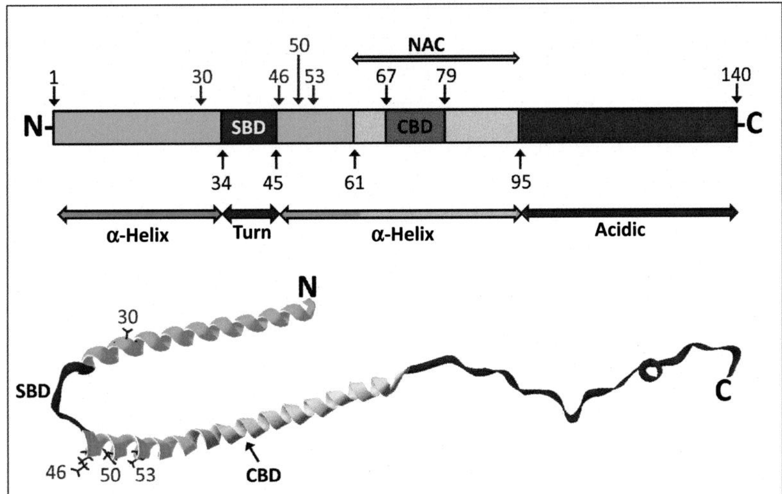

FIGURE 10.1 **Structure of α-synuclein.** Mutations associated with Parkinson's disease (A30P, E46K, H50Q, and A53T) are indicated in red. The N-terminal domain 1–60 containing the SBD (34–45 turn, blue frame) is in green. The NAC domain (61–95, nonamyloid component) is framed in yellow. The main cholesterol-binding domain (CBD 67–79) is in orange. The C-terminal acidic domain 96–140 is framed in red. The helix, turn, and disordered structures of micelle-bound α-synuclein (retrieved from PDB entry # 1XQ8) are shown in the lower panel.

and negatively charged at pH 7.[9] Parkinson's disease is essentially sporadic, often associated to environmental toxicants (pesticides), and, more rarely, familial. Both mutations (Fig. 10.1) and duplication/triplication of the SNCA gene leading to overexpression of wild-type α-synuclein have been associated with familial forms of the disease.[23] The main Parkinson's associated mutations of α-synuclein are A30P,[24] E46K,[25,26] A53T,[8] and the more recently identified H50Q.[27,28]

Despite two decades of intense research efforts, the exact biological function of α-synuclein is still unknown. As discussed by Bendor et al.,[29] α-synuclein has been in fact independently discovered on several occasions. Originally, the protein was identified in purified preparations of cholinergic vesicles of the Torpedo electric organ.[30] Apart from this presynaptic localization, the protein was also detected at the nuclear envelope and was accordingly designated synuclein, (*syn* for synapse, *nuclein* for the nucleus). As indicated earlier, α-synuclein was also identified as the precursor protein for the NAC (α-synuclein61–95) found in senile plaques in Alzheimer's disease.[18] Nevertheless, it is likely that the main function of α-synuclein is related to its presence at presynaptic terminals.[31] Mice lacking α-synuclein (SNCA-knockout mice) are fully viable, yet they exhibit an impairment in replenishment of docked pools from the reserve pool of synaptic vesicles.[32,33] In contrast, transgenic mice overexpressing α-synuclein showed an impairment in synaptic vesicle exocytosis[34] and a reduction in neurotransmitter release.[35] Overall, these studies suggested that the physiological function of α-synuclein is to regulate synaptic vesicle mobilization at nerve terminals.[36] More generally, because triple-knockout mice lacking all synucleins (α, β, and γ) exhibited age-dependent decreased SNARE-complex assembly, it is likely that synucleins may, collectively, function to sustain normal SNARE-complex assembly in a presynaptic terminal during aging.

10.3 INTRACELLULAR α-SYNUCLEIN BINDS TO SYNAPTIC VESICLES AND REGULATES VESICLE TRAFFICKING, DOCKING, AND RECYCLING

As shown in Fig. 10.1, α-synuclein does not contain a transmembrane domain that could account for a specific interaction with synaptic vesicles. Hence, the binding of α-synuclein to the membrane of synaptic vesicles is rather weak, which probably explains why the majority of α-synuclein behaves as a soluble protein.[37,38] Nevertheless, it provides a decisive advantage because in this case the protein can freely and reversibly move from the cytoplasm to the membrane of synaptic vesicles. A schematic description of the role of α-synuclein in neurotransmitter release and synaptic vesicle trafficking is shown in Fig. 10.2.

Two important properties of α-synuclein explain why this protein localizes specifically to presynaptic boutons rather than other cell membranes.[29] First, α-synuclein has a marked preference for membranes with a high curvature.[39,40] As a matter of fact, synaptic vesicles have a small radius that induces particularly high membrane curvature. On the other hand, α-synuclein has a high affinity for acidic glycerophospholipids[41] that are present on the cytoplasmic leaflet of intracellular membranes. Because only synaptic vesicles combine a high curvature and the presence of acidic lipids, then we may admit that α-synuclein can manage the feat of exhibiting a high specificity for synaptic vesicles over other membranes while

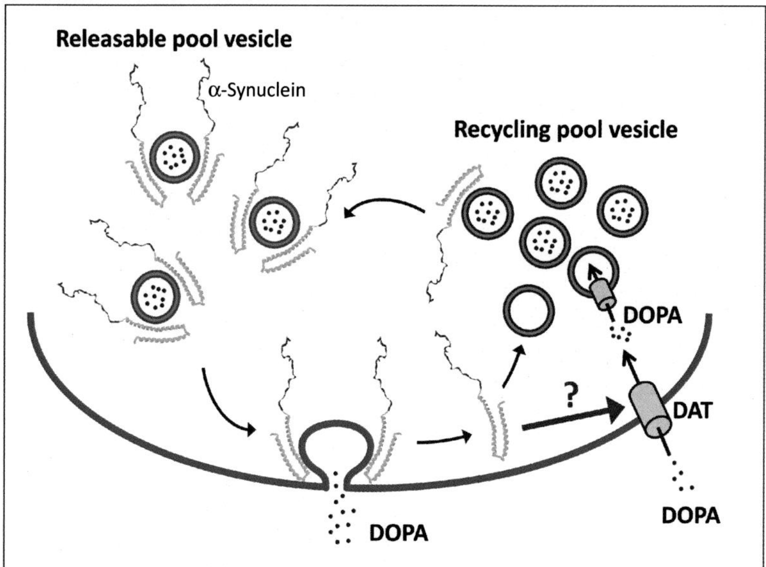

FIGURE 10.2 **Putative role of α-synuclein in synaptic vesicles cycle.** Intracellular α-synuclein binds to synaptic vesicles (either from the releasable or recycling pools) and regulates vesicle docking and fusion. The high curvature of synaptic vesicles is consistent with the curved structure that the protein acquires upon binding to the membrane of synaptic vesicles. Dopamine (DOPA) released from dopaminergic neurons can be recaptured via the dopamine transporter (DAT). A direct effect of α-synuclein on plasma membrane DAT has been suggested (red arrow). In Parkinson's disease, the overexpression of α-synuclein or the presence of mutant proteins such as E46K or A53T may affect the exocytosis of synaptic vesicles and their recycling, resulting in a decrease of dopamine release.

being only loosely and transiently associated to these vesicles. Obviously, α-synuclein has to adopt a slightly curved structure to optimally match the shape of the vesicle membrane. Mutational studies have suggested that a series of threonine residues in the N-terminal part of α-synuclein could allow the protein to adopt a shape compatible with small vesicles.[42] Replacing those residues with less polar and larger amino acids (leucine and phenylalanine) resulted in a loss of specificity for both acidic lipids and small vesicles. Physicochemical studies with synthetic lipids also revealed that α-synuclein interacts with acidic glycerophospholipids displaying a specific combination of acyl chains.[41] These data indicated that α-synuclein binds to membranes with precise fluidity characteristics.

Whatever the exact molecular mechanism of synaptic vesicle–α-synuclein interactions, it is now admitted that α-synuclein may regulate vesicle availability in the different pools as well as vesicle docking and fusion.[36,43] More generally, it has been proposed that one important physiological role of α-synuclein could be to protect presynaptic terminals against neurodegeneration.[43] In this respect, it should be underscored that dopamine is a potential hazard for dopaminergic neurons because its oxidation can generate reactive oxygen species (ROS) and reactive quinones in the cytoplasm.[43] Correspondingly, a defect in synaptic vesicles function (due to overexpression or mutation of α-synuclein) could result in the cytoplasmic accumulation of dopamine, which in turn could be responsible for the oxidative damage observed in Parkinson's disease. It has also been suggested that α-synuclein could interact with the dopamine transporter (DAT) and control its activity[44–46] (Fig. 10.2). However, these *in vitro* studies are challenged by studies showing that mice lacking α-synuclein have no observable changes in DAT function.[47,48] Finally, α-synuclein may regulate the biosynthesis of dopamine through a direct interaction with tyrosine hydroxylase.[49] Overall, we can see that most if not all biological activities of α-synuclein are intimately connected to the control of dopamine synthesis, storage, release, and recycling in dopaminergic neurons. In contrast, any dysfunction of α-synuclein caused by mutation, overexpression, and/or aggregation could result in severe functional impairments of these neurons, as observed in Parkinson's disease.[1]

10.4 α-SYNUCLEIN IS SECRETED, EXTRACELLULAR, AND TAKEN UP BY SEVERAL BRAIN CELL TYPES

The presynaptic localization of α-synuclein and its association with synaptic vesicles initially suggested that α-synuclein is mainly expressed in the cytoplasm of neurons.[50] However, this concept was challenged by the unexpected detection of α-synuclein in human cerebrospinal fluid and blood plasma from both normal and Parkinson's disease patients.[51] In line with this finding, it has been shown that α-synuclein is secreted in the extracellular space both *in vitro* and *in vivo*.[52] These studies prompted a consideration of whether the secretion of α-synuclein from neurons could contribute to the amplification and spread of neurodegenerative changes in brain microenvironment.[53] Later on, it was proposed that Parkinson's disease and related synucleopathies involve prion-like propagation of α-synuclein aggregates.[54–56] The prion-like hypothesis for Parkinson's disease postulates that α-synuclein, in a misfolded conformation, is released from a donor neuron and then is taken up by a neighboring brain cell, which can be either neuronal or a glial.[57] Following this cell-to-cell transfer, the misfolded α-synuclein acts as a template to convert an endogenous and "correctly folded" α-synuclein into a new misfolded unit, and so on until large aggregates of insoluble α-synuclein are formed in the recipient cell. As discussed in Chapter 8, the term *misfolding* is confusing because at the molecular level, protein folding results from a combination of mechanisms involving both the amino acid sequence of the protein and its environment. Thus, we should not consider that the protein is "misfolded" but rather that a given folding of the protein, resulting from a cascade of physicochemical interactions in a specific microenvironment, leads to a functional defect of the protein or to direct neurotoxic effects. Nevertheless, beyond this semantic reservation, the prion-like concept of synucleopathies is particularly attractive because it explains how α-synuclein aggregates may arise in brain cells that do not synthesize α-synuclein. For instance in multiple system atrophy, oligodendrocytes display α-synuclein-positive cytoplasmic inclusions although they do not express α-synuclein mRNA. This enigma was solved when it was established that cultured oligodendrocytes could take up α-synuclein proteins.[57] These data supported the notion that prion-like replication of α-synuclein may play a role in the pathogenesis of Parkinson's disease and other synucleopathies.

Several mechanisms have been proposed to explain the extracellular presence of α-synuclein in the brain.[58] Although α-synuclein lacks an endoplasmic reticulum/Golgi-targeting signal peptide, it appeared that a small portion of the protein is secreted into the extracellular space through an unconventional exocytosis process.[59] In fact, it is likely that α-synuclein is secreted via an exosomal, Ca^{2+}-dependent mechanism.[60] Exosomes are small vesicles (100 nM diameter, or approximately the size of an enveloped virus such as HIV-1) that can interact with recipient cells and mediate intercellular communications in various ways including endocytosis, receptor-ligand binding, attachment, or fusion with the plasma membrane.[61–63] Both Alzheimer's β-amyloid peptides and infectious prion proteins have been found extracellularly in association with exosomes.[64,65] Therefore, it is now considered that exosomes may play a central role in the pathogenesis of various neurodegenerative diseases, including Alzheimer's, Creutzfeldt–Jakob, Parkinson's, and Huntington's diseases, and ALS (amyotrophic lateral sclerosis).[66] From the physiological point of view, exosomes have been shown to mediate a novel mode of bidirectional neuron–glia communication[62] and to actively participate in interneuronal networks linked to neural plasticity.[61] In this respect, the association of α-synuclein with exosomes in the extracellular space is highly significant because it may contribute to the spreading and pathogenesis of α-synuclein in Parkinson's disease and other synucleopathies. For this reason, α-synuclein can no longer be considered as a cytoplasmic protein capable of interacting with synaptic vesicles at neural ends, but as an ubiquitous protein present in various aqueous and membrane environments, inside, on the surface and around different brain cell types. Because α-synuclein has the weird property to adapt its shape to each of these microenvironments, it has been referred to as a "chameleon" protein (i.e., a protein with exceptional conformational plasticity).[67–69]

10.5 α-SYNUCLEIN: A MULTIFACETED PROTEIN WITH EXCEPTIONAL CONFORMATIONAL PLASTICITY

α-synuclein is a typical example of an intrinsically disordered protein (IDP), that is, a protein that lacks a precise 3D structure and thus fluctuates between energetically similar extended conformations.[70–72] As discussed in Chapter 8, the folding of proteins in water rely

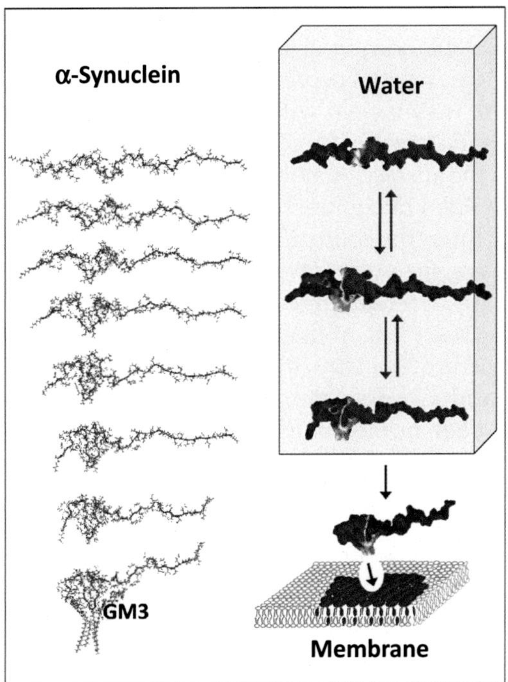

FIGURE 10.3 **Various structures of α-synuclein in solution and in membranes.** In aqueous solutions, α-synuclein fluctuates between a series of extended conformations but can also adopt a more condensed structure resulting from the local folding of specific regions. This can be seen in the molecular models (left panel) and in the surface rendering of the corresponding conformers (right panel). Biological membranes, especially lipid raft domains enriched in glycosphingolipids and cholesterol act as a selection filter able to stabilize condensed α-synuclein conformers once bound to raft lipids (e.g., ganglioside GM3).

on the formation of an apolar core that self-organizes into a water-free central domain for both entropic (hydrophobic effect) and enthalpic (van der Waals interactions) reasons. This mechanism requires a proper balance of apolar versus polar residues in the protein, so that proteins without enough apolar residues have no reason to adopt a globular shape in water. Such is the case of α-synuclein that remains largely unstructured in aqueous solution.[67] Examples of conformational fluctuation of α-synuclein monomers in water solution are shown in Fig. 10.3. Nevertheless, various physicochemical approaches have shown that α-synuclein can locally adopt a more compact state that can be stabilized by long-range intramolecular interactions.[73–76] Moreover, studies with short synthetic peptides have indicated that specific turn structures in α-synuclein could be secured by a network of hydrogen bonds.[77] Turn structures are illustrated in the models of Fig. 10.3 (left panel) that show the progressive folding of a region in the N-terminal part of α-synuclein into a hairpin domain.

This process was confirmed by both circular dichroism and Fourier transform infrared spectroscopy, suggesting that α-synuclein has a slightly collapsed conformation, with 14% of its secondary structure corresponding to turns.[74] In line with these data, the determination of the hydrodynamic dimensions of α-synuclein revealed that the protein was indeed more compact than expected from an extended random coil devoid of secondary structures.[73] In the models of Fig. 10.3, the hairpin structure corresponded to the sphingolipid-binding domain (SBD) of the protein, which encompasses the amino acid residues 34–45.[20] This SBD mediates the interaction of α-synuclein with various glycosphingolipids, including ganglioside GM3.[20,78] Thus, when the prestructured α-synuclein comes close to a cellular membrane expressing GM3 (such as the surface of an oligodendrocyte or an astrocyte[79]), the hairpin domain is already partially folded into a functional SBD. In this case, the plasma membrane acts as a chaperone that finely tunes the conformation of the SBD so that it adapts its final shape to the ganglioside. At this stage, the process becomes irreversible and the whole α-synuclein will lie on the membrane surface. The repetitive lipid head group will induce the helical folding of the N-terminal domain on each side of the SBD, and α-synuclein will soon acquire its classical membrane-bound structure.[80] *In vitro*, the α-helical structuration of α-synuclein can occur with[81] or without gangliosides.[82] However, *in vivo*, all brain cells display abundant amounts of glycosphingolipids on their surface, suggesting a role for these lipids in α-synuclein structuration.[83] Moreover, glycosphingolipid expression in the brain is cell type-specific.[79] Therefore, it is likely that the SBD, already present in soluble α-synuclein monomers as a hydrogen bond-stabilized motif, helps the extracellular protein to select a cellular target. In contrast, the extended α-helices of the N-terminal region are probably not present in α-synuclein before its interaction with membrane lipids, but are readily induced soon after the protein has landed on the cell surface.[84] Nevertheless, it should be mentioned that an endogenous native form of α-synuclein isolated from various cell types, including neural cells, consists of a folded tetramer.[85] This finding raises the interesting possibility that a part of cell-derived α-synuclein could spontaneously acquire an α-helical structure, at least in the cytoplasm of brain cells. However, this issue is still under debate after other groups failed to identify such folded tetramers in brain tissues and found instead that most α-synuclein consists of a largely unstructured monomer.[86,87] Future studies will help to clarify this important issue.

In all cases, a growing body of scientific evidence indicates that membrane lipids exert a chaperone activity that both induces and stabilizes a large helix-turn-helix domain accounting for half of the protein.[80,88] The remaining part of α-synuclein, which correspond to the acidic C-terminus, remains unfolded even when the protein is bound to brain membranes. In this respect, it can be considered as the larger intrinsically disordered fragment of α-synuclein. This domain probably plays a prominent role in the chaperone activity of α-synuclein on the proteins of the SNARE-complex.[89] The exceptional conformational plasticity of α-synuclein is further illustrated in Fig. 10.4.

Under certain conditions, the physical contact between soluble α-synuclein monomers may lead to the formation of oligomers. Then these oligomers can follow two distinct pathways: (1) the fibrillar pathway leading to β-sheet enriched protofibrils that will progressively grow and form the large α-synuclein fibrils found in Lewy bodies[90,91]; and (2) the off-fibril pathway leading to annular oligomers or other forms of oligomers but not fibrils.[92–94] Annular oligomers may form ion-permeable pores in the plasma membrane of brain cells.[82,95,96] It has not been determined yet whether the β-sheet structure of α-synuclein in Lewy bodies is neurotoxic of protective; therefore, annular oligomeric pores could be in fact the main neurotoxic form of α-synuclein.[58,88] Understanding how α-synuclein penetrates the membrane of brain cells and forms an oligomeric channel is thus of primary importance, together with the assessment of the impact of mutations associated with familial Parkinson's disease in

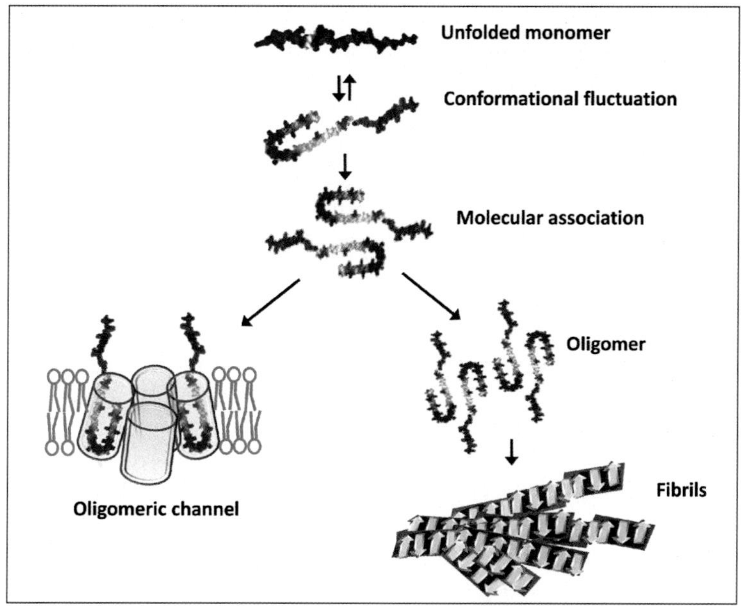

FIGURE 10.4 **Oligomerization and aggregation pathways of α-synuclein.** In aqueous solution, α-synuclein is a disordered protein that fluctuates between a myriad of extended conformations. Sometimes the protein can locally fold in a more condensed structure prone to oligomerization. At this stage, two pathways are possible. When α-synuclein oligomers remain attached to the membrane, they can form Ca^{2+}-permeable pores referred to as amyloid pores or (oligomeric channels) with a typical donut-like shape (left panel). In solution and at high concentration of α-synuclein, a conformational change will occur. The protein will further aggregate into β-rich amyloid fibrils (right panel) that are the main components of Lewy bodies.

this process. To achieve this goal, we first have to carefully study the molecular mechanisms controlling the interaction of α-synuclein with membrane lipids.

10.6 HOW α-SYNUCLEIN INTERACTS WITH MEMBRANE LIPIDS

α-synuclein can interact with three main types of lipids: acidic phospholipids (cytoplasmic leaflet of the plasma membrane, synaptic vesicles), glycosphingolipids (extracellular leaflet of the plasma membrane), and cholesterol (inside the membrane). The first lipids shown to bind α-synuclein were acidic glycerophospholipids[97–100]: phosphatidic acid, phosphatidylserine, and phosphatidylglycerol. All these studies converged to the common observation that upon binding to lipid vesicles containing these acidic phospholipids, the N-terminal part of α-synuclein readily underwent α-helical structuration. At this time, the finding that lipid raft integrity is required for the normal, presynaptic localization of α-synuclein[30] strongly suggested that the protein recognized some lipid raft components from the cytoplasmic leaflet of the plasma membrane. For this reason, great attention was given to decipher the molecular mechanisms associated with the binding of α-synuclein to phosphatidylserine, a typical lipid of the inner leaflet of lipid rafts.[101] Strikingly, it was observed that a variety of synthetic phosphatidylserine with defined acyl chains did not support binding when used individually. Indeed, it appeared that α-synuclein binding requires a precise combination of phosphatidylserine with oleic (C18:1ω9) and polyunsaturated (e.g., arachidonic acid C20:4ω6) fatty acyl chains. Clearly, these data demonstrated that the head group of phosphatidylserine is not, by itself, sufficient to create a functional binding site for α-synuclein. Other determinants such as lipid–lipid interactions and membrane fluidity of the artificial vesicles used for the binding assay also played a critical role in the recognition of α-synuclein. This is a perfect example of the complexity of interpreting lipid–protein binding experiments using artificial bilayer membrane systems.

As the complexity of head group increases, the binding parameters appear more clear-cut, or at least less dependent on the various physicochemical cofactors that affect the binding of α-synuclein to acidic phospholipids. An exhaustive study of α-synuclein/glycosphingolipid interactions allowed to unravel the molecular mechanisms controlling the binding of α-synuclein to the extracellular side of lipid rafts.[20] Langmuir monolayers of reconstituted monolayers of highly purified glycosphingolipids assist in determining an order of preference of α-synuclein for these lipids: GM3 > Gb3 > GalCer-NFA > GM1 > sulfatide > GalCer-HFA > LacCer > GM4 > GM2 > asialo-GM1 > GD3, indicating a marked preference for glycosphingolipid with one, three, or five sugar units.[20] Although α-synuclein could interact with GM1,[81,102] one of the most abundant gangliosides of neurons,[103] the protein has a significantly higher affinity for GM3, the main astrocytic ganglioside.[104] α-Synuclein also recognized the glycosphingolipids expressed by oligodendrocytes (i.e., GalCer and sulfatide). In marked contrast with Aβ, which recognized only GalCer with a hydroxylated fatty acid (GalCer-HFA),[105] α-synuclein made little difference between GalCer-HFA and GalCer-NFA (GalCer with a nonhydroxylated fatty acid). These data indicated that the galactose head group of GalCer constituted by itself a sufficient binding motif for α-synuclein, independently of its conformation. Monolayer studies conducted with a minimal SBD peptide (α-synuclein34–45), combined with studies with mutant peptides and molecular modeling approaches, which allowed to precise at the atomic scale how α-synuclein interacts with all these glycosphingolipids.[20] For instance, residues Tyr-39, Lys-34, and Lys-45 were important for GM3 binding, whereas only Tyr-39 appeared critical for GM1 recognition. Molecular models of the α-synuclein–GM3 complex are shown in Fig. 10.5.

The stoichiometry of the complex is two GM3 per α-synuclein protein. Specifically, the protein inserts its SBD in the chalice-shaped dimer of GM3. It precisely positions the aromatic side chain of Tyr-39 at the center of the chalice, close to the apolar phase of the plasma membrane. At this stage, the aromatic ring of Tyr-39 forms a transient OH–π bond[106] with an OH group carried by the first sugar (glucose) of a GM3 head group (green GM3 in Fig. 10.5). A detailed geometric analysis of this OH–π bond involved in the α-synuclein–GM3 complex has been recently published.[107] The cationic groups of Lys-34 and Lys-45, located at each end of the SBD, interact with the sialic acid residue of the GM3 head groups. The energy of interaction of α-synuclein for the GM3 dimer has been estimated to –116 kJ mol^{-1} on the basis of molecular modeling simulations.[107] For comparison, α-synuclein interacts with only one GM1 molecule, and the energy of interaction in this case is significantly lower (–40 kJ mol^{-1}).[78] By itself, the OH–π bond between Tyr-39 and GM3 has an energy of interaction of –6 kJ mol^{-1}. This bond is rather weak and definitely not sufficient to stabilize the complex in its initial topology. Indeed, molecular dynamics simulations of the α-synuclein–GM3 complex in a membrane matrix indicated that the OH–π bond no longer existed after one nanosecond (ns). In fact,

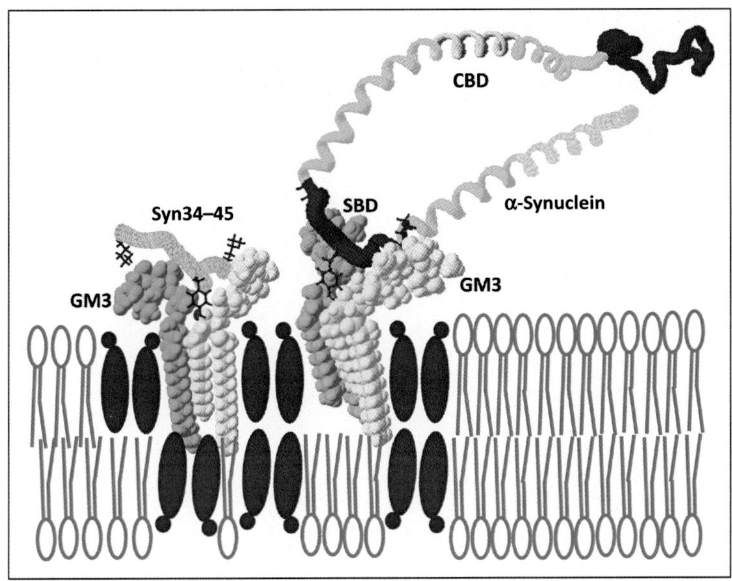

FIGURE 10.5 **Interaction of α-synuclein with gangliosides: role of the SBD.** The SBD of α-synuclein (linear fragment 34–45) has a high affinity for glycosphingolipids, especially ganglioside GM3. The SBD recognizes a typical dimer of GM3 whose overall conformation is stabilized by GM3–cholesterol interactions. Both the synthetic SBD (syn34–45) and the whole α-synuclein (1–140) recognize the GM3 dimer. In both cases, the aromatic residue Tyr-39 (in purple) is at a center of the complex, whereas the basic residues Lys-34 and Lys-45 occupy lateral positions. The aromatic ring of Tyr-39 interacts with an OH group belonging to the sugar head group of the green GM3, thereby forming a typical OH–π bond. The GM3 molecules colored green and yellow; cholesterol is red and the other membrane lipids (chiefly glycerophospholipids) are outlined in green. The C-terminal acidic domain of α-synuclein is in red and the main cholesterol-binding domain (CBD) in orange.

disrupting the OH–π bond allowed the peptide to pursue its journey through the apolar phase of the membrane, thereby starting an insertion process (Fig. 10.6).

After 10 ns, the peptide had traveled ~20 Å and was totally dipped into the membrane. This rapid progression is due to the polar/apolar clash between the OH group of Tyr-39 and the aliphatic chains of membrane lipids. To be more precise, the polar phenolic OH group of Tyr-39 repulses the apolar groups of the lipids as the insertion gradually progresses along the aliphatic chains of membrane lipids. In this respect, one could consider that α-synuclein could remain perfectly bound to GM3 if the ganglioside–α-synuclein interaction was strong enough. However, this is not the case because the OH–π interaction, which controls the formation of the complex, is a weak hydrogen bond. When the α-synuclein–GM3 complex is formed, the polar OH group of Tyr-39 is deeply dipped in the membrane, so that it faces an unfriendly apolar milieu. This conflict of polarity generates a clash that forces α-synuclein to pursue its journey across the membrane until the whole SBD eventually finds a more comfortable polar environment, the cytosol in this case. This mechanism of membrane insertion has been described extensively in a recent publication.[107]

The ganglioside-driven insertion of α-synuclein in a lipid raft domain will allow the protein to meet cholesterol, which is abundantly expressed in lipid raft domains.[108] Several

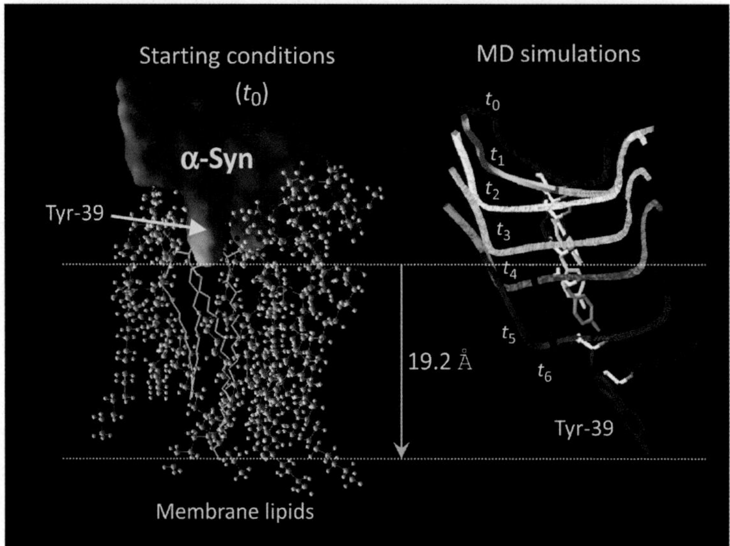

FIGURE 10.6 **Role of Tyr-39 in the insertion of α-helical α-synuclein.** The SBD of α-synuclein complex with two GM3 molecules was merged with a model membrane containing cholesterol and phosphatidylcholine. Energy minimization of the whole system was achieved with the Polak–Ribiere algorithm in order to obtain realistic starting conditions (left panel) referred t_0 , or base time. The OH–π bond between GM3 and Tyr-39 of α-synuclein was still present at this step. Molecular dynamics simulations were then conducted for 10 ns. The OH–π bond appeared to be labile, allowing the peptide to gradually penetrate in the lipid membrane. Snapshots of the insertion process were taken at six different times (from t_1 to t_6) corresponding to 1, 2, 4, 6, 8, and 10 ns, respectively (right panel). *(Reproduced from Fantini and Yahi[107] with kind permission from Springer Science and Business Media.)*

characteristics of cholesterol should be mentioned before we can detail the molecular mechanisms controlling α-synuclein–cholesterol interactions. First, in lipid raft microdomains, cholesterol is not accessible to extracellular proteins because it is totally masked by the large polar head groups of sphingolipids, including both sphingomyelin and glycosphingolipids.[108] However, cholesterol has a dramatic impact on the conformation of these head groups so that sphingolipids interacting with cholesterol are geometrically "prepared" to be recognized by extracellular proteins.[21,105] The masking effect of sphingolipids is highly specific: indeed, it does not exist with phosphatidylcholine. A simple experiment can illustrate this point. Methyl-β-cyclodextrin is a cylindrical protein than can extract cholesterol from lipid membranes.[21] When a mixed monolayer of phosphatidylcholine and cholesterol is spread at the air–water interface, the injection of methyl-β-cyclodextrin in the aqueous subphase is immediately followed by a drop of the surface pressure, indicating that the dextrin has efficiently taken cholesterol molecules from the monolayer.[21] In contrast, when cholesterol is mixed with gangliosides (e.g., GM3), it is not accessible to the dextrin and, in this case, there is no decrease in the surface pressure. Because cholesterol tunes the conformation of the ganglioside, the binding of external ligands is greatly accelerated. Such is the case for Aβ peptides and α-synuclein, which interact with gangliosides more rapidly in presence than in absence of cholesterol. However, the affinity of these proteins for gangliosides is not affected by cholesterol, so that the cholesterol effect is definitely kinetic.[109] As already discussed, a simple way to interpret the cholesterol effect on protein–ganglioside interactions is to consider that although cholesterol is totally masked by the sphingolipid head groups, it "signals" its presence through a fine-tuning effect on ganglioside conformation. Hence, the interaction of extracellular proteins with glycosphingolipids will more efficiently occur in presence of cholesterol, despite the fact that cholesterol, at this stage, does not physically interact with the protein.[88] The second remark pertaining to the role of cholesterol in the interaction of α-synuclein in the plasma membrane of brain cells is that the protein has to penetrate inside the membrane to find cholesterol. The OH–π driven mechanism described earlier has given some clues on this insertion process, and a 20 Å-length insertion of the protein is sufficient to reach the cholesterol area (Fig. 10.6). Again, the tight association of cholesterol with glycosphingolipids in lipid rafts is a warranty that cholesterol will be there for the protein.

As discussed in Chapter 6, several types of cholesterol-binding motifs have been characterized in membrane proteins,[110] including the consensus cholesterol recognition/interaction amino acid consensus sequence CRAC[111] and the more recently discovered inversion of CRAC referred to as CARC.[112] The injection of α-synuclein monomers in the aqueous phase facing a monolayer of cholesterol triggered a rapid increase of the surface pressure, which indicated that the protein interacted with cholesterol.[21] A rapid glance at the amino acid sequence of α-synuclein revealed that the protein displays a CRAC motif spanning amino acid residues 37–43 (**V**L**Y**VGS**K**, with the amino acids defining the CRAC motif underscored). Correspondingly, the recombinant α-synuclein fragment 1–60, which contains this CRAC motif, was also able to interact with cholesterol monolayers. Surprisingly, this interaction was of low affinity. Moreover, the complementary fragment of α-synuclein (i.e., 61–140), although lacking any conventional cholesterol-binding motif, appeared to bind cholesterol with high affinity, and definitely more efficiently than the 1–60 fragment.[21]

Computational approaches coupled with monolayer studies with synthetic peptides led to the identification of a second cholesterol-binding domain in α-synuclein. This high-affinity cholesterol-binding motif mapped to 67–79 region, with the following amino acid sequence: GGAVVTGVTAVAQ. Interestingly, this novel and unconventional cholesterol-binding domain corresponded to the previously characterized fusogenic tilted peptide of α-synuclein.[113] Incidentally, one can remark that this cholesterol-binding domain (CBD) belongs to the NAC region (61–95) of α-synuclein (Fig. 10.1). Both the low-affinity (CRAC) and high-affinity (tilted/fusogenic, CBD) cholesterol-binding domains of α-synuclein are shown in Fig. 10.7.

It is interesting to note that the OH group of cholesterol can form a hydrogen bond with the OH of Tyr-39 (CRAC domain) and with the amide group of Gln-79 (tilted/fusogenic domain). Thus, the difference in the affinity for cholesterol was due to van der Waals interactions that the apolar part of cholesterol establishes with the apolar amino acid side chains of each domain. Indeed, only two apolar residues contributed to the cholesterol–CRAC interaction (Val-37 and Leu-38), for a total energy of interaction of interaction of –16 kJ mol^{-1}. In contrast, the van der Waals interactions between cholesterol and the tilted/fusogenic domain of α-synuclein involved five apolar residues[21] and accounted for as much as –36 kJ mol^{-1}.

Two important features of cholesterol–α-synuclein interactions warrant mention here. First, the CRAC and the tilted/fusogenic domains have opposite orientations in the α-helical membrane-bound conformation of α-synuclein (Fig. 10.7). Thus α-synuclein cannot interact simultaneously with both cholesterol molecules. If both molecules are involved in the insertion of α-synuclein in cholesterol-rich membrane, the protein binds sequentially to one

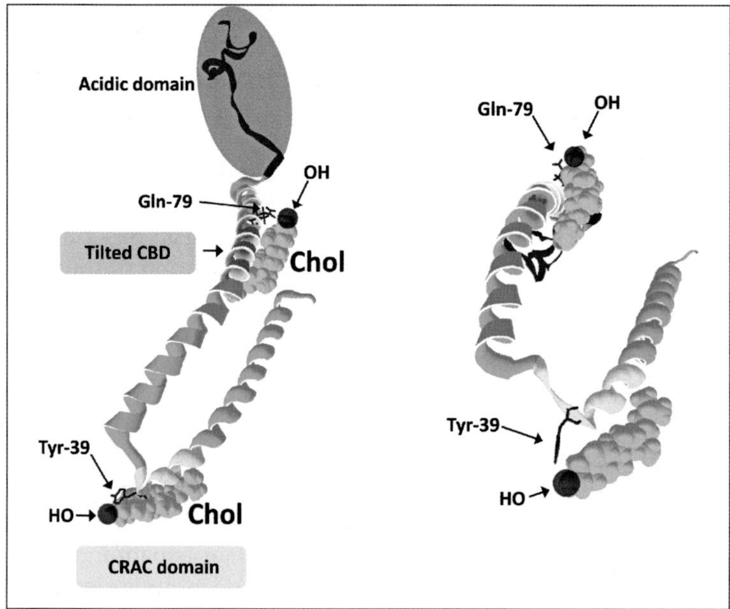

FIGURE 10.7 **Cholesterol-binding domains in α-synuclein.** The protein contains a CRAC domain that binds cholesterol with low affinity and a tilted cholesterol-binding domain (CBD) that displays a high affinity for cholesterol. The OH group of cholesterol can form a hydrogen bond with the OH of Tyr-39 (CRAC domain) and with the amide group of Gln-79 (tilted CBD). Note α-synuclein cannot interact simultaneously with both cholesterol molecules because that they have opposite orientations. Cholesterol is in blue, with its OH group in red.

of these domains and then to the other one, and it also changes its orientation in between. Second, molecular modeling simulations of the cholesterol–α-synuclein complexes indicated that the 67–79 fragment fully retained its tilted orientation, with an angle of 46°, upon binding to cholesterol (in the left panels of Fig. 10.8). The surface views of the complex (in the right panels of Fig. 10.8) show how α-synuclein and cholesterol can mutually adapt their shape during the binding process, resulting in a perfect fit. This tilted orientation is mandatory for the oligomerization of α-synuclein into an annular channel-like pore.

10.7 OLIGOMERIZATION OF α-SYNUCLEIN INTO Ca^{2+}-PERMEABLE ANNULAR CHANNELS

The overall topology of annular pores formed by oligomers of α-synuclein is shown in Fig. 10.9. The electropotential surface of the pore (negative in red, apolar in white, and positive in blue) illustrates the high level of polarization of the oligomer formed by six α-synuclein units in a cholesterol-rich membrane (left panel of Fig. 10.9). At the entry of the pore, the cluster of negative charges displayed by acidic residues will attract cations such as Ca^{2+} into the channel. By constraining a tilted orientation of the transmembrane segment of α-synuclein, cholesterol plays a critical chaperone-like effect on the oligomerization of α-synuclein into annular pores.

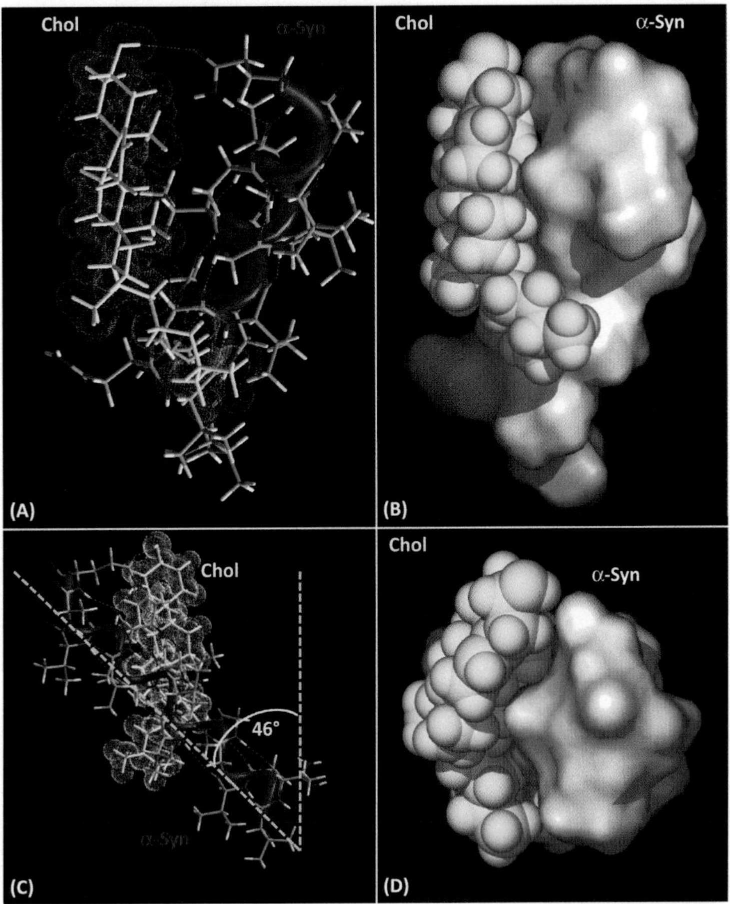

FIGURE 10.8 **Tilted orientation of α-synuclein upon cholesterol binding.** The models were obtained by molecular dynamics simulations of the cholesterol–α-synuclein complex. (**A**) The α-helix of α-synuclein (α-syn) is colored in red and cholesterol (chol) in yellow. (**B**) Surface view of the complex (cholesterol in yellow). (**C**) Tilted topology of the complex. The angle between the helix axis of the tilted cholesterol-binding domain of α-synuclein and the main axis of cholesterol is 46°. (**D**) Surface view illustrates the close contact between cholesterol and α-synuclein (cholesterol wraps around the α-helix of the tilted cholesterol-binding domain). *(Reproduced from Fantini et al.*[21] *with permission from Elsevier.)*

The involvement of cholesterol in α-synuclein pore formation is further illustrated in the right panel of Fig. 10.9. The optimal interactions between the transmembrane domains of α-synuclein are guided by cholesterol that provides a "wall" of rigid lipids onto which the channel is progressively built. When the channel is formed, cholesterol molecules remain tightly associated with the α-synuclein subunits. In this case, cholesterol is considered as a nonannular lipid (i.e., a lipid that is not exchangeable with other lipids from the bulk membrane).[114] A similar effect of cholesterol has been demonstrated for various Aβ peptides, which can also form Ca^{2+}-permeable channels in presence of cholesterol.[115] In both cases (Aβ and α-synuclein),

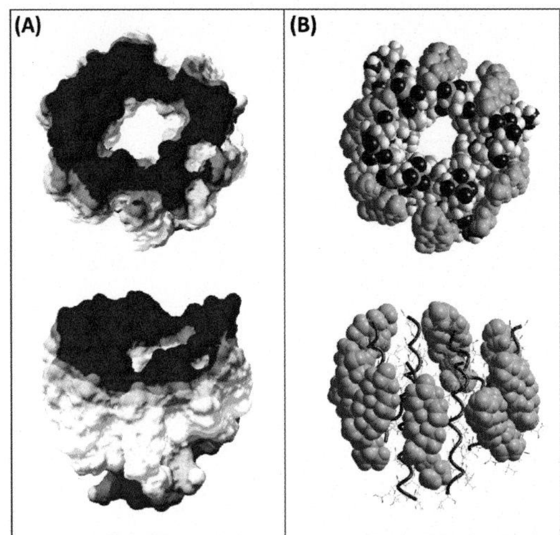

FIGURE 10.9 **α-Synuclein pore formation driven by cholesterol.** (**A**) Surface potential representation (negative in red, positive in blue, and apolar in white) of α-synuclein oligomers forming an annular pore. The models show both a top and a lateral view of the pore. (**B**) Cholesterol molecules (sphere rendering in blue) stabilize the oligomer formed by six α-synuclein proteins (67–79 fragment) and eight cholesterol molecules. In the top view, the peptide is represented with atom spheres. In the lateral view, the helical structure of the 67–79 fragment is represented by a black coil (atoms in stick rendering).

the tilted orientation of transmembrane segments of the amyloid proteins favors the functional protein–protein interactions that control the oligomerization process.[116]

The sequence of events that leads extracellular disordered α-synuclein monomers to form highly organized annular channels in the plasma membrane of brain cells[107] is summarized in Fig. 10.10. The first step is the attraction of α-synuclein by lipid rafts microdomains for which the protein has a marked preference over more fluid regions of the membrane enriched in phosphatidylcholine.[37] Once bound to the surface of lipid rafts, the protein will definitively adopt a α-helical structure with a central turn domain containing the SBD.

A functional interaction between this SBD and a chalice-shaped dimer of ganglioside GM3 will then occur, after which the protein rapidly penetrates the membrane. At this stage, α-synuclein will meet cholesterol and stick to it with its tilted/fusogenic CBD. The interaction of α-synuclein with cholesterol will stop the insertion process and lock the transmembrane segment of α-synuclein in the tilted orientation that is required for the oligomerization into an annular channel. Moreover, the formation of a regular channel implies that the mode of insertion of α-synuclein in the membrane is the same for all monomers. This "leveling" effect is controlled by cholesterol, which acts as a wedge, so that the tilted CBD is immersed at a precise and constant distance from the membrane surface. Downstream from the CBD, the C-terminal region of α-synuclein contains numerous acidic residues that confers a large electrically negative field. This negative field is naturally repulsive and could, in theory, preclude the oligomerization process. However, the tilted configuration of the CBD ensures that the negative domain of α-synuclein is rejected sufficiently far from each transmembrane domain in the extracellular space. In fact, the concentration of acidic amino acid residues in

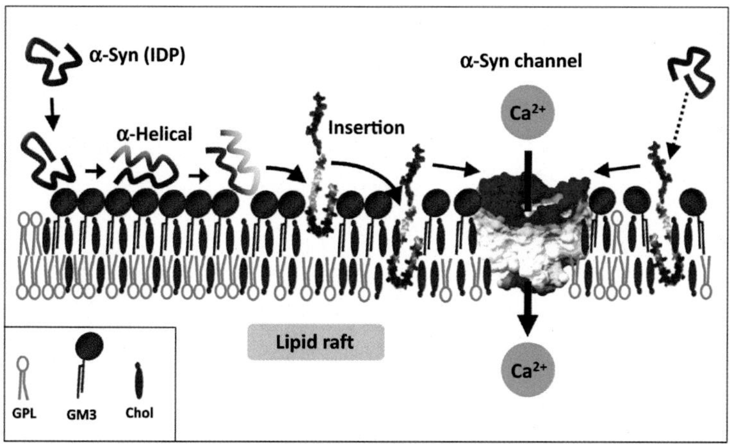

FIGURE 10.10 **Role of lipid rafts in the formation of amyloid pores by helical α-synuclein.** In the aqueous extracellular milieu, α-synuclein is a disordered protein (IDP). Following binding to lipid rafts, α-synuclein is constrained to adopt an α-helical structure. The α-helical α-synuclein penetrates the membrane, interacts with cholesterol (chol), and eventually forms Ca^{2+}-permeable oligomeric channels. Thus, both GM3 (initial binding and helical folding) and cholesterol (membrane insertion and pore formation) act as chaperones during this process. GPL, glycerophospholipid.

the extracellular face of the oligomeric α-synuclein channel constitutes an attractive crater for Ca^{2+} ions. The polarity of the channel (negative outside) is consistent with the vectorial transport of Ca^{2+} from the extracellular space to the cytosol, in agreement with the calcium channel hypothesis of Parkinson's[58] and Alzheimer's[117,118] diseases. This further underscores the common mechanisms of neurotoxicity associated with different neurodegenerative diseases.[88,119–121] In summary, both GM3 (initial binding and α-helical folding) and cholesterol (membrane insertion and pore formation) act as chaperones during the formation of annular pores of α-synuclein in lipid raft domains of brain cells (Fig. 10.10). In full agreement with this model of channel formation, it has been shown that both the N-terminal domain (1–60), which contains the SBD, and the NAC fragment (61–95), which contains the CBD, are responsible for the membrane translocation of α-synuclein.[122]

10.8 ELECTROPHYSIOLOGICAL STUDIES OF OLIGOMERIC α-SYNUCLEIN CHANNELS

Artificial planar lipid membranes, such as the black lipid membrane, are wonderful systems for studying ion channels in minimal but fully controlled experimental conditions.[123] As shown in Fig. 10.11, the setup is particularly well suited for electrophysiological analysis of ion transport. The bilayer is reconstituted by "painting" the hole that separates two aqueous compartments. Both compartments are easily accessible so that one can modulate the composition of the bathing solutions. Coupling the setup with an electrophysiological recording system helps ensure that the bilayer is leakproof, and it also allows to measure the passage of ions in real time.[77] Hence, black lipid membranes are a method of choice for studying the

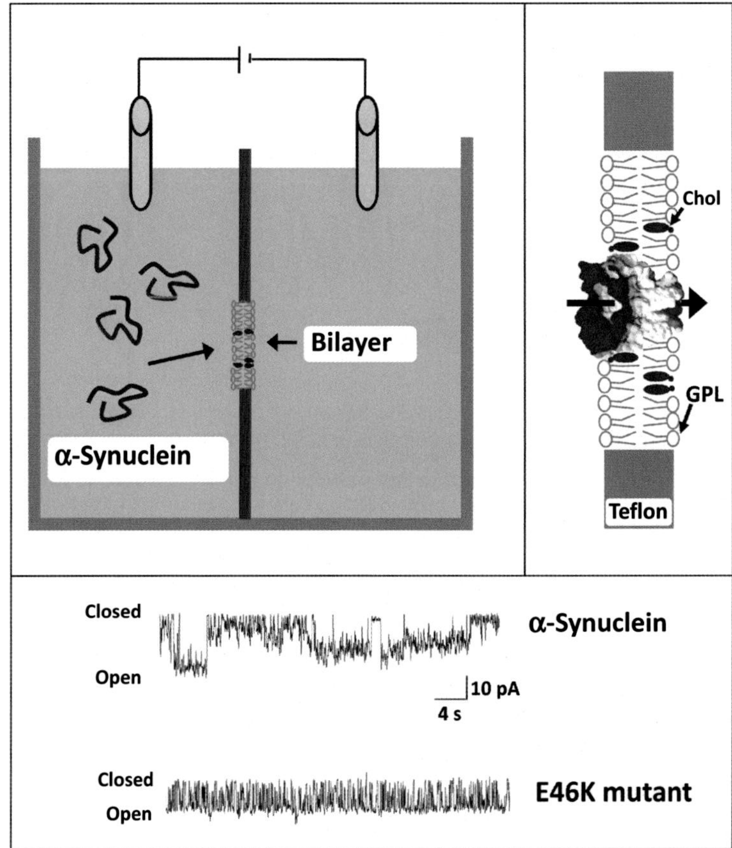

FIGURE 10.11 **Electrophysiological studies of α-synuclein channels in model membranes.** Planar lipid bilayer (black lipid membranes, BLM) are formed by injecting a lipid mixture into the hole of a Delrin chamber separating two compartments (upper left panel). The protein (here α-synuclein) is added in the *cis* compartment. The electrical properties of the reconstituted bilayer and the passage of ions across amyloid pores (black arrow in the upper right panel) can be measured by coupling the device with an electrophysiological recording system capable of resolving currents in the low pA range. Typical recordings of wild-type and mutant α-synuclein channels reconstituted in a phosphatidylcholine bilayer are shown in the lower panel. *(Reproduced from Di Pasquale et al.[77] Copyright © 2010, with permission from Elsevier.)*

formation and properties of oligomeric pores formed by amyloid proteins in lipid bilayers. The force of this system is its simplicity: a lipid bilayer, the protein injected as monomers in the *cis* compartment, and ions (Fig. 10.11). Several reports indicate the capability of soluble α-synuclein monomers to interact with black lipid membranes of various lipid composition and to form a uniform and distinct oligomeric pore.[82,124,125] A model of an oligomeric pore of α-synuclein immersed in phospholipid–cholesterol bilayer and some representative examples of single-channel currents are shown in Fig. 10.11. One of these recordings is typical of the E46K mutant of α-synuclein.[77] Interestingly, the E46K mutant formed channels that are less conductive than those formed by the wild-type protein exhibited higher selectivity for cations and showed an asymmetrical response to voltage and nonstop channel activity.[77] This channelopathy was functionally corrected when ganglioside GM3 was incorporated in the lipid bilayer. This correction required the interaction of the mutant α-synuclein with GM3 and was not observed with ganglioside GM1. However, the corrective effect was also detected in presence of both GM3 and GM1 in the bilayer. Incidentally, GM1 was found to interact with α-synuclein and to inhibit fibrillation.[81] Moreover, the binding of α-synuclein to GM1 induced substantial α-helical structure.[81] In several animal models of Parkinson's disease, treatment with GM1 induced a partial recovery of both neurochemical and behavioral parameters and, concomitantly, reversed the dopaminergic deficits in nigrostriatal neurons of aged rats.[120] Overall, these data support the notion that the interaction of α-synuclein with cell surface gangliosides is a complex process that may lead to either increased or reduced neurotoxicity. More specifically, the data reported by Di Pasquale et al.[77] raised the interesting possibility that ganglioside GM3, through a specific regulation of mutant α-synuclein channels, could play a key protective role in the pathogenesis of familial Parkinson's disease.[77] Nevertheless, other data suggested that GM3 could promote the insertion of α-synuclein monomers in brain cell membranes, leading to oligomeric pore formation in cholesterol-rich domains.[107,116] Clearly, this issue – the effect of cholesterol/GM3 complexes on the electrophysiological activity of α-synuclein – warrants further consideration and new studies, especially with cholesterol-containing bilayers. A comparison of all α-synuclein mutants associated with familial Parkinson's disease should also be informative (Fig. 10.1). This issue is important because high membrane ion permeability caused by mutant α-synuclein (A30P and A53T) has been considered as a possible mechanism of degeneration of neurons in familial forms of Parkinson's disease.[126]

In any case, the data obtained with black lipid membranes clearly showed that the perturbations of membrane permeability induced by α-synuclein are due to the formation of a uniform and highly ordered pore species that most likely correspond to the annular oligomers visualized by atomic force microscopy.[92,121,127] Thus, these studies rule out diffuse bilayer damage due to the stochastic insertion of heterogeneous aggregates of α-synuclein.[125] The similarity between α-synuclein and bacterial pore-forming toxins such as colicins has been emphasized by several authors.[77,82,125] Indeed, colicin E1, like α-synuclein, inserts into anionic areas of the plasma membrane via an α-helix-dependent process and forms channel-like structures.[128] The identification of annular pores as the main neurotoxic species of α-synuclein oligomers in various forms of Parkinson's disease could lead to new therapeutic approaches, as discussed in Chapter 14. The key role of membrane lipids, especially cholesterol and glycosphingolipids, in the formation of α-synuclein channels gives a unique opportunity to limit the spreading of the disease at its earliest stages by disrupting the interaction of α-synuclein with these lipids.

10.9 CELLULAR TARGETS FOR α-SYNUCLEIN IN THE BRAIN: THE LIPID CONNECTION

Because brain lipids play a key role in the pathogenesis of Parkinson's disease and related synucleopathies, it is important to identify which lipids are recognized by α-synuclein and the consequences of α-synuclein binding to these lipids. The exhaustive analysis of α-synuclein–glycosphingolipid interactions by the Langmuir monolayer approach indicated

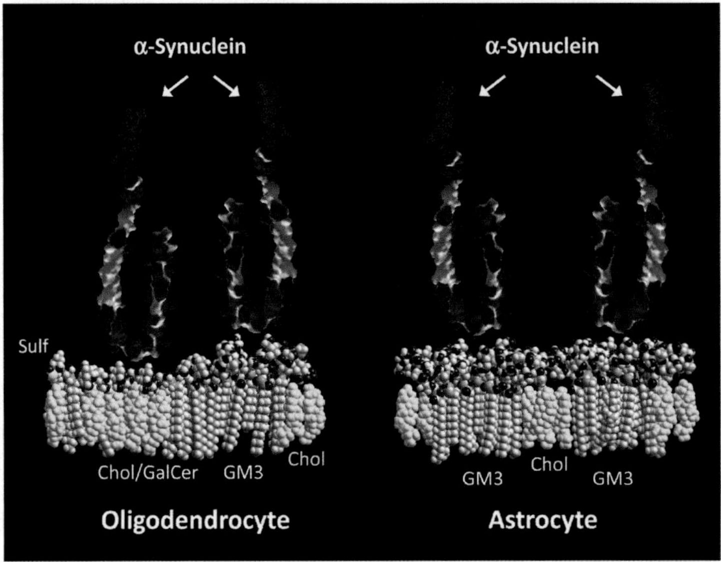

FIGURE 10.12 **Molecular modeling of the interaction between α-synuclein and oligodendrocytes and astrocytes.** In both cases it is assumed that the protein preferentially binds to lipid rafts. Upon membrane binding, α-synuclein acquires a α-helical structure that facilitates membrane penetration.

that, in addition to GM3, α-synuclein also recognized GalCer and sulfatide.[20] These data strongly suggested that α-synuclein is able to interact with the external surface of both astrocytes (enriched in GM3[104]) and oligodendrocytes (enriched in GalCer, sulfatide, and GM3[79]) through direct binding to these sphingolipids. It should be underscored that this does not exclude other modes of binding that involve, for instance, membrane glycerophospholipids or proteins. Nevertheless, the abundant and specific expression of the previously mentioned glycosphingolipids on the surface of astrocytes and oligodendrocytes is consistent with a facilitated interaction of α-synuclein with these brain cells. In agreement with this notion, functional interactions of α-synuclein with astrocytes and oligodendrocytes have been reported consistently.[57,129,130] To visualize how α-synuclein could bind to these cells, molecular models of α-synuclein–GalCer and α-synuclein–GM3 complexes on astrocytes and oligodendrocytes are presented in Fig. 10.12.

The interaction of α-synuclein with GalCer is by no way similar to the interaction of Aβ with this particular glycosphingolipid. In the case of Aβ, only GalCer with a α-hydroxylated fatty acid in its ceramide moiety (GalCer-HFA) is recognized by the amyloid protein.[105] Thus, GalCer-NFA (with a nonhydroxylated fatty acid) is totally ignored by Aβ. However, in presence of cholesterol, GalCer-NFA changes the orientation of its sugar head so that its conformation resembles GalCer-HFA. The tuning of GalCer-NFA by cholesterol effect is mediated by a hydrogen bond network involving the OH group of cholesterol, the glycosidic bond and the –NH group of GalCer-NFA, which forces the sugar to adopt a topology parallel to the membrane. This L-shape of GalCer is spontaneously acquired by GalCer-HFA because the internal OH group of the fatty acid plays the same role as the external OH of cholesterol.[105] The consequence is that in absence of cholesterol, Aβ selects GalCer-HFA and ignores GalCer-NFA, whereas in

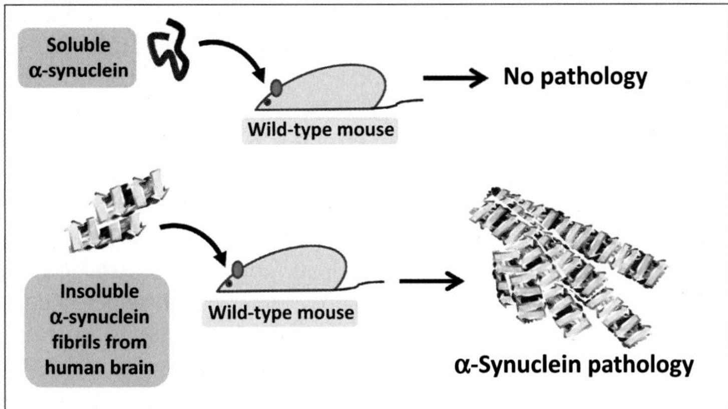

FIGURE 10.13 **Prion-like propagation of α-synuclein.** Intracerebral injections of α-synuclein fibrils from brains of patients with dementia with Lewy bodies induced α-synuclein pathology in wild-type mice (lower panel). In contrast, mice injected with soluble α-synuclein did not develop such pathologies (upper panel). This illustration summarizes data published by Masuda-Suzukake et al.[131]

presence of cholesterol Aβ indistinctively interacts with both types of GalCer.[105] Notably, the interaction of α-synuclein with GalCer does not follow this scheme: the protein makes little difference between GalCer-HFA and GalCer-NFA.[20] In fact, α-synuclein recognizes both types of GalCer and even has a slight preference for GalCer-NFA. The ability to bind indistinctively to all forms GalCer could facilitate the interaction of α-synuclein with the surface of oligodendrocytes, which have been shown to accumulate Lewy bodies in spite of the fact these cells do not synthesize α-synuclein.[57] An experimental demonstration of such a propagation of α-synuclein pathology in wild-type mice[131] is summarized in Fig. 10.13.

Binding of α-synuclein to cells surface glycosphingolipids expressed by various brain cell types could facilitate the spreading of synucleopathies through a prion-like mechanism.[54] In this respect, it is worthy to note GalCer is recognized by all three major amyloid proteins (i.e., PrP,[132] Aβ peptides,[105] and α-synuclein).[20] Moreover, GalCer remains tightly attached to infectious prion particles.[133] Thus, this glycosphingolipid probably plays a common role in the prion-like dissemination of neurodegenerative diseases, and this role could be to transfer the "infection" to glial cells. This further illustrates how complex may be the involvement of glycosphingolipids in these diseases, because GalCer also acts as a protective lipid able to maintain the cellular PrP protein in its nonpathological α-helical conformation.[88,134,135]

10.10 CONCLUSION OF THE INVESTIGATION: WHO IS GUILTY, WHO IS INNOCENT?

These considerations lead us to ask the question of the culpability of α-synuclein in Parkinson's disease. Is protein the culprit? Without any doubt, yes. Is it the only culprit? No, because Parkinson's disease is somewhat like a gang affair, because α-synuclein would probably not do any harm without the complicity of lipids. Among the bad associates, cholesterol and glycosphingolipids should be blamed as strongly as α-synuclein, because they greatly facilitate the binding, insertion, and oligomerization of the protein into Ca^{2+}-permeable amyloid pores that are likely the main neurotoxic species in Parkinson's disease.[58,88] There are some mitigating circumstances for α-synuclein because the harmful conformation of the protein is totally controlled by membrane lipids. In water, α-synuclein is mostly disordered and innocuous. Finally, it should be mentioned that gangliosides such as GM3 may exert a corrective function of the channelopathy induced by mutant α-synuclein in artificial bilayers.[77] Moreover, deficiency of GM1 has been shown to correlate with Parkinson's disease in rodents and humans.[136] Correspondingly, GM1 treatment of Parkinson's disease patients seemed to induce some clinical benefit.[137] Even if the mechanisms by which selected gangliosides might improve the patients' symptoms of Parkinson' s disease remain to be fully characterized, these new tracks deserve to be explored.

10.11 KEY EXPERIMENT: PESTICIDES AND ANIMAL MODELS OF PARKINSON'S DISEASE[138,139]

Two possible genetic causes of Parkinson's disease include (1) mutations of the SNCA gene, and (2) duplication or triplication of this gene, which suggests that the level of expression of the protein is also a causal factor of the disease.[140] Correspondingly, α-synuclein overexpression induced some dopaminergic neuron loss in rats.[141] Nevertheless, although transgenic models could reproduce some features of Parkinson's disease, rare are the models displaying both motor symptoms and selective dopaminergic neuronal death in the substantia nigra.[142]

In 2000, Betarbet et al. reported that chronic exposure of rats to rotenone (Fig. 10.14), a mitochondrial complex I inhibitor, could reproduce most of the anatomical, neurochemical,

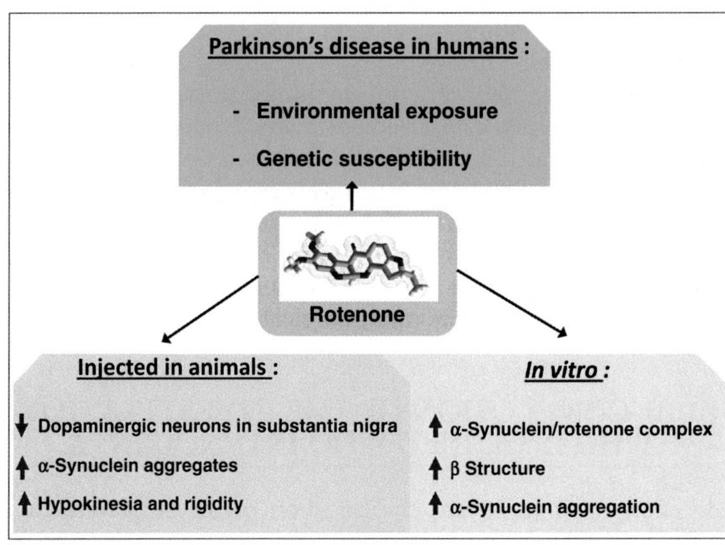

FIGURE 10.14 **Epidemiological and experimental link between Parkinson's disease and the pesticide rotenone.**

behavioral, and neuropathological features of Parkinson's disease.[138] A remarkable feature of rotenone-treated rats was the detection of α-synuclein-positive cytoplasmic inclusions, similar to Lewy bodies, in nigral dopaminergic neurons.[143] As stated by Greenamyre et al.,[143] the rotenone model was initially developed to test the mitochondrial hypothesis of Parkinson's disease (i.e., the hypothesis that the disease might be associated with mitochondrial defects). It turned out that because rotenone is a pesticide, epidemiological studies began to look at the potential role of exposure to rotenone *per se* as a risk factor for Parkinson's disease in humans.[143] In this respect, it is quite remarkable that the rotenone model sparked subsequent epidemiological studies, which indeed have suggested a potential role for rotenone in some cases of Parkinson's disease.[144,145]

Despite its merits, the rotenone model for Parkinson's disease has been criticized,[146,147] some authors argued that it does not recapitulate the motor symptoms of Parkinson's disease.[148] An improved model based on intraperitoneal injections of rotenone was developed in 2009 by the Greenamyre group.[139] This time, all rotenone-treated animals developed bradykinesia, postural instability, and/or rigidity. α-synuclein aggregates were observed in dopamine neurons of the substantia nigra. A 45% loss of dopaminergic neurons was also observed in the substantia nigra, associated with significant loss of striatal dopamine (Fig. 10.14).

Little doubt remains that rotenone and Parkinson's disease are associated by a causative link (Fig. 10.14). Genetic factors may obviously increase the susceptibility of individuals for developing Parkinson's disease following a chronic exposure to rotenone and other pesticides,[149] but the risk is now well recognized.[145] Physicochemical studies revealed that rotenone, in addition to its mitochondrial effects, can physically interact with α-synuclein and induce its aggregation mediated by β-rich structures.[150] The rotenone binding site on α-synuclein has not been identified yet, but the apolar NAC domain, known to play a role in α-synuclein aggregation,[22] could be a good candidate. In any case, animal models of Parkinson's disease based on neurotoxicants[151] or viral vectors[152] are indispensable for studying stage-specific pathologic mechanisms and testing new drugs active at the earliest stages of the disease.

References

1. Recchia A, Debetto P, Negro A, Guidolin D, Skaper SD, Giusti P. Alpha-synuclein and Parkinson's disease. *FASEB J*. 2004;18(6):617–626.
2. Wooten GF, Currie LJ, Bovbjerg VE, Lee JK, Patrie J. Are men at greater risk for Parkinson's disease than women? *J Neurol Neurosurg Psychiatry*. 2004;75(4):637–639.
3. Baldereschi M, Di Carlo A, Rocca WA, et al. Parkinson's disease and parkinsonism in a longitudinal study: two-fold higher incidence in men. ILSA Working Group. Italian Longitudinal Study on Aging. *Neurology*. 2000;55(9):1358–1363.
4. Siderowf A, Stern M. Update on Parkinson disease. *Ann Intern Med*. 2003;138(8):651–658.
5. Lotharius J, Brundin P. Pathogenesis of Parkinson's disease: dopamine, vesicles and alpha-synuclein. *Nat Rev Neurosci*. 2002;3(12):932–942.
6. Goedert M. Alpha-synuclein and neurodegenerative diseases. *Nat Rev Neurosci*. 2001;2(7):492–501.
7. Spillantini MG, Schmidt ML, Lee VM, Trojanowski JQ, Jakes R, Goedert M. Alpha-synuclein in Lewy bodies. *Nature*. 1997;388(6645):839–840.
8. Polymeropoulos MH, Lavedan C, Leroy E, et al. Mutation in the alpha-synuclein gene identified in families with Parkinson's disease. *Science*. 1997;276(5321):2045–2047.
9. Breydo L, Wu JW, Uversky VN. Alpha-synuclein misfolding and Parkinson's disease. *Biochim Biophys Acta*. 2012;1822(2):261–285.

10. Vekrellis K, Xilouri M, Emmanouilidou E, Rideout HJ, Stefanis L. Pathological roles of alpha-synuclein in neurological disorders. *Lancet Neurol*. 2011;10(11):1015–1025.
11. McKeith IG, Burn DJ, Ballard CG, et al. Dementia with Lewy bodies. *Semin Clin Neuropsychiatry*. 2003;8(1):46–57.
12. Simard M, van Reekum R. Dementia with Lewy bodies in Down's syndrome. *Int J Geriatr Psychiatry*. 2001;16(3):311–320.
13. Arai K, Kato N, Kashiwado K, Hattori T. Pure autonomic failure in association with human alpha-synucleinopathy. *Neurosci Lett*. 2000;296(2–3):171–173.
14. Saito Y, Kawai M, Inoue K, et al. Widespread expression of alpha-synuclein and tau immunoreactivity in Hallervorden-Spatz syndrome with protracted clinical course. *J Neurol Sci*. 2000;177(1):48–59.
15. Larson ME, Sherman MA, Greimel S, et al. Soluble alpha-synuclein is a novel modulator of Alzheimer's disease pathophysiology. *J Neurosci*. 2012;32(30):10253–10266.
16. Chen X, de Silva HA, Pettenati MJ, et al. The human NACP/alpha-synuclein gene: chromosome assignment to 4q21.3-q22 and TaqI RFLP analysis. *Genomics*. 1995;26(2):425–427.
17. Touchman JW, Dehejia A, Chiba-Falek O, et al. Human and mouse alpha-synuclein genes: comparative genomic sequence analysis and identification of a novel gene regulatory element. *Genome Res*. 2001;11(1):78–86.
18. Ueda K, Fukushima H, Masliah E, et al. Molecular cloning of cDNA encoding an unrecognized component of amyloid in Alzheimer disease. *Proc Natl Acad Sci USA*. 1993;90(23):11282–11286.
19. George JM. The synucleins. *Genome Biol*. 2002;3(1):Reviews3002.
20. Fantini J, Yahi N. Molecular basis for the glycosphingolipid-binding specificity of alpha-synuclein: key role of tyrosine 39 in membrane insertion. *J Mol Biol*. 2011;408(4):654–669.
21. Fantini J, Carlus D, Yahi N. The fusogenic tilted peptide (67-78) of alpha-synuclein is a cholesterol binding domain. *Biochim Biophys Acta*. 2011;1808(10):2343–2351.
22. El-Agnaf OM, Bodles AM, Guthrie DJ, Harriott P, Irvine GB. The N-terminal region of non-A beta component of Alzheimer's disease amyloid is responsible for its tendency to assume beta-sheet and aggregate to form fibrils. *Eur J Biochem*. 1998;258(1):157–163.
23. Kasten M, Klein C. The many faces of alpha-synuclein mutations. *Mov Disord*. 2013;28(6):697–701.
24. Kruger R, Kuhn W, Muller T, et al. Ala30Pro mutation in the gene encoding alpha-synuclein in Parkinson's disease. *Nat Genet*. 1998;18(2):106–108.
25. Zarranz JJ, Alegre J, Gomez-Esteban JC, et al. The new mutation, E46K, of alpha-synuclein causes Parkinson and Lewy body dementia. *Ann Neurol*. 2004;55(2):164–173.
26. Choi W, Zibaee S, Jakes R, et al. Mutation E46K increases phospholipid binding and assembly into filaments of human alpha-synuclein. *FEBS Lett*. 2004;576(3):363–368.
27. Appel-Cresswell S, Vilarino-Guell C, Encarnacion M, et al. Alpha-synuclein p.H50Q, a novel pathogenic mutation for Parkinson's disease. *Mov Disord*. 2013;28(6):811–813.
28. Proukakis C, Dudzik CG, Brier T, et al. A novel alpha-synuclein missense mutation in Parkinson disease. *Neurology*. 2013;80(11):1062–1064.
29. Bendor JT, Logan TP, Edwards RH. The function of alpha-synuclein. *Neuron*. 2013;79(6):1044–1066.
30. Maroteaux L, Campanelli JT, Scheller RH. Synuclein: a neuron-specific protein localized to the nucleus and presynaptic nerve terminal. *J Neurosci*. 1988;8(8):2804–2815.
31. Iwai A, Masliah E, Yoshimoto M, et al. The precursor protein of non-A beta component of Alzheimer's disease amyloid is a presynaptic protein of the central nervous system. *Neuron*. 1995;14(2):467–475.
32. Abeliovich A, Schmitz Y, Farinas I, et al. Mice lacking alpha-synuclein display functional deficits in the nigrostriatal dopamine system. *Neuron*. 2000;25(1):239–252.
33. Cabin DE, Shimazu K, Murphy D, et al. Synaptic vesicle depletion correlates with attenuated synaptic responses to prolonged repetitive stimulation in mice lacking alpha-synuclein. *J Neurosci*. 2002;22(20):8797–8807.
34. Scott DA, Tabarean I, Tang Y, Cartier A, Masliah E, Roy S. A pathologic cascade leading to synaptic dysfunction in alpha-synuclein-induced neurodegeneration. *J Neurosci*. 2010;30(24):8083–8095.
35. Nemani VM, Lu W, Berge V, et al. Increased expression of alpha-synuclein reduces neurotransmitter release by inhibiting synaptic vesicle reclustering after endocytosis. *Neuron*. 2010;65(1):66–79.
36. Lashuel HA, Overk CR, Oueslati A, Masliah E. The many faces of alpha-synuclein: from structure and toxicity to therapeutic target. *Nat Rev. Neurosci*. 2013;14(1):38–48.
37. Fortin DL, Troyer MD, Nakamura K, Kubo S, Anthony MD, Edwards RH. Lipid rafts mediate the synaptic localization of alpha-synuclein. *J Neurosci*. 2004;24(30):6715–6723.

38. Kahle PJ, Neumann M, Ozmen L, et al. Subcellular localization of wild-type and Parkinson's disease-associated mutant alpha-synuclein in human and transgenic mouse brain. *J Neurosci*. 2000;20(17):6365–6373.
39. Middleton ER, Rhoades E. Effects of curvature and composition on alpha-synuclein binding to lipid vesicles. *Biophys J*. 2010;99(7):2279–2288.
40. Jensen MB, Bhatia VK, Jao CC, et al. Membrane curvature sensing by amphipathic helices: a single liposome study using alpha-synuclein and annexin B12. *J Biol Chem*. 2011;286(49):42603–42614.
41. Kubo S, Nemani VM, Chalkley RJ, et al. A combinatorial code for the interaction of alpha-synuclein with membranes. *J Biol Chem*. 2005;280(36):31664–31672.
42. Pranke IM, Morello V, Bigay J, et al. alpha-synuclein and ALPS motifs are membrane curvature sensors whose contrasting chemistry mediates selective vesicle binding. *J Cell Biol*. 2011;194(1):89–103.
43. Venda LL, Cragg SJ, Buchman VL, Wade-Martins R. alpha-synuclein and dopamine at the crossroads of Parkinson's disease. *Trends Neurosci*. 2010;33(12):559–568.
44. Fountaine TM, Wade-Martins R. RNA interference-mediated knockdown of alpha-synuclein protects human dopaminergic neuroblastoma cells from MPP(+) toxicity and reduces dopamine transport. *J Neurosci Res*. 2007;85(2):351–363.
45. Sidhu A, Wersinger C, Vernier P. Does alpha-synuclein modulate dopaminergic synaptic content and tone at the synapse? *FASEB J*. 2004;18(6):637–647.
46. Lee FJ, Liu F, Pristupa ZB, Niznik HB. Direct binding and functional coupling of alpha-synuclein to the dopamine transporters accelerate dopamine-induced apoptosis. *FASEB J*. 2001;15(6):916–926.
47. Yavich L, Tanila H, Vepsalainen S, Jakala P. Role of alpha-synuclein in presynaptic dopamine recruitment. *J Neurosci*. 2004;24(49):11165–11170.
48. Senior SL, Ninkina N, Deacon R, et al. Increased striatal dopamine release and hyperdopaminergic-like behaviour in mice lacking both alpha-synuclein and gamma-synuclein. *Eur J Neurosci*. 2008;27(4):947–957.
49. Perez RG, Waymire JC, Lin E, Liu JJ, Guo F, Zigmond MJ. A role for alpha-synuclein in the regulation of dopamine biosynthesis. *J Neurosci*. 2002;22(8):3090–3099.
50. Lee HJ, Choi C, Lee SJ. Membrane-bound alpha-synuclein has a high aggregation propensity and the ability to seed the aggregation of the cytosolic form. *J Biol Chem*. 2002;277(1):671–678.
51. Borghi R, Marchese R, Negro A, et al. Full length alpha-synuclein is present in cerebrospinal fluid from Parkinson's disease and normal subjects. *Neurosci Lett*. 2000;287(1):65–67.
52. El-Agnaf OM, Salem SA, Paleologou KE, et al. Alpha-synuclein implicated in Parkinson's disease is present in extracellular biological fluids, including human plasma. *FASEB J*. 2003;17(13):1945–1947.
53. Lee SJ. Origins and effects of extracellular alpha-synuclein: implications in Parkinson's disease. *J Mol Neurosci*. 2008;34(1):17–22.
54. Angot E, Steiner JA, Hansen C, Li JY, Brundin P. Are synucleinopathies prion-like disorders? *Lancet Neurol*. 2010;9(11):1128–1138.
55. Frost B, Diamond MI. Prion-like mechanisms in neurodegenerative diseases. *Nat Rev Neurosci*. 2010;11(3):155–159.
56. Goedert M, Clavaguera F, Tolnay M. The propagation of prion-like protein inclusions in neurodegenerative diseases. *Trends Neurosci*. 2010;33(7):317–325.
57. Reyes JF, Rey NL, Bousset L, Melki R, Brundin P, Angot E. Alpha-synuclein transfers from neurons to oligodendrocytes. *Glia*. 2014;62(3):387–398.
58. Pacheco C, Aguayo LG, Opazo C. An extracellular mechanism that can explain the neurotoxic effects of alpha-synuclein aggregates in the brain. *Front Physiol*. 2012;3:297.
59. Lee HJ, Suk JE, Bae EJ, Lee SJ. Clearance and deposition of extracellular alpha-synuclein aggregates in microglia. *Biochem Biophys Res Commun*. 2008;372(3):423–428.
60. Emmanouilidou E, Melachroinou K, Roumeliotis T, et al. Cell-produced alpha-synuclein is secreted in a calcium-dependent manner by exosomes and impacts neuronal survival. *J Neurosci*. 2010;30(20):6838–6851.
61. Chivet M, Hemming F, Pernet-Gallay K, Fraboulet S, Sadoul R. Emerging role of neuronal exosomes in the central nervous system. *Front Physiol*. 2012;3:145.
62. Fruhbeis C, Frohlich D, Kuo WP, et al. Neurotransmitter-triggered transfer of exosomes mediates oligodendrocyte-neuron communication. *PLoS Biol*. 2013;11(7):e1001604.
63. Chivet M, Javalet C, Hemming F, et al. Exosomes as a novel way of interneuronal communication. *Biochem Soc Trans*. 2013;41(1):241–244.

64. Rajendran L, Honsho M, Zahn TR, et al. Alzheimer's disease beta-amyloid peptides are released in association with exosomes. *Proc Natl Acad Sci USA*. 2006;103(30):11172–11177.
65. Arellano-Anaya ZE, Huor A, Leblanc P, et al. Prion strains are differentially released through the exosomal pathway. *Cell Mol Life Sci*. 2014.
66. Bellingham SA, Guo BB, Coleman BM, Hill AF. Exosomes: vehicles for the transfer of toxic proteins associated with neurodegenerative diseases? *Front Physiol*. 2012;3:124.
67. Uversky VN. A protein-chameleon: conformational plasticity of alpha-synuclein, a disordered protein involved in neurodegenerative disorders. *J Biomol Struct Dyn*. 2003;21(2):211–234.
68. Drescher M, Huber M, Subramaniam V. Hunting the chameleon: structural conformations of the intrinsically disordered protein alpha-synuclein. *Chembiochem*. 2012;13(6):761–768.
69. Silva BA, Uversky VN. Targeting the chameleon: a focused look at alpha synuclein and its roles in neurodegeneration. *Mol Neurobiol*. 2013;47:446–459.
70. Uversky VN. Intrinsic disorder in proteins associated with neurodegenerative diseases. *Front Biosci*. 2009;14:5188–5238.
71. Xue B, Mizianty MJ, Kurgan L, Uversky VN. Protein intrinsic disorder as a flexible armor and a weapon of HIV-1. *Cell Mol Life Sci*. 2012;69(8):1211–1259.
72. Uversky VN. The mysterious unfoldome: structureless, underappreciated, yet vital part of any given proteome. *J Biomed Biotechnol*. 2010;2010:568068.
73. Uversky VN, Li J, Souillac P, et al. Biophysical properties of the synucleins and their propensities to fibrillate: inhibition of alpha-synuclein assembly by beta- and gamma-synucleins. *J Biol Chem*. 2002;277(14):11970–11978.
74. Morar AS, Olteanu A, Young GB, Pielak GJ. Solvent-induced collapse of alpha-synuclein and acid-denatured cytochrome c. *Protein Sci*. 2001;10(11):2195–2199.
75. Ruf RA, Lutz EA, Zigoneanu IG, Pielak GJ. Alpha-synuclein conformation affects its tyrosine-dependent oxidative aggregation. *Biochemistry*. 2008;47(51):13604–13609.
76. Bertoncini CW, Jung YS, Fernandez CO, et al. Release of long-range tertiary interactions potentiates aggregation of natively unstructured alpha-synuclein. *Proc Natl Acad Sci USA*. 2005;102(5):1430–1435.
77. Di Pasquale E, Fantini J, Chahinian H, Maresca M, Taieb N, Yahi N. Altered ion channel formation by the Parkinson's-disease-linked E46K mutant of alpha-synuclein is corrected by GM3 but not by GM1 gangliosides. *J Mol Biol*. 2010;397(1):202–218.
78. Yahi N, Fantini J. Deciphering the glycolipid code of Alzheimer's and Parkinson's amyloid proteins allowed the creation of a universal ganglioside-binding Peptide. *PloS One*. 2014;9(8):e104751.
79. van Echten-Deckert G, Walter J. Sphingolipids: critical players in Alzheimer's disease. *Prog Lipid Res*. 2012;51(4):378–393.
80. Ulmer TS, Bax A, Cole NB, Nussbaum RL. Structure and dynamics of micelle-bound human alpha-synuclein. *J Biol Chem*. 2005;280(10):9595–9603.
81. Martinez Z, Zhu M, Han S, Fink AL. GM1 specifically interacts with alpha-synuclein and inhibits fibrillation. *Biochemistry*. 2007;46(7):1868–1877.
82. Zakharov SD, Hulleman JD, Dutseva EA, Antonenko YN, Rochet JC, Cramer WA. Helical alpha-synuclein forms highly conductive ion channels. *Biochemistry*. 2007;46(50):14369–14379.
83. Schnaar RL, Gerardy-Schahn R, Hildebrandt H. Sialic acids in the brain: gangliosides and polysialic acid in nervous system development, stability, disease, and regeneration. *Physiol Rev*. 2014;94(2):461–518.
84. Pfefferkorn CM, Jiang Z, Lee JC. Biophysics of alpha-synuclein membrane interactions. *Biochim Biophys Acta*. 2012;1818(2):162–171.
85. Bartels T, Choi JG, Selkoe DJ. Alpha-synuclein occurs physiologically as a helically folded tetramer that resists aggregation. *Nature*. 2011;477(7362):107–110.
86. Fauvet B, Mbefo MK, Fares MB, et al. Alpha-synuclein in central nervous system and from erythrocytes, mammalian cells, and *Escherichia coli* exists predominantly as disordered monomer. *J Biol Chem*. 2012;287(19):15345–15364.
87. Burre J, Vivona S, Diao J, Sharma M, Brunger AT, Sudhof TC. Properties of native brain alpha-synuclein. *Nature*. 2013;498(7453):E4–E6:discussion E6–E7.
88. Fantini J, Yahi N. Molecular insights into amyloid regulation by membrane cholesterol and sphingolipids: common mechanisms in neurodegenerative diseases. *Expert Rev Mol Med*. 2010;12:e27.
89. Burre J, Sharma M, Tsetsenis T, Buchman V, Etherton MR, Sudhof TC. Alpha-synuclein promotes SNARE-complex assembly *in vivo* and *in vitro*. *Science*. 2010;329(5999):1663–1667.

90. Conway KA, Lee SJ, Rochet JC, Ding TT, Williamson RE, Lansbury Jr PT. Acceleration of oligomerization, not fibrillization, is a shared property of both alpha-synuclein mutations linked to early-onset Parkinson's disease: implications for pathogenesis and therapy. *Proc Natl Acad Sci USA*. 2000;97(2):571–576.
91. Waxman EA, Giasson BI. Molecular mechanisms of alpha-synuclein neurodegeneration. *Biochim Biophys Acta*. 2009;1792(7):616–624.
92. Kostka M, Hogen T, Danzer KM, et al. Single particle characterization of iron-induced pore-forming alpha-synuclein oligomers. *J Biol Chem*. 2008;283(16):10992–11003.
93. Glabe CG. Structural classification of toxic amyloid oligomers. *J Biol Chem*. 2008;283(44):29639–29643.
94. Kayed R, Sokolov Y, Edmonds B, et al. Permeabilization of lipid bilayers is a common conformation-dependent activity of soluble amyloid oligomers in protein misfolding diseases. *J Biol Chem*. 2004;279(45):46363–46366.
95. Volles MJ, Lansbury PT. Vesicle permeabilization by protofibrillar α-synuclein is sensitive to Parkinson's disease-linked mutations and occurs by a pore-like mechanism. *Biochemistry*. 2002;41(14):4595–4602.
96. Volles MJ, Lansbury Jr PT. Zeroing in on the pathogenic form of alpha-synuclein and its mechanism of neurotoxicity in Parkinson's disease. *Biochemistry*. 2003;42(26):7871–7878.
97. Davidson WS, Jonas A, Clayton DF, George JM. Stabilization of alpha-synuclein secondary structure upon binding to synthetic membranes. *J Biol Chem*. 1998;273(16):9443–9449.
98. Eliezer D, Kutluay E, Bussell Jr R, Browne G. Conformational properties of alpha-synuclein in its free and lipid-associated states. *J Mol Biol*. 2001;307(4):1061–1073.
99. Jo E, Fuller N, Rand RP, St George-Hyslop P, Fraser PE. Defective membrane interactions of familial Parkinson's disease mutant A30P alpha-synuclein. *J Mol Biol*. 2002;315(4):799–807.
100. Chandra S, Chen X, Rizo J, Jahn R, Sudhof TC. A broken alpha-helix in folded alpha-synuclein. *J Biol Chem*. 2003;278(17):15313–15318.
101. Pike LJ, Han X, Chung KN, Gross RW. Lipid rafts are enriched in arachidonic acid and plasmenylethanolamine and their composition is independent of caveolin-1 expression: a quantitative electrospray ionization/mass spectrometric analysis. *Biochemistry*. 2002;41(6):2075–2088.
102. Bartels T, Kim NC, Luth ES, Selkoe DJ. *N*-alpha-acetylation of alpha-synuclein increases its helical folding propensity, GM1 binding specificity and resistance to aggregation. *PloS One*. 2014;9(7):e103727.
103. Ledeen RW, Diebler MF, Wu G, Lu ZH, Varoqui H. Ganglioside composition of subcellular fractions, including pre- and postsynaptic membranes, from Torpedo electric organ. *Neurochem Res*. 1993;18(11):1151–1155.
104. Asou H, Hirano S, Uyemura K. Ganglioside composition of astrocytes. *Cell Struct Funct*. 1989;14(5):561–568.
105. Yahi N, Aulas A, Fantini J. How cholesterol constrains glycolipid conformation for optimal recognition of Alzheimer's beta amyloid peptide (Abeta1-40). *PloS One*. 2010;5(2):e9079.
106. Steiner T, Koellner G. Hydrogen bonds with pi-acceptors in proteins: frequencies and role in stabilizing local 3D structures. *J Mol Biol*. 2001;305(3):535–557.
107. Fantini J, Yahi N. The driving force of alpha-synuclein insertion and amyloid channel formation in the plasma membrane of neural cells: key role of ganglioside- and cholesterol-binding domains. *Adv Exp Med Biol*. 2013;991:15–26.
108. Fantini J, Garmy N, Mahfoud R, Yahi N. Lipid rafts: structure, function and role in HIV. Alzheimer's and prion diseases. *Expert Rev Mol Med*. 2002;4(27):1–22.
109. Fantini J, Yahi N, Garmy N. Cholesterol accelerates the binding of Alzheimer's beta-amyloid peptide to ganglioside GM1 through a universal hydrogen-bond-dependent sterol tuning of glycolipid conformation. *Front Physiol*. 2013;4:120.
110. Fantini J, Barrantes FJ. How cholesterol interacts with membrane proteins: an exploration of cholesterol-binding sites including CRAC, CARC, and tilted domains. *Front Physiol*. 2013;4:31.
111. Jamin N, Neumann JM, Ostuni MA, et al. Characterization of the cholesterol recognition amino acid consensus sequence of the peripheral-type benzodiazepine receptor. *Mol Endocrinol*. 2005;19(3):588–594.
112. Baier CJ, Fantini J, Barrantes FJ. Disclosure of cholesterol recognition motifs in transmembrane domains of the human nicotinic acetylcholine receptor. *Sci Rep*. 2011;1:69.
113. Crowet JM, Lins L, Dupiereux I, et al. Tilted properties of the 67-78 fragment of alpha-synuclein are responsible for membrane destabilization and neurotoxicity. *Proteins*. 2007;68(4):936–947.
114. Fantini J, Barrantes FJ. Sphingolipid/cholesterol regulation of neurotransmitter receptor conformation and function. *Biochim Biophys Acta*. 2009;1788(11):2345–2361.
115. Di Scala C, Troadec JD, Lelievre C, Garmy N, Fantini J, Chahinian H. Mechanism of cholesterol-assisted oligomeric channel formation by a short Alzheimer beta-amyloid peptide. *J Neurochem*. 2013;128(1):186–195.

116. Di Scala C, Chahinian H, Yahi N, Garmy N, Fantini J. Interaction of Alzheimer's beta-amyloid peptides with cholesterol: mechanistic insights into amyloid pore formation. *Biochemistry*. 2014;53(28):4489–4502.
117. Arispe N, Diaz JC, Simakova O. Abeta ion channels. Prospects for treating Alzheimer's disease with Abeta channel blockers. *Biochim Biophys Acta*. 2007;1768(8):1952–1965.
118. Lal R, Lin H, Quist AP. Amyloid beta ion channel: 3D structure and relevance to amyloid channel paradigm. *Biochim Biophys Acta*. 2007;1768(8):1966–1975.
119. Kayed R, Head E, Thompson JL, et al. Common structure of soluble amyloid oligomers implies common mechanism of pathogenesis. *Science*. 2003;300(5618):486–489.
120. Piccinini M, Scandroglio F, Prioni S, et al. Deregulated sphingolipid metabolism and membrane organization in neurodegenerative disorders. *Mol Neurobiol*. 2010;41(2–3):314–340.
121. Quist A, Doudevski I, Lin H, et al. Amyloid ion channels: a common structural link for protein-misfolding disease. *Proc Natl Acad Sci USA*. 2005;102(30):10427–10432.
122. Ahn KJ, Paik SR, Chung KC, Kim J. Amino acid sequence motifs and mechanistic features of the membrane translocation of alpha-synuclein. *J Neurochem*. 2006;97(1):265–279.
123. Grewer C, Gameiro A, Mager T, Fendler K. Electrophysiological characterization of membrane transport proteins. *Ann Rev Biophys*. 2013;42:95–120.
124. Tosatto L, Andrighetti AO, Plotegher N, et al. Alpha-synuclein pore forming activity upon membrane association. *Biochim Biophys Acta*. 2012;1818(11):2876–2883.
125. Schmidt F, Levin J, Kamp F, Kretzschmar H, Giese A, Botzel K. Single-channel electrophysiology reveals a distinct and uniform pore complex formed by alpha-synuclein oligomers in lipid membranes. *PloS One*. 2012;7(8):e42545.
126. Furukawa K, Matsuzaki-Kobayashi M, Hasegawa T, et al. Plasma membrane ion permeability induced by mutant alpha-synuclein contributes to the degeneration of neural cells. *J Neurochem*. 2006;97(4):1071–1077.
127. Lashuel HA, Hartley D, Petre BM, Walz T, Lansbury Jr PT. Neurodegenerative disease: amyloid pores from pathogenic mutations. *Nature*. 2002;418(6895):291.
128. Stroud RM, Reiling K, Wiener M, Freymann D. Ion-channel-forming colicins. *Curr Opin Struct Biol*. 1998;8(4):525–533.
129. Wakabayashi K, Hayashi S, Yoshimoto M, Kudo H, Takahashi H. NACP/alpha-synuclein-positive filamentous inclusions in astrocytes and oligodendrocytes of Parkinson's disease brains. *Acta Neuropathol*. 2000;99(1):14–20.
130. Kovacs GG, Breydo L, Green R, et al. Intracellular processing of disease-associated alpha-synuclein in the human brain suggests prion-like cell-to-cell spread. *Neurobiol Dis*. 2014;69:76–92.
131. Masuda-Suzukake M, Nonaka T, Hosokawa M, et al. Prion-like spreading of pathological alpha-synuclein in brain. *Brain*. 2013;136(Pt 4):1128–1138.
132. Mahfoud R, Garmy N, Maresca M, Yahi N, Puigserver A, Fantini J. Identification of a common sphingolipid-binding domain in Alzheimer, prion, and HIV-1 proteins. *J Biol Chem*. 2002;277(13):11292–11296.
133. Klein TR, Kirsch D, Kaufmann R, Riesner D. Prion rods contain small amounts of two host sphingolipids as revealed by thin-layer chromatography and mass spectrometry. *Biol Chem*. 1998;379(6):655–666.
134. Fantini J. How sphingolipids bind and shape proteins: molecular basis of lipid-protein interactions in lipid shells, rafts and related biomembrane domains. *Cell Mol Life Sci*. 2003;60(6):1027–1032.
135. Sarnataro D, Campana V, Paladino S, Stornaiuolo M, Nitsch L, Zurzolo C. PrP(C) association with lipid rafts in the early secretory pathway stabilizes its cellular conformation. *Mol Biol Cell*. 2004;15(9):4031–4042.
136. Wu G, Lu ZH, Kulkarni N, Ledeen RW. Deficiency of ganglioside GM1 correlates with Parkinson's disease in mice and humans. *J Neurosci Res*. 2012;90(10):1997–2008.
137. Schneider JS, Sendek S, Daskalakis C, Cambi F. GM1 ganglioside in Parkinson's disease: results of a five year open study. *J Neurol Sci*. 2010;292(1–2):45–51.
138. Betarbet R, Sherer TB, MacKenzie G, Garcia-Osuna M, Panov AV, Greenamyre JT. Chronic systemic pesticide exposure reproduces features of Parkinson's disease. *Nat Neurosci*. 2000;3(12):1301–1306.
139. Cannon JR, Tapias V, Na HM, Honick AS, Drolet RE, Greenamyre JT. A highly reproducible rotenone model of Parkinson's disease. *Neurobiol Dis*. 2009;34(2):279–290.
140. Blandini F, Armentero MT. Animal models of Parkinson's disease. *FEBS J*. 2012;279(7):1156–1166.
141. Yamada M, Mizuno Y, Mochizuki H. Parkin gene therapy for alpha-synucleinopathy: a rat model of Parkinson's disease. *Hum Gene Ther*. 2005;16(2):262–270.
142. Fernagut PO, Chesselet MF. Alpha-synuclein and transgenic mouse models. *Neurobiol Dis*. 2004;17(2):123–130.

143. Greenamyre JT, Cannon JR, Drolet R, Mastroberardino PG. Lessons from the rotenone model of Parkinson's disease. *Trends Pharmacol Sci*. 2010;31(4):141–142, author reply 142–143.
144. Dhillon AS, Tarbutton GL, Levin JL, et al. Pesticide/environmental exposures and Parkinson's disease in East Texas. *J Agromedicine*. 2008;13(1):37–48.
145. Tanner CM, Kamel F, Ross GW, et al. Rotenone, paraquat, and Parkinson's disease. *Environ Health Perspect*. 2011;119(6):866–872.
146. Cicchetti F, Drouin-Ouellet J, Gross RE. Environmental toxins and Parkinson's disease: what have we learned from pesticide-induced animal models? *Trends Pharmacol Sci*. 2009;30(9):475–483.
147. Schmidt WJ, Alam M. Controversies on new animal models of Parkinson's disease pro and con: the rotenone model of Parkinson's disease (PD). *J Neural Transm Suppl*. 2006;70:273–276.
148. Hoglinger GU, Oertel WH, Hirsch EC. The rotenone model of parkinsonism – the five years inspection. *J Neural Transm Suppl*. 2006;70:269–272.
149. Cannon JR, Greenamyre JT. Gene-environment interactions in Parkinson's disease: specific evidence in humans and mammalian models. *Neurobiol Dis*. 2013;57:38–46.
150. Silva BA, Einarsdottir O, Fink AL, Uversky VN. Biophysical characterization of alpha-synuclein and rotenone interaction. *Biomolecules*. 2013;3(3):703–732.
151. Uversky VN. Neurotoxicant-induced animal models of Parkinson's disease: understanding the role of rotenone, maneb and paraquat in neurodegeneration. *Cell Tissue Res*. 2004;318(1):225–241.
152. Decressac M, Mattsson B, Lundblad M, Weikop P, Bjorklund A. Progressive neurodegenerative and behavioural changes induced by AAV-mediated overexpression of alpha-synuclein in midbrain dopamine neurons. *Neurobiol Dis*. 2012;45(3):939–953.

CHAPTER

11

Alzheimer's Disease

Jacques Fantini, Nouara Yahi

OUTLINE

11.1 ALZHEIMER'S DISEASE: A RAPID SURVEY, FROM 1906 TO 2014

The three major neurodegenerative disorders are Alzheimer's, Creutzfeldt–Jakob, and Parkinson's diseases. All these diseases are referred to by the name of the physicians who discovered them or described them in detail. Of course it is not the doctors who decided to self-designate these diseases with their own name, but their successors in the study of brain pathologies, in honor of these pioneers.

Parkinson's disease was named by the French neurologist Jean Martin Charcot in recognition of the importance of London doctor James Parkinson who, in 1817, published a detailed medical description of "shaking palsy." In fact, the disease was known since antiquity, and it was previously described by the famous physician Galen in 175 C.E.

The first description of the disease that was later named Creutzfeldt–Jakob disease was given in 1920 by Hans Creutzfeldt, a German neurologist, although it is now considered that his case was highly atypical. A year later another German neurologist, Alfons Jakob, described four new cases of the disorder and thought, probably erroneously, that these cases

Brain Lipids in Synaptic Function and Neurological Disease. http://dx.doi.org/10.1016/B978-0-12-800111-0.00011-4

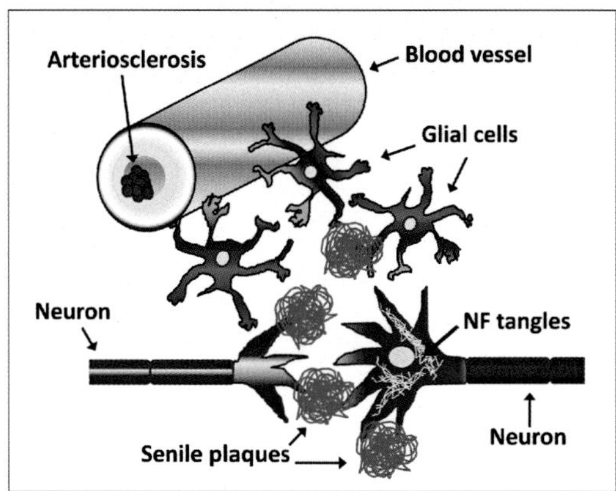

FIGURE 11.1 **What Alois Alzheimer saw in the autopsied brain of Auguste D. in 1906.** In the histological description of the first documented case of Alzheimer's disease, Alois Alzheimer mentioned three main abnormalities: arteriosclerotic lesions of blood vessels, intracellular fibrils (now referred to as neurofibrillar tangles), and senile plaques.

were similar to the one reported by Creutzfeldt. In 1922, German neuropathologist Walther Spielmeyer proposed the eponym of Creutzfeldt–Jakob disease to describe this new neurodegenerative disorder.

The history of Alzheimer's disease also began with one peculiar case. In 1906, German neurologist Alois Alzheimer described the illness of a 51-year-old woman named Auguste D. who exhibited severe cognitive disorders pertaining to memory, language, and social interaction.[1] This patient died 5 years after her admittance to Frankfurt's hospital. After the patient's death, Alzheimer performed an autopsy on her brain, using a silver staining technique (the Bielschwsky method) that allowed him to detect unusual formations, now known as neurofibrillary tangles and senile plaques (Fig. 11.1). He also noted a regularly atrophied brain and signs of arteriosclerosis.[1] In 1909, Alzheimer's colleague Gaetano Perusini published by his own four cases (three new cases plus the already known and documented case of Auguste D.) of patients with "clinically and histologically peculiar mental illnesses in advanced age." Most importantly, Perusini emphasized the presence of a "common main finding" in all four cases, namely ganglial cell fibrils and peculiar plaques. However, it is not Perusini who coined the name *Alzheimer's disease,* but Emil Kraepelin, the director of the Royal Psychiatric Clinic in Munich who had offered Alzheimer a position in the clinic's anatomical laboratory. (It is interesting to note that after his death in 1915, Alzheimer was succeeded by Spielmeyer, the one who coined the name *Creutzfeldt–Jakob disease,* as director of this laboratory.) In the 1910 edition of his *Handbook of Psychiatry,* Kraepelin associated the disease characterized by both cognitive defects and histological abnormalities with Alzheimer's name for the first time. The recognition of these disorders as a new disease should be considered as a decisive breakthrough, because at that time, dementia was considered a natural progression of age. Correspondingly, senility was widely accepted as a part of aging. Hence, Alzheimer's disease was not clearly differentiated from other types of age-induced dementia or senility. This probably explains why Alzheimer's disease did not become a serious issue until the late 1970s. Forty years after, the number of diagnosed cases of Alzheimer's disease over the world is ~36 million. In the United States, one in nine individuals over age 65 has Alzheimer's disease, and one-third over age 85 are afflicted with the illness. The disease disproportionately affects women (two-thirds of cases) versus men (one-third), which cannot simply be attributed to the higher longevity of women compared to men.[2] Indeed, it is likely that the higher incidence of Alzheimer's disease in women results from specific pathogenic mechanisms (one of these potential mechanisms associated with gender-specific lipid expression is discussed in this chapter).

The extracellular senile plaques and the intracellular neurofibrillary tangles discovered by Alzheimer (Fig. 11.1) have remained for more than a century the hallmarks of Alzheimer's disease and the main targets of therapeutic strategies.[3] The discovery of two proteins associated with these abnormal histological structures (i.e., tau for the tangles[4,5] and β-amyloid peptide (Aβ) for the plaques)[6,7] have designated the molecular culprits to hunt. Unfortunately, despite decades of research efforts, Alzheimer's disease is still a fatal disorder for which we are still desperately seeking a cure.[3,8] In this chapter, we will neither follow the chronological order of scientific findings about Alzheimer's disease, nor the chronological sequence of events starting from the proteolytic cleavage of the amyloid precursor protein (APP) leading to Aβ release,[9] oligomerization,[10] aggregation,[11] or tau phosphorylation.[12] Instead, we will try to identify the main mechanisms of neurotoxicity at work in the disease, with the eye of an investigator at the scene of a crime. Then we will patiently disentangle all these threads back to the source of guilt. At the time of writing this book, PubMed contains more than 100,000 references to Alzheimer's disease, 20% of which are reviews. Thus, it is totally unrealistic to attempt to cover all aspects of this disease. Our selective approach is essence biased, yet in accordance to the main topic of this book (i.e., brain lipids). In this subtopic, we have selected cholesterol and glycosphingolipids because, as discussed in Chapters 8, 9, and 10, these lipids are heavily involved in the common mechanisms of neurotoxicity associated with neurodegenerative diseases.[13,14]

11.2 THE AMYLOID PARADIGM

Alzheimer's disease is characterized by the progressive dysfunction of synapses and neuronal loss leading to dementia. Logically, the histological abnormalities found in the brain of Alzheimer's disease patients (i.e., tangles and plaques) have prompted researchers to link these structures with the disease, which has led to the formulation of the amyloid hypothesis according to the accumulation of the β-amyloid peptide (Aβ) that triggers a cascade of events resulting in the deposit of neurofibrillary tangles and amyloid plaques that directly induce neuronal loss.[3] In fact, the observations of Alois Alzheimer were focused on neurofibrillary tangles rather than amyloid plaques, but the interest in amyloid plaques was renewed when several researchers (including Stanley Prusiner) realized that prion aggregates could also form amyloid-like material.[15,16] This parallel between Alzheimer's and Creutzfeldt–Jakob diseases was both unexpected and exciting. Shortly after, the plaques from Alzheimer's disease patients were shown to contain an aggregated peptide of about 40 amino acids, now referred to as Aβ.[6,7,17] Subsequently, the microtubule-associated protein tau was identified as the main component of the neurofibrillary tangles.[18] Finally it was shown that the Aβ peptide was neurotoxic by itself[19] and that it could also induce the abnormal phosphorylation of the tau protein[20] that resulted in its aggregation together with microtubule fragilization.[21] In 1991, J. Hardy and D. Allsop summarized the formulation of the amyloid hypothesis in a sentence: "*The pathological cascade for the disease process is most likely to be: beta-amyloid deposition – tau phosphorylation and tangle formation – neuronal death.*"[22] And in 1992, the "amyloid cascade hypothesis" was published in *Science*.[11]

Aβ is produced by the proteolytic cleavage of a membrane protein called APP, a 770 amino acid protein expressed by neurons.[23] Two mechanisms of APP processing have been characterized: the nonamyloidogenic pathway leading to the release of the secreted APP α (sAPPα) and the amyloidogenic pathway generating various Aβ peptides with 38, 40, 42, or 43 amino acids.[24] Only the amyloid pathway is considered pathogenic, because sAPPα has been shown to control and modulate key neural functions, including neuronal excitability, synaptic plasticity, neurite outgrowth, synaptogenesis, and cell survival.[25] The physiological functions of Aβ peptides have not been identified yet, apart from a possible role in antimicrobial defense[26] and, at low concentrations, a positive modulatory effect on synaptic plasticity and memory.[27,28] Mutations in the APP gene have been linked to rare familial forms of Alzheimer's disease,[29] including for instance D678N (D7N in Aβ),[25] E693G (E22G in Aβ),[30] V717I,[31] or V717F.[32] Transgenic mice overexpressing the V171F human APP protein displayed Alzheimer-type neuropathology, including Aβ deposits, neuritic plaques, and synaptic loss.[33] This finding confirmed the causative link between APP and Alzheimer's disease and reinforced the notion that aberrant APP processing and/or overexpression of Aβ peptides were the primary cause of Alzheimer's disease.

Like α-synuclein (Chapter 10), Aβ is an intrinsically disordered protein (IDP)[34] that may adopt various types of conformations, depending on the environment. In the extracellular aqueous milieu, Aβ will fold into β-rich structures that may further aggregate into amyloid fibrils.[35] These amyloid fibrils are the major component of amyloid plaques.[36,37] The NAC fragment of α-synuclein is also found in these extracellular deposits,[38] which indicates that, at the molecular level, the proteins involved the pathogenesis of Alzheimer's and Parkinson's diseases are interconnected (see Chapter 10).

The responsibility of amyloid plaques in the pathogenesis of Alzheimer's disease has long been the central focus of drug development strategy. Indeed, a solid body of experimental evidence obtained both *in vitro* and *in vivo* showed that Aβ fibrils could induce synapse dysfunction and neuronal loss.[3] The weak point is that amyloid plaques are also found in the brain of cognitively normal individuals.[39–41] Thus amyloid plaques are not sufficient by themselves to trigger the synaptic dysfunctions that lead to memory loss and other cognitive deficits in Alzheimer's disease patients. Moreover, it is now obvious that the clinical strategies targeting amyloid plaques have failed.[3,8,42] Overall, these facts have divided the scientific community into two schools of thought: those who consider Alzheimer's disease as caused independent of Aβ,[3,43] and those who still consider Aβ to be involved, yet independent of amyloid plaques.[8] This issue is of primary importance because identification of the culprit will orient future therapeutic strategies. After the false accusation against amyloid plaques, a new error would significantly delay the development of anti-Alzheimer's drugs. Ongoing inquiries strongly suggest that Aβ is still the culprit, but not in the form of amyloid plaques. An alternative statement of the amyloid hypothesis is that various types of Aβ oligomers are the main neurotoxic species at work in Alzheimer's disease.[44,45] Among these oligomers, those forming Ca^{2+}-permeable channels in the plasma membrane of brain cells have captured the attention of numerous laboratories.[8,46–48]

11.3 THE CALCIUM HYPOTHESIS OF ALZHEIMER'S DISEASE

The calcium hypothesis of Alzheimer's disease is based on the recognition that the homeostasis of cytosolic calcium concentration $[Ca^{2+}]_i$ plays a key role in brain aging, so that sustained changes in $[Ca^{2+}]_i$ homeostasis would be responsible for age-associated brain changes.[49] This claim led several researchers to postulate that Aβ peptides could alter calcium fluxes in brain cells. Three possibilities have been considered: Aβ-induced release of intracellular calcium stores,[50] direct effect of Aβ peptide on plasma membrane calcium channels,[51,52] and plasma membrane permeabilization induced by Aβ.[53] It turned out that the latter proposal gradually led to a new paradigm of Alzheimer's disease. Twenty years ago, Arispe et al. have elegantly demonstrated that Aβ peptide forms electrophysiologically active calcium channels in planar bilayer membranes. These data suggested that the neurotoxicity of Aβ was due to its capability to form such calcium-permeable channels. These channels were selectively blocked by Zn^{2+} ions, which suggested the presence of histidine residues lining the pore mouth.[54,55] Then it was shown that Aβ channels could also form in membrane patches excised from hypothalamic neurons[56] and in cultured neuron-like cells.[57,58] Atomic force microscopy (AFM) studies of Aβ1–42 in solution showed globular structures that did not appear to form fibers.[58] In planar bilayers, multimeric channel-like structures were consistently detected.[58] Finally, soluble oligomers of Aβ were detected in the brain of Alzheimer's disease patients, and, unlike amyloid plaques, the oligomer burden seemed to correlate with cognitive symptoms.[40] Overall, everything fits to charge Aβ oligomers, and especially those forming oligomeric channels in neuron membranes, as the main culprits of Alzheimer's disease.[8] We will now try to figure out how such a small peptide could form highly structured pore-like assemblies in the plasma membrane of brain cells and determine the role of cholesterol and sphingolipids in this process. The main steps of our investigation are summarized in Fig. 11.2.

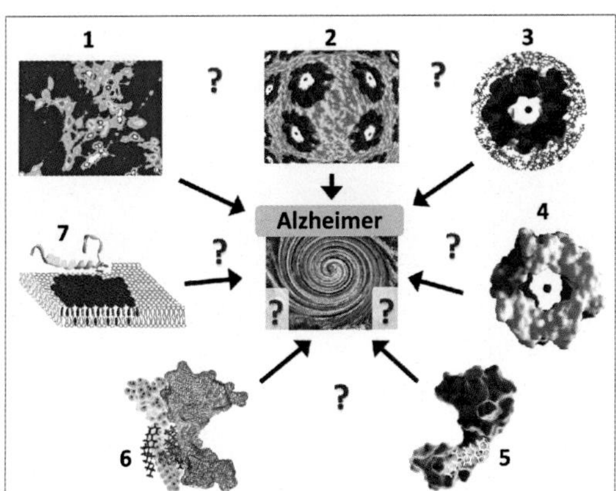

FIGURE 11.2 **The calcium hypothesis of Alzheimer's disease: from cells to molecules. 1.** Calcium fluxes are visualized in neural cells incubated with Aβ monomers. **2.** Calcium enters the cells via plasma membrane pores. **3.** These annular pores formed by Aβ oligomers can be reconstituted in artificial lipid membranes for electrophysiological studies. **4.** Both *in silico* (molecular modeling simulations) and microscopic (AFM) approaches may be used to determine the 3D structure of membrane-bound Aβ oligomers. **5.** Membrane cholesterol is required for the formation of α-helical Aβ pores. **6.** Glycosphingolipids (e.g., ganglioside GM1) facilitate the initial interaction of Aβ monomers with the plasma membrane. **7.** This interaction occurs in lipid raft microdomains of the plasma membrane.

11.4 AMYLOID PORES: β, α, OR BOTH?

Amyloid pores do exist, because we can see them through microscopic approaches.[59] In fact, amyloid pores have progressively supplanted amyloid fibrils in the toxicological studies of various amyloid proteins, including Aβ, α-synuclein, and amylin.[59,60] How are these pores formed? Aβ is by essence an IDP that is expected to fluctuate between numerous equally probable conformations in aqueous solution.[61,62] Indeed, the apolar–polar balance of amino acids in Aβ does not allow the formation of an apolar core that would induce the folding of the protein into a globular shape. Nevertheless, Aβ contains a short stretch (16-KLVFFAED-23) that has a high propensity to form a β-rich structure in water.[63] Indeed, NMR studies of amyloid oligomers in solution showed a typical organization in a double-layered β sheet in which each monomer is folded into a β-strand/turn/β-strand motif.[45] Combining these NMR data with AFM structural studies and molecular dynamics simulations has made it possible to obtain high-resolution models of Aβ channels with a β-barrel topology.[45,48,64] The formation of β-barrel channels in the plasma membrane of a neuron is schematized in Fig. 11.3. Because the folding of Aβ into β-sheet structures may occur in an aqueous environment,[45] we surmised in this schematic rendition of Aβ pore formation that the β-barrel channel is formed in the extracellular milieu prior to its insertion in the membrane. This issue was previously discussed by Jang et al.[45] who reasoned that preformed β-structures, thus potentially β-barrel oligomers of Aβ, can penetrate the membrane with great efficiency.[65,66] Thus, the following sequence of events can be described (Fig. 11.3): (1) Aβ is first released in the extracellular milieu as a disordered protein fluctuating between several extended conformations; (2) above a critical concentration, Aβ starts to fold into a β-structure that progressively forms a β-hairpin motif; (3) this β-hairpin is the building block of a large β-barrel oligomer that is readily built in the extracellular milieu; (4) the resulting β-barrel pore has a high energetic potential allowing it to mechanically disrupt the membrane through a "punching" mechanism.

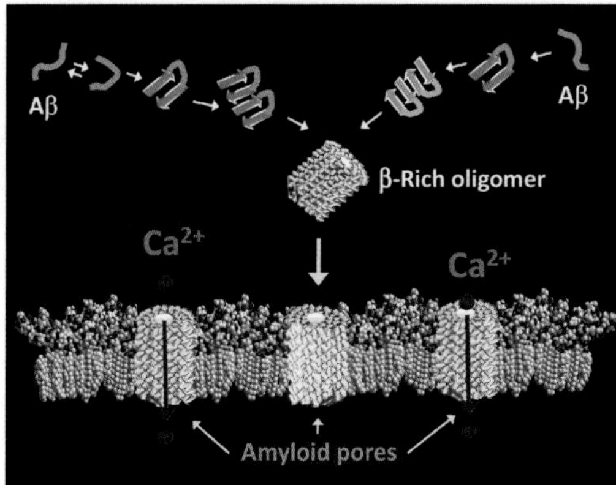

FIGURE 11.3 **Formation of β-barrel ion channels by Aβ.** In the extracellular milieu, Aβ monomers are mostly disordered. Above a critical concentration, Aβ spontaneously folds into a typical β-hairpin and starts to aggregate. A large β-barrel oligomer is generated through the ordered recruitment and assembly of numerous β-structured Aβ peptides. Such β-barrels can penetrate the plasma membrane of neural cells through a punching mechanism, thereby forming Ca^{2+}-permeable amyloid pores.

We do not know whether specific lipids are required for the perforation process or whether the β-barrel is strong enough to bore a hole in the membrane on its own. However, lipids may be necessary to seal the contacts between the amyloid pore and the membrane, and this late step may involve specific lipid–protein interactions. In this respect, the case of cholesterol-dependent cytolysins (CDCs) warrants mention because amyloid pores may share some structural similarity with bacterial pore-forming toxins.[67–69] These CDCs are a family of protein toxins secreted by bacteria (e.g., *Clostridium perfringens*) as water-soluble monomers. Upon binding to eukaryotic cell membranes, the toxin molecules form circular β-rich oligomeric complexes that subsequently penetrate the lipid bilayer to form aqueous pores.[70] The perforation of host cellular membranes by CDCs and subsequent cytolysis play an important role in bacterial pathogenesis.[71] In this case, pore formation is conditioned by the availability of free cholesterol at the membrane surface.[72] Correspondingly, the binding of *C. perfringens* toxin perfringolysin to membranes is triggered when the concentration of cholesterol exceeds the association capacity of membrane lipids, so that this cholesterol excess is free to associate with the toxin.[72] It will be interesting to assess whether β-rich Aβ oligomers can recognize cholesterol in a similar manner and determine in this case whether an excess of cholesterol could stimulate the insertion of these preformed Aβ pores.

It should be pointed out that because Aβ has a high affinity for membrane lipids, the formation of Aβ oligomers in the extracellular milieu will generally occur at high peptide concentrations. In this case, the high extracellular concentration of Aβ will favor peptide–peptide contacts rather than the higher energetic Aβ–lipid interactions. This tendency is in line with the notion that molecular crowding is a key parameter that can control the structuration of IDPs.[73] If this notion is proved correct, it would definitely exonerate brain lipids from the accusation that they trigger the formation of ion channels with β-barrel topology. However, the investigation must proceed with the examination of another important mechanism of pore formation.

The other possibility for the formation of an amyloid pore is the insertion of Aβ monomers in the membrane, which could be followed by an oligomerization process inside the membrane. Like α-synuclein,[74] Aβ is constrained to adopt an α-helical structure when bound to lipids.[75] Thus, the first step in the formation of an α-structured Aβ pore is the interaction of Aβ with selected membrane lipids (e.g., ganglioside GM1, which is concentrated in lipid raft domains).[76] In this case, Aβ will readily adopt an α-helical conformation (Fig. 11.4). The insertion of α-helical Aβ may require the involvement of additional lipids, such as cholesterol, which will also promote the oligomerization process leading to the formation of a functional Ca^{2+}-permeable pore.[77] As noted by Jang et al.,[45] the α-helical structure of Aβ is first formed on the membrane surface.[78] Thus, the insertion of α-helical Aβ may require an energy input that can be furnished by the interaction of Aβ with cholesterol.[77] In this respect, it is clear that the formation of an α-helical oligomeric pore of Aβ is strictly dependent upon the presence of specific lipids, especially cholesterol, in the plasma membrane of brain cells.

It should be noted that both β-barrel and α-helical pores models of oligomeric Aβ are compatible with the dimensions of the oligomers visualized by AFM studies.[48,60,64,77] However, as previously discussed, no objective reason causes us to suspect a role for membrane lipids in the formation of ion channels with a β-barrel topology. We will thus pursue our investigations by examining the role played by these lipids during the formation of α-helical Aβ channels. The first lipid to assess is cholesterol.

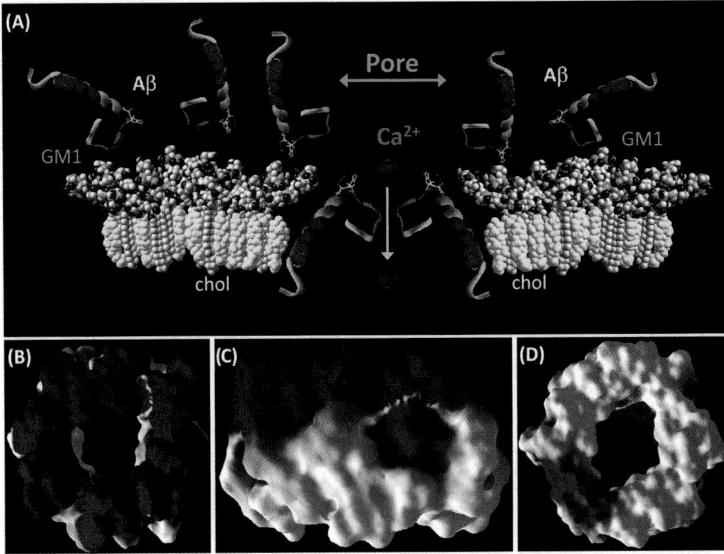

FIGURE 11.4 **Formation of α-helical ion channels by Aβ.** At appropriate Aβ–GM1 ratios, Aβ monomers may bind to cell surface GM1 and subsequently fold into an α-helical structure. The insertion of α-helical Aβ in lipid raft domains enriched in cholesterol (chol) is followed by the oligomerization of Aβ into an oligomeric annular pore permeable to calcium. This sequence of events is summarized in panel A. The structure of these oligomeric Aβ channels is shown B (top view), C (lateral view), and D (bottom view). The surface potential of the models is colored in red (negative), blue (positive), or white (apolar regions).

11.5 CHOLESTEROL

Determining whether Aβ can interact with cholesterol is important not only for deciphering the lipid-assisted mechanism of Aβ pore formation[79] but also for understanding how cholesterol controls the neurotoxicity of Aβ.[80] Docking studies indicated that the C-terminal region of the membrane-bound α-helical Aβ conformers contains a cholesterol-binding domain encompassing amino acid residues 20–35.[81] Interestingly, molecular modeling simulations with a series of peptides derived from this domain showed that the cholesterol-binding activity required the minimal core fragment Aβ25–35.[81,82] The different cholesterol–Aβ complexes obtained in silico are illustrated in Fig. 11.5. Because, in all cases, the interaction relied mostly on van der Waals forces between apolar residues, it is not surprising that the energy of interaction is correlated to the length of the cholesterol-binding motif: –76.9, –67.8, and –37.1 kJ mol^{-1} for Aβ1–40, Aβ22–35, and Aβ25–35, respectively.[79] Among the amino acids that play a key role in cholesterol binding of these Aβ fragments, Lys-28 is remarkable. Indeed, Lys-28 is the sole common residue systematically involved in cholesterol binding in all these motifs.[79] A common involvement does not necessarily imply a unique molecular mechanism of interaction. In fact, the side chain of Lys-28 has a specific orientation in each cholesterol–peptide complex. It interacts with the isooctyl chain of cholesterol in the case of Aβ1–40 and with the sterane backbone in the case of Aβ22–35. In the shorter peptide (Aβ25–35), it forms a hydrogen bond with the OH group of cholesterol.[79] These different topologies of the Aβ–cholesterol complex have a profound impact on the oligomerization properties of Aβ in cholesterol-rich membranes.

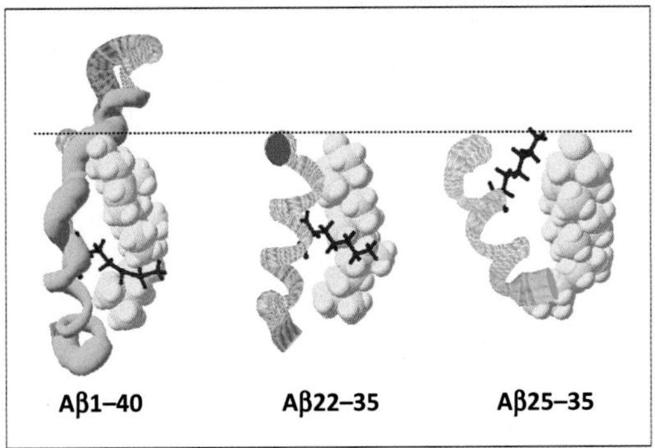

FIGURE 11.5 **Molecular interactions between cholesterol and Aβ peptides.** Cholesterol is in yellow, the peptide backbone in green, and the side chain of Lys-28 in blue. Note that each complex has its own geometry and a specific orientation of Lys-28. The dotted line indicates the position of the OH group of cholesterol.

Physicochemical studies with both wild-type and mutant Aβ peptides were conducted to validate the information given by molecular modeling approaches. In these experiments, a cholesterol monolayer was prepared at the air–water interface, and the Aβ peptide, in monomeric form, was injected in the aqueous subphase underneath the monolayer.[81] Peptides encompassing the N-terminal part of Aβ showed a weak affinity for cholesterol monolayers, confirming that the cholesterol-binding domain is located outside this region.[81] In contrast, all peptides derived from the C-terminal region of Aβ could interact with cholesterol monolayers, providing that they contained the minimal core motif 25–35.[81,82] Thus, these data remarkably confirmed the molecular modeling studies of Aβ–cholesterol interactions. The Aβ22–35 peptide was selected as a reference cholesterol-binding domain, and the cholesterol-binding capability of Aβ22–35 mutants was analyzed.[81] The data showed that the amino acid residues identified as critical for cholesterol binding in modeling studies were indeed involved in the physical interaction with cholesterol. The amino acids implicated in cholesterol binding were Ala-21, Val-24, Lys-28, and Ile-32.[81] Replacing any of these residues by a glycine induced a significant decrease in cholesterol binding. Mutating other amino acids such as Gly-29 and Gly-33 did not affect cholesterol binding, in line with the modeling data.[81]

An interesting feature of all cholesterol–Aβ complexes is the tilted topology of the peptide with respect to the main axis of cholesterol. In fact, the geometry of the Aβ–cholesterol complex is quite similar to the α-synuclein–cholesterol complex.[83] By constraining the conformation of Aβ and α-synuclein in a tilted α-helical domain, cholesterol prevents the dilution of these peptides in the apolar phase of the membrane. Instead, the α-helices remain firmly attached to cholesterol and are encouraged to form an oligomeric channel through the establishment of specific peptide–peptide interactions. In the case of Aβ, the interpeptide contacts are stabilized by a network of hydrogen bonds involving Asn-27 and Lys-28.[77] Without the help of cholesterol, it is likely that this hydrogen bond would not form because of the propensity of the lysine side chain to "snorkel" at the membrane–water interface (i.e., to remove the ε-amino group from the apolar phase of the membrane and let it "breathe" at the cell surface).[84] It is only because cholesterol is tightly bound to the methylene groups of Lys-28 that in Aβ22–35 (and in longer Aβ peptides) the ε-amino group is blocked inside the membrane where it can find Asn-27 on a neighboring peptide (Fig. 11.6). In the case of

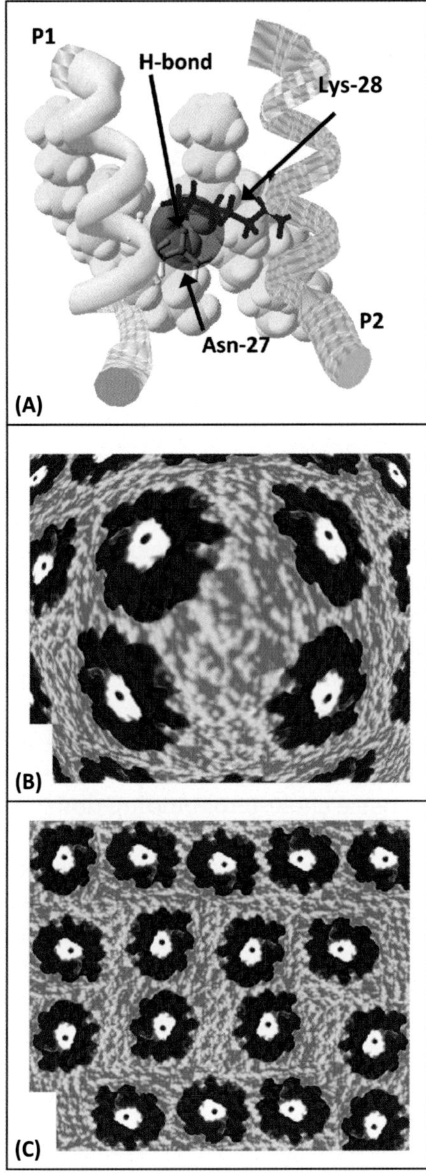

FIGURE 11.6 **Cholesterol-assisted formation of α-helical Aβ channels.** (A) Molecular interactions stabilizing a dimer of Aβ peptides (P1 and P2, in green) in the presence of cholesterol (in yellow). The hydrogen bond between Asn-27 of P1 and Lys-28 of P2 is indicated (red disk). (B, C) Top views of Aβ channels inserted in the plasma membrane at two different scales (obtained by molecular modeling simulations of Aβ in a cholesterol-containing lipid matrix).

Aβ25–35 (Fig. 11.5), the ε-amino group of Lys-28 forms a hydrogen bond with the OH group of cholesterol. Under these conditions, the mechanism of oligomerization cannot be driven by the formation of hydrogen bonds between Lys-28 and Asn-27. A network of van der Waals interactions between apolar residues replaces the lost hydrogen bond and ensures the oligomerization of Aβ25–35 in an annular channel.[82] Cholesterol renders possible these van der Waals interactions because in the Aβ25–35/cholesterol complex, the residues controlling the oligomerization process through peptide–peptide interactions are fully accessible. The mechanism of cholesterol-assisted ion channel formation through the oligomerization of tilted by α-helical Aβ peptides has been described in detail in a recent review.[79]

Analysis of calcium fluxes in neural cells incubated with monomers of Aβ peptides (including Aβ1–42, Aβ22–35, and Aβ25–35) confirmed that these peptides could form Ca^{2+}-permeable channels. These channels required the presence of cholesterol, because pretreatment of the cells with the cholesterol-depleting agent methyl-β-cyclodextrin abrogated the increase in calcium fluxes evoked by Aβ peptides.[77,82] The involvement of extracellular Ca^{2+} in the calcium fluxes elicited by Aβ pores was demonstrated by either removing calcium from the extracellular medium or chelating it with EGTA. In both cases, Aβ no longer affected the Ca^{2+} homeostasis of cultured neural cells.[77] The "amyloid" character of these channels was assessed by their sensitivity to Zn^{2+} cations.[77,82] Most importantly, the residues that appeared critical for cholesterol binding were also important for the formation of Aβ channels. Moreover, a double mutant of Aβ22–35 at positions 27 and 28 (involved in both cholesterol binding and oligomeric pore formation) was totally unable to form Ca^{2+}-permeable channels in neural cells.[77] Overall, these data fit with several reports indicating that cholesterol facilitates the insertion of Aβ peptides in bilayer membranes and induces an α-helical conformation.[85,86] In fact, the balance between α-helices and β-strands in Aβ peptides is controlled by the peptide–lipid ratio.[87] Schematically, at low peptide concentrations, the lipids win and force Aβ to form an α-helix. At high peptide concentrations, peptide–peptide interactions are preferred over peptide binding to lipids, so that the peptide is free to adopt the structure it prefers in water (i.e., β-strands)[87] that may further self-aggregate into preformed ion channels with a β-barrel topology.[48,64] Hence, these data support the notion that cholesterol plays a critical role in the insertion, α-helical folding, and oligomerization of Aβ monomers into functional amyloid pore channels (Fig. 11.6). In this respect, it is interesting to note that the first models of Aβ channels were based on the α-helical topology of the amyloid peptide.[88]

Now we will pursue our investigation to understand how Aβ peptides can have access to cholesterol even though this lipid is totally masked by the voluminous head groups of sphingolipids such as ganglioside GM1.

11.6 GM1

For years, it has been known that Aβ has a high affinity for glycosphingolipids, especially gangliosides.[89] Among the gangliosides recognized by Aβ the most important is probably GM1, which forms highly neurotoxic aggregates when bound to Aβ·[90] The molecular mechanisms controlling the binding of Aβ to GM1 have been studied through various experimental approaches, including NMR and Langmuir monolayers.[91,92] Overall, these data suggest that both the N-terminal[93–95] and the C-terminal[96] parts of GM1 are involved in Aβ binding, which involved both hydrogen bonding and van der Waals forces.[97] This involvement would be possible only if a significant part of the Aβ peptide penetrates the GM1-containing membrane, possibly through an insertion process. A minimal peptide encompassing residues 5–16 of Aβ has been shown to recognize GM1 monolayers and to bind to chalice-shaped dimers of GM1.[92,95] Aβ5–16 fulfils the criteria of a sphingolipid-binding domain (SBD), including lateral charged residues (Arg-5, Lys-16) and a central aromatic amino acid (Tyr-10). This topology is strikingly similar to the SBD of α-synuclein, which has been delineated to amino acids 34–45.[94] However, the SBD of α-synuclein has a marked preference for GM3 versus GM1, whereas the SBD of Aβ has a high affinity for GM1 and a lower affinity for GM3.[94] Both computational and mutational studies reveal that the binding to GM1 involves a pair of contiguous histidine residues in the SBD of Aβ, namely His-13 and His-14.[94] Notably, single mutations of either His-13 or His-14 in Aβ5–16 resulted in a total loss of the GM1-binding capability.[95] In fact, each histidine residue binds to one of two GM1 gangliosides, forming a chalice-shaped dimer in the membrane (Fig. 11.7A). This unique topology of Aβ is induced by cholesterol, whose OH group forms a hydrogen bond with the glycosydic bond linking the glycone part of GM1 to its ceramide anchor.[92,98] In absence of cholesterol, the binding of Aβ5–16 to GM1 occurs with a lag.[92] In other words, cholesterol does not increase the affinity of Aβ for GM1 but accelerates the binding reaction through a conformational effect on GM1. This situation is similar to the effect of cholesterol on the interaction between α-synuclein and GM3.[83] In both cases, cholesterol tunes the conformation of the ganglioside and facilitates the recruitment of functional ganglioside dimers, which is typically a kinetic effect.[92] The universality of this cholesterol effect, both at the molecular level (hydrogen bond-assisted conformational tuning) and at the membrane level (acceleration of the binding of amyloid proteins to lipid raft domains), makes it a hallmark of membrane–amyloid interactions. In a larger perspective, it has been suggested that it could represent one of the facets of cholesterol regulation of membrane receptors.[92,98,99]

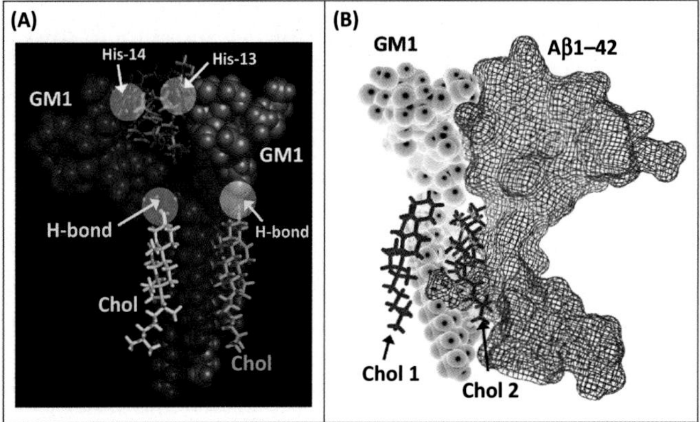

FIGURE 11.7 **Molecular interaction between Aβ and GM1.** (A) Interaction of Aβ5–16 with a chalice-shaped dimer of GM1. GM1 molecules are in blue, and cholesterol (chol) bound to GM1 (hydrogen bonds) in yellow and green. The positions of His-13 and His-14 of Aβ5–16 are indicated. (B) Interaction of Aβ1–42 with GM1 and cholesterol. The cholesterol in purple (chol 1) interacts with GM1 and the one in pink (chol 2) with Aβ.

Our interpretation of this phenomenon is that when cholesterol affects the conformation of sphingolipids, it signals its presence to extracellular proteins that need cholesterol to perform their physiological or pathological functions.[14,83] Indeed, the voluminous head groups of sphingolipids, especially in complex gangliosides, recover the small polar group of cholesterol so that it is totally masked. By constraining a specific ganglioside conformation recognized by amyloid proteins such as Aβ, cholesterol accelerates the interaction of the protein in a zone of the membrane where the sterol is present in large quantities, just underneath the sphingolipids.[76]

At this stage of our investigation, we can link both cholesterol and GM1, which intimately cooperate to attract Aβ in a lipid raft area of the plasma membrane. The attachment of Aβ to GM1, under the control of cholesterol, is the first step in a cascade of events that will lead to the insertion, oligomerization, and formation of a Ca^{2+}-permeable annular pore. At high Aβ–GM1 ratios, the ganglioside induces a β-structuration of Aβ that serves as a seed for further aggregation and generation of highly neurotoxic aggregates.[91,100] At low Aβ–GM1, Aβ will adopt the typical α-helical fold of membrane-bound amyloid proteins.[101] At this stage, the helical cholesterol-binding domain[81] is still extracellular. The insertion of Aβ in the membrane requires a significant rearrangement of the GM1 dimer that should dissociate to let the peptide interact with cholesterol. A similar mechanism has been proposed for the ganglioside–cholesterol-dependent insertion of α-synuclein.[83,102] The topology of the Aβ1–42/cholesterol complex is shown in Fig. 11.7B. It should be noted that this complex can still bind to a GM1 molecule and that this GM1 still interacts with cholesterol. Overall, the GM1/Aβ1–42/cholesterol complex involves two cholesterol molecules, one bound to the peptide and the other one to the ganglioside. The orientation of this multimolecular lipid/Aβ1–42 complex is compatible with the oligomerization of Aβ into an annular Ca^{2+}-permeable channel. The histidine residues that were initially interacting with the sugar head groups of the ganglioside dimer were relocated at the entry of the channel (Fig. 11.8). The presence of histidine residues at the pore mouth is consistent with the inhibitory effect of Zn^{2+} on Aβ-induced

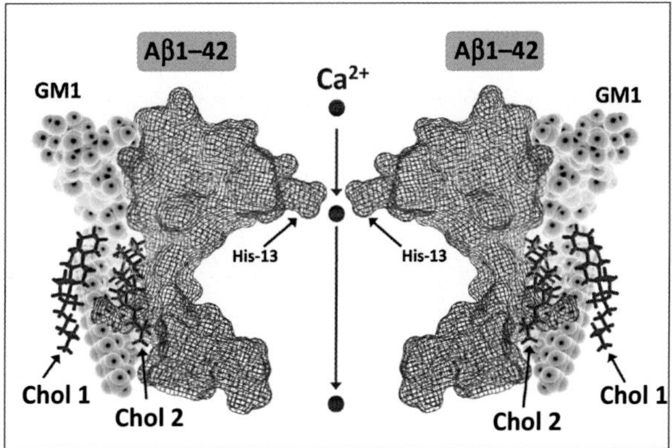

FIGURE 11.8 **How α-helical Aβ1–42 bound to both GM1 and cholesterol can form a Ca^{2+}-permeable amyloid pore.** The topology of Aβ1–42 constrained by both GM1 and cholesterol is the same as in Fig. 11.7B. Note that once bound to Aβ1–42, GM1 and cholesterol determine the orientation of Aβ inside the membrane. His-13, located at the pore mouth, may attract Ca^{2+} ions in the channel. The interaction of Zn^{2+} ions with these residues blocks the channel.

calcium fluxes.[54,55] As discussed in Chapter 14, amyloid pore formation can be inhibited by disrupting Aβ–ganglioside[95] or Aβ–cholesterol[82] interactions. Thus these findings open new possibilities of therapy aimed at preventing Aβ binding and insertion in lipid raft domains of neuronal plasma membranes.[95]

11.7 LIPID RAFTS: MATRIX FOR APP PROCESSING AND FACTORY FOR Aβ PRODUCTION

The next question to ask is where does brain Aβ come from? At first glance, the case is closed: Aβ is produced by the posttranslational proteolytic processing of a precursor protein called APP. Then why don't all people develop Alzheimer's disease? As stated earlier, proteolytic processing of Aβ follows two distinct pathways.[9] In the "physiological" pathway (referred to as *nonamyloidogenic*), APP is sequentially cleaved by two proteases called α- and γ-secretases (Fig. 11.9). The reason why this pathway is nonamyloidogenic is that α-secretase cuts APP at the 17th amino acid of the Aβ sequence,[9] thereby eliminating any further possibility to generate Aβ. Most importantly, this α cleavage produces a large secreted extracellular domain (s-APPα) whereas the C-terminal fragment (C83), now with 83 amino acids, remains associated with the membrane. Then C83 is cleaved by γ-secretase into two fragments, p3 and AICD (APP intracellular domain) that are both rapidly degraded. In the amyloidogenic pathway, APP is first cleaved by β-secretase, which generates a membrane-associated C-terminal fragment (C99) whose N-terminal amino acid corresponds to the first of the Aβ peptides (Asp-1) and a secreted extracellular domain referred to as sAPPβ·[103] Of all these proteolytic fragments, only Aβ is pathogenic, whereas sAPPα (but not sAPPβ) appears to have neuroprotective functions.[104] Therefore, the production of Aβ through the amyloidogenic pathway of APP processing generates a neurotoxic product (Aβ) and decreases the production of a neuroprotective factor (sAPPα). Does it mean that Aβ is always neurotoxic? In fact it may heavily depend on its concentration. Indeed, some studies have shown that at picomolar

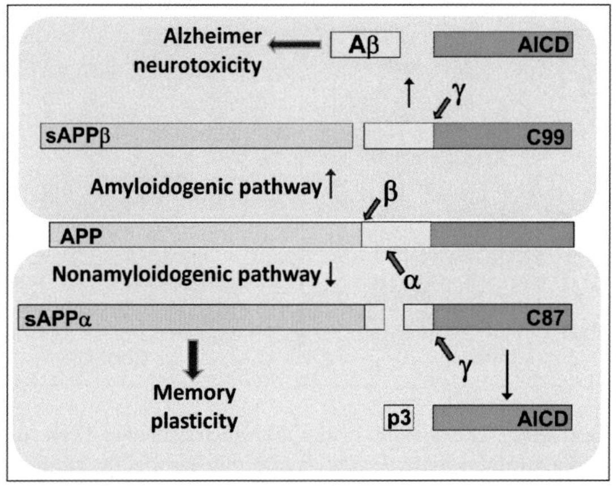

FIGURE 11.9 **The amyloidogenic and nonamyloidogenic pathways of APP processing.**

concentrations, Aβ could positively modulate synaptic plasticity and memory by increasing long-term potentiation.[27,28]

If the amyloidogenic route of APP processing becomes preponderant over the nonamyloidogenic pathway, then the brain will have to manage large amounts of Aβ peptides. In physiological conditions, the excess of Aβ can be cleared by microglial cells, yet this important function may progressively decrease with aging.[105] Microglial cells, which are the brain's equivalent of macrophages colonizing peripheral tissues, fulfill immune, and cleaning functions in the central nervous system. In particular, these cells have been reported to mediate the clearance of both fibrillar[106] and soluble[107] forms of Aβ peptides through distinct mechanisms (receptor-mediated phagocytosis for Aβ fibrils, macropinocytosis for soluble Aβ). This function is crucial because in physiological conditions, the amyloid peptide is cleared at a rate equivalent to its production (~8% of total brain Aβ per hr), so that microglial cells efficiently prevent the accumulation and deposition of Aβ in the normal brain.[108] Affecting this balance by either increasing Aβ production of diminishing Aβ clearance would result in a pathological accumulation of Aβ in the brain, which, according to the amyloid hypothesis, would lead to the development Alzheimer's disease in its most common form (i.e., sporadic in more than 99% of cases). The microglia-dependent clearance of Aβ also involves apoE-containing lipoprotein particles that have been reported to bind to Aβ, thereby facilitating its endocytosis by microglial cells.[109] More recently, an alternative hypothesis has been proposed. Instead of a direct interaction between Aβ and apoE, both proteins would in fact compete for the same low-density lipoprotein receptor-related protein 1 (LRP1) expressed on the surface of astrocytes.[110] Whatever the exact mechanism by which lipoproteins control Aβ metabolism in brain tissues, and especially the involvement of apoE in Aβ clearance, the interest in apoE has been independently motivated by genetic studies. Indeed, individuals with the e4 allele of apoE (apoE4) have an increased risk of developing Alzheimer's disease.[111] In contrast, people with the apoE2 allele may have a risk reduced by 50% because this particular version of the gene is neuroprotective against Alzheimer's disease.[112]

After this short survey of the mechanisms that link lipoprotein metabolism to Aβ clearance, we will examine the amyloidogenic pathway of APP processing in order to understand why the processing of Aβ becomes out of control in Alzheimer's disease. Because it is a membrane protein, the lipids surrounding APP in the membrane are at the forefront. Under physiological conditions, only a small proportion of APP is present in lipid rafts.[113] However, APP can be attracted in lipid rafts through binding of its intracellular domain to the raft proteins flotillin-1[114] or flotillin-2.[115] The possibility of moving APP from nonraft to raft domains logically suggests a switch-like mechanism of processing. The presence of APP in two membrane pools could lead to distinct processing mechanisms.[116,117] According to this model, amyloidogenic processing occurred in lipid rafts.[116,118] In agreement with this notion, the site of Aβ production in vivo was also localized in raft membranes.[119] Thus, the amyloidogenic pathway is stimulated when both APP and β-secretase are located in the same membrane compartment.[116] Reducing cholesterol levels markedly decreased Aβ production, which further confirmed the role of raft lipids in the amyloidogenic pathway of APP.[116] In contrast, when APP is present outside rafts (which is probably the "normal" situation), it is cleaved by α-secretase through the nonamyloidogenic process.[116] Overall, these data indicate that the access of secretases to APP, and therefore Aβ production, is controlled by dynamic interactions of APP with raft lipids, especially cholesterol.[120]

The molecular mechanisms of APP–cholesterol interactions have been unraveled by Sanders and colleagues who reported a series of advanced NMR data obtained with the C99 fragment of APP incorporated in various types of micelles.[121–123] These studies indicated that C99 interacts with cholesterol through a domain encompassing amino acid residues 689–711 of APP, which correspond to amino acids 18–40 in Aβ· The cholesterol-binding domain of APP contains a typical Gly-XXX-Gly motif that confers a high flexibility to the membrane-spanning region of C99.[123] As discussed in a recent review,[79] the avidity of APP for cholesterol needs to be finely tuned so that Aβ peptides can be liberated from the membrane and not trapped by membrane cholesterol. The presence of glycine residues at the binding site is consistent with this notion because the side chain of Gly residues (a single hydrogen atom) offers only a small surface for cholesterol together with a high degree of freedom, so that cholesterol can easily detach from APP.[79] In contrast, the binding of cholesterol to Aβ peptides is mediated by large apolar residues that, together with the long aliphatic chain of Lys-28, confer a high energy of interaction to the Aβ–cholesterol complex.[81] In this respect, it is interesting to note that although partially overlapping, the cholesterol-binding domain of APP and Ab peptides markedly differ in the molecular mechanisms that control cholesterol binding.[79] Finally, high cholesterol, especially after midlife, has been shown to increase the risk of developing Alzheimer's disease.[124–126] In the light of these data, one can hypothesize that the cholesterol–Alzheimer's connection might, at least in part, reflect a cholesterol-induced shift of APP processing from the nonamyloidogenic to the amyloidogenic pathway.

11.8 GENDER-SPECIFIC MECHANISMS

At this stage we have already incriminated cholesterol for its role in the oligomerization of Aβ channels and in the stimulation of Aβ production via the amyloidogenic processing of APP. The case is not closed, however, and cholesterol may also play a role in the higher risk for women versus men in developing Alzheimer's disease. The implication of cholesterol in such gender-specific mechanisms may be linked to its mode of interaction with glycosphingolipids. In this chapter, we have described the interaction of Aβ with GM1, yet Aβ can also interact with several other glycosphingolipids,[89] including galactosylceramide (GalCer).[98] As described in Chaper 1, GalCer exists in two forms that differ in their the absence or presence of an OH group in the acyl chain of the ceramide moiety. GalCer with this OH group is named GalCer-HFA (HFA for hydroxylated fatty acid), whereas its companion devoid of OH group is named GalCer-NFA (NFA for nonhydroxylated fatty acid). Interestingly, Aβ1–40 recognized GalCer-HFA, but not GalCer-NFA.[98] Cholesterol inhibited the interaction of Aβ with GalCer-HFA, but rendered GalCer-NFA competent for Aβ binding. Molecular dynamics simulations and X-ray diffraction studies of GalCer-NFA and GalCer-HFA shed some light on this significant phenomenon.[98,127,128] In fact, the galactose ring of GalCer-NFA protrudes at 180° with respect to the main axis of the ceramide backbone[128] (Fig. 11.10A). Thus, it does not offer any planar surface onto which Aβ can spread.[98] This explains why Aβ does bind to GalCer-NFA. However, in presence of cholesterol, the galactose head group of GalCer-NFA came closer to the ceramide and the

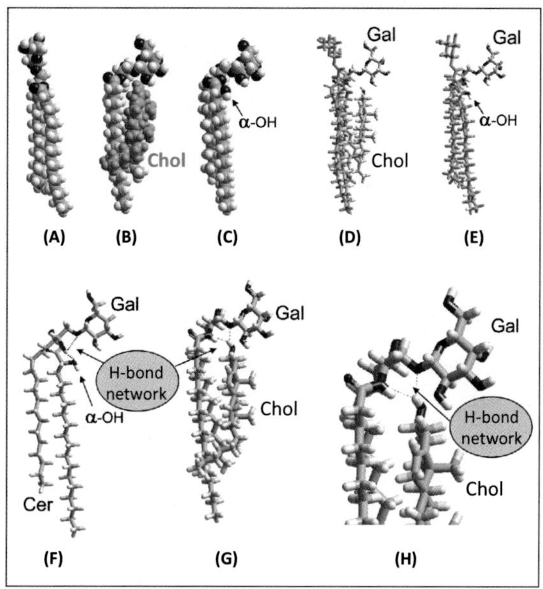

FIGURE 11.10 **Molecular modeling simulations of GalCer-HFA and GalCer-NFA (alone or complexed with cholesterol).** (A) GalCer-NFA. (B) GalCer-NFA complexed with cholesterol (in green). (C) GalCer-HFA (the α-OH group is indicated). (D) Superposition of GalCer-NFA (same model as in subpart A, colored in green) with the GalCer-NFA/cholesterol complex (identical to subpart B). Note the distinct orientation of the galactose head group (Gal) induced by cholesterol. (E) Superposition of GalCer-NFA (in green) with GalCer-HFA. (F) In GalCer-HFA, the galactose head group is maintained in a shovel-like conformation by a network of H-bonds. (G) In GalCer-NFA complexed with cholesterol, the galactose head group is also maintained in a typical shovel-like conformation through a H-bond network. (H) Higher magnification of the H-bond network in GalCer-NFA complexed with cholesterol. *(Reproduced with permission from Fantini et al.[98])*

glycosphingolipid adopted a shovel-like L-shape structure Fig. 11.10B,D,E that was recognized by Aβ·[98] This cholesterol-constrained conformation of GalCer-NFA was stabilized by a hydrogen-bond network involving the OH group of cholesterol, the NH of sphingosine, and the oxygen atom of the glycosidic bond Fig. 11.10H. In GalCer-HFA, the α-OH group of the acyl chain restricted the conformation of the galactose ring so that the molecule spontaneously adopted, without the help of cholesterol, the L-shape structure that is required for Aβ binding Fig. 11.10C. This conformation of GalCer-HFA has been confirmed by crystallographic data that demonstrated the orientation of the galactose head group GalCer-HFA is stabilized by a network of hydrogen bonds Fig. 11.10F involving the α-OH group of the acyl chain, the NH of sphingosine, and the oxygen atom of the glycosidic bond.[127,128] In presence of cholesterol, the OH group of the sterol may form a hydrogen bond with the OH group of the sterol and this new bond will affect the conformation of GalCer, explaining why cholesterol inhibited the binding of Aβ to GalCer-HFA. Overall, these data indicated that the α-OH group of the acyl chain and the OH of cholesterol have a comparable conformational effect on GalCer, and that cholesterol can either inhibit or facilitate membrane-Aβ interactions through fine tuning of glycosphingolipid conformation.[98] This modulatory tuning-like effect of cholesterol was also observed for other NFA and HFA glycosphingolipids, including LacCer and GM1.[98]

The discrimination of Aβ between HFA and NFA glycosphingolipids and the cholesterol-driven modulation of Aβ binding to these membrane lipids may be linked to the neuropathology of Alzheimer's disease in men and women. It has been reported that the balance between NFA and HFA-ceramides is markedly altered in an animal model (the transgenic APPSL/PS1Ki mouse) of Alzheimer's disease.[129] Specifically, female mice exhibited a strong increase in HFA-ceramides (those recognized by Aβ), whereas males showed a significant elevation of NFA-ceramide species (which do not interact with Aβ). The authors of this study concluded that these gender differences in sphingolipid accumulation could contribute to the increased propensity of females to develop an Alzheimer's disease.[129] How cholesterol affects Aβ–sphingolipid interactions in men and women suffering from Alzheimer's disease has not been determined yet.

11.9 CONCLUSIONS OF THE INQUIRY

In this chapter, we covered a select part of the complex relationship between Alzheimer's disease and brain membranes. Related topics are covered in other chapters of this book. For instance, the involvement of the nicotinic acetylcholine receptor in the cascade of events that links Aβ to the hyperphosphorylation of tau[130] is discussed in Chapter 7 after the structural description of this receptor and its interaction with nonannular lipids. It should also be mentioned that membrane lipids are now believed to favor the aggregation of the tau protein through specific lipid–protein interactions[131,132] as described in Chapter 6. The connection among cholesterol metabolism, statins, and Alzheimer's disease[126] is treated in Chapter 3. In the present chapter, we chose to focus our discussion on glycosphingolipids and cholesterol, both types of lipids that play a critical role in the neurotoxicity of Aβ.[133] The charges against these lipids are summarized as follows. Cholesterol stimulates Aβ production via the amyloidogenic pathway of APP processing[116] and facilitates the oligomerization of Aβ peptides into annular pores that allow Ca^{2+} ions uncontrolled entry into brain cells.[77] Ganglioside GM1 may induce the formation of highly toxic Aβ fibrils[134] and, depending on Aβ concentrations, may also favor α-helical structuration[101] that leads to membrane insertion and amyloid pore formation.[77] Moreover, Aβ oligomers bind to GM1 in vivo, and blocking GM1–Aβ interactions decreased Aβ oligomer-mediated long-term potentiation (LTP) impairment.[135] Aβ binding to GalCer[98] may be related to the lethal interaction of the amyloid peptide to oligodendrocytes[136,137] and contribute to the glial inflammation in Alzheimer's disease.[138] Finally, differential expression of HFA and NFA sphingolipids may account for women's higher risk for developing an Alzheimer's disease.

If we still consider the amyloid paradigm of Alzheimer's disease as valid, it might be difficult to exonerate the principal actor of this scenario (i.e., Aβ).[11] Nevertheless, transferring the responsibility of neurotoxicity from amyloid plaques to Aβ oligomers have allowed membrane lipids to attain a "best supporting role." Without the help of sphingolipids and cholesterol, it is likely that Aβ would not, by itself, induce the important synaptic dysfunction and neuronal loss that occurs during the course of Alzheimer's disease. In summary, Aβ and membrane lipids, the two culprits, share an equal culpability.

11.10 KEY EXPERIMENT: A BLOOD-BASED TEST TO PREDICT ALZHEIMER'S DISEASE?

The need to develop reliable test for predicting the development of Alzheimer's disease before the cognitive manifestations of the disease is imperative. The mechanisms of pore formation described in this chapter are probably operative in the early asymptomatic stages. When the disease is eventually diagnosed, it might be too late to target amyloid pores. Thus it is important to diagnose the disease as soon as possible and to try to control the neurotoxicity of Aβ peptides before significant synaptic dysfunction and neuronal loss occurs. A large number of proteins, peptides, and amino acids, including Aβ, have been examined in plasma in order to develop a blood-based test to predict Alzheimer's disease.[139] For instance, it has been suggested that high concentrations of Aβ1–40 with low concentrations of Aβ1–42 in the plasma are indicative of an increased risk of dementia. Unfortunately, proteomic approaches have not yet allowed the development a reliable blood detection test.[140] Recently Mapstone et al.[141] published the results of a lipidomic approach to the detection of preclinical Alzheimer's disease in a group of cognitively normal adults aged about 80 years. On the basis of a metabolomic analysis, they identified and validated a set of 10 lipids from peripheral blood that predicted the development of Alzheimer's disease within 2–3 years before the symptoms with greater than 90% accuracy. The originality of this biomarker panel is that it is based on lipid metabolites that may reflect cell membrane integrity, especially phosphatidylcholines and lysophosphatidylcholine (i.e., phosphatidylcholine with only one acyl chain). This association is not totally surprising because membrane phospholipid abnormalities have been detected in the brains of Alzheimer's disease patients.[142] Moreover, changes in lysophosphatidylcholine to phosphatidylcholine ratios have been consistently detected in the cerebrospinal fluid of patients with Alzheimer's disease.[143] Therefore, it is likely that the amounts and ratios of phospholipids in the plasma may reflect abnormalities of lipid metabolism in brain cell membranes that are likely to occur in the earliest stages of Alzheimer's disease. The quantitative detection of these phospholipids in blood samples may thus be highly predictive of the development of Alzheimer's disease within 2–3 years as proposed by Mapstone et al.[141] Incidentally, these data strongly support the notion that Alzheimer's disease is primarily a disorder of the plasma membrane.[13,144]

References

1. Maurer K. Historical background of Alzheimer's research done 100 years ago. *J Neural Transm*. 2006;113(11): 1597–1601.
2. Vina J, Lloret A. Why women have more Alzheimer's disease than men: gender and mitochondrial toxicity of amyloid-beta peptide. *J Alzheimers Dis*. 2010;20(suppl 2):S527–S533.
3. Morris GP, Clark IA, Vissel B. Inconsistencies and controversies surrounding the amyloid hypothesis of Alzheimer's disease. *Acta Neuropathol Commun*. 2014;2(1):135.
4. Kosik KS, Joachim CL, Selkoe DJ. Microtubule-associated protein tau (tau) is a major antigenic component of paired helical filaments in Alzheimer disease. *Proc Natl Acad Sci USA*. 1986;83(11):4044–4048.
5. Wood JG, Mirra SS, Pollock NJ, Binder LI. Neurofibrillary tangles of Alzheimer disease share antigenic determinants with the axonal microtubule-associated protein tau (tau). *Proc Natl Acad Sci USA*. 1986;83(11):4040–4043.
6. Allsop D, Landon M, Kidd M. The isolation and amino acid composition of senile plaque core protein. *Brain Res*. 1983;259(2):348–352.
7. Masters CL, Simms G, Weinman NA, Multhaup G, McDonald BL, Beyreuther K. Amyloid plaque core protein in Alzheimer disease and Down syndrome. *Proc Natl Acad Sci USA*. 1985;82(12):4245–4249.

8. Jang H, Connelly L, Arce FT, et al. Alzheimer's disease: which type of amyloid-preventing drug agents to employ? *Phys Chem Chem Phys*. 2013;15(23):8868–8877.
9. Zhang H, Ma Q, Zhang YW, Xu H. Proteolytic processing of Alzheimer's beta-amyloid precursor protein. *J Neurochem*. 2012;120(suppl 1):9–21.
10. Glabe CG. Structural classification of toxic amyloid oligomers. *J Biol Chem*. 2008;283(44):29639–29643.
11. Hardy JA, Higgins GA. Alzheimer's disease: the amyloid cascade hypothesis. *Science*. 1992;256(5054):184–185.
12. Zheng WH, Bastianetto S, Mennicken F, Ma W, Kar S. Amyloid beta peptide induces tau phosphorylation and loss of cholinergic neurons in rat primary septal cultures. *Neuroscience*. 2002;115(1):201–211.
13. van Echten-Deckert G, Walter J. Sphingolipids: critical players in Alzheimer's disease. *Prog Lipid Res*. 2012;51(4):378–393.
14. Fantini J, Yahi N. Molecular insights into amyloid regulation by membrane cholesterol and sphingolipids: common mechanisms in neurodegenerative diseases. *Expert Rev Mol Med*. 2010;12:e27.
15. Masters CL, Gajdusek DC, Gibbs Jr CJ. The familial occurrence of Creutzfeldt-Jakob disease and Alzheimer's disease. *Brain*. 1981;104(3):535–558.
16. Prusiner SB, McKinley MP, Bowman KA, et al. Scrapie prions aggregate to form amyloid-like birefringent rods. *Cell*. 1983;35(2 Pt 1):349–358.
17. Glenner GG, Wong CW. Alzheimer's disease: initial report of the purification and characterization of a novel cerebrovascular amyloid protein. *Biochem Biophys Res Commun*. 1984;120(3):885–890.
18. Grundke-Iqbal I, Iqbal K, Tung YC, Quinlan M, Wisniewski HM, Binder LI. Abnormal phosphorylation of the microtubule-associated protein tau (tau) in Alzheimer cytoskeletal pathology. *Proc Natl Acad Sci USA*. 1986;83(13):4913–4917.
19. Yankner BA, Dawes LR, Fisher S, Villa-Komaroff L, Oster-Granite ML, Neve RL. Neurotoxicity of a fragment of the amyloid precursor associated with Alzheimer's disease. *Science*. 1989;245(4916):417–420.
20. Busciglio J, Lorenzo A, Yeh J, Yankner BA. Beta-amyloid fibrils induce tau phosphorylation and loss of microtubule binding. *Neuron*. 1995;14(4):879–888.
21. Alonso AC, Zaidi T, Grundke-Iqbal I, Iqbal K. Role of abnormally phosphorylated tau in the breakdown of microtubules in Alzheimer disease. *Proc Natl Acad Sci USA*. 1994;91(12):5562–5566.
22. Hardy J, Allsop D. Amyloid deposition as the central event in the aetiology of Alzheimer's disease. *Trends Pharmacol Sci*. 1991;12(10):383–388.
23. Guo Q, Li H, Gaddam SS, Justice NJ, Robertson CS, Zheng H. Amyloid precursor protein revisited: neuron-specific expression and highly stable nature of soluble derivatives. *J Biol Chem*. 2012;287(4):2437–2445.
24. Dimitrov M, Alattia JR, Lemmin T, et al. Alzheimer's disease mutations in APP but not gamma-secretase modulators affect epsilon-cleavage-dependent AICD production. *Nat Commun*. 2013;4:2246.
25. Mattson MP. Cellular actions of beta-amyloid precursor protein and its soluble and fibrillogenic derivatives. *Physiol Rev*. 1997;77(4):1081–1132.
26. Soscia SJ, Kirby JE, Washicosky KJ, et al. The Alzheimer's disease-associated amyloid beta-protein is an antimicrobial peptide. *PloS One*. 2010;5(3):e9505.
27. Morley JE, Farr SA, Banks WA, Johnson SN, Yamada KA, Xu L. A physiological role for amyloid-beta protein: enhancement of learning and memory. *J Alzheimers Dis*. 2010;19(2):441–449.
28. Puzzo D, Privitera L, Leznik E, et al. Picomolar amyloid-beta positively modulates synaptic plasticity and memory in hippocampus. *J Neurosci*. 2008;28(53):14537–14545.
29. Levy E, Carman MD, Fernandez-Madrid IJ, et al. Mutation of the Alzheimer's disease amyloid gene in hereditary cerebral hemorrhage, Dutch type. *Science*. 1990;248(4959):1124–1126.
30. Nilsberth C, Westlind-Danielsson A, Eckman CB, et al. The 'Arctic' APP mutation (E693G) causes Alzheimer's disease by enhanced Abeta protofibril formation. *Nat Neurosci*. 2001;4(9):887–893.
31. Yoshioka K, Miki T, Katsuya T, Ogihara T, Sakaki Y. The 717Val—Ile substitution in amyloid precursor protein is associated with familial Alzheimer's disease regardless of ethnic groups. *Biochem Biophys Res Commun*. 1991;178(3):1141–1146.
32. Murrell J, Farlow M, Ghetti B, Benson MD. A mutation in the amyloid precursor protein associated with hereditary Alzheimer's disease. *Science*. 1991;254(5028):97–99.
33. Games D, Adams D, Alessandrini R, et al. Alzheimer-type neuropathology in transgenic mice overexpressing V717F beta-amyloid precursor protein. *Nature*. 1995;373(6514):523–527.
34. Uversky VN. Intrinsic disorder in proteins associated with neurodegenerative diseases. *Front Biosci*. 2009;14:5188–5238.

35. Luhrs T, Ritter C, Adrian M, et al. 3D structure of Alzheimer's amyloid-beta(1-42) fibrils. *Proc Natl Acad Sci USA*. 2005;102(48):17342–17347.
36. Serpell LC. Alzheimer's amyloid fibrils: structure and assembly. *Biochim Biophys Acta*. 2000;1502(1):16–30.
37. Friedrich RP, Tepper K, Ronicke R, et al. Mechanism of amyloid plaque formation suggests an intracellular basis of Abeta pathogenicity. *Proc Natl Acad Sci USA*. 2010;107(5):1942–1947.
38. Masliah E, Iwai A, Mallory M, Ueda K, Saitoh T. Altered presynaptic protein NACP is associated with plaque formation and neurodegeneration in Alzheimer's disease. *Am J Pathol*. 1996;148(1):201–210.
39. Rentz DM, Locascio JJ, Becker JA, et al. Cognition, reserve, and amyloid deposition in normal aging. *Ann Neurol*. 2010;67(3):353–364.
40. Esparza TJ, Zhao H, Cirrito JR, et al. Amyloid-beta oligomerization in Alzheimer dementia versus high-pathology controls. *Ann Neurol*. 2013;73(1):104–119.
41. Chetelat G, La Joie R, Villain N, et al. Amyloid imaging in cognitively normal individuals, at-risk populations and preclinical Alzheimer's disease. *Neuroimage Clin*. 2013;2:356–365.
42. Rosenblum WI. Why Alzheimer trials fail: removing soluble oligomeric beta amyloid is essential, inconsistent, and difficult. *Neurobiol Aging*. 2014;35(5):969–974.
43. Chetelat G. Alzheimer disease: Abeta-independent processes-rethinking preclinical AD. *Nat Rev Neurol*. 2013;9(3):123–124.
44. Shankar GM, Li S, Mehta TH, et al. Amyloid-beta protein dimers isolated directly from Alzheimer's brains impair synaptic plasticity and memory. *Nat Med*. 2008;14(8):837–842.
45. Jang H, Arce FT, Ramachandran S, Capone R, Lal R, Nussinov R. Beta-barrel topology of Alzheimer's beta-amyloid ion channels. *J Mol Biol*. 2010;404(5):917–934.
46. Ono K, Yamada M. Low-n oligomers as therapeutic targets of Alzheimer's disease. *J Neurochem*. 2011;117(1): 19–28.
47. Zhao LN, Long H, Mu Y, Chew LY. The Toxicity of Amyloid beta Oligomers. *Int J Mol Sci*. 2012;13(6):7303–7327.
48. Shafrir Y, Durell S, Arispe N, Guy HR. Models of membrane-bound Alzheimer's Abeta peptide assemblies. *Proteins*. 2010;78(16):3473–3487.
49. Khachaturian ZS. The role of calcium regulation in brain aging: reexamination of a hypothesis. *Aging*. 1989;1(1):17–34.
50. Berridge MJ. Calcium hypothesis of Alzheimer's disease. *Pflugers Arch*. 2010;459(3):441–449.
51. Thinnes FP. On a solution to the riddle of "amyloid made" versus "amyloid regulated" channels in cell membranes. Letter to the Editor. *J Alzheimer Dis*. 2013. Available from: http://www.j-alz.com/node/309 (September 1, 2013).
52. Thinnes FP. Concerning a clue on the enigma of "amyloid made" versus "amyloid regulated" channels in cell membranes: a comment on M. A. Mukhamedyarov et al. NBP. 43: 479–484. 2013. *Neurosci Behav Physiol*. 2014;44(1):125–126.
53. Kayed R, Sokolov Y, Edmonds B, et al. Permeabilization of lipid bilayers is a common conformation-dependent activity of soluble amyloid oligomers in protein misfolding diseases. *J Biol Chem*. 2004;279(45):46363–46366.
54. Arispe N, Pollard HB, Rojas E. Zn2+ interaction with Alzheimer amyloid beta protein calcium channels. *Proc Natl Acad Sci USA*. 1996;93(4):1710–1715.
55. Rhee SK, Quist AP, Lal R. Amyloid beta protein-(1-42) forms calcium-permeable, Zn2+-sensitive channel. *J Biol Chem*. 1998;273(22):13379–13382.
56. Kawahara M, Arispe N, Kuroda Y, Rojas E. Alzheimer's disease amyloid beta-protein forms Zn(2+)-sensitive, cation-selective channels across excised membrane patches from hypothalamic neurons. *Biophys J*. 1997;73(1): 67–75.
57. Sanderson KL, Butler L, Ingram VM. Aggregates of a beta-amyloid peptide are required to induce calcium currents in neuron-like human teratocarcinoma cells: relation to Alzheimer's disease. *Brain Res*. 1997;744(1):7–14.
58. Lin H, Bhatia R, Lal R. Amyloid beta protein forms ion channels: implications for Alzheimer's disease pathophysiology. *FASEB J*. 2001;15(13):2433–2444.
59. Lashuel HA, Hartley D, Petre BM, Walz T, Lansbury Jr PT. Neurodegenerative disease: amyloid pores from pathogenic mutations. *Nature*. 2002;418(6895):291.
60. Quist A, Doudevski I, Lin H, et al. Amyloid ion channels: a common structural link for protein-misfolding disease. *Proc Natl Acad Sci USA*. 2005;102(30):10427–10432.
61. Uversky VN, Dave V, Iakoucheva LM, et al. Pathological unfoldomics of uncontrolled chaos: intrinsically disordered proteins and human diseases. *Chem Rev*. 2014;114(13):6844–6879.
62. Uversky VN. Introduction to intrinsically disordered proteins (IDPs). *Chem Rev*. 2014;114(13):6557–6560.

63. Kallberg Y, Gustafsson M, Persson B, Thyberg J, Johansson J. Prediction of amyloid fibril-forming proteins. *J Biol Chem*. 2001;276(16):12945–12950.
64. Jang H, Arce FT, Ramachandran S, et al. Truncated beta-amyloid peptide channels provide an alternative mechanism for Alzheimer's disease and Down syndrome. *Proc Natl Acad Sci USA*. 2010;107(14):6538–6543.
65. de Planque MR, Raussens V, Contera SA, et al. Beta-sheet structured beta-amyloid(1-40) perturbs phosphatidylcholine model membranes. *J Mol Biol*. 2007;368(4):982–997.
66. Ege C, Lee KY. Insertion of Alzheimer's A beta 40 peptide into lipid monolayers. *Biophys J*. 2004;87(3):1732–1740.
67. Lashuel HA, Lansbury Jr PT. Are amyloid diseases caused by protein aggregates that mimic bacterial pore-forming toxins? *Q Rev Biophys*. 2006;39(2):167–201.
68. Kagan BL, Thundimadathil J. Amyloid peptide pores and the beta sheet conformation. *Adv Exp Med Biol*. 2010;677:150–167.
69. Butterfield SM, Lashuel HA. Amyloidogenic protein-membrane interactions: mechanistic insight from model systems. *Angew Chem Int Ed Engl*. 2010;49(33):5628–5654.
70. Ramachandran R, Tweten RK, Johnson AE. Membrane-dependent conformational changes initiate cholesterol-dependent cytolysin oligomerization and intersubunit beta-strand alignment. *Nat Struct Mol Biol*. 2004;11(8):697–705.
71. Gilbert RJ. Cholesterol-dependent cytolysins. *Adv Exp Med Biol*. 2010;677:56–66.
72. Flanagan JJ, Tweten RK, Johnson AE, Heuck AP. Cholesterol exposure at the membrane surface is necessary and sufficient to trigger perfringolysin O binding. *Biochemistry*. 2009;48(18):3977–3987.
73. Breydo L, Reddy KD, Piai A, Felli IC, Pierattelli R, Uversky VN. The crowd you're in with: effects of different types of crowding agents on protein aggregation. *Biochim Biophys Acta*. 2014;1844(2):346–357.
74. Ulmer TS, Bax A, Cole NB, Nussbaum RL. Structure and dynamics of micelle-bound human alpha-synuclein. *J Biol Chem*. 2005;280(10):9595–9603.
75. Coles M, Bicknell W, Watson AA, Fairlie DP, Craik DJ. Solution structure of amyloid beta-peptide(1-40) in a water-micelle environment. Is the membrane-spanning domain where we think it is? *Biochemistry*. 1998;37(31):11064–11077.
76. Fantini J, Garmy N, Mahfoud R, Yahi N. Lipid rafts: structure, function and role in HIV, Alzheimer's and prion diseases. *Expert Rev Mol Med*. 2002;4(27):1–22.
77. Di Scala C, Troadec JD, Lelievre C, Garmy N, Fantini J, Chahinian H. Mechanism of cholesterol-assisted oligomeric channel formation by a short Alzheimer beta-amyloid peptide. *J Neurochem*. 2013.
78. Shao H, Jao S, Ma K, Zagorski MG. Solution structures of micelle-bound amyloid beta-(1-40) and beta-(1-42) peptides of Alzheimer's disease. *J Mol Biol*. 1999;285(2):755–773.
79. Di Scala C, Chahinian H, Yahi N, Garmy N, Fantini J. Interaction of Alzheimer's beta-amyloid peptides with cholesterol: mechanistic insights into amyloid pore formation. *Biochemistry*. 2014;53(28):4489–4502.
80. Arispe N, Doh M. Plasma membrane cholesterol controls the cytotoxicity of Alzheimer's disease AbetaP (1-40) and (1-42) peptides. *FASEB J*. 2002;16(12):1526–1536.
81. Di Scala C, Yahi N, Lelievre C, Garmy N, Chahinian H, Fantini J. Biochemical identification of a linear cholesterol-binding domain within Alzheimer's beta amyloid peptide. *ACS Chem Neurosci*. 2013;4(3):509–517.
82. Fantini J, Di Scala C, Yahi N, et al. Bexarotene blocks calcium-permeable ion channels formed by neurotoxic Alzheimer's β-amyloid peptides. *ACS Chem Neurosci*. 2014.
83. Fantini J, Carlus D, Yahi N. The fusogenic tilted peptide (67-78) of alpha-synuclein is a cholesterol binding domain. *Biochim Biophys Acta*. 2011;1808(10):2343–2351.
84. Strandberg E, Killian JA. Snorkeling of lysine side chains in transmembrane helices: how easy can it get? *FEBS Lett*. 2003;544(1–3):69–73.
85. Ji SR, Wu Y, Sui SF. Cholesterol is an important factor affecting the membrane insertion of beta-amyloid peptide (A beta 1-40), which may potentially inhibit the fibril formation. *J Biol Chem*. 2002;277(8):6273–6279.
86. Yu X, Zheng J. Cholesterol promotes the interaction of Alzheimer beta-amyloid monomer with lipid bilayer. *J Mol Biol*. 2012;421(4–5):561–571.
87. Wong PT, Schauerte JA, Wisser KC, et al. Amyloid-beta membrane binding and permeabilization are distinct processes influenced separately by membrane charge and fluidity. *J Mol Biol*. 2009;386(1):81–96.
88. Durell SR, Guy HR, Arispe N, Rojas E, Pollard HB. Theoretical models of the ion channel structure of amyloid beta-protein. *Biophys J*. 1994;67(6):2137–2145.
89. Ariga T, Kobayashi K, Hasegawa A, Kiso M, Ishida H, Miyatake T. Characterization of high-affinity binding between gangliosides and amyloid beta-protein. *Archiv Biochem Biophys*. 2001;388(2):225–230.

90. Yanagisawa K, Odaka A, Suzuki N, Ihara Y. GM1 ganglioside-bound amyloid beta-protein (A beta): a possible form of preamyloid in Alzheimer's disease. *Nat Med*. 1995;1(10):1062–1066.
91. Matsuzaki K, Kato K, Yanagisawa K. Abeta polymerization through interaction with membrane gangliosides. *Biochim Biophys Acta*. 2010;1801(8):868–877.
92. Fantini J, Yahi N, Garmy N. Cholesterol accelerates the binding of Alzheimer's beta-amyloid peptide to ganglioside GM1 through a universal hydrogen-bond-dependent sterol tuning of glycolipid conformation. *Front Physiol*. 2013;4:120.
93. Mahfoud R, Garmy N, Maresca M, Yahi N, Puigserver A, Fantini J. Identification of a common sphingolipid-binding domain in Alzheimer, prion, and HIV-1 proteins. *J Biol Chem*. 2002;277(13):11292–11296.
94. Fantini J, Yahi N. Molecular basis for the glycosphingolipid-binding specificity of alpha-synuclein: key role of tyrosine 39 in membrane insertion. *J Mol Biol*. 2011;408(4):654–669.
95. Yahi N, Fantini J. Deciphering the glycolipid code of Alzheimer's and Parkinson's amyloid proteins allowed the creation of a universal ganglioside-binding peptide. *PloS One*. 2014;9(8):e104751.
96. Utsumi M, Yamaguchi Y, Sasakawa H, Yamamoto N, Yanagisawa K, Kato K. Up-and-down topological mode of amyloid beta-peptide lying on hydrophilic/hydrophobic interface of ganglioside clusters. *Glycoconj J*. 2009;26(8):999–1006.
97. Ikeda K, Matsuzaki K. Driving force of binding of amyloid beta-protein to lipid bilayers. *Biochem Biophys Res Commun*. 2008;370(3):525–529.
98. Yahi N, Aulas A, Fantini J. How cholesterol constrains glycolipid conformation for optimal recognition of Alzheimer's beta amyloid peptide (Abeta1-40). *PloS One*. 2010;5(2):e9079.
99. Lingwood D, Binnington B, Rog T, et al. Cholesterol modulates glycolipid conformation and receptor activity. *Nat Chem Biol*. 2011;7(5):260–262.
100. Kakio A, Nishimoto S, Yanagisawa K, Kozutsumi Y, Matsuzaki K. Interactions of amyloid beta-protein with various gangliosides in raft-like membranes: importance of GM1 ganglioside-bound form as an endogenous seed for Alzheimer amyloid. *Biochemistry*. 2002;41(23):7385–7390.
101. McLaurin J, Franklin T, Fraser PE, Chakrabartty A. Structural transitions associated with the interaction of Alzheimer beta-amyloid peptides with gangliosides. *J Biol Chem*. 1998;273(8):4506–4515.
102. Fantini J, Yahi N. The driving force of alpha-synuclein insertion and amyloid channel formation in the plasma membrane of neural cells: key role of ganglioside- and cholesterol-binding domains. *Adv Exp Med Biol*. 2013;991: 15–26.
103. Cole SL, Vassar R. The Alzheimer's disease beta-secretase enzyme, BACE1. *Mol Neurodegener*. 2007;2:22.
104. Mattson MP, Cheng B, Culwell AR, Esch FS, Lieberburg I, Rydel RE. Evidence for excitoprotective and intraneuronal calcium-regulating roles for secreted forms of the beta-amyloid precursor protein. *Neuron*. 1993;10(2):243–254.
105. Hickman SE, Allison EK, El Khoury J. Microglial dysfunction and defective beta-amyloid clearance pathways in aging Alzheimer's disease mice. *J Neurosci*. 2008;28(33):8354–8360.
106. Bamberger ME, Harris ME, McDonald DR, Husemann J, Landreth GE. A cell surface receptor complex for fibrillar beta-amyloid mediates microglial activation. *J Neurosci*. 2003;23(7):2665–2674.
107. Mandrekar S, Jiang Q, Lee CY, Koenigsknecht-Talboo J, Holtzman DM, Landreth GE. Microglia mediate the clearance of soluble Abeta through fluid phase macropinocytosis. *J Neurosci*. 2009;29(13):4252–4262.
108. Bateman RJ, Munsell LY, Morris JC, Swarm R, Yarasheski KE, Holtzman DM. Human amyloid-beta synthesis and clearance rates as measured in cerebrospinal fluid in vivo. *Nat Med*. 2006;12(7):856–861.
109. Kim J, Basak JM, Holtzman DM. The role of apolipoprotein E in Alzheimer's disease. *Neuron*. 2009;63(3):287–303.
110. Verghese PB, Castellano JM, Garai K, et al. ApoE influences amyloid-beta (Abeta) clearance despite minimal apoE/Abeta association in physiological conditions. *Proc Natl Acad Sci USA*. 2013;110(19):E1807–1816.
111. Strittmatter WJ, Saunders AM, Schmechel D, et al. Apolipoprotein E: high-avidity binding to beta-amyloid and increased frequency of type 4 allele in late-onset familial Alzheimer disease. *Proc Natl Acad Sci USA*. 1993;90(5):1977–1981.
112. Chouraki V, De Bruijn RF, Chapuis J, et al. A genome-wide association meta-analysis of plasma Abeta peptides concentrations in the elderly. *Mol Psychiatry*. 2014.
113. Parkin ET, Turner AJ, Hooper NM. Amyloid precursor protein, although partially detergent-insoluble in mouse cerebral cortex, behaves as an atypical lipid raft protein. *Biochem J*. 1999;344(Pt 1):23–30.
114. Chen TY, Liu PH, Ruan CT, Chiu L, Kung FL. The intracellular domain of amyloid precursor protein interacts with flotillin-1, a lipid raft protein. *Biochem Biophys Res Commun*. 2006;342(1):266–272.

115. Schneider A, Rajendran L, Honsho M, et al. Flotillin-dependent clustering of the amyloid precursor protein regulates its endocytosis and amyloidogenic processing in neurons. *J Neurosci*. 2008;28(11):2874–2882.
116. Ehehalt R, Keller P, Haass C, Thiele C, Simons K. Amyloidogenic processing of the Alzheimer beta-amyloid precursor protein depends on lipid rafts. *J Cell Biol*. 2003;160(1):113–123.
117. Hicks DA, Nalivaeva NN, Turner AJ. Lipid rafts and Alzheimer's disease: protein-lipid interactions and perturbation of signaling. *Front Physiol*. 2012;3:189.
118. Rushworth JV, Hooper NM. Lipid rafts: linking Alzheimer's amyloid-beta production, aggregation, and toxicity at neuronal membranes. *Int J Alzheimers Dis*. 2010;2011:603052.
119. Lee SJ, Liyanage U, Bickel PE, Xia W, Lansbury Jr PT, Kosik KS. A detergent-insoluble membrane compartment contains A beta *in vivo*. *Nat Med*. 1998;4(6):730–734.
120. Lahdo R, Coillet-Matillon S, Chauvet JP, de La Fourniere-Bessueille L. The amyloid precursor protein interacts with neutral lipids. *Eur J Biochem*. 2002;269(8):2238–2246.
121. Beel AJ, Mobley CK, Kim HJ, et al. Structural studies of the transmembrane C-terminal domain of the amyloid precursor protein (APP): does APP function as a cholesterol sensor? *Biochemistry*. 2008;47(36):9428–9446.
122. Beel AJ, Sakakura M, Barrett PJ, Sanders CR. Direct binding of cholesterol to the amyloid precursor protein: an important interaction in lipid-Alzheimer's disease relationships? *Biochim Biophys Acta*. 2010;1801(8):975–982.
123. Barrett PJ, Song Y, Van Horn WD, et al. The amyloid precursor protein has a flexible transmembrane domain and binds cholesterol. *Science*. 2012;336(6085):1168–1171.
124. Kivipelto M, Helkala EL, Laakso MP, et al. Midlife vascular risk factors and Alzheimer's disease in later life: longitudinal, population based study. *BMJ*. 2001;322(7300):1447–1451.
125. Pappolla MA, Bryant-Thomas TK, Herbert D, et al. Mild hypercholesterolemia is an early risk factor for the development of Alzheimer amyloid pathology. *Neurology*. 2003;61(2):199–205.
126. Wood WG, Li L, Muller WE, Eckert GP. Cholesterol as a causative factor in Alzheimer's disease: a debatable hypothesis. *J Neurochem*. 2013.
127. Lofgren H, Pascher I. Molecular arrangements of sphingolipids. The monolayer behaviour of ceramides. *Chem Phys Lipids*. 1977;20(4):273–284.
128. Nyholm PG, Pascher I, Sundell S. The effect of hydrogen bonds on the conformation of glycosphingolipids. Methylated and unmethylated cerebroside studied by X-ray single crystal analysis and model calculations. *Chem Phys Lipids*. 1990;52(1):1–10.
129. Barrier L, Ingrand S, Fauconneau B, Page G. Gender-dependent accumulation of ceramides in the cerebral cortex of the APP(SL)/PS1Ki mouse model of Alzheimer's disease. *Neurobiol Aging*. 2010;31(11):1843–1853.
130. Oz M, Lorke DE, Yang KH, Petroianu G. On the interaction of beta-amyloid peptides and alpha7-nicotinic acetylcholine receptors in Alzheimer's disease. *Curr Alzheimer Res*. 2013;10(6):618–630.
131. Kunze G, Barre P, Scheidt HA, Thomas L, Eliezer D, Huster D. Binding of the three-repeat domain of tau to phospholipid membranes induces an aggregated-like state of the protein. *Biochim Biophys Acta*. 2012;1818(9): 2302–2313.
132. Georgieva ER, Xiao S, Borbat PP, Freed JH, Eliezer D. Tau binds to lipid membrane surfaces via short amphipathic helices located in its microtubule-binding repeats. *Biophys J*. 2014;107(6):1441–1452.
133. Eckert GP, Wood WG, Muller WE. Lipid membranes and beta-amyloid: a harmful connection. *Curr Protein Pept Sci*. 2010;11(5):319–325.
134. Yanagisawa K. Pathological significance of ganglioside clusters in Alzheimer's disease. *J Neurochem*. 2011; 116(5):806–812.
135. Hong S, Ostaszewski BL, Yang T, et al. Soluble Abeta oligomers are rapidly sequestered from brain ISF in vivo and bind GM1 ganglioside on cellular membranes. *Neuron*. 2014;82(2):308–319.
136. Xu J, Chen S, Ahmed SH, et al. Amyloid-beta peptides are cytotoxic to oligodendrocytes. *J Neurosci*. 2001; 21(1):Rc118.
137. Lee JT, Xu J, Lee JM, et al. Amyloid-beta peptide induces oligodendrocyte death by activating the neutral sphingomyelinase-ceramide pathway. *J Cell Biol*. 2004;164(1):123–131.
138. Roth AD, Ramirez G, Alarcon R, Von Bernhardi R. Oligodendrocytes damage in Alzheimer's disease: beta amyloid toxicity and inflammation. *Biol Res*. 2005;38(4):381–387.
139. Kawarabayashi T, Shoji M. Plasma biomarkers of Alzheimer's disease. *Curr Opin Psychiatr*. 2008;21(3):260–267.
140. Thambisetty M, Lovestone S. Blood-based biomarkers of Alzheimer's disease: challenging but feasible. *Biomark Med*. 2010;4(1):65–79.

141. Mapstone M, Cheema AK, Fiandaca MS. Plasma phospholipids identify antecedent memory impairment in older adults. *Nat Med*. 2014;20(4):415–418.
142. Nitsch RM, Blusztajn JK, Pittas AG, Slack BE, Growdon JH, Wurtman RJ. Evidence for a membrane defect in Alzheimer disease brain. *Proc Natl Acad Sci USA*. 1992;89(5):1671–1675.
143. Mulder C, Wahlund LO, Teerlink T, et al. Decreased lysophosphatidylcholine/phosphatidylcholine ratio in cerebrospinal fluid in Alzheimer's disease. *J Neural Transm*. 2003;110(8):949–955.
144. Lukiw WJ. Alzheimer's disease (AD) as a disorder of the plasma membrane. *Front Physiol*. 2013;4:24.

CHAPTER

12

Viral and Bacterial Diseases

Jacques Fantini, Nouara Yahi

OUTLINE

12.1 OVERVIEW OF BRAIN PATHOGENS

The brain is a territory that must maintain a high level of protection against infectious agents just like an impregnable fortress. Therefore, the brain relies primarily on the blood–brain barrier that, under normal conditions, efficiently insulates the neurons of the central nervous system from the bloodstream and effectively controls blood–brain exchanges. An important feature of the brain is that it seems to have adopted a strategy to avoid the harmful effects of the immune response to pathogens, especially inflammation.[1,2] In particular, brain neurons do not express class I histocompatibility antigens and thus are not killed by antigen-specific cytotoxic T cells that normally recognize infectious cells and kill them. Although the lack of expression of class I antigens by brain neurons has been challenged,[3,4] neurons are nevertheless generally considered as an ideal sanctuary allowing microbes to hide from the immune system.[1] In any case, the fortress is undoubtedly robust, and most of the pathogens that infect the brain must develop sophisticated strategies to enter this sanctuary. These mechanisms include blood–brain barrier disruption, infection of white blood cells used as Trojan horse-like carriers, and propagation along axons coupled with retrograde synaptic transport,

Brain Lipids in Synaptic Function and Neurological Disease. http://dx.doi.org/10.1016/B978-0-12-800111-0.00012-6

via either peripheral or olfactory/trigeminal nerves.[1] Representative examples of each route of brain infection used by bacteria and/or viruses are described in the following sections.

12.2 PATHOGEN TRAFFIC TO THE BRAIN

12.2.1 Rabies Virus: A Traveler to the Brain via Retrograde Axonal Routes

The first mechanism we will describe is used by rabies virus, a lethal pathogen transmitted to humans by the bite of infected animals. Rabies virus belongs to the large family of Rhabdoviridae,[5] which also includes the vesicular stomatitis virus (VSV). Among the wide variety of pathogens that infect the brain, rabies virus is certainly one of the few to be authentically neurotropic. Indeed, this virus classically enters a motor neuron at the neuromuscular junction and is then subsequently transported in the retrograde direction through the axon of the infected neuron.[6,7] The genome of rabies virus consists of a negative-stranded RNA of about 12 kb, which encodes five viral proteins referred to as N (nucleoprotein), P (phosphoprotein), M (matrix protein), G (glycoprotein), and L (large protein corresponding to the polymerase).[8] The glycoprotein self-associates into trimeric spikes that are regularly arranged on the surface of the virus[9] (Fig. 12.1).

These spikes are required for the attachment of the virus to the surface of the host cell through specific interactions with a cell surface receptor.[7] The exact nature of this receptor is still a matter of debate, although a number of candidates have been suggested, including the neural cell adhesion molecule N-CAM[10] and the low-affinity nerve growth factor receptor p75NTR.[11] As for most other enveloped viruses, it is likely that cell surface glycolipids,[12]

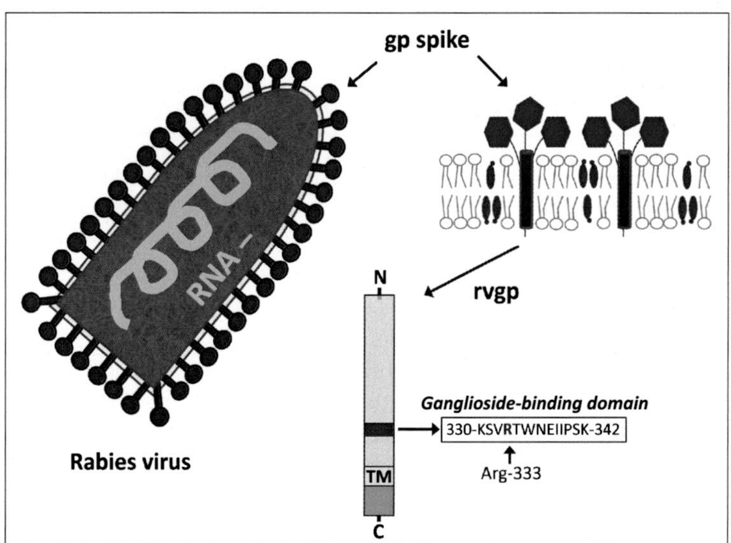

FIGURE 12.1 **Rabies virus.** Rabies virus has a bullet-like morphology. Its envelope consists of a lipid bilayer taken from the plasma membrane of an infected host cell during the budding process. Its surface is decorated by a regular arrangement of spikes formed by trimers of the rabies virus glycoprotein (rvgp). This glycoprotein is a class I transmembrane protein displaying a potential ganglioside-binding domain in a highly antigenic region.

especially gangliosides,[13] may serve as surface-binding sites for rabies virus. The involvement of gangliosides in virus binding to host cells has been demonstrated by a series of experiments performed with chicken embryo-related (CER) cells that are particularly sensitive to rabies virus infection.[14] When these cells were treated with various concentrations of neuraminidase, an enzyme that cleaves the sialic acids of cell surface gangliosides and glycoproteins, a dose-dependent decrease of rabies infection was observed.[13] Then, the insertion of exogenous gangliosides into the plasma membrane of neuraminidase-treated cells allowed these cells to recover their initial susceptibility to rabies virus infection. A structure–activity relationship indicated that highly sialylated gangliosides, including GT1b and GQ1b were the most active "recovery" gangliosides, whereas monosialylated species such as GM1 or GM3 had little or no activity. Finally, the disialylated ganglioside GD1a had an intermediate effect. These spectacular findings were corroborated by other experimental approaches. Ganglioside-enriched membrane fractions prepared from rat brain suspensions were shown to compete with virus receptors on fibroblasts and neurons.[15] In these experiments, host proteins did not seem to be involved.[5] Overall these data strongly suggested that the rabies virus interacted with selected gangliosides on the presynaptic neuronal membrane. Interestingly, the presynaptic membrane is enriched with trisialylated gangliosides and, in marked contrast with its postsynaptic counterpart, lacks GM1.[16] This differential ganglioside expression in synaptic membranes may favor the binding of rabies virus to the presynaptic neuron. Given the current uncertainty about the exact nature of rabies virus receptors, it is regrettable that this mechanism has not been explored further.

Mutational studies have shown that the infectivity of rabies virus is highly dependent on its glycoprotein.[5] The rabies virus glycoprotein contains 505 amino acids (after cleavage of the signal peptide) that form three domains: a large N-terminal exofacial domain (1–439), a transmembrane domain (440–461), and a small cytosolic domain (462–505). One of the main antigenic sites, which extends from amino acid 333 to 338,[17] has captured our attention because it belongs to a region of the protein displaying a potential ganglioside-binding domain (Fig. 12.1). Because this region is antigenic, it is located on the surface of the protein and thus fully accessible to external ligands. Indeed, our algorithm for the detection of sphingolipid-binding domains (SBDs)[18] identified a linear domain of rabies virus glycoprotein (330–342) with high ganglioside-binding potential. This domain contains Arg-333, a residue that has been shown to be highly critical for virus pathogenicity.[19,20] Replacement of this arginine residue by either leucine, isoleucine, cysteine, or methionine resulted in a totally avirulent virus.[21] In contrast, the substitution of Arg-333 with lysine did not decrease the pathogenicity of the virus, indicating that it is not the exact nature of the side chain but the presence of a positive charge at this position that is necessary for viral virulence. It is tempting to speculate that the role of this positive charge is to interact electrostatically with the negative charge of ganglioside receptors expressed on the presynaptic membrane of infected neurons. To test this hypothesis, we conducted molecular modeling simulations of the 330–342 fragment of rabies virus glycoprotein and searched a fit for GT1b gangliosides. The results of this modeling exercise are presented in Fig. 12.2.

The energy minimization of the 330–342 fragment showed that this motif could adopt a loop structure stabilized by a network of hydrogen and electrostatic bonds involving the side chains of Arg-333 and Glu-337. When this conformer was merged with a cluster of GT1b gangliosides organized into chalice-shaped dimers,[22] an important conformational rearrangement of the 330–342 fragment allowed a reorientation of the side chain of Arg-333 toward the head group of GT1b. The interaction of the 330–342 fragment with GT1b dimers was

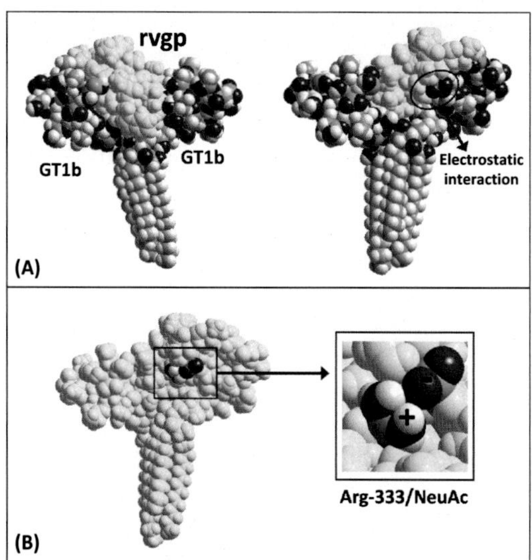

FIGURE 12.2 **Interaction of the rabies virus glycoprotein with GT1b.** The structure of the ganglioside-binding domain (330–342) of the rabies virus glycoprotein (rvgp, in yellow) was obtained by molecular modeling simulations with the HyperChem program. When merged with a cluster of gangliosides GT1b, the protein fragment fit quite well with a GT1b dimer forming a chalice-like receptacle (A). The interaction was stabilized by an electrostatic bridge between the cationic side chain (guanidinium group) of Arg-333 and the carboxylate negative charge of a proximal sialic acid (NeuAc) residue of the glycone head group of the ganglioside. In panel B, the whole complex is in yellow except the atoms involved in this electrostatic interaction.

thus stabilized by an electrostatic bridge between the cationic group of Arg-333 and the closest anionic carboxylate group of the ganglioside head group (Fig. 12.2). Lysine could form this electrostatic bridge, but not amino acids with a neutral lateral chain. Therefore, these results are in full agreement with the mutational studies of rabies virus infectivity.[21] Moreover, our molecular modeling approach strongly suggested that the rabies virus glycoprotein is a ganglioside-binding protein and that the antigenic site 333–338 is physically involved in viral attachment to host cell gangliosides. Finally we showed that the rabies virus glycoprotein can bind to GT1b with high affinity, in agreement with the ability of this particular ganglioside to restore the infectivity of rabies virus in ganglioside-negative cells.[5]

Although these data obtained *in silico* require experimental confirmation, they suggest an interesting analogy between the rabies virus glycoprotein and the HIV-1 surface envelope glycoprotein gp120. Initially, the concept of a common SBD in unrelated sphingolipid-binding proteins has been developed on the basis of a structural homology between the V3 loop of HIV-1 gp120 and the amyloid proteins Aβ and PrP.[23] The V3 loop is also the main antigenic site of gp120 that induces neutralizing antibody responses in AIDS patients.[24,25] The exceptional variability of V3 loop sequences, at the scale of an infected individual and also within the whole population of infected patients, generates myriads of HIV-1 quasi-species that cannot be eradicated by classical vaccination procedures.[26] It also explains why the HIV-1 gp120 from various virus isolates can bind to distinct glycosphingolipids.[27] The finding of a structurally related V3-like SBD in an antigenic region of the rabies virus glycoprotein suggests that

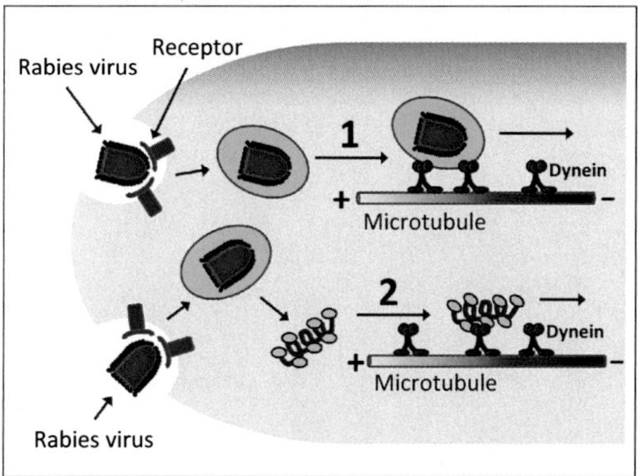

FIGURE 12.3 **Entry and retrograde transport of rabies virus in a peripheral neuron.** After a bite by a rabid dog, rabies viruses are concentrated at the neuromuscular synapse and attracted by cell surface receptors (probably gangliosides) expressed by the neuron. The virus is endocyted in a vesicle. Two scenarios have been proposed to explain the retrograde transport of the virus across the axon. **1.** The virus may remain in the vesicle, and this vesicle interacts with the transport protein dynein, which ensures the migration of the package toward the minus end of microtubules. **2.** A variant of this process postulates that the virus loses its envelope, and its nucleocapsid is transported by dynein along the microtubules.

ganglioside recognition and antigenicity might involve the same molecular determinants in solvent-accessible domains of infectious proteins. This fascinating relationship could be exploited for the development of innovative prophylactic and/or therapeutic strategies against various neurological disorders and infectious diseases (see Chapters 8, 13, and 14).

The next step is the pH-dependent fusion of the viral envelope with intracellular vesicles.[7] At this stage, the virus is inside the cell but in a region lacking the cellular facilities for protein biosynthesis. To circumvent this problem, the virus migrates through the axon toward the neuronal cell body (Fig. 12.3). Two mechanisms of retrograde transport involving microtubular motor molecules have been considered and are still debated: direct attachment of the nucleocapsid to the dynein motor via an interaction between rabies virus phosphoprotein and the dynein light chain 8,[28,29] or encapsulation of the virus in vesicles attached to the dynein motor.[30] Dynein is a cytosolic protein that transports various cellular cargo along microtubules toward the minus-end of the microtubule, which is usually oriented toward the cell center.[31] In the nervous system, dynein is thus a motor for retrograde axonal transport along the tracks of axonal microtubules.[32,33]

The transcription and replication of rabies virus occurs in cytoplasmic inclusions (Negri bodies) of infected neurons.[34] The last phase of the life cycle is the assembly of the viral components, immediately followed by the release of infectious virions through a budding mechanism.[35] At this stage, the detection of virions in the extracellular milieu indicated that rabies virus in the human central nervous system could spread through the intercellular spaces.[36] After intraocular inoculation of the virus in mice, viral antigens were detected first in the trigeminal nerve ganglia, then in the cerebral cortex and cerebellum.[37] The lack of inflammatory reaction suggested that the lethal outcome of this experimental inoculation was due to a direct effect of the virus on the functions of neural cells. In fact, the mechanisms of neuropathogenicity of rabies virus are still poorly understood, and the initial hypothesis that the virus could induce neuronal death through apoptotic mechanisms[38] has been dismissed.[7]

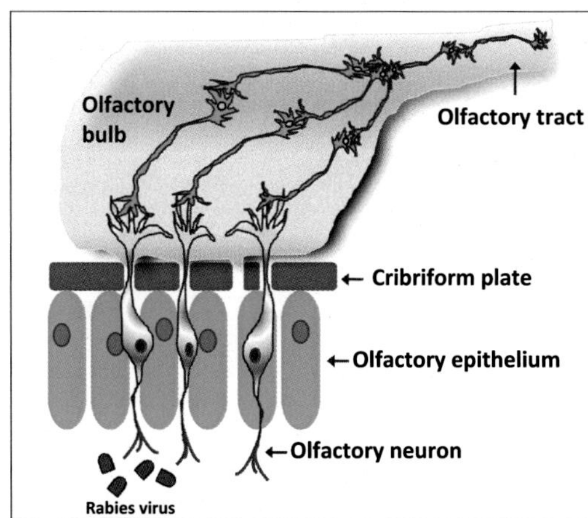

FIGURE 12.4 **Entry of rabies virus into the brain through the nasal cavity.** The nasal cavity is innervated by the olfactory and trigeminal nerves, and it is thus a portal of virus entry into the central nervous system. Several cases of laboratory workers contaminated by rabies virus through aerosols have been reported. This route of infection is also used by other pathogens (e.g., measles virus) to gain entry into the brain. Olfactory neurons in the nasal epithelium are the primary target for these viruses. Then they migrate along a nerve bundle through the olfactory bulb until they reach the brain parenchyma after intracellular microtubule transport and transsynaptic spread.

It is important to note that the rabies virus can also be transmitted through the colonization of the olfactory bulbs via the nasal cavity[39,40] (Fig. 12.4). Indeed, animal models of infection have shown that viruses could migrate from the olfactory bulb to higher brain regions, including the basal nuclei, hypothalamus, and cerebellum.[41–43]

Cases of contamination of laboratory workers by aerosols have been reported.[5] Finally, the trigeminal nerves innervate the nasal olfactory epithelium and could thus be used as an alternative brain entry route for pathogens.[44] This could explain the transmission of rabies virus from undiagnosed infected donors to healthy recipients.[45]

At the end of this brief survey of rabies virus infections, it is striking to realize how poor our knowledge of this virus is. Rabies is a threatening lethal disease that has long been associated with terrible images of friendly dogs suddenly changed into aggressive and deadly beasts.[5] The development of a vaccine strategy by Pasteur considerably decreased the incidence of the disease in western countries, but the virus still kills about 50,000 humans in the world each year, which most likely underestimates the actual incidence.[7] As a matter of fact, we know much more about HIV, which emerged during the last quarter of the twentieth century, than about rabies virus, which is the transmissible agent of a disease known for more than 4000 years. However, knowing is not curing and despite our ignorance of several important cellular and molecular mechanisms associated with rabies virus infection and pathogenesis, at least we can, by empiric means, save contaminated people from an automatic death.[7] Unfortunately, our encyclopedic knowledge of the transmissible agent of AIDS has not yet allowed us to reach the same level for HIV-associated diseases. Like rabies virus, HIV can infect the brain and induce several neurological disorders including dementia and memory deficits.[46] However, HIV does not travel through peripheral axons but instead has developed a complex strategy to cross the blood–brain barrier.

12.2.2 HIV: Strategies for Crossing the Blood–Brain Barrier

Attacking the blood–brain barrier is a second type of brain-invading strategy. This mechanism is chiefly used by pathogens present in the bloodstream, especially HIV.[47] HIV is a retrovirus that primarily infects $CD4^+$ T cells and monocytes/macrophages, resulting in a progressive and eventually fatal immunodeficiency that left the organism without defense against opportunistic infections. Neurological complications of AIDS[46] suggest that the virus could pass from blood to brain and, once in place, perturb neural functions either through direct viral effects (infection of brain cells, neurotoxicity of HIV proteins) or via indirect mechanisms (endogenous neurotoxic/inflammatory factors).[46,48] In any case, the responsibility of HIV in these perturbations is supported by the global beneficial impact of antiretroviral therapy on HIV-associated dementia.[49,50]

To gain entry into the brain, HIV must disrupt or at least fragilize this barrier whose function is to regulate the passage of compounds from the bloodstream to the brain, and vice versa. The viral strains isolated from the brain are generally monocyte/macrophage-tropic strains, indicating that the virus can use infected monocytes as Trojan horses to invade the brain.[51] Infected lymphocytes[48] and free virions[52] may also cross the blood–brain barrier. As described in Chapter 14, the blood–brain barrier consists of an endothelium surrounded by different cell types (pericytes, astrocytes, neurons) that all contribute to the barrier function.[47] The tightness of the endothelium is ensured by the establishment of tight junctions between endothelial cells (Fig. 12.5). A network of protein–protein interactions involving both transmembrane proteins (claudins and occludins) and the actin cytoskeleton participate in these

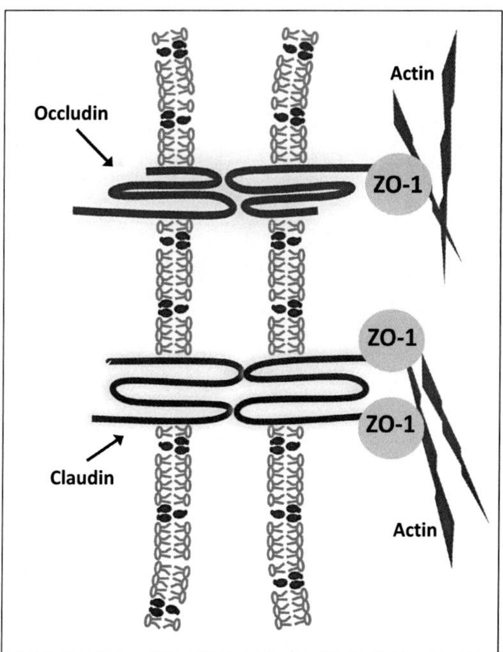

FIGURE 12.5 **Schematic structure and organization of tight junction proteins.**

junctional complexes.[53,54] Claudins and occludins are tetraspan membrane proteins with both their N-terminal and C-terminal parts located in the cytosol. Their extracellular loop domains ensure a tight contact between two adjacent endothelial cells through homologous claudin–claudin and occludin–occludin *trans* interactions (Fig. 12.5).

In the cytosol, zonula occludens proteins (ZO-1, ZO-2) link these transmembrane proteins to the actin cytoskeleton. Compared to other tissues, brain endothelial cells have a particularly high level of expression of occludins[55] so that brain capillaries are tighter than peripheral blood vessels. Correspondingly, the brain endothelium efficiently restricts the passive "paracellular" diffusion of compounds through the tight junctions.[56] Nevertheless, many lipophilic compounds can diffuse passively through tight junctions.[57,58] Apart from this paracellular route, endothelial cells express a series of transporters for the transcellular delivery of essential polar molecules (sugars, amino acids, ATP).[59,60] Therefore, even though the brain endothelium, sealed by tight junctions, would seem to provide an absolute barrier preventing the entry of blood cells into brain, this notion has been thoroughly challenged. It has been shown that leukocyte traffic in the central nervous system is a physiological mechanisms ensuring the routine immune surveillance of the brain,[61] leading to the "Trojan horse" hypothesis in which HIV enters the brain as a "passenger" of immune cells that are normally trafficking to the brain[62] (Fig. 12.6). In other words, these cells introduce a pathogen in the brain while

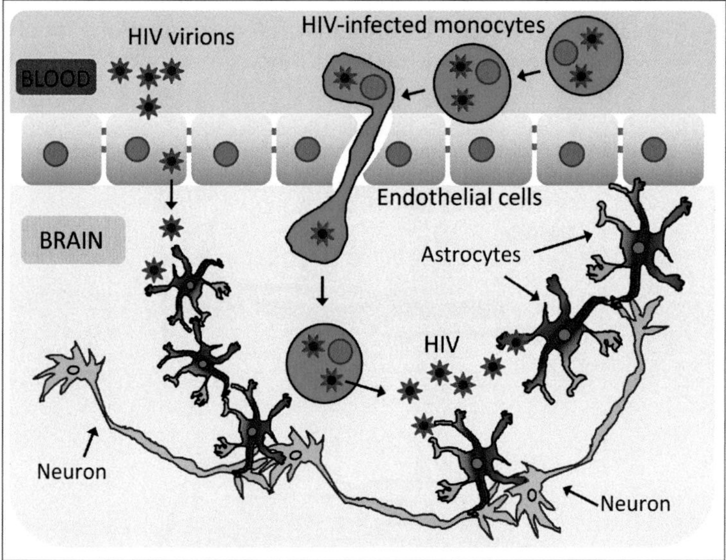

FIGURE 12.6 **How HIV crosses the blood–brain barrier.** The tightness of the blood–brain barrier is warranted by tight junctions formed between endothelial cells (blue cells). However, the barrier is not absolute and brain pathogens can make their way to the brain. Conceptually, both free HIV virions and HIV-infected leukocytes can theoretically cross the blood–brain barrier. Free viruses can be taken up by endothelial cells from the bloodstream and delivered to brain tissues by a transcellular mechanism. HIV-infected lymphocytes and monocytes can cross the endothelium through the paracellular pathway provided that tight junctions are (at least transiently) disrupted. The entry of leukocytes into the brain occurs physiologically to ensure an immune surveillance of the brain. Productive infection of monocytes in the brain generates HIV virions that can secondarily infect astrocytes. Both HIV virions and secreted HIV proteins can perturb brain functions and lead to the neurological disorders associated with HIV infection.

checking that the brain is not infected through one of numerous hijacking strategies used by pathogens.

The cellular transmigration of immune cells through the brain endothelium requires the transient and reversible opening of tight junctions that results from a highly regulated leukocyte–endothelial cell cross-talk.[63] The process of leukocyte extravasation is easier for infected cells that express an activated phenotype, which renders them particularly prone to transendothelial migration.[64] High levels of proinflammatory cytokines such as IL-1 and TNF-α in HIV infection may also perturb the assembly/disassembly cycles of tight junctions, thereby increasing the permeability of the endothelium to infected cells.[47] Finally, HIV-1 proteins may also directly affect the blood–brain barrier through specific effects on endothelial cells. The HIV-1 transactivator Tat protein, which is secreted by infected cells,[65] has been shown to decrease the expression of tight junction proteins (occludins, ZO-1) in brain endothelial cells.[66] This Tat-induced disruption of tight junctions may greatly facilitate the access of HIV-infected cells to brain tissues. The surface envelope glycoprotein gp120 can also increase the permeability of the blood–brain barrier by disrupting the connection between tight junctions and the actin cytoskeleton.[67,68] Overall these data strongly suggest that through the specific effects of its proteins Tat and gp120 on tight junction proteins, HIV sets up favorable conditions that facilitate the transmigration of infected immune cells into the brain.

HIV enters the brain early after systemic infection[69] and may thus induce neurological disorders soon after contamination. The study of a case of iatrogenic transmission of HIV-1 showed that the virus could be detected in the brain as soon as 15 days after the mistaken inoculation of the virus.[69] Thus although HIV is not a typical neurotropic virus that, like rabies virus, primarily infects neurons, it is a "brain-resident" pathogen that rapidly colonizes brain tissues and induces neurological disorders that are a hallmark of HIV-associated disease. Nevertheless, the mechanisms by which HIV induces neurological damage are not clearly understood, especially because no evidence supports HIV infection of neurons. Instead, the replication of HIV in the brain must be sustained by macrophages and/or glial cells. The production of HIV proteins (e.g., Tat and gp120) by these infected cells may directly affect neuronal functions (HIV-mediated injury mechanism). Alternatively, neurons may be damaged by indirect inflammatory processes triggered in response to HIV infection (bystander mechanism).[48] In the latter case, the role of HIV would be essentially to switch on a self-sustained inflammatory process that would rapidly become an HIV-independent process. Nevertheless, in both cases the injury process requires – at least in the beginning for the bystander mechanism – the productive infection of HIV in brain tissues. It is at this stage that brain lipids enter the scene.

As a retrovirus, HIV requires a cellular receptor to fuse with the plasma membrane of a host cell. In T-lymphocytes and monocytes/macrophages, this receptor is the transmembrane protein CD4.[70] In addition to CD4, coreceptors are also required for the fusion reaction.[71] Two coreceptors are widely used by most HIV-1 strains, defining two main tropisms of HIV-1: CCR5, used by R5 virus, and CXCR4, used by X4 strains. CCR5 and CXCR4 are members of the family of chemokine receptors that are G protein-coupled proteins with seven transmembrane domains.[72] The selection of one coreceptor instead of another is determined by the amino acid sequence of the V3 loop of gp120.[73] According to a widely accepted model, the binding of HIV-1 gp120 to CD4 induces conformational changes in gp120 that lead to recognition of either CCR5 or CXCR4, depending on the virus tropism.[71,74] As schematized in Fig. 12.7, the

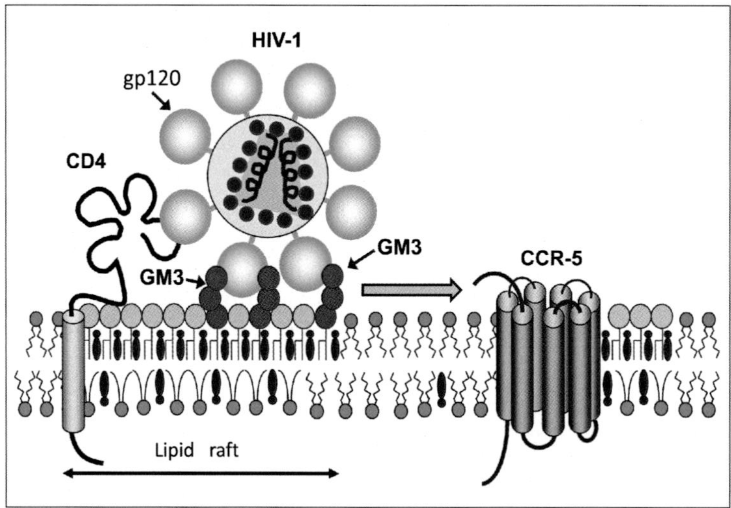

FIGURE 12.7 **Role of glycosphingolipids in HIV-1 fusion mechanisms.** In $CD4^+$ lymphocytes, HIV-1 binds to CD4, a transmembrane protein, via a specific interaction with its surface envelope glycoprotein gp120. Because CD4 is associated with lipid rafts, this primary interaction occurs in a membrane environment enriched in glycosphingolipids. The V3 loop of g120 can recognize different types of glycosphingolipids, including ganglioside GM3. Following binding to CD4, the V3 loop of gp120 binds to GM3 so that the virus remains firmly attached to the surface of the raft. After this initial binding step, the virus will sail on the cell surface until reaching the CCR5 coreceptor.

transfer of the viral particle from CD4 to the coreceptor is facilitated by glycosphingolipids that play an accessory but indispensable role in the fusion process.[74–76]

These glycosphingolipids are recognized by the V3 domain of gp120, so that a glycosphingolipid tropism is superposed to the coreceptor tropism.[77] It has been suggested that R5 strains interact with GM3 and X4 viruses with Gb_3.[78] In addition, HIV-1 gp120 can interact with several other glycosphingolipids such as GalCer and sulfatide.[79,80] Apart from microglia and perivascular macrophages that fulfill immune functions in the brain, the vast majority of brain cells, including neurons, astrocytes, oligodendrocytes, and endothelial cells do not express CD4.[48] Under these circumstances, the selective cell surface expression of glycosphingolipids may attract both infectious HIV-1 particles and free gp120 in the vicinity of brain cells. Historically, GalCer has been the first glycolipid receptor shown to mediate both gp120 binding and CD4-independent infection of cultured brain cells.[79] HIV infection of oligodendrocytes has been reported *in vitro*[81] but not *in vivo*.[82] In fact, *in situ* PCR/hybridization studies of brain tissues from AIDS patients revealed that only macrophages, microglial, and, to a lesser extent, astrocytes were infected by HIV.[82] Nevertheless, the incubation of oligodendrocytes with gp120 impaired myelin formation through a GalCer-dependent mechanism.[83] Hence, GalCer-mediated binding of gp120 to oligodendrocytes rather than direct infection of these cells[84] could explain the demyelination (myelin pallor) observed in patients with HIV-associated encephalopathy.[85] Interestingly, it has been shown that HIV-1 gp120 glycoprotein induced Ca^{2+} responses in oligodendrocytes of the cerebellum,[86] suggesting that these glial cells play a critical role in the neurological manifestations of AIDS.[87] Incidentally, this illustrates once again that both viral and amyloid proteins involved in neurological disorders share similar mechanisms of pathogenesis, including binding to glycosphingolipids and perturbation of calcium fluxes.

In contrast with oligodendrocytes, astrocytes are infected by HIV *in vivo* and these cells may serve as an important reservoir for the virus. The simultaneous expression of the CCR5 coreceptor and of GM3 by human astrocytes is consistent with the selective infection of astrocytes by R5 strains of HIV-1.[88] It should be noted that astrocytes express other HIV-1 coreceptors, such as CXCR4.[89] Thus, it is surprising that X4 strains of HIV-1 do not infect these cells, most likely due to a restriction imposed by GM3, which is recognized by the gp120 from R5 but not X4 strains.[27] Because R5 strains predominate in the early stages of primary HIV-1 infection, these data also suggest that astrocytes can be infected by HIV-1 shortly after the contamination, basically as soon as the virus is present in the brain. Apart from direct infection, free gp120 has been shown to perturb astrocyte functions by inducing inflammatory responses,[90] oxidative stress,[91] calcium signaling,[92] or apoptosis.[93]

At the end of this survey of HIV infection and pathogenesis in the brain, it is interesting to note that all the glycosphingolipids recognized by gp120 (GalCer, sulfatide, and GM3) have a relatively small head group. Moreover, these glycolipids are strongly expressed by glial cells, but not (or in small amounts) by neurons. This aspect of HIV–glycolipid interactions in the brain has not been studied and thus warrants further consideration. More generally, we would like to emphasize that the strategies used by pathogens to invade host cells and tissues may be diverse, but, when it comes to the cell surface, glycosphingolipids are always at the frontline. Identifying the glycosphingolipid(s) recognized by a pathogen should be a priority for both fundamental knowledge and therapeutic strategies. Too often, and this is typically the case for rabies virus, the discovery a protein receptor for a pathogen overshadows previous characterizations of glycolipid receptors. This discrimination of glycolipids to the benefit of proteins (considered by most researchers as "noble" molecules) is not only totally puzzling and poorly scientific, it may also significantly delay the development of antimicrobial therapeutics. We hope that this book will convince the readers, especially young scientists, that glycosphingolipids, as well as other brain lipids, are fascinating objects of study for biologists.

12.2.3 Listeria Monocytogenes

We will now describe the mechanisms of brain invasion used by a bacterium that causes meningitis, *Listeria monocytogenes.*[94] This Gram-positive bacterium is a food-borne pathogen that can invade the central nervous systems by three distinct mechanisms[95]: (1) direct infection of endothelial cells, (2) passage of infected monocytes through the blood–brain barrier, and (3) infection of peripheral neurons and retrograde transport to the brain. In this respect, and although *L. monocytogenes* is far from a brain-exclusive pathogen, this bacterium is particularly efficient at invading brain tissues. Correspondingly, central nervous system infections are present in about 50% of infected patients.[96] Moreover, *L. monocytogenes* is one of the most common causes of bacterial meningitis in western countries.[97–99] As a food contaminant, *L. monocytogenes* enters the body through the gastrointestinal mucosa.[100] It has thus developed adaptive strategies to resist to the low pH of the stomach, and to the detergent activity of bile salts in the upper small intestine.[101] *L. monocytogenes* penetrates the intestinal mucosa directly via enterocytes, or indirectly via invasion of Peyer's patches.[102]

Following this passage through the gastrointestinal barrier, the bacterium spreads via the lymph and the blood to distant tissues. In the peripheral blood, *L. monocytogenes* is found in a subpopulation of monocytes that sustain a systemic infection status.[103] These infected monocytes can adhere to the wall of brain capillaries and migrate into brain tissues by diapedesis through the endothelium.[104] This Trojan horse mechanism mediated by infected monocytes is quite similar to the one described for HIV (Fig. 12.6). Alternatively, free *L. monocytogenes* can infect brain endothelial cells and replicate in these cells.[105] Thus, direct infection of endothelial cells is a second route of entry of this bacterium from the blood stream to the brain. The third route of brain invasion by *L. monocytogenes* involves peripheral neurons and retrograde axonal transport, a mechanism also used by rabies virus (Fig. 12.3). This neural route of infection has been suspected for several cases of brain encephalitis associated with *L. monocytogenes.*[106] Correspondingly, inflammatory lesions associated with *L. monocytogenes* infection have been observed in cranial nerves V, VII, IX, and XII that innervate the oropharynx.[107]

The pathogenicity of *L. monocytogenes* is mediated by a pore-forming hemolysin toxin, listeriolysin O. It is interesting to note that this toxin is used by the bacterium to escape the intracellular vacuoles of infected monocytes during the course of cell-to-cell transmission.[108] Listeriolysin O belongs to the family of cholesterol-dependent cytolysins that contains more than 20 pore-forming toxins produced by various bacteria including *Bacillus anthracis* (anthrolysin), *Clostridium perfringens* (perfringolysin), and *Streptococcus pneumoniae* (pneumolysin).[109] All these toxins are secreted by bacteria as soluble monomers but, following binding to cholesterol in host cell membranes, oligomerize into large pores (up to 35 nm in diameter). The choice of cholesterol among other lipids in the membrane of host cells is important, because it assures that the toxin cannot be harmful for the bacterium that produces it, given that bacterial membranes lack cholesterol.[109] Mutational studies have indicated that cholesterol is recognized by two consecutive amino acids, Thr-515 and Leu-516,[110,111] which form a "sensor" at the tip of a short loop joining two β-strands (505–514 and 517–525) in the C-terminal domain of listeriolysin O[112] (Fig. 12.8).

The conservation of this pair of residues among all cholesterol-dependent cytolysins suggests a common mechanism of cholesterol binding.[111] Obviously only two amino acid residues cannot account for a binding process involving the whole cholesterol molecule. In fact, Thr-515/Leu-516 may recognize the OH group of cholesterol and the first ring of its sterane backbone, as illustrated in Fig. 12.8. In other words, the pair of "'cholesterol sensor" residues may allow the C-terminal part of the soluble listeriolysin O monomer to detect cholesterol in the plasma membrane of the target cell and to promote the first toxin–membrane contact. A vicinal loop consisting of a conserved apolar undecapeptide (483–493) (Fig. 12.8) will then be responsible for the initial insertion of listeriolysin O within the host cell membrane (Fig. 12.9A). However, the formation of a pore requires deeper plasma membrane insertion of the toxin.

Because the attachment of listeriolysin occurs in a cholesterol-enriched domain of the membrane, it is logical to search for a cholesterol-binding domain in the C-terminal part of the toxin. A first approach is to determine whether the amino acid sequence contains a linear cholesterol-binding motif fulfilling the criteria of the CRAC[113] or the CARC[114] algorithms. We found that the 507–516 segment of listeriolysin O (**R**NISI**W**GTT**L**) is a CARC-like motif (the amino acids important for the definition of the motif are bold and underscored). This motif

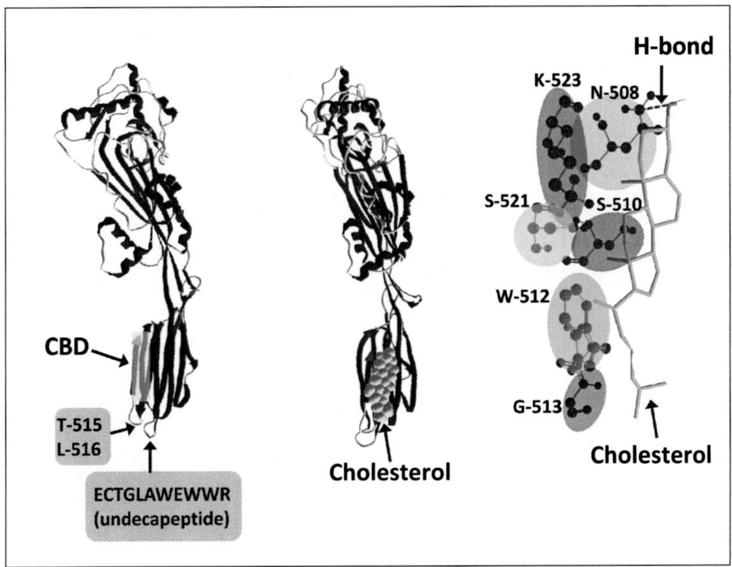

FIGURE 12.8 **Structure and cholesterol-binding domain of listeriolysin O.** The C-terminal region of listeriolysin O (PDB entry # 4CDB) contains a series of β-strands linked by short loops. One of these loops corresponds to a conserved undecapeptide involved in membrane insertion. A pair of residues (Thr-516/Leu-516) has been identified as critical for cholesterol binding by mutational studies. We find a cholesterol-binding domain (CBD, shaded in yellow in the model on the left) in the structure of listeriolysin O. Molecular dynamics simulations show that this domain displays a high affinity for cholesterol (model in the middle panel). The amino acid residues of the CBD involved in cholesterol binding are shown on the right panel.

is notable because it contains the pair of residues (Thr-515/Leu-516) that ensure the initial binding to cholesterol. The sliding of the loop containing both residues could thus ensure a stronger interaction with cholesterol, not only on the cell surface but inside the membrane, as illustrated in Fig. 12.9A. Molecular dynamics simulations were then performed to assess whether the CARC domain of listeriolysin O could actually interact with cholesterol. In fact, we found that both β-strands joined by Thr-515/Leu-516 form a 3D binding domain displaying a high affinity for cholesterol (calculated energy of interaction of -65 kJ mol^{-1}). In our model the amide group of Asn-508 formed a hydrogen bond with the OH group of cholesterol (Fig. 12.8). This asparagine residue belongs to the β-strand 505–514, which also contains Ser-510, Trp-512, and Gly-513, all involved in cholesterol binding. The next β-strand (517–525) contributes also to cholesterol binding through contacts with residues Ser-521 and Lys-523 (Fig. 12.8). Two important features of these cholesterol–listeriolysin O interactions should be underscored. First, one should note that the cholesterol bound to the C-terminal domain of the toxin has an extended linear conformation (Fig. 12.9B), which is totally distinct from the tilted conformers involved in binding to amyloid proteins.[115] Indeed, cholesterol does not induce any tilt in the monomers of listeriolysin O inserted into the membrane of host cells. The orientation of the toxin remains perpendicular to the plane of membrane, which is required for the formation of a large oligomeric pore[112] (Fig. 12.9C). Secondly, the binding of cholesterol is fully consistent with the oligomerization process of listeriolysin O monomers,

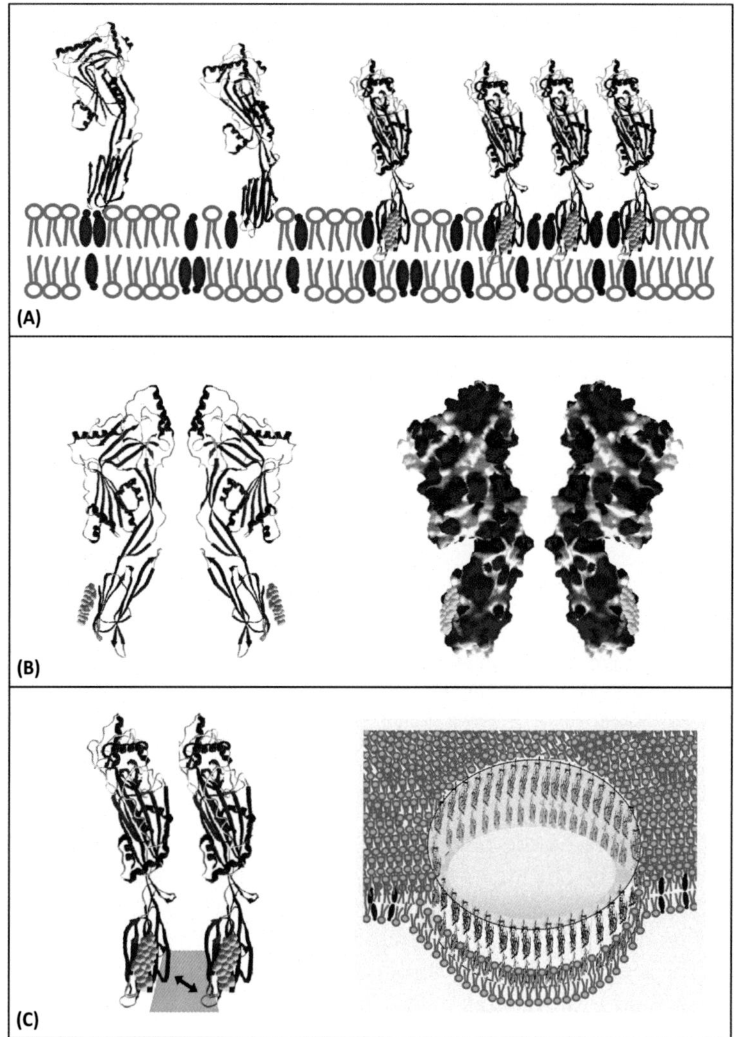

FIGURE 12.9 **Insertion and oligomeric pore formation by listeriolysin O.** (A) Step-by-step insertion (from left to right) of listeriolysin in cholesterol-rich domain of the plasma membrane. Schematic bulk lipids are in green and cholesterol in red, yet the real cholesterol molecules bound to the toxin are shown in yellow. The initial contact of the soluble listeriolysin O monomer with the membrane is mediated by the surface binding of the Thr-515/Leu-516 "sensor" with the accessible parts of cholesterol. Then the undecapeptide penetrates into the membrane, followed by the C-terminal β-strands displaying the cholesterol-binding domain. Only the earliest steps of the insertion process are shown. Further membrane insertion may concern the membrane-inserting helix bundles located in the middle of the toxin.[112] (B) Secondary structure (left panel) and surface potential (right panel) of listeriolysin O bound to membrane cholesterol (in yellow) (PDB entry # 4CDB). (C) Schematic drawing (not a molecular model) of a ring formed by 36 monomers of listeriolysin O. The oligomerization process involves close contacts between Leu-461 and Trp-489 (arrow) belonging to two vicinal monomers.[112] A spectacular rendition of such pores has been published by Köster et al.[112]

a coordinated process that involves, in the C-terminal region, an interaction between Leu-461 on one monomer and Trp-489 on its immediate neighbor.[112] As illustrated in Fig. 12.9C, cholesterol is clearly outside these junction points.

When applied extracellularly, listerolysin O forms, in the plasma membrane of recipient cells, oligomeric pores that mediate Ca^{2+} fluxes from the extracellular milieu.[116] In this respect the bacterial toxin behaves exactly as amyloid proteins that also self-assemble into Ca^{2+}-permeable pores through a cholesterol-dependent mechanism.[115,117] Once again, we can see that at the molecular level that these mechanisms of toxicity rely on specific lipid–protein interactions and that among the lipids concerned, cholesterol often (if not always) plays a prominent role. This is not unexpected. Cholesterol combines several features that render this lipid unique among all other membrane lipids.[118,119] First, it is the only membrane lipid that is not formed by condensation of a fatty acid. Acyl chains display a high biochemical diversity (length of the carbon chain, absence or presence of double bonds, absence or presence of α–OH groups). Therefore, in contrast with acyl-containing lipids, cholesterol has a unique structure. Cholesterol has a relatively mobile isooctyl chain that can adapt its shape and orientation to protein ligands. In contrast, the sterane backbone made of four contiguous rings is rigid enough to provide a solid wall for proteins that penetrates the membrane. This capability is particularly important for the formation of structured assemblies such as oligomeric pores that require a precise adjustment of the monomeric bricks. Cholesterol has two opposite faces, one smooth and the other rough, spiked by methyl groups. This "bifacial" lipid can thus interact with two distinct types of ligands, most often sphingolipids for the smooth side and transmembrane domains of proteins on the rough one. This later feature allows several combinations of lipid–protein interactions, resulting in fine conformational adjustments and functional control of membrane receptors through coordinated cholesterol–receptor and cholesterol–sphingolipid interactions.[118,119]

12.3 OVERVIEW OF BRAIN PATHOGENS

The variety of mechanisms allowing pathogen organisms to enter the brain explains why the brain is not a sanctuary efficiently protected from these invaders. We will briefly list the main viruses and bacteria able to penetrate the brain parenchyma and perturb neural functions in the central nervous system. Among these pathogens, those causing meningitis are particularly threatening.

12.3.1 Bacteria

Apart from *L. monocytogenes* (discussed earlier), several other bacterium species can cause meningitis, including *Neisseria meningitidis*, *Haemophilus influenzae*, and *Streptococcus pneumoniae*. Although people with bacterial meningitis usually recover, these bacteria can cause irreversible damage of the central nervous systems. It has been estimated that more than 4000 cases and approximately 500 deaths from bacterial meningitis occurred annually in the United States during 2003–2007.[120] Bacterial meningitis cases are commonly reported in Europe as well. In France, about 5000 bacterial meningitis cases were reported in 2001–2012.[121]

12.3.1.1 N. meningitidis

N. meningitidis is a strictly human, Gram-negative, diplococcal pathogenic bacterium responsible for septicemia and meningococcal meningitis, a severe infection resulting in coma and death within a few hours.[122] The disease is highly contagious especially for children and young adults living in close promiscuity (e.g., in military barracks and college dormitories). In fact, *N. meningitidis* is generally an asymptomatic colonizer of the human nasopharynx.[123] Only a small proportion of these bacteria are able to disseminate from the oropharyngeal cavity to the bloodstream and cross the blood–brain barrier to eventually invade the meninges. In marked contrast with *L. monocytogenes*, *N. meningitidis* does not need a leukocyte Trojan horse to pass the blood–brain barrier but disrupts the tight junctions of the brain endothelium.[123] Specifically, the adhesion of *N. meningitidis* to endothelial cells promotes the cleavage of occludin by the metalloproteinase MMP-8.[124] The cascade of events leading to this deleterious effect on brain endothelial cells is mediated a by signal transduction pathway involving lipid rafts.[125] In line with these findings, cholesterol depletion with methyl-β-cyclodextrin efficiently blocked the initiation of the cellular response triggered by the bacterium in endothelial cells.[126]

N. meningitidis has been observed around the capillaries of the subarachnoidal space, in the brain parenchyma and the choroidal plexuses, and inside brain vessels.[123] *In vitro* studies with experimental models of the human meninges showed that *N. meningitidis* readily binds to the leptomeninges and meningeal blood vessels.[127] These data suggested that the dissemination of *N. meningitidis* through the meningeal space requires the adhesion of the bacteria to meningeal cells. The participation of plasma membrane gangliosides in the adhesion of *N. meningitidis* to human cells has been suggested.[128] Hugosson et al. have studied the binding of radiolabeled *N. meningitidis* bacteria to glycosphingolipids on thin-layer chromatograms.[129] Interesting similarities were found with the glycolipid binding specificity of *H. influenzae*, another nasopharyngeal bacterium responsible for acute meningitis. These bacteria did not bind to brain gangliosides (GM1, GM3, GD1a) but to a series of neutral glycosphingolipids, some of which are expressed by the human oropharyngeal epithelium.[129] One should note that human brain endothelial cells may also express neutral glycosphingolipids (e.g., LacCer)[130,131] that can serve as attachment sites for circulating *N. meningitidis*.

As discussed previously, *N. meningitidis* does not penetrates the blood–brain barrier hidden in "Trojan horse" leukocytes, but interacts directly with brain endothelial cells. *In vivo*, the blood flow generates mechanical forces that limit the binding of bacteria to the endothelium surface.[123] Interestingly, an immunohistological study of a meningococcal sepsis case revealed that the adhesion of *N. meningitidis* was restricted to capillaries located in low blood flow regions.[132] Experimental studies using a flow chamber assay indicated that after their initial attachment to the surface of endothelial cells, the bacteria could resist high-velocity blood flow conditions.[132] Because they form large membrane areas decorated with high numbers of glycosphingolipid molecules, lipid rafts represent ideal landing platforms for inducing the adhesion of *N. meningitidis* on the brain endothelium surface. Once in the brain, the bacteria are not trapped by neural and glial cells because these brain cells express a pattern of glycosphingolipids (including gangliosides) that are not recognized by *N. meningitidis,* leaving the bacteria free to rapidly invade meningeal cells, which probably express adequate glycolipid and/or protein receptors. Taken together, all these data show that the plasma membrane of host cells displays a wide repertory of glycosphingolipids that are used as primary attachment sites by most bacteria.[133–135]

12.3.1.2 H. influenzae

H. influenzae is a Gram-negative rod-shaped bacterium present in the nasopharynx of approximately 75% of healthy children and adults. It can be either encapsulated (thus typable) or unencapsulated (nontypable). Encapsulated *H. influenzae* is the leading cause of acute bacterial meningitis in children under age 5. The development of vaccine against *H. influenzae* type B has spectacularly decreased the incidence of the disease.[136] Nowadays, this type of meningitis occurs in approximately 1 in 100,000 children in the United States, and the disease starts to disappear worldwide.[137–140] This reduction is one of the most remarkable achievements of an antibacterial vaccine. Incidentally, *H. influenzae* is the first living organism to have its genomic DNA sequenced in totality.[141]

Although the pathogenesis of *H. influenzae* infections is not completely understood, the type B polysaccharide capsule is considered as a major virulence factor.[142] Encapsulated bacteria penetrate the epithelium of the nasopharynx and invade the blood capillaries directly. Then the bacteria cross the blood–brain barrier and invade the meninges. Overall, the route of infection of *H. influenzae* is similar to the one used by *N. meningitidis*. Like *N. meningitidis*, *H. influenzae* can increase the permeability of the blood–brain barrier through disrupting effects on tight junctions.[143] However, the mechanisms used by *H. influenzae* to permeabilize the barrier are various and complex.[108] In particular, it has been suggested that a porin produced by *H. influenzae* could trigger a signaling cascade, leading to the biosynthesis of proinflammatory cytokines (IL-1, TNF-α) that correlate with injury of the blood–brain barrier.[144] A remarkable convergence in glycosphingolipid recognition by *H. influenzae* and *N. meningitides* also occurs. These bacteria recognize a similar pattern of neutral glycosphingolipids, but not the major gangliosides (GM1, GD1a, or GM3) expressed by brain neurons and glial cells[129,145,146] As discussed previously, this rare glycosphingolipid-binding pattern may allow these bacteria to reach meningeal cells without being glued by neurons or glial cells during their travel to the meninges. Alternatively, it has been proposed that the bacteria could directly reach the subarachnoid space via the nasal lymphatics.[147]

Finally an intriguing aspect of *H. influenzae* infections warrants mention. These bacteria express a lipopolysaccharide whose exposed surface is decorated with phosphorylcholine. As described in Chapter 1, phosphorylcholine is a building block of phosphatidylcholine, a major membrane lipid of eukaryotic cells. However, most bacteria lack phosphatidylcholine[148] and thus *H. influenzae* has to rely on host cell lipids to obtain phosphorylcholine.[149] The transfer of phosphorylcholine from the host to the bacterial cell surface is an adaptative process that allows the pathogen to mimic characteristics of host cells and to become more resistant to host immune responses.[150] Another astonishing case of "lipid mimicry" has been reported for *H. influenzae*. A particular strain of nontypable *H. influenzae* seemed to express a ganglioside GM1-like structure that has been suspected as a generator of anti-GM1 antibodies responsible for an axonal Guillain–Barré syndrome.[151]

12.3.1.3 S. pneumoniae

A third case of bacterial meningitis that we will describe is caused by *S. pneumoniae*, a Gram-positive bacterium commonly referred to as pneumococcus. Pneumococcus infections are the most common cause of bacterial meningitis in adults. After the initial colonization of the nasopharynx, *S. pneumoniae* can enter the brain through the blood–brain barrier via a combination of transcellular and paracellular mechanisms.[108] It has also been reported that pneumococci may penetrate the brain directly from the nasal cavity by axonal transport mediated by olfactory neurons.[152] In this case, the bacteria are not detected in the circulating blood. The mechanisms controlling the adhesion of *S. pneumoniae* to the surface of host cells are particularly interesting. Like *H. influenzae,* the surface of pneumococci displays a polysaccharide (C-polysaccharide) decorated by phosphorylcholine units. This polysaccharide–phosphorylcholine combination binds to the glycosphingolipid receptor of *S. pneumoniae,* asialo-GM1.[153] Indeed, pneumococci deprived of choline did not bind to asialo-GM1, indicating that phosphorylcholine is essential for the interaction between C-polysaccharide and the glycosphingolipid receptor.[153] In line with these data, the incubation of pneumococci with micelles of asialo-GM1 reduced the colonization of the nasal mucosa, olfactory system, and brain.[152] Thus, host glycosphingolipids play a key role in the blood-independent invasion of pneumococci in brain tissues. A molecular model of the asialo-GM1–phosphorylcholine complex in relation with *S. pneumonia* adhesion is presented in Fig. 12.10.

Apart from glycosphingolipids, cholesterol is also involved in the pathogenesis of pneumococci-associated meningitis. Like *L. monocytogenes, S. pneumoniae* produces a cholesterol-dependent cytolysin, referred to as pneumolysin, that is critical for bacterial virulence. Soluble pneumolysin binds cholesterol with high affinity.[154] Once bound to cholesterol in the membrane of host cells, pneumolysin monomers form ring-shaped oligomeric pores.[155] The 3D structure of these pores has not been fully deciphered, but structural homology models coupled with cryoelectron microscopy have assisted in reconstituting a scenario of pore formation by 44 toxin monomers inserting a total of 176 β-strands in the bilayer. This giant

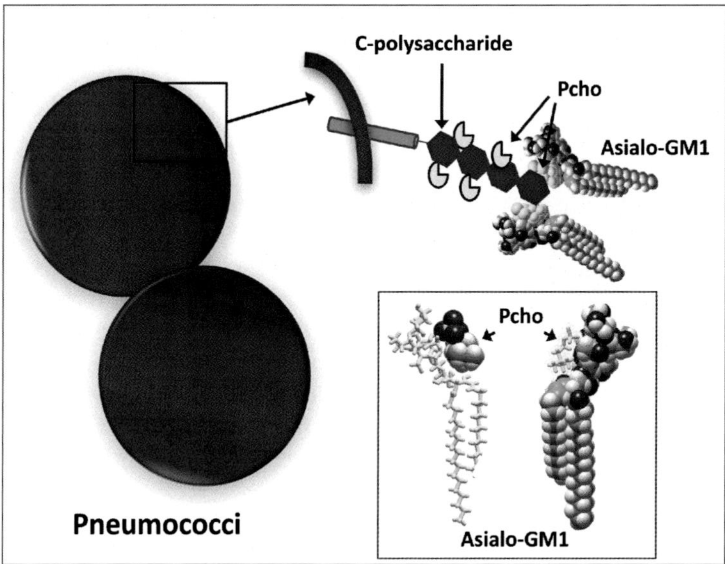

FIGURE 12.10 **Binding of *S. pneumoniae* to host cell glycosphingolipids.** Pneumococci are often observed under the microscope as bacteria pairs (diplococci). The major surface component of pneumococci, the C-polysaccharide, is decorated by phosphorylcholine (Pcho) residues (in yellow). Phosphorylcholine interacts with the sugar head group of asialo-GM1. In the inset, two renditions of the asialo-GM1–phosphorylcholine complex are shown. The models were obtained by molecular dynamics simulations with the HyperChem program.

assembly created a transmembrane channel of 26 nm in diameter. Pneumolysin affects the integrity of the barrier function of human brain endothelial cells, suggesting that this toxin may contribute to the entry of pneumococci into the cerebral compartment and favor the development of brain edema in pneumococcal meningitis.[156] Therefore, whatever the site of entry of pneumococci into the brain, membrane lipids of host cells play a critical cofactor function that facilitates the microbial invasion: on one side, cholesterol helps pneumolysin to perforate the blood–brain barrier, and on the other side asialo-GM1 allows the bacteria to reach the brain through the nasal cavity. Hence, these lipids are a target for therapeutic approaches.

12.3.2 Viruses

Meningitis can also be caused by viral infections of the central nervous system, yet the symptoms are generally less severe than bacterial meningitis. Enteroviruses are the most common cause of viral meningitis,[157,158] but several other viruses may cause the disease, including mumps, herpes, measles, influenza, and mosquito-transmitted arboviruses. Moreover, apart from meningitis, virus infection of the central nervous system may evoke various neurological symptoms such as encephalitis, myelin damage, cognitive changes, or neuronal destruction.[159,160]

12.3.2.1 Enteroviruses

Enteroviruses are icosahedral, nonenveloped viruses of approximately 30 nm in diameter that can survive the low pH of the stomach following fecal/oral transmission.[161] Enteroviruses can use a wide variety of cell surface receptors to gain entry into host cells, including gangliosides.[162] Three routes of brain invasion have been proposed, which indicates that the exact mechanism has not been deciphered yet. In the first model, the virus is transported through the endothelial cells of the blood–brain barrier presumably via a transcellular process.[163] A variant of this mechanism could be the passage of enterovirus-infected immune cells through the blood–brain barrier (Trojan horse strategy).[164] Finally, on the basis of experiments performed in mice with poliovirus (a member of the enterovirus family), Ren and Racaniello proposed that the virus could spread from muscle to the central nervous system along nerve pathways.[165] Of high interest is the observation that enteroviruses display a selective tropism for brain cells (neurons, astrocytes, macrophages, microglia, and/or oligodendrocytes) and for brain areas (choroid plexus, hippocampus, and cortex).[161] The mechanisms controlling these overlapping tropic parameters expressed at two distinct scales (individual cells and brain regions) are largely unknown. Because enteroviruses are known to recognize some gangliosides,[162] it would be particularly interesting to assess whether membrane lipids, which are differentially expressed according to cell type, brain region, and the age, may be involved in this phenomenon.

12.3.2.2 JC Virus

JC virus is an interesting case of a neurotropic virus that may bind to cell surface glycosphingolipids to enter specific brain cell types. This polyomavirus is responsible for a fatal demyelinating disease of the central nervous system called progressive multifocal leukoencephalopathy.[166] *In vivo*, JC virus infection has been detected in neurons, astrocytes, and oligodendrocytes.[167] Two main types of cell surface receptors for JC virus have been described:

sialic acid-containing glycoconjugates[168] and the serotonin receptor 5-HT2R.[169] Virus-like particles displaying the major capsid protein VP1 selectively interacted with gangliosides GM3, GD2, GD3, GD1b, GT1b, and GQ1b, but did not bind to GM1a or GM2.[170] The specificity of glycosphingolipid expression in these cells is consistent with the pattern of glycosphingolipid recognition by JC virus: neurons express GD1b, GT1b, and GQ1b; astrocytes and oligodendrocytes express GM3.[171] The preferential interaction of JC virus particles with gangliosides versus neutral glycosphingolipids suggests a specific binding mechanism involving sialic acid. In agreement with this notion, a striking correlation is evident between the expression of the JC virus receptor-type sialic acid on cells and their susceptibility to infection by the virus.[172]

Thus, the role of sialic acid in JC virus infection has been firmly established.[168] Correspondingly, a linear sialylated pentasaccharide with the sequence NeuAc–α2,6–Gal–β1,4–GlcNAc–β1,3–Gal–β1,4–Glc present on host glycoproteins and glycolipids has been characterized as a specific JC virus recognition motif.[173] The crystal structure of the major JC virus capsid protein VP1, previously shown to interact with some gangliosides,[170] has recently been solved.[173] In these experiments, recombinant VP1 proteins oligomerized into pentameric structures that resemble capsid assemblies[174] (Fig. 12.11).

The structure of a complex between the VP1 pentamer and the sialylated pentasaccharide revealed interesting features of protein–ganglioside interactions. Firstly, the oligosaccharide in solution can adopt a variety of conformations allowed by the possibilities of rotation of the glycosidic bonds.[175] However, the binding of the oligosaccharide to the protein proceeds through an induced-fit mechanism during which both the glycone and the protein mutually adapt their shape to their molecular partner.[176] In the VP1–glycone complex, the bound sialylated oligosaccharide adopts an L-like structure, with the shorter leg chiefly formed by the sialic acid (NeuNAc) and the longer leg consisting of the Gal–β1,4–GlcNAc–β1,3–Gal–β1,4–Glc tetrasaccharide[173] (Fig. 12.12).

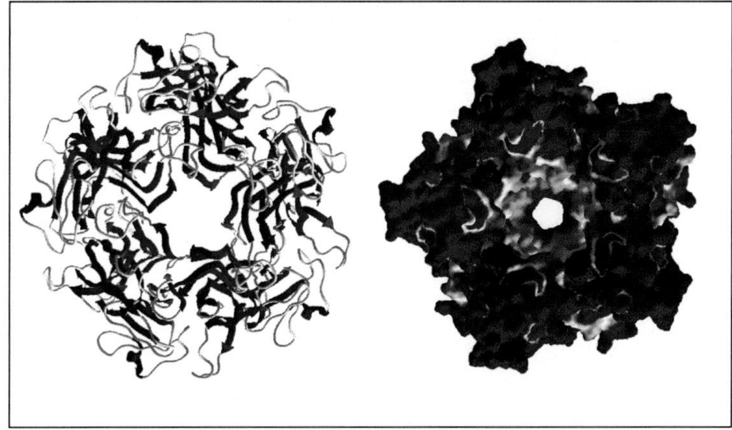

FIGURE 12.11 **Capsid-like pentameric assembly of recombinant JC virus protein VP1.** The secondary structure of the pentamer (left panel) and the surface electropotential (right panel) have been obtained from PDB entry # 1VPN.[177]

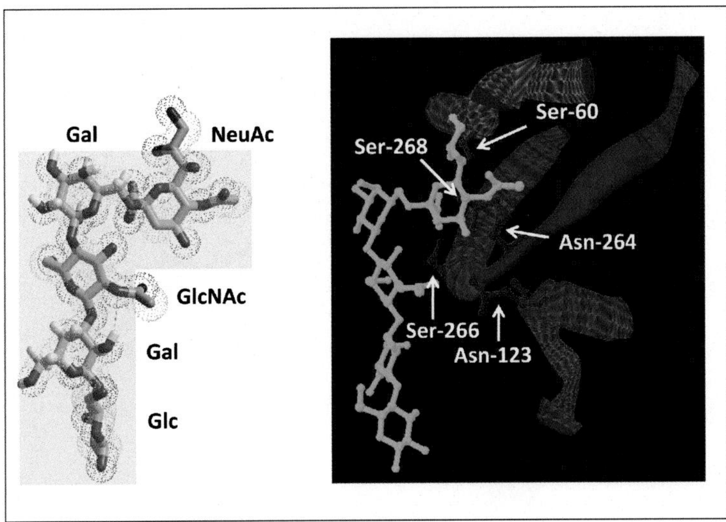

FIGURE 12.12 **Binding of a sialylated pentasaccharide to capsid protein VP1 of JC virus.** The typical L-shaped structure adopted by the sialylated pentasaccharide NeuAc–α2,6–Gal–β1,4–GlcNAc–β1,3–Gal–β1,4–Glc upon binding to the VP1 protein is show in the left panel. The main amino acid residues involved in the binding are shown in the right panel. The structures were obtained from PDB entry 3NXD.

The binding of the sialylated oligosaccharide receptor to VP1 is mostly driven by a network of sugar–protein hydrogen bonds involving serine (Ser-60, Ser-266, Ser-268) and asparagine (Asn-123, Asn-264) (Fig. 12.12). As expected, the sialic acid residue of the oligosaccharide plays a major role in these interactions. Most importantly, this structural study was validated by mutational studies of the sialic acid binding site of VP1.[173] Mutations of the serine residues involved in the binding of NeuNAc (Ser-266 and Ser-268) completely abolished JC virus spread and infectivity,[173] showing that the sugar–protein interactions mediated by these residues are required for a productive interaction with sialic acid-containing receptors, and thus for JC virus replication. A structural study of the interaction between VP1 pentamers and another minimal receptor-mimicking ligand, disialylated hexasaccharide, has also been published.[177] In this case, the authors did not detect any conformational rearrangement of the capsid protein.[177] To further illustrate the complexity of these virus–receptor interactions, the complex with the disialylated oligosaccharide involved a specific set of acidic residues distinct from those defining the binding site for the sialylated pentasaccharide. A model of the pentameric capsid-like VP1 assembly bound to the disialylated oligosaccharide is shown in Fig. 12.13A. It is interesting to note that both oligosaccharides bind to the same region of VP1, yet only the topology of the VP1–pentasaccharide complex is consistent with a virus–glycolipid interaction (Fig. 12.13B). The mode of association of VP1 with the disialylated hexasaccharide, which occurs without conformational adjustment, may be representative of a virus–glycoprotein rather than a virus–glycolipid interaction. Indeed, it is difficult to figure out how the VP1 domain located below the hexasaccharide-binding site could be accommodated by a membrane glycolipid.

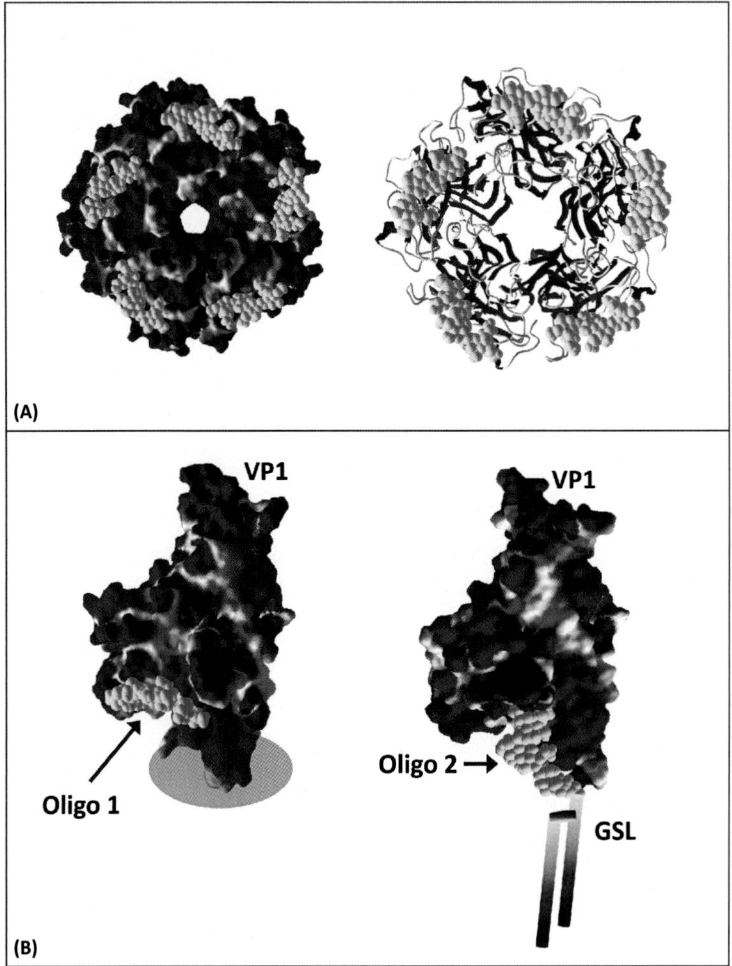

FIGURE 12.13 **Two modes of sialylated oligosaccharide binding on VP1 pentamers.** (A) Binding of disialylated hexasaccharide (in yellow) to the VP1 pentamer shown in secondary structure (right) or surface electropotential (left) rendering. The structures were obtained from PDB entry 1VPN.[177] (B) Although both the disialylated hexasaccharide[177] (oligo 1, left panel) and the sialylated pentasaccharide[173] (oligo 2, right panel) bind to the same region of VP1, each oligosaccharide has its own mechanism of interaction. The binding of pentasaccharide induces a conformational change in VP1 that is consistent with an interaction of the oligosaccharide motif linked to a membrane glycolipid (right panel). In contrast, the hexasaccharide does not induce any conformational change in VP1. In the latter case, the topology of the complex suggests an interaction with a glycoprotein decorated with the oligosaccharide motif rather than binding to a membrane glycolipid. The region of VP1–oligo 1 complex that would probably evoke a steric clash with a lipid bilayer is highlighted (left panel).

The second type of receptor used by JC virus is the serotonin receptor 5-HT2A, a G protein-coupled receptor with seven transmembrane domains.[169] The involvement of this cell surface receptor in JV virus entry has been demonstrated by the inhibitory effect of 5-HT2A receptor antagonists and monoclonal antibodies directed against 5-HT2A receptors. Oligodendrocyte progenitor cells, which express this particular serotonin receptor, are susceptible to JC virus infection.[178] The concomitant requirement of a ganglioside and a proteic receptor is a common mechanism shared by various viruses[77,179] and toxins[180] to gain entry into host cells (see Chapter 13).

12.3.2.3 *Herpes Viruses*

Several species of herpes viruses (*Herpesviridae*), including Epstein–Barr, herpes simplex, and varicella-zoster viruses, may cause meningitis and other neurological disorders. These DNA viruses are transmitted to the brain through the olfactory and/or trigeminal nerve pathway.[108,181] Herpes viruses bind to heparan sulfate and to nectin-1 expressed on the apical side of olfactory epithelial cells.[182,183] Virions first attach the heparan sulfate moieties of cell surface proteoglycans.[184,185] These sulfated membrane molecules serve as low-affinity attachment sites for the virus that is subsequently delivered to high-affinity protein receptors. As with most other viruses, lipid rafts are involved in the cellular attachment of various types of herpes virus.[186–188] The presence of cholesterol in lipid raft facilitates the fusion of herpes simplex virus (HSV-1) and varicella-zoster virus with the plasma membrane of host cells.[189,190] Moreover, during the maturation of herpes viruses, the assembly of viral components also occurs in lipid rafts so that host cell glycosphingolipids and cholesterol can be included in the viral envelope.[191] The incorporation of host gangliosides may thus extend the repertory of virus tropism through either specific glycolipid–glycolipid (transcarbohydrate) interactions[192] or glycolipid binding to adhesion molecules, a process referred to as the "glycosynapse" by Hakomori.[193,194] For instance, it has been reported that the incorporation of host cell-derived GM3 gangliosides into the envelope of HIV-1 virions could mediate the capture of these "humanized" virions by dendritic cells.[195] It is likely that several other viruses use such mimicry mechanisms for facilitating virus capture and dissemination *in vivo*.

Apart from being responsible for specific neurological disorders, *Herpesviridae* may also be linked to the etiology of Alzheimer's disease. Indeed, an intriguing relationship has been observed between herpes simplex virus (HSV-1) infection and some aspects of Alzheimer's disease pathogenesis.[196] First, the transsynaptic spread of Tau pathology *in vivo*[197] has been compared to the dissemination of a viral infection from brain cell to brain cell.[196] Second, HSV-1 has been shown to promote both neurotoxin accumulation of Aβ peptides[198] and Tau phosphorylation.[199] Third, the propagation of abnormal Tau proteins via paracrine or cell-to-cell mechanisms can be legitimately questioned whereas this mode of transmission is routine for HSV-1.[196] Overall, the notion that "reactivated HSV-1 could account for the intracerebral propagation of Alzheimer's disease changes in the human brain"[196] compels our attention.

12.4 KEY EXPERIMENT: WHAT IS A VIRUS RECEPTOR?

A virus receptor can be defined as a host cell surface component recognized by the virus as a gateway to entry into the cell. Ideally, a virus receptor would fulfill three main characteristics: (1) a physical interaction between the virus and the receptor should be demonstrated; (2) occupying the virus-binding site of the receptor (e.g., with an antibody directed against the receptor, should inhibit virus infection); and (3) the cellular sensitivity to virus infection should correlate with receptor expression. Therefore, cells lacking the receptor should not be infected, and transfection with the gene coding for the receptor would confer sensitivity to infection.

The CD4 transmembrane protein expressed by a subset of human T-lymphocytes is widely considered as the main receptor for HIV-1. However, CD4 does not fulfill all the characteristics listed. CD4 interacts with the HIV-1 surface envelope glycoprotein gp120 with high affinity[200,201] and some anti-CD4 antibodies, directed against the gp120 binding site, inhibit HIV-1 infection of $CD4^+$ lymphocytes.[70] The ability of HIV-1 to infect several types of CD4-negative cells, including fibroblasts,[202] neural,[203] and intestinal epithelial cells[204] led to the identification of an alternative receptor for HIV-1, the oligodendrocyte differentiation marker GalCer.[79,80] In this respect, HIV-1 behaved as an atypic retrovirus able to use either a protein (CD4) or a glycolipid (GalCer) to gain entry into various cell types.

Murine fibroblasts do not express CD4 or GalCer and are not sensitive to HIV-1 infection. Therefore, they provided a good model to determine whether the expression of human CD4 in these cells would confer sensitivity to infection. This was not the case. HIV-1 could attach to the surface of murine cells transfected with the gene coding for human CD4, but it was the only property that the transfection conferred. These data suggested that in addition to CD4, auxiliary factors, referred to as *fusion cofactors,* were required for the post-binding fusion step to ensure the penetration of the virus genome in the host cell. After a long search, the cofactor was identified as *fusin,*[205,206] a G protein-coupled transmembrane receptor that proved to be a member of the family of chemokine receptors. The main coreceptors allowing the postbinding fusion events of HIV-1 were CXCR4 (formerly referred to as fusin) and CCR5.[72] This time, the cotransfection of human CD4 and either CXCR4 or CCR5 conferred to initially resistant cells susceptibility to HIV-1 infection.[205] Independently of these coreceptor studies, Puri et al. showed that metabolic depletion of glycosphingolipid levels in human $CD4^+$ cells rendered these cells incompetent for gp120-dependent fusion.[207] These intriguing data indicated that in addition to the coreceptors, HIV-1 fusion also required a glycolipid.[207] A series of elegant reconstitution experiments demonstrated that the missing element was present in human erythrocyte membranes[207] and was finally identified as the neutral glycosphingolipid Gb_3.[208] The strain-dependent interaction of HIV-1 gp120 with specific glycolipids (chiefly Gb_3 and GM3) provided a biochemical basis to these data and further supported the notion that these glycolipids might serve as fusion cofactors.[78]

This was not the end of the story. The requirement of specific human glycosphingolipids in the transfer of HIV-1 from CD4 to a functional coreceptor (CCR5 or CXCR4) suggested a lipid raft-dependent mechanism of HIV-1 binding and fusion (Fig. 12.7). According to this model, the virus binds to CD4 in a lipid raft domain. This binding exposes the V3 loop of gp120 that becomes available to interact with an adequate coreceptor. Because the coreceptor is not initially associated with lipid rafts, the V3 loop is stabilized by multiple interactions with raft glycolipids (Gb3 or GM3). The virus, still bound to CD4 and to the glycolipids, sails on the raft until it finds its coreceptor to start the fusion process (Fig. 12.7). But what would happen if a cell overexpresses the glycolipid? It can be anticipated that the virus would be stuck on lipid rafts without possibility of engaging functional interactions with the coreceptors. Indeed, it has been reported that a high level of expression of GM3 conferred resistance to HIV-1 fusion.[209] These data have been obtained *in vitro* with transfected murine cells. However, the same kind of situation may exist *in vivo*. The minor P blood group phenotype is characterized by a high level of expression of Gb_3 (also referred to as CD77 or Pk antigen) in the plasma membrane of human lymphocytes.[210] *In vitro*, these cells appeared to be highly resistant to

HIV-1 infection, which may suggest that the blood group Pk could confer a natural resistance to HIV-1 infection.[211] If this proved to be the case, it would add to the extreme variety of biological functions involving glycosphingolipids. A parallel case has been described in birds that express unusually high levels of the ganglioside receptors used by tetanus toxin and are naturally resistant to tetanus.[180] In both cases (tetanus and HIV-1 infections), the overexpression of glycolipids may glue the toxin and the virus on the surface of target cells without any possibility for transfer to their protein receptors. Finally, it is interesting to note that synthetic soluble analogs of Gb_3 (adamantyl-Gb_3) inhibit HIV-1 infection *in vitro*[212] and could be thus considered an alternative therapeutic approach to antiretroviral agents. Similarly, oligosaccharide dendrimers based on the glycone moiety of Gb_3 and GM3 are efficient blockers of HIV-1 infection.[213] Such lipid-based antiviral strategies could also be used to prevent the formation of Ca^{2+}-permeable amyloid pores for the treatment of neurodegenerative diseases, as discussed in Chapter 14.

References

1. Kristensson K. Microbes' roadmap to neurons. *Nat Rev Neurosci*. 2011;12(6):345–357.
2. Niederkorn JY. See no evil, hear no evil, do no evil: the lessons of immune privilege. *Nat Immunol*. 2006;7(4): 354–359.
3. Huh GS, Boulanger LM, Du H, Riquelme PA, Brotz TM, Shatz CJ. Functional requirement for class I MHC in CNS development and plasticity. *Science*. 2000;290(5499):2155–2159.
4. Goddard CA, Butts DA, Shatz CJ. Regulation of CNS synapses by neuronal MHC class I. *Proc Natl Acad Sci USA*. 2007;104(16):6828–6833.
5. Tsiang H. Pathophysiology of rabies virus infection of the nervous system. *Adv Virus Res*. 1993;42:375–412.
6. Kelly RM, Strick PL. Rabies as a transneuronal tracer of circuits in the central nervous system. *J Neurosci Methods*. 2000;103(1):63–71.
7. Schnell MJ, McGettigan JP, Wirblich C, Papaneri A. The cell biology of rabies virus: using stealth to reach the brain. *Nat Rev Microbiol*. 2010;8(1):51–61.
8. Conzelmann KK, Cox JH, Schneider LG, Thiel HJ. Molecular cloning and complete nucleotide sequence of the attenuated rabies virus SAD B19. *Virology*. 1990;175(2):485–499.
9. Gaudin Y, Ruigrok RW, Tuffereau C, Knossow M, Flamand A. Rabies virus glycoprotein is a trimer. *Virology*. 1992;187(2):627–632.
10. Thoulouze MI, Lafage M, Schachner M, Hartmann U, Cremer H, Lafon M. The neural cell adhesion molecule is a receptor for rabies virus. *J Virol*. 1998;72(9):7181–7190.
11. Tuffereau C, Benejean J, Blondel D, Kieffer B, Flamand A. Low-affinity nerve-growth factor receptor (P75NTR) can serve as a receptor for rabies virus. *EMBO J*. 1998;17(24):7250–7259.
12. Wunner WH, Reagan KJ, Koprowski H. Characterization of saturable binding sites for rabies virus. *J Virol*. 1984;50(3):691–697.
13. Superti F, Hauttecoeur B, Morelec MJ, Goldoni P, Bizzini B, Tsiang H. Involvement of gangliosides in rabies virus infection. *J Gen Virol*. 1986;67(Pt 1):47–56.
14. Cardoso TC, da Silva LH, da Silva SE, et al. Chicken embryo related (CER) cell line for quantification of rabies neutralizing antibody by fluorescent focus inhibition test. *Biologicals*. 2006;34(1):29–32.
15. Conti C, Superti F, Tsiang H. Membrane carbohydrate requirement for rabies virus binding to chicken embryo related cells. *Intervirology*. 1986;26(3):164–168.
16. Ledeen RW, Diebler MF, Wu G, Lu ZH, Varoqui H. Ganglioside composition of subcellular fractions, including pre- and postsynaptic membranes, from Torpedo electric organ. *Neurochem Res*. 1993;18(11):1151–1155.
17. Schubert M, Harmison GG, Meier E. Primary structure of the vesicular stomatitis virus polymerase (L) gene: evidence for a high frequency of mutations. *J Virol*. 1984;51(2):505–514.
18. Fantini J, Garmy N, Yahi N. Prediction of glycolipid-binding domains from the amino acid sequence of lipid raft-associated proteins: application to HpaA, a protein involved in the adhesion of *Helicobacter pylori* to gastrointestinal cells. *Biochemistry*. 2006;45(36):10957–10962.

19. Dietzschold B, Wunner WH, Wiktor TJ, et al. Characterization of an antigenic determinant of the glycoprotein that correlates with pathogenicity of rabies virus. *Proc Natl Acad Sci USA*. 1983;80(1):70–74.
20. Seif I, Coulon P, Rollin PE, Flamand A. Rabies virulence: effect on pathogenicity and sequence characterization of rabies virus mutations affecting antigenic site III of the glycoprotein. *J Virol*. 1985;53(3):926–934.
21. Tuffereau C, Leblois H, Benejean J, Coulon P, Lafay F, Flamand A. Arginine or lysine in position 333 of ERA and CVS glycoprotein is necessary for rabies virulence in adult mice. *Virology*. 1989;172(1):206–212.
22. Yahi N, Fantini J. Deciphering the glycolipid code of Alzheimer's and Parkinson's amyloid proteins allowed the creation of a universal ganglioside-binding peptide. *PloS One*. 2014;9(8):e104751.
23. Mahfoud R, Garmy N, Maresca M, Yahi N, Puigserver A, Fantini J. Identification of a common sphingolipid-binding domain in Alzheimer, prion, and HIV-1 proteins. *J Biol Chem*. 2002;277(13):11292–11296.
24. Javaherian K, Langlois AJ, LaRosa GJ, et al. Broadly neutralizing antibodies elicited by the hypervariable neutralizing determinant of HIV-1. *Science*. 1990;250(4987):1590–1593.
25. Krachmarov CP, Kayman SC, Honnen WJ, Trochev O, Pinter A. V3-specific polyclonal antibodies affinity purified from sera of infected humans effectively neutralize primary isolates of human immunodeficiency virus type 1. *AIDS Res Hum Retroviruses*. 2001;17(18):1737–1748.
26. Wolfs TF, de Jong JJ, Van den Berg H, Tijnagel JM, Krone WJ, Goudsmit J. Evolution of sequences encoding the principal neutralization epitope of human immunodeficiency virus 1 is host dependent, rapid, and continuous. *Proc Natl Acad Sci USA*. 1990;87(24):9938–9942.
27. Hammache D, Pieroni G, Yahi N, et al. Specific interaction of HIV-1 and HIV-2 surface envelope glycoproteins with monolayers of galactosylceramide and ganglioside GM3. *J Biol Chem*. 1998;273(14):7967–7971.
28. Jacob Y, Badrane H, Ceccaldi PE, Tordo N. Cytoplasmic dynein LC8 interacts with lyssavirus phosphoprotein. *J Virol*. 2000;74(21):10217–10222.
29. Raux H, Flamand A, Blondel D. Interaction of the rabies virus P protein with the LC8 dynein light chain. *J Virol*. 2000;74(21):10212–10216.
30. Klingen Y, Conzelmann KK, Finke S. Double-labeled rabies virus: live tracking of enveloped virus transport. *J Virol*. 2008;82(1):237–245.
31. Roberts AJ, Kon T, Knight PJ, Sutoh K, Burgess SA. Functions and mechanics of dynein motor proteins. *Nat Rev Mol Cell Biol*. 2013;14(11):713–726.
32. Schnapp BJ, Reese TS. Dynein is the motor for retrograde axonal transport of organelles. *Proc Natl Acad Sci USA*. 1989;86(5):1548–1552.
33. Brady ST. Molecular motors in the nervous system. *Neuron*. 1991;7(4):521–533.
34. Lahaye X, Vidy A, Pomier C, et al. Functional characterization of Negri bodies (NBs) in rabies virus-infected cells: Evidence that NBs are sites of viral transcription and replication. *J Virol*. 2009;83(16):7948–7958.
35. Okumura A, Harty RN. Rabies virus assembly and budding. *Adv Virus Res*. 2011;79:23–32.
36. Iwasaki Y, Liu DS, Yamamoto T, Konno H. On the replication and spread of rabies virus in the human central nervous system. *J Neuropathol Exp Neurol*. 1985;44(2):185–195.
37. Sinchaisri TA, Nagata T, Yoshikawa Y, Kai C, Yamanouchi K. Immunohistochemical and histopathological study of experimental rabies infection in mice. *J Vet Med Sci*. 1992;54(3):409–416.
38. Fu ZF, Jackson AC. Neuronal dysfunction and death in rabies virus infection. *J Neurovirol*. 2005;11(1):101–106.
39. Conomy JP, Leibovitz A, McCombs W, Stinson J. Airborne rabies encephalitis: demonstration of rabies virus in the human central nervous system. *Neurology*. 1977;27(1):67–69.
40. Lafay F, Coulon P, Astic L, et al. Spread of the CVS strain of rabies virus and of the avirulent mutant AvO1 along the olfactory pathways of the mouse after intranasal inoculation. *Virology*. 1991;183(1):320–330.
41. Munster VJ, Prescott JB, Bushmaker T, et al. Rapid Nipah virus entry into the central nervous system of hamsters via the olfactory route. *Sci Rep*. 2012;2:736.
42. Schrauwen EJ, Herfst S, Leijten LM, et al. The multibasic cleavage site in H5N1 virus is critical for systemic spread along the olfactory and hematogenous routes in ferrets. *J Virol*. 2012;86(7):3975–3984.
43. Phillips AT, Stauft CB, Aboellail TA, et al. Bioluminescent imaging and histopathologic characterization of WEEV neuroinvasion in outbred CD-1 mice. *PloS One*. 2013;8(1):e53462.
44. Schaefer ML, Bottger B, Silver WL, Finger TE. Trigeminal collaterals in the nasal epithelium and olfactory bulb: a potential route for direct modulation of olfactory information by trigeminal stimuli. *J Comp Neurol*. 2002;444(3):221–226.
45. Baer GM, Shaddock JH, Houff SA, Harrison AK, Gardner JJ. Human rabies transmitted by corneal transplant. *Archiv Neurol*. 1982;39(2):103–107.

46. Spudich S, Gonzalez-Scarano F. HIV-1-related central nervous system disease: current issues in pathogenesis, diagnosis, and treatment. *Cold Spring Harb Perspect Med*. 2012;2(6):a007120.
47. Strazza M, Pirrone V, Wigdahl B, Nonnemacher MR. Breaking down the barrier: the effects of HIV-1 on the blood–brain barrier. *Brain Res*. 2011;1399:96–115.
48. Gonzalez-Scarano F, Martin-Garcia J. The neuropathogenesis of AIDS. *Nat Rev Immunol*. 2005;5(1):69–81.
49. d'Arminio Monforte A, Cinque P, Mocroft A, et al. Changing incidence of central nervous system diseases in the EuroSIDA cohort. *Ann Neurol*. 2004;55(3):320–328.
50. Heaton RK, Clifford DB, Franklin Jr DR, et al. HIV-associated neurocognitive disorders persist in the era of potent antiretroviral therapy: CHARTER Study. *Neurology*. 2010;75(23):2087–2096.
51. Persidsky Y, Stins M, Way D, et al. A model for monocyte migration through the blood–brain barrier during HIV-1 encephalitis. *J Immunol*. 1997;158(7):3499–3510.
52. Banks WA, Freed EO, Wolf KM, Robinson SM, Franko M, Kumar VB. Transport of human immunodeficiency virus type 1 pseudoviruses across the blood–brain barrier: role of envelope proteins and adsorptive endocytosis. *J Virol*. 2001;75(10):4681–4691.
53. Nico B, Ribatti D. Morphofunctional aspects of the blood–brain barrier. *Curr Drug Metabol*. 2012;13(1):50–60.
54. Tamura A, Tsukita S. Paracellular barrier and channel functions of TJ claudins in organizing biological systems: Advances in the field of barriology revealed in knockout mice. *Semin Cell Dev Biol*. 2014;36C:177–185.
55. Hirase T, Staddon JM, Saitou M, et al. Occludin as a possible determinant of tight junction permeability in endothelial cells. *J Cell Sci*. 1997;110(Pt 14):1603–1613.
56. Abbott NJ, Dolman DE, Patabendige AK. Assays to predict drug permeation across the blood–brain barrier, and distribution to brain. *Curr Drug Metabol*. 2008;9(9):901–910.
57. Rapoport SI, Levitan H. Neurotoxicity of X-ray contrast media. Relation to lipid solubility and blood–brain barrier permeability. *Am J Roentgenol Radium Ther Nucl Med*. 1974;122(1):186–193.
58. Ramsay RE, Hammond EJ, Perchalski RJ, Wilder BJ. Brain uptake of phenytoin, phenobarbital, and diazepam. *Archiv Neurol*. 1979;36(9):535–539.
59. Ohtsuki S, Terasaki T. Contribution of carrier-mediated transport systems to the blood–brain barrier as a supporting and protecting interface for the brain; importance for CNS drug discovery and development. *Pharm Res*. 2007;24(9):1745–1758.
60. Abbott NJ. Blood–brain barrier structure and function and the challenges for CNS drug delivery. *J Inherit Metab Dis*. 2013;36(3):437–449.
61. Hickey WF. Leukocyte traffic in the central nervous system: the participants and their roles. *Semin Immunol*. 1999;11(2):125–137.
62. Peluso R, Haase A, Stowring L, Edwards M, Ventura P. A Trojan horse mechanism for the spread of visna virus in monocytes. *Virology*. 1985;147(1):231–236.
63. Greenwood J, Heasman SJ, Alvarez JI, Prat A, Lyck R, Engelhardt B. Review: leucocyte-endothelial cell crosstalk at the blood–brain barrier: a prerequisite for successful immune cell entry to the brain. *Neuropathol Appl Neurobiol*. 2011;37(1):24–39.
64. Persidsky Y, Ghorpade A, Rasmussen J, et al. Microglial and astrocyte chemokines regulate monocyte migration through the blood–brain barrier in human immunodeficiency virus-1 encephalitis. *Am J Pathol*. 1999;155(5):1599–1611.
65. Ensoli B, Buonaguro L, Barillari G, et al. Release, uptake, and effects of extracellular human immunodeficiency virus type 1 Tat protein on cell growth and viral transactivation. *J Virol*. 1993;67(1):277–287.
66. Pu H, Hayashi K, Andras IE, Eum SY, Hennig B, Toborek M. Limited role of COX-2 in HIV Tat-induced alterations of tight junction protein expression and disruption of the blood–brain barrier. *Brain Res*. 2007;1184:333–344.
67. Kanmogne GD, Schall K, Leibhart J, Knipe B, Gendelman HE, Persidsky Y. HIV-1 gp120 compromises blood–brain barrier integrity and enhances monocyte migration across blood–brain barrier: implication for viral neuropathogenesis. *J Cereb Blood Flow Metab*. 2007;27(1):123–134.
68. Nakamuta S, Endo H, Higashi Y, et al. Human immunodeficiency virus type 1 gp120-mediated disruption of tight junction proteins by induction of proteasome-mediated degradation of zonula occludens-1 and -2 in human brain microvascular endothelial cells. *J Neurovirol*. 2008;14(3):186–195.
69. Davis LE, Hjelle BL, Miller VE, et al. Early viral brain invasion in iatrogenic human immunodeficiency virus infection. *Neurology*. 1992;42(9):1736–1739.
70. Klatzmann D, Champagne E, Chamaret S, et al. T-lymphocyte T4 molecule behaves as the receptor for human retrovirus LAV. *Nature*. 1984;312(5996):767–768.

71. Blumenthal R, Durell S, Viard M. HIV entry and envelope glycoprotein-mediated fusion. *J Biol Chem*. 2012;287(49):40841–40849.
72. Berger EA, Murphy PM, Farber JM. Chemokine receptors as HIV-1 coreceptors: roles in viral entry, tropism, and disease. *Ann Rev Immunol*. 1999;17:657–700.
73. Speck RF, Wehrly K, Platt EJ, et al. Selective employment of chemokine receptors as human immunodeficiency virus type 1 coreceptors determined by individual amino acids within the envelope V3 loop. *J Virol*. 1997;71(9):7136–7139.
74. Fantini J, Garmy N, Mahfoud R, Yahi N. Lipid rafts: structure, function and role in HIV, Alzheimer's and prion diseases. *Expert Rev Mol Med*. 2002;4(27):1–22.
75. Rawat SS, Viard M, Gallo SA, Rein A, Blumenthal R, Puri A. Modulation of entry of enveloped viruses by cholesterol and sphingolipids (Review). *Mol Membr Biol*. 2003;20(3):243–254.
76. Viard M, Parolini I, Rawat SS, et al. The role of glycosphingolipids in HIV signaling, entry and pathogenesis. *Glycoconj J*. 2004;20(3):213–222.
77. Fantini J, Hammache D, Pieroni G, Yahi N. Role of glycosphingolipid microdomains in CD4-dependent HIV-1 fusion. *Glycoconj J*. 2000;17(3–4):199–204.
78. Hammache D, Yahi N, Maresca M, Pieroni G, Fantini J. Human erythrocyte glycosphingolipids as alternative cofactors for human immunodeficiency virus type 1 (HIV-1) entry: evidence for CD4-induced interactions between HIV-1 gp120 and reconstituted membrane microdomains of glycosphingolipids (Gb3 and GM3). *J Virol*. 1999;73(6):5244–5248.
79. Harouse JM, Bhat S, Spitalnik SL, et al. Inhibition of entry of HIV-1 in neural cell lines by antibodies against galactosyl ceramide. *Science*. 1991;253(5017):320–323.
80. Yahi N, Baghdiguian S, Moreau H, Fantini J. Galactosyl ceramide (or a closely related molecule) is the receptor for human immunodeficiency virus type 1 on human colon epithelial HT29 cells. *J Virol*. 1992;66(8):4848–4854.
81. Albright AV, Strizki J, Harouse JM, Lavi E, O'Connor M, Gonzalez-Scarano F. HIV-1 infection of cultured human adult oligodendrocytes. *Virology*. 1996;217(1):211–219.
82. Takahashi K, Wesselingh SL, Griffin DE, McArthur JC, Johnson RT, Glass JD. Localization of HIV-1 in human brain using polymerase chain reaction/in situ hybridization and immunocytochemistry. *Ann Neurol*. 1996;39(6):705–711.
83. Kimura-Kuroda J, Nagashima K, Yasui K. Inhibition of myelin formation by HIV-1 gp120 in rat cerebral cortex culture. *Archiv Virol*. 1994;137(1-2):81–99.
84. Bernardo A, Agresti C, Levi G. HIV-gp120 affects the functional activity of oligodendrocytes and their susceptibility to complement. *J Neurosci Res*. 1997;50(6):946–957.
85. Johnson RT. The virology of demyelinating diseases. *Ann Neurol*. 1994;36(suppl):S54–S60.
86. Codazzi F, Menegon A, Zacchetti D, Ciardo A, Grohovaz F, Meldolesi J. HIV-1 gp120 glycoprotein induces [Ca2+]i responses not only in type-2 but also type-1 astrocytes and oligodendrocytes of the rat cerebellum. *Eur J Neurosci*. 1995;7(6):1333–1341.
87. Esiri MM, Morris CS. Cellular basis of HIV infection of the CNS and the AIDS dementia complex: oligodendrocyte. *J NeuroAIDS*. 1996;1(1):133–160.
88. Thompson KA, Churchill MJ, Gorry PR, et al. Astrocyte specific viral strains in HIV dementia. *Ann Neurol*. 2004;56(6):873–877.
89. Bajetto A, Bonavia R, Barbero S, et al. Glial and neuronal cells express functional chemokine receptor CXCR4 and its natural ligand stromal cell-derived factor 1. *J Neurochem*. 1999;73(6):2348–2357.
90. Ronaldson PT, Bendayan R. HIV-1 viral envelope glycoprotein gp120 triggers an inflammatory response in cultured rat astrocytes and regulates the functional expression of P-glycoprotein. *Mol Pharmacol*. 2006;70(3):1087–1098.
91. Saha RN, Pahan K. Differential regulation of Mn-superoxide dismutase in neurons and astroglia by HIV-1 gp120: implications for HIV-associated dementia. *Free Rad Biol Med*. 2007;42(12):1866–1878.
92. Banerjee S, Walseth TF, Borgmann K, et al. CD38/cyclic ADP-ribose regulates astrocyte calcium signaling: implications for neuroinflammation and HIV-1-associated dementia. *J Neuroimmune Pharmacol*. 2008;3(3):154–164.
93. Shah A, Kumar S, Simon SD, Singh DP, Kumar A. HIV gp120- and methamphetamine-mediated oxidative stress induces astrocyte apoptosis via cytochrome P450 2E1. *Cell Death Dis*. 2013;4:e850.
94. Doganay M. Listeriosis: clinical presentation. *FEMS Immunol Med Microbiol*. 2003;35(3):173–175.
95. Drevets DA, Bronze MS. *Listeria monocytogenes*: epidemiology, human disease, and mechanisms of brain invasion. *FEMS Immunol Med Microbiol*. 2008;53(2):151–165.

96. Siegman-Igra Y, Levin R, Weinberger M, et al. *Listeria monocytogenes* infection in Israel and review of cases worldwide. *Emerg Infect Dis*. 2002;8(3):305–310.
97. Durand ML, Calderwood SB, Weber DJ, et al. Acute bacterial meningitis in adults. A review of 493 episodes. *N Engl J Med*. 1993;328(1):21–28.
98. Sigurdardottir B, Bjornsson OM, Jonsdottir KE, Erlendsdottir H, Gudmundsson S. Acute bacterial meningitis in adults. A 20-year overview. *Archiv Int Med*. 1997;157(4):425–430.
99. Kyaw MH, Christie P, Jones IG, Campbell H. The changing epidemiology of bacterial meningitis and invasive non-meningitic bacterial disease in Scotland during the period 1983-99. *Scand J Infect Dis*. 2002;34(4): 289–298.
100. Sleator RD, Watson D, Hill C, Gahan CG. The interaction between *Listeria monocytogenes* and the host gastrointestinal tract. *Microbiology*. 2009;155(Pt 8):2463–2475.
101. Gahan CG, Hill C. *Listeria monocytogenes*: survival and adaptation in the gastrointestinal tract. *Front Cell Infect Microbiol*. 2014;4:9.
102. Schuppler M, Loessner MJ. The opportunistic pathogen *Listeria monocytogenes*: pathogenicity and interaction with the mucosal immune system. *Int J Inflam*. 2010;2010:704321.
103. Drevets DA, Dillon MJ, Schawang JS, et al. The Ly-6Chigh monocyte subpopulation transports *Listeria monocytogenes* into the brain during systemic infection of mice. *J Immunol*. 2004;172(7):4418–4424.
104. Join-Lambert OF, Ezine S, Le Monnier A, et al. *Listeria monocytogenes*-infected bone marrow myeloid cells promote bacterial invasion of the central nervous system. *Cell Microbiol*. 2005;7(2):167–180.
105. Drevets DA, Leenen PJ, Greenfield RA. Invasion of the central nervous system by intracellular bacteria. *Clin Microbiol Rev*. 2004;17(2):323–347.
106. Antal EA, Dietrichs E, Loberg EM, Melby KK, Maehlen J. Brain stem encephalitis in listeriosis. *Scand J Infect Dis*. 2005;37(3):190–194.
107. Antal EA, Loberg EM, Dietrichs E, Maehlen J. Neuropathological findings in 9 cases of *Listeria monocytogenes* brain stem encephalitis. *Brain Pathol*. 2005;15(3):187–191.
108. Dando SJ, Mackay-Sim A, Norton R, et al. Pathogens penetrating the central nervous system: infection pathways and the cellular and molecular mechanisms of invasion. *Clin Microbiol Rev*. 2014;27(4):691–726.
109. Hamon MA, Ribet D, Stavru F, Cossart P. Listeriolysin O: the Swiss army knife of Listeria. *Trends Microbiol*. 2012;20(8):360–368.
110. Soltani CE, Hotze EM, Johnson AE, Tweten RK. Structural elements of the cholesterol-dependent cytolysins that are responsible for their cholesterol-sensitive membrane interactions. *Proc Natl Acad Sci USA*. 2007;104(51):20226–20231.
111. Farrand AJ, LaChapelle S, Hotze EM, Johnson AE, Tweten RK. Only two amino acids are essential for cytolytic toxin recognition of cholesterol at the membrane surface. *Proc Natl Acad Sci USA*. 2010;107(9):4341–4346.
112. Koster S, van Pee K, Hudel M, et al. Crystal structure of listeriolysin O reveals molecular details of oligomerization and pore formation. *Nat Commun*. 2014;5:3690.
113. Jamin N, Neumann JM, Ostuni MA, et al. Characterization of the cholesterol recognition amino acid consensus sequence of the peripheral-type benzodiazepine receptor. *Mol Endocrinol*. 2005;19(3):588–594.
114. Baier CJ, Fantini J, Barrantes FJ. Disclosure of cholesterol recognition motifs in transmembrane domains of the human nicotinic acetylcholine receptor. *Sci Rep*. 2011;1:69.
115. Di Scala C, Chahinian H, Yahi N, Garmy N, Fantini J. Interaction of Alzheimer's beta-amyloid peptides with cholesterol: mechanistic insights into amyloid pore formation. *Biochemistry*. 2014;53(28):4489–4502.
116. Repp H, Pamukci Z, Koschinski A, et al. Listeriolysin of *Listeria monocytogenes* forms Ca2+-permeable pores leading to intracellular Ca2+ oscillations. *Cell Microbiol*. 2002;4(8):483–491.
117. Di Scala C, Troadec JD, Lelievre C, Garmy N, Fantini J, Chahinian H. Mechanism of cholesterol-assisted oligomeric channel formation by a short Alzheimer beta-amyloid peptide. *J Neurochem*. 2014;128(1):186–195.
118. Fantini J, Barrantes FJ. Sphingolipid/cholesterol regulation of neurotransmitter receptor conformation and function. *Biochim Biophys Acta*. 2009;1788(11):2345–2361.
119. Fantini J, Barrantes FJ. How cholesterol interacts with membrane proteins: an exploration of cholesterol-binding sites including CRAC, CARC, and tilted domains. *Front Physiol*. 2013;4:31.
120. Thigpen MC, Whitney CG, Messonnier NE, et al. Bacterial meningitis in the United States, 1998-2007. *N Engl J Med*. 2011;364(21):2016–2025.
121. Levy C, Varon E, Taha MK, et al. Change in French bacterial meningitis in children resulting from vaccination. *Arch Pediatr*. 2014;21(7):736–744.

122. van Deuren M, Brandtzaeg P, van der Meer JW. Update on meningococcal disease with emphasis on pathogenesis and clinical management. *Clin Microbiol Rev*. 2000;13(1):144–166:table of contents.
123. Coureuil M, Join-Lambert O, Lecuyer H, Bourdoulous S, Marullo S, Nassif X. Mechanism of meningeal invasion by *Neisseria meningitidis*. *Virulence*. 2012;3(2):164–172.
124. Schubert-Unkmeir A, Konrad C, Slanina H, Czapek F, Hebling S, Frosch M. *Neisseria meningitidis* induces brain microvascular endothelial cell detachment from the matrix and cleavage of occludin: a role for MMP-8. *PLoS Pathog*. 2010;6(4):e1000874.
125. Coureuil M, Mikaty G, Miller F, et al. Meningococcal type IV pili recruit the polarity complex to cross the brain endothelium. *Science*. 2009;325(5936):83–87.
126. Soyer M, Charles-Orszag A, Lagache T, et al. Early sequence of events triggered by the interaction of *Neisseria meningitidis* with endothelial cells. *Cell Microbiol*. 2014;16(6):878–895.
127. Hardy SJ, Christodoulides M, Weller RO, Heckels JE. Interactions of *Neisseria meningitidis* with cells of the human meninges. *Mol Microbiol*. 2000;36(4):817–829.
128. Rumiantsev SN, Avrova NF, Pospelov VF, Denisova NA. The effect of gangliosides on the adhesive interaction of *Neisseria meningitidis* with human cells. *Zh Mikrobiol Epidemiol Immunobiol*. 1990;10:29–32.
129. Hugosson S, Angstrom J, Olsson BM, et al. Glycosphingolipid binding specificities of *Neisseria meningitidis* and *Haemophilus influenzae*: detection, isolation, and characterization of a binding-active glycosphingolipid from human oropharyngeal epithelium. *J Biochem*. 1998;124(6):1138–1152.
130. Kanda T, Ariga T, Kubodera H, et al. Glycosphingolipid composition of primary cultured human brain microvascular endothelial cells. *J Neurosci Res*. 2004;78(1):141–150.
131. Betz J, Bielaszewska M, Thies A, et al. Shiga toxin glycosphingolipid receptors in microvascular and macrovascular endothelial cells: differential association with membrane lipid raft microdomains. *J Lipid Res*. 2011;52(4):618–634.
132. Mairey E, Genovesio A, Donnadieu E, et al. Cerebral microcirculation shear stress levels determine *Neisseria meningitidis* attachment sites along the blood–brain barrier. *J Exp Med*. 2006;203(8):1939–1950.
133. Bock K, Karlsson KA, Stromberg N, Teneberg S. Interaction of viruses, bacteria and bacterial toxins with host cell surface glycolipids. Aspects on receptor identification and dissection of binding epitopes. *Adv Exp Med Biol*. 1988;228:153–186.
134. Karlsson KA. Animal glycosphingolipids as membrane attachment sites for bacteria. *Ann Rev Biochem*. 1989;58: 309–350.
135. Taieb N, Yahi N, Fantini J. Rafts and related glycosphingolipid-enriched microdomains in the intestinal epithelium: bacterial targets linked to nutrient absorption. *Adv Drug Deliv Rev*. 2004;56(6):779–794.
136. MacNeil JR, Cohn AC, Farley M, et al. Current epidemiology and trends in invasive *Haemophilus influenzae* disease–United States, 1989-2008. *Clin Infect Dis*. 2011;53(12):1230–1236.
137. McIntyre PB, O'Brien KL, Greenwood B, van de Beek D. Effect of vaccines on bacterial meningitis worldwide. *Lancet*. 2012;380(9854):1703–1711.
138. Fleming JA, Dieye Y, Ba O, et al. Effectiveness of *Haemophilus influenzae* type B conjugate vaccine for prevention of meningitis in Senegal. *PediatrInfect Dis J*. 2011;30(5):430–432.
139. Kabore NF, Poda GE, Barro M, et al. Impact of vaccination on admissions for *Haemophilus influenzae* b meningitis from 2004 to 2008 in Bobo Dioulasso, Burkina Faso. *Med Sante Trop*. 2012;22(4):425–429.
140. Sigauque B, Vubil D, Sozinho A, et al. *Haemophilus influenzae* type b disease among children in rural Mozambique: impact of vaccine introduction. *J Pediatr*. 2013;163(suppl 1):S19–S24.
141. Fleischmann RD, Adams MD, White O, et al. Whole-genome random sequencing and assembly of *Haemophilus influenzae* Rd. *Science*. 1995;269(5223):496–512.
142. Moxon ER, Kroll JS. Type b capsular polysaccharide as a virulence factor of *Haemophilus influenzae*. *Vaccine*. 1988;6(2):113–115.
143. Quagliarello VJ, Long WJ, Scheld WM. Morphologic alterations of the blood–brain barrier with experimental meningitis in the rat. Temporal sequence and role of encapsulation. *J Clin Invest*. 1986;77(4):1084–1095.
144. Galdiero M, D'Amico M, Gorga F, et al. *Haemophilus influenzae* porin contributes to signaling of the inflammatory cascade in rat brain. *Infect Immun*. 2001;69(1):221–227.
145. Fakih MG, Murphy TF, Pattoli MA, Berenson CS. Specific binding of *Haemophilus influenzae* to minor gangliosides of human respiratory epithelial cells. *Infect Immun*. 1997;65(5):1695–1700.
146. Berenson CS, Sayles KB, Huang J, Reinhold VN, Garlipp MA, Yohe HC. Nontypeable *Haemophilus influenzae*-binding gangliosides of human respiratory (HEp-2) cells have a requisite lacto/neolacto core structure. *FEMS Immunol Med Microbiol*. 2005;45(2):171–182.

147. Filippidis A, Fountas KN. Nasal lymphatics as a novel invasion and dissemination route of bacterial meningitis. *Med Hypotheses*. 2009;72(6):694–697.
148. Martinez-Morales F, Schobert M, Lopez-Lara IM, Geiger O. Pathways for phosphatidylcholine biosynthesis in bacteria. *Microbiology*. 2003;149(Pt 12):3461–3471.
149. Fan X, Goldfine H, Lysenko E, Weiser JN. The transfer of choline from the host to the bacterial cell surface requires glpQ in *Haemophilus influenzae*. *Mol Microbiol*. 2001;41(5):1029–1036.
150. Clark SE, Snow J, Li J, Zola TA, Weiser JN. Phosphorylcholine allows for evasion of bactericidal antibody by *Haemophilus influenzae*. *PLoS Pathog*. 2012;8(3):e1002521.
151. Mori M, Kuwabara S, Miyake M, et al. *Haemophilus influenzae* infection and Guillain-Barre syndrome. *Brain*. 2000;123(Pt 10):2171–2178.
152. van Ginkel FW, McGhee JR, Watt JM, Campos-Torres A, Parish LA, Briles DE. Pneumococcal carriage results in ganglioside-mediated olfactory tissue infection. *Proc Natl Acad Sci USA*. 2003;100(24):14363–14367.
153. Sundberg-Kovamees M, Holme T, Sjogren A. Interaction of the C-polysaccharide of *Streptococcus pneumoniae* with the receptor asialo-GM1. *Microb Pathog*. 1996;21(4):223–234.
154. Nollmann M, Gilbert R, Mitchell T, Sferrazza M, Byron O. The role of cholesterol in the activity of pneumolysin, a bacterial protein toxin. *Biophys J*. 2004;86(5):3141–3151.
155. Rossjohn J, Gilbert RJ, Crane D, et al. The molecular mechanism of pneumolysin, a virulence factor from *Streptococcus pneumoniae*. *J Mol Biol*. 1998;284(2):449–461.
156. Zysk G, Schneider-Wald BK, Hwang JH, et al. Pneumolysin is the main inducer of cytotoxicity to brain microvascular endothelial cells caused by *Streptococcus pneumoniae*. *Infect Immun*. 2001;69(2):845–852.
157. Rotbart HA. Enteroviral infections of the central nervous system. *Clin Infect Dis*. 1995;20(4):971–981.
158. Desmond RA, Accortt NA, Talley L, Villano SA, Soong SJ, Whitley RJ. Enteroviral meningitis: natural history and outcome of pleconaril therapy. *Antimicrob Agents Chemother*. 2006;50(7):2409–2414.
159. Zhou L, Miranda-Saksena M, Saksena NK. Viruses and neurodegeneration. *Virol J*. 2013;10:172.
160. Studahl M, Lindquist L, Eriksson BM, et al. Acute viral infections of the central nervous system in immunocompetent adults: diagnosis and management. *Drugs*. 2013;73(2):131–158.
161. Rhoades RE, Tabor-Godwin JM, Tsueng G, Feuer R. Enterovirus infections of the central nervous system. *Virology*. 2011;411(2):288–305.
162. Yang B, Chuang H, Yang KD. Sialylated glycans as receptor and inhibitor of enterovirus 71 infection to DLD-1 intestinal cells. *Virol J*. 2009;6:141.
163. Yang WX, Terasaki T, Shiroki K, et al. Efficient delivery of circulating poliovirus to the central nervous system independently of poliovirus receptor. *Virology*. 1997;229(2):421–428.
164. Tabor-Godwin JM, Ruller CM, Bagalso N, et al. A novel population of myeloid cells responding to coxsackievirus infection assists in the dissemination of virus within the neonatal CNS. *J NeurosciV 30*. 2010;(25):8676–8691.
165. Ren R, Racaniello VR. Poliovirus spreads from muscle to the central nervous system by neural pathways. *J Infect Dis*. 1992;166(4):747–752.
166. Seth P, Diaz F, Major EO. Advances in the biology of JC virus and induction of progressive multifocal leukoencephalopathy. *J Neurovirol*. 2003;9(2):236–246.
167. Wuthrich C, Dang X, Westmoreland S, et al. Fulminant JC virus encephalopathy with productive infection of cortical pyramidal neurons. *Ann Neurol*. 2009;65(6):742–748.
168. Gee GV, Dugan AS, Tsomaia N, Mierke DF, Atwood WJ. The role of sialic acid in human polyomavirus infections. *Glycoconj J*. 2006;23(1–2):19–26.
169. Elphick GF, Querbes W, Jordan JA, et al. The human polyomavirus, JCV, uses serotonin receptors to infect cells. *Science*. 2004;306(5700):1380–1383.
170. Komagome R, Sawa H, Suzuki T, et al. Oligosaccharides as receptors for JC virus. *J Virol*. 2002;76(24):12992–13000.
171. van Echten-Deckert G, Walter J. Sphingolipids: critical players in Alzheimer's disease. *Prog Lipid Res*. 2012;51(4):378–393.
172. Eash S, Tavares R, Stopa EG, Robbins SH, Brossay L, Atwood WJ. Differential distribution of the JC virus receptor-type sialic acid in normal human tissues. *Am J Pathol*. 2004;164(2):419–428.
173. Neu U, Maginnis MS, Palma AS, et al. Structure-function analysis of the human JC polyomavirus establishes the LSTc pentasaccharide as a functional receptor motif. *Cell Host Microbe*. 2010;8(4):309–319.
174. Salunke DM, Caspar DL, Garcea RL. Self-assembly of purified polyomavirus capsid protein VP1. *Cell*. 1986;46(6):895–904.

175. Breg J, Kroon-Batenburg LM, Strecker G, Montreuil J, Vliegenthart JF. Conformational analysis of the sialyl alpha(2–3/6)N-acetyllactosamine structural element occurring in glycoproteins, by two-dimensional NOE 1H-NMR spectroscopy in combination with energy calculations by hard-sphere exo-anomeric and molecular mechanics force-field with hydrogen-bonding potential. *Eur J Biochem*. 1989;178(3):727–739.
176. Fantini J, Yahi N. Molecular insights into amyloid regulation by membrane cholesterol and sphingolipids: common mechanisms in neurodegenerative diseases. *Expert Rev Mol Med*. 2010;12:e27.
177. Stehle T, Harrison SC. High-resolution structure of a polyomavirus VP1-oligosaccharide complex: implications for assembly and receptor binding. *EMBO J*. 1997;16(16):5139–5148.
178. Schaumburg C, O'Hara BA, Lane TE, Atwood WJ. Human embryonic stem cell-derived oligodendrocyte progenitor cells express the serotonin receptor and are susceptible to JC virus infection. *J Virol*. 2008;82(17): 8896–8899.
179. Hammache D, Yahi N, Pieroni G, Ariasi F, Tamalet C, Fantini J. Sequential interaction of CD4 and HIV-1 gp120 with a reconstituted membrane patch of ganglioside GM3: implications for the role of glycolipids as potential HIV-1 fusion cofactors. *Biochem Biophys Res Commun*. 1998;246(1):117–122.
180. Montecucco C. How do tetanus and botulinum toxins bind to neuronal membranes? *Trends Biochem Sci*. 1986;11:314–317.
181. Shivkumar M, Milho R, May JS, Nicoll MP, Efstathiou S, Stevenson PG. Herpes simplex virus 1 targets the murine olfactory neuroepithelium for host entry. *J Virol*. 2013;87(19):10477–10488.
182. Milho R, Frederico B, Efstathiou S, Stevenson PG. A heparan-dependent herpesvirus targets the olfactory neuroepithelium for host entry. *PLoS Pathog*. 2012;8(11):e1002986.
183. Lu G, Zhang N, Qi J, et al. Crystal structure of HSV-2 gD bound to nectin-1 reveals a conserved mode of receptor recognition. *J Virol*. 2014.
184. Shieh MT, WuDunn D, Montgomery RI, Esko JD, Spear PG. Cell surface receptors for herpes simplex virus are heparan sulfate proteoglycans. *J Cell Biol*. 1992;116(5):1273–1281.
185. Heldwein EE, Krummenacher C. Entry of herpesviruses into mammalian cells. *Cell Mol Life Sci*. 2008;65(11): 1653–1668.
186. Bender FC, Whitbeck JC, Ponce de Leon M, Lou H, Eisenberg RJ, Cohen GH. Specific association of glycoprotein B with lipid rafts during herpes simplex virus entry. *J Virol*. 2003;77(17):9542–9552.
187. Suzuki T, Suzuki Y. Virus infection and lipid rafts. *Biol Pharm Bull*. 2006;29(8):1538–1541.
188. Takahashi T, Suzuki T. Function of membrane rafts in viral lifecycles and host cellular response. *Biochem Res Int*. 2011;2011:245090.
189. Hambleton S, Steinberg SP, Gershon MD, Gershon AA. Cholesterol dependence of varicella-zoster virion entry into target cells. *J Virol*. 2007;81(14):7548–7558.
190. Wudiri GA, Pritchard SM, Li H, et al. Molecular requirement for sterols in herpes simplex virus entry and infectivity. *J Virol*. 2014;88(23):13918–13922.
191. Kawabata A, Tang H, Huang H, Yamanishi K, Mori Y. Human herpesvirus 6 envelope components enriched in lipid rafts: evidence for virion-associated lipid rafts. *Virol J*. 2009;6:127.
192. Handa K, Hakomori SI. Carbohydrate to carbohydrate interaction in development process and cancer progression. *Glycoconj J*. 2012;29(8–9):627–637.
193. Hakomori S. Glycosynapses: microdomains controlling carbohydrate-dependent cell adhesion and signaling. *An Acad Bras Cienc*. 2004;76(3):553–572.
194. Hakomori SI. The glycosynapse. *Proc Natl Acad Sci USA*. 2002;99(1):225–232.
195. Puryear WB, Yu X, Ramirez NP, Reinhard BM, Gummuluru S. HIV-1 incorporation of host-cell-derived glycosphingolipid GM3 allows for capture by mature dendritic cells. *Proc Natl Acad Sci USA*. 2012;109(19):7475–7480.
196. Ball MJ, Lukiw WJ, Kammerman EM, Hill JM. Intracerebral propagation of Alzheimer's disease: strengthening evidence of a herpes simplex virus etiology. *Alzheimers Dement*. 2013;9(2):169–175.
197. Liu L, Drouet V, Wu JW, et al. Trans-synaptic spread of tau pathology *in vivo*. *PloS One*. 2012;7(2):e31302.
198. Piacentini R, Civitelli L, Ripoli C, et al. HSV-1 promotes Ca2+-mediated APP phosphorylation and Abeta accumulation in rat cortical neurons. *Neurobiol Aging*. 2011;32(12):2323:e2313-e2326.
199. Zambrano A, Solis L, Salvadores N, Cortes M, Lerchundi R, Otth C. Neuronal cytoskeletal dynamic modification and neurodegeneration induced by infection with herpes simplex virus type 1. *J Alzheimers Dis*. 2008;14(3):259–269.
200. Landau NR, Warton M, Littman DR. The envelope glycoprotein of the human immunodeficiency virus binds to the immunoglobulin-like domain of CD4. *Nature*. 1988;334(6178):159–162.

201. Moore JP, McKeating JA, Weiss RA, Sattentau QJ. Dissociation of gp120 from HIV-1 virions induced by soluble CD4. *Science*. 1990;250(4984):1139–1142.
202. Ikeuchi K, Kim S, Byrn RA, Goldring SR, Groopman JE. Infection of nonlymphoid cells by human immunodeficiency virus type 1 or type 2. *J Virol*. 1990;64(9):4226–4231.
203. Harouse JM, Kunsch C, Hartle HT, et al. CD4-independent infection of human neural cells by human immunodeficiency virus type 1. *J Virol*. 1989;63(6):2527–2533.
204. Fantini J, Yahi N, Chermann JC. Human immunodeficiency virus can infect the apical and basolateral surfaces of human colonic epithelial cells. *Proc Natl Acad Sci USA*. 1991;88(20):9297–9301.
205. Feng Y, Broder CC, Kennedy PE, Berger EA. HIV-1 entry cofactor: functional cDNA cloning of a seven-transmembrane, G protein-coupled receptor. *Science*. 1996;272(5263):872–877.
206. Endres MJ, Clapham PR, Marsh M, et al. CD4-independent infection by HIV-2 is mediated by fusin/CXCR4. *Cell*. 1996;87(4):745–756.
207. Puri A, Hug P, Munoz-Barroso I, Blumenthal R. Human erythrocyte glycolipids promote HIV-1 envelope glycoprotein-mediated fusion of CD4+ cells. *Biochem Biophys Res Commun*. 1998;242(1):219–225.
208. Puri A, Hug P, Jernigan K, et al. The neutral glycosphingolipid globotriaosylceramide promotes fusion mediated by a CD4-dependent CXCR4-utilizing HIV type 1 envelope glycoprotein. *Proc Natl Acad Sci USA*. 1998;95(24):14435–14440.
209. Rawat SS, Gallo SA, Eaton J, et al. Elevated expression of GM3 in receptor-bearing targets confers resistance to human immunodeficiency virus type 1 fusion. *J Virol*. 2004;78(14):7360–7368.
210. Hellberg A, Steffensen R, Yahalom V, et al. Additional molecular bases of the clinically important p blood group phenotype. *Transfusion*. 2003;43(7):899–907.
211. Lund N, Olsson ML, Ramkumar S, et al. The human P(k) histo-blood group antigen provides protection against HIV-1 infection. *Blood*. 2009;113(20):4980–4991.
212. Lund N, Branch DR, Mylvaganam M, et al. A novel soluble mimic of the glycolipid, globotriaosyl ceramide inhibits HIV infection. *AIDS*. 2006;20(3):333–343.
213. Rosa Borges A, Wieczorek L, Johnson B, et al. Multivalent dendrimeric compounds containing carbohydrates expressed on immune cells inhibit infection by primary isolates of HIV-1. *Virology*. 2010;408(1):80–88.

CHAPTER

13

A Unifying Theory

Jacques Fantini, Nouara Yahi

OUTLINE

13.1 WHY DO WE NEED A UNIFYING THEORY?

In face of complex phenomena, we need sense. Biology is an experimental science whose conceptual frames evolve perpetually, depending on the advance of knowledge. New theories arise whereas others are abandoned, which is how science goes on. If an experiment contradicts theory, we should trust the experiment and challenge the theory. Sometimes we may hesitate between two interpretations of the same biological fact. In this case, it may be wise to use Occam's razor (also referred to as the law of economy or the parsimony principle) as a problem-solving guideline. This rule says: When you have two competing explanations for the same phenomenon, the simpler one is more likely to be true. It is not our purpose to decide whether the use of Occam's razor always leads (or not) to a correct interpretation. What we take from this principle is an encouragement to identify common molecular mechanisms accounting for various and different causes of neurological disorders. As striking as it may seem, a broad range of unrelated viral, bacterial, and amyloid proteins use in fact the same types of lipids (glycosphingolipids and cholesterol) to bind to neural cells and penetrate

Brain Lipids in Synaptic Function and Neurological Disease. http://dx.doi.org/10.1016/B978-0-12-800111-0.00013-8

their plasma membrane. At the molecular level, all these protein–membrane interactions look similar and can thus be described with a unique model. We do not need multiple mechanisms explaining how each pathogen interacts with brain cells because one general mechanism applies for all pathogens. This notion would not be dismissed by Occam's razor.

The objective of this chapter is thus to show that brain pathogens and/or their proteins have independently developed the same membrane burglary strategy. As we will see, there are no multiple ways to interact with glycosphingolipids or with cholesterol, and the protein domains devoted to these crucial functions are indeed remarkably convergent. In the same way that acetyl-coA is at the crossroads of many metabolic pathways, cholesterol is the common, universal, and almost unique passageway for entering brain membranes. Therefore, what applies for a given protein may generally be extrapolated to the others, with minor variations to accommodate some degree of lipid-binding specificity, especially when glycosphingolipids are involved in the initial steps of membrane attachment. The frequent coinvolvement of cholesterol and sphingolipids suggests that these common mechanisms take place in lipid rafts. This is obviously often the case, but it is not an absolute requirement. Above all, we are talking about molecular mechanisms that concern two partners, a pathogenic protein and a lipid that bind together through well-defined interactions described at the atomic resolution. We emphasized two salient features of these interactions (i.e., the importance of the geometry of the lipid–protein complexes and the prominent role of protein disorder in the mutual adaptation of proteins to lipid structures). This unifying theory of protein pathogenicity in the central nervous system may give some clues for developing innovative therapeutic strategies for brain disorders, as discussed in the last chapter of this book.

13.2 BACTERIA, VIRUSES, AND AMYLOIDS CONVERGE AT BRAIN MEMBRANES

As discussed in the previous chapters, the brain is not a sanctuary protected from the invasion of pathogens, but rather subject to a wide variety of microbial assaults including bacterial, virus, and prion infections. In addition, the brain may generate a series of amyloid proteins whose propagation, neurotoxicity, and disease-associated effects are based on self-aggregation properties. Viral proteins, bacterial toxins, prions, and amyloid proteins begin to be dangerous when they leave the aqueous extracellular milieu in order to adhere to plasma membrane of host cells. This rule applies to pathogens infecting peripheral tissues and to brain pathogens as well. At first glance, the wide variety of pathogens and toxins able to perturb the central nervous system should discourage everyone trying to find common mechanisms of invasion or toxicity. This might be true at the tissue and cellular levels, but not at the molecular level. In fact, the molecular mechanisms controlling the interaction of pathogen-associated and pathogenic proteins are remarkably convergent. To illustrate this point, we will first consider a given pathogenic protein, which can be a bacterial toxin, a microbial adhesin, or an amyloid protein. In the beginning, the pathogenic protein is present in the extracellular milieu in a form that does not induce any toxicity. For instance, in the case of amyloid proteins, this inoffensive form is generally a disordered monomer.[1–4] In all cases, the pathogenic protein needs a membrane to express its neurotoxic potential. The brain offers a large choice of plasma membranes lining neurons, astrocytes, oligodendrocytes, microglia, macrophages, and endothelial cells. The biochemical composition of these membranes may greatly vary from cell to cell, in function of the geographical zone of the brain, and as the result of aging (see Chapter 4). Brain lipids, which represent 50% of the dry weight of this organ,[5] are at the frontline facing pathogenic agents. Therefore, it is not surprising that the first step of the adhesion of a pathogenic protein to a brain cell generally involves a molecular interaction of the protein with membrane lipids. Moreover, the broad variety of brain lipid structures has allowed each pathogen to choose a preferred target for driving its adhesion to the cell surface. When a lipid receptor is identified, we should keep in mind that the choice of this lipid by the pathogen (one also could reversibly say "the choice of a pathogen by the lipid") reflects a long evolutionary process. This molecular adaptation is attested by the high quality of the fit generally displayed by pathogenic proteins for their lipid receptors.[6–11] For instance, the HIV-1 surface envelope glycoprotein has the same nanomolar affinity for CD4 than for its glycolipid receptor GalCer expressed on the surface of brain oligodendrocytes.[12] More generally, two types of lipids appeared to play a critical role in the adhesion process of pathogenic proteins to brain membranes: glycosphingolipids and cholesterol. These lipids often cooperate to ensure both binding and postbinding mechanisms that allow a broad range of pathogens and pathogenic proteins to invade and/or destroy brain cells.

13.3 GLYCOSPHINGOLIPIDS AND CHOLESTEROL IN BRAIN MEMBRANES: "UN PAS DE DEUX"

In the plasma membrane, glycosphingolipids are exclusively located in the exofacial leaflet, whereas cholesterol is found in both leaflets (Fig. 13.1).

Glycosphingolipids are always associated with cholesterol, forming specific condensed complexes[13] that are physically separated from the bulk membrane. These entities are alternatively referred to as lipid rafts,[14] membrane rafts,[15–18] or sphingolipid microdomains.[19] Cholesterol is more fickle because in addition to sphingolipids of the outer plasma membrane leaflet, it can also interact with glycerophospholipids in both outer and inner leaflets.[20] Glycosphingolipids have a large head group that totally covers cholesterol. In contrast, glycerophospholipids of the exofacial leaflet (e.g., phosphatidycholine) do not mask cholesterol. Thus, in the hemimembrane facing the extracellular milieu, the two distinct pools of membrane cholesterol include a nonaccessible pool hidden by glycosphingolipids in lipid rafts and an accessible pool outside lipid rafts (Fig. 13.1). For a pathogenic protein, targeting a raft domain through binding to glycosphingolipids indicates that cholesterol is actually present underneath the sugar head group. The reverse is not true: binding to cholesterol outside lipid rafts is in fact a shortcut for a rapid access to the apolar phase of the membrane. In this case, glycolipids are totally ignored.

We will first treat the adhesion of a pathogenic protein via the long route crossing both glycosphingolipids and cholesterol. In the beginning, the protein "sees" only the glycone part of the glycolipid, which extrudes from the membrane (Fig. 13.2).

Here the glycolipid can be compared to the tip of an iceberg floating on the water surface. However this comparison is not especially good because in the case of the glycolipid, the global shape of the sugar head group is controlled by a functional interaction with cholesterol.[21,22] Thus, the immersed part of the glycolipid is in fact a glycolipid–cholesterol complex.

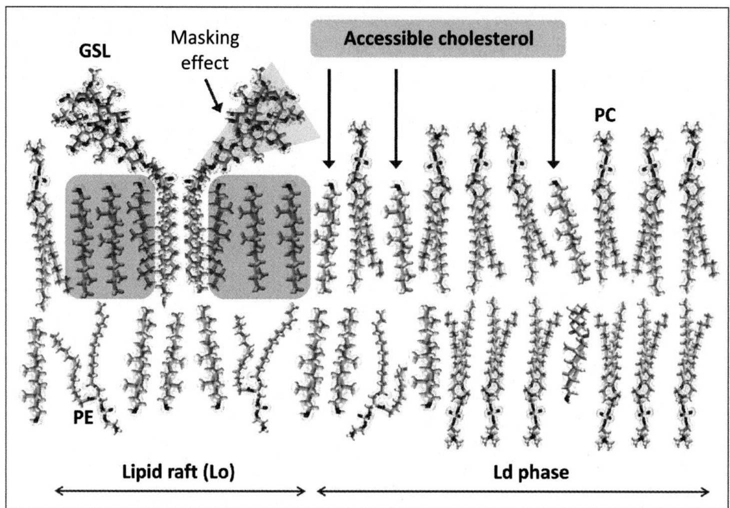

FIGURE 13.1 **Glycosphingolipid and cholesterol organization in the plasma membrane.** Glycosphingolipids (GSL) are exclusively found in the exofacial leaflet of the plasma membrane, whereas cholesterol is found on both sides. Cholesterol is concentrated in lipid raft domains (liquid-ordered, Lo phase) but is also found in the liquid-disordered (Ld) phase, yet in lower amounts. Cholesterol in Ld phase is accessible to extracellular ligands because it is not masked by phosphatidylcholine (PC). In contrast, the large head groups of glycosphingolipids totally cover the raft-associated pool of cholesterol (purple areas). In the cytoplasmic leaflet, cholesterol can be associated with phosphatidylethanolamine (PE) or phosphatidylserine (not shown) in lipid rafts, and with phosphatidylcholine outside raft domains.

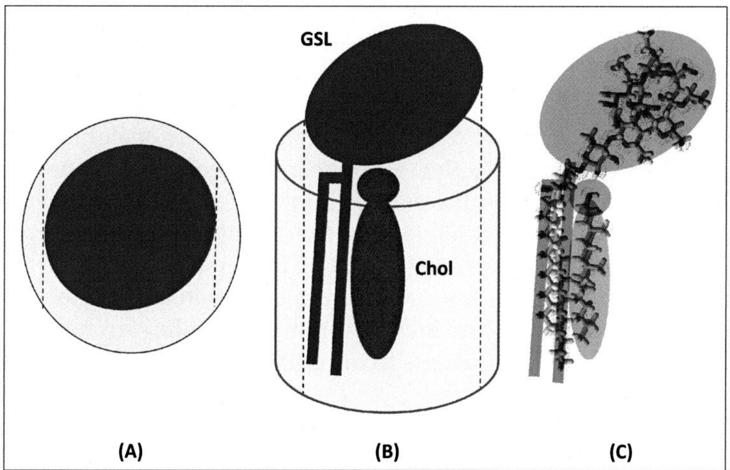

FIGURE 13.2 **What an extracellular protein sees of a glycosphingolipid–cholesterol complex.** An extracellular protein coming close to the membrane may detect the sugar head group of a glycosphingolipid (GSL), but not cholesterol (A). In fact, cholesterol is there, closely associated with the GSL, yet totally masked by the large head group of the glycolipid (B). In panel C, a real molecular model of a GSL–cholesterol complex is represented at a scale identical to that used for the cartoon in panel B.

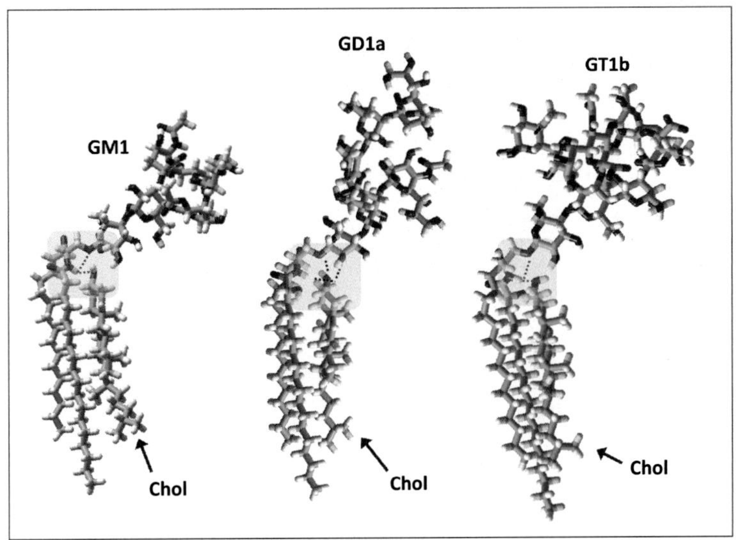

FIGURE 13.3 **Conformational effect of cholesterol on gangliosides.** The OH group of cholesterol binds to glycosphingolipids at the ceramide–sugar junction. A network of hydrogen bonds (dashed lines in orange areas) stabilizes each ganglioside–cholesterol complex. These hydrogen bonds generally involve the OH group of cholesterol, the NH group of sphingosine in the ceramide part, and the glycosidic bond linking the glycone to the ceramide anchor. In some cases (e.g., for GD1a), an OH group belonging to the first sugar may also be involved.

In this complex, the OH group of cholesterol is ideally located to interact with the glycosidic bond, joining the glycone domain of the glycolipids to its ceramide anchor. This interaction, mediated by a hydrogen bond, attracts the first sugar of the glycolipid close to the membrane. The conformation of the sugar head group is constrained by a network of hydrogen bonds involving the OH of cholesterol, the amide group of ceramide, the glycosidic bond, and the first sugar. Overall, both the membrane orientation and the 3D structure conformation of the whole head group are concerned by this conformational effect of cholesterol. We described the molecular mechanisms of this cholesterol-driven tuning of glycolipid structure for GalCer in Chapter 11. Here we show that this effect applies to more complex glycosphingolipids such as GM1, GD1a, or GT1b (Fig. 13.3).

It is important to note that in these cholesterol–sphingolipid interactions, the oxygen atom of the OH group of cholesterol always plays the role of hydrogen bond acceptor. This same is not true for an association of cholesterol with phosphatidylcholine. Indeed, this latter lipid lacks hydrogen bond donor groups located at the interfacial zone where the OH group of cholesterol lies. In contrast, the NH group carried by the ceramide backbone of sphingolipids is an efficient hydrogen donor group, and it is ideally placed to contact the OH group of cholesterol (Fig. 13.3). This phenomenon contributes to render sphingolipids more attractive for cholesterol than phosphatidylcholine. However, the consequence of this preference is that cholesterol is so deeply attached to the glycosphingolipid that it is not visible from the exterior of the membrane. Reciprocally, the glycosphingolipid does not remain insensitive to this mark of attachment, and it moves its head group according to the structural instructions given by the sterol. Overall, what the extracellular protein sees when it comes close to the

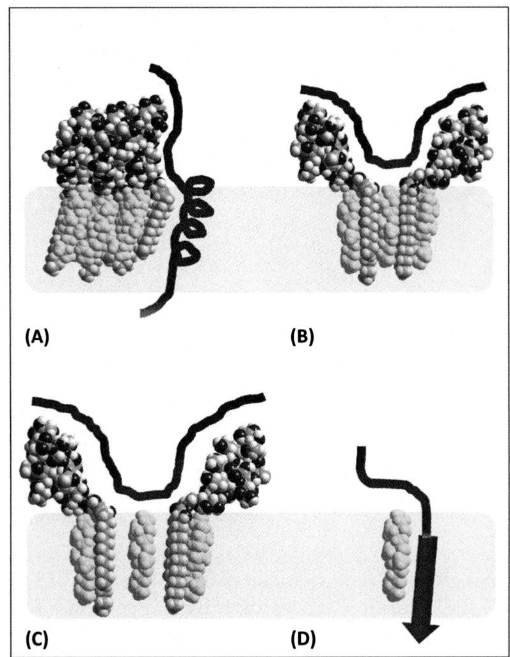

FIGURE 13.4 **Protein–lipid interactions in lipid rafts.** (A) Cluster of ganglioside–cholesterol complexes stabilized by sugar–sugar interactions through a dense network of hydrogen bonds. In this case, the available surface of contact for an extracellular protein is quite small, and the numerous hydrogen bonds prevent any adjustment of the sugar head groups. Therefore, this organization of cell surface gangliosides is not particularly adapted to the binding of extracellular proteins. However, transmembrane proteins expressed by the host could perfectly interact with the periphery of the cluster. (B) A chalice-shaped dimer of ganglioside stabilized by three cholesterol molecules. In this case, the gangliosides form a chalice-like receptacle with a large surface of contact for extracellular proteins. (C) Once the protein is bound to the chalice, the ganglioside dimer dissociates, allowing a free access to membrane cholesterol. (D) Secondary interactions with cholesterol facilitate membrane insertion. In all four models, cholesterol is in yellow and the gangliosides in atom colors (carbon green, oxygen red, nitrogen blue, and hydrogen white).

glycolipid is not just a sugar head group but a structured motif whose conformation is dictated by cholesterol. As discussed in Chapters 10 and 11 for amyloid proteins, we can interpret this conformational tuning as a means for cholesterol to report its presence to the invader. If the pathogenic protein sticks to the glycolipid, it is sure to find an accomplice (a traitor in this case) who will help break the membrane.

The high concentration of glycosphingolipids in membrane microdomains forces these molecules to interact with each other. Given that the glycosphingolipids are already associated with cholesterol, the possibilities for glycolipid–glycolipid interactions are limited. The head group of glycosphingolipids displays numerous hydrogen bond-forming atoms, including OH, *N*-acetyl, and carboxylate groups carried by neutral, acetylated sugars, and sialic acids. A first type of glycolipid cluster topology is obtained by combining a maximal number of hydrogen bonds between the sugar head groups of two vicinal glycolipids (Fig. 13.4A).

The structure obtained is a rigid cluster that let little space for the protein, as illustrated in Fig. 13.4A. In fact, this type of protein–glycolipid interactions can be used by transmembrane proteins located at the periphery of a membrane raft. In this case, the transmembrane domain may interact with the ceramide part of the sphingolipid or with cholesterol. The juxtamembrane domain of the protein may interact with the sugar head group of the glycolipid. A second possible topology of glycosphingolipid clusters is the formation of chalice-shaped dimers (Fig. 13.4B). In this case, a large accessible area can accommodate the pathogenic protein. For the protein, this particular topology of glycolipid clusters is advantageous for three reasons. First, the landing surface is curved and its deepest level is near the apolar phase of the membrane (Fig. 13.4B). Second, the separation of the glycolipid dimer (Fig. 13.4C) gives immediate access to cholesterol, which may subsequently facilitate membrane insertion (Fig. 13.4D). In this respect, it is likely that the binding of the pathogenic protein in the chalice could induce the dissociation of the glycolipid dimer through a minimal conformational change (e.g., by destabilizing cholesterol–glycolipid interactions). Third, dimeric glycolipids may have more degrees of freedom in the membrane than condensed clusters. Therefore, such glycolipid dimers could rapidly engage simultaneous interactions with oligomeric proteins (e.g., bacterial toxins). Molecular modeling studies of various glycolipid–pathogen interactions support this model of protein–glycolipid interactions.

As recurrently discussed in this book, pathogenic proteins have developed strikingly convergent strategies for invading and/or damaging host target cells.[23] The common and almost universal use of glycosphingolipids as primary attachment receptors on the host cell membrane relies on a short protein domain referred to as the sphingolipid-binding domain (SBD). Several SBDs have been described in this book (see Chapters 6, 7, 9, 10, 11, and 12) so we will only briefly address here the most salient structural and functional properties of these domains. SBDs are short motifs (12–25 amino acid residues) forming a structurally conserved loop fold.[20,24,25] The dimensions of this loop are compatible with the insertion of the SBD in a chalice-shaped glycolipid dimer. Little (or no) sequence homology occurs between the SBDs from unrelated proteins. However, the glycolipid-binding properties of the motif are determined by a specific combination of amino acids: (1) a central aromatic residue that can stack onto a sugar ring and engage CH–π interactions; (2) charged residues, most often lysine or arginine, at each end of the motif (this ensures that the whole loop is readily accessible at the surface of the protein); (3) turn-inducing residues (glycine and/or proline); and (4) small residues (glycine and/or serine) conferring a sufficient mobility to the loop so that it can adapt its shape to the chalice receptacle of glycolipid dimers. Overall, this particular combination of residues in a short motif has rendered feasible the accurate prediction of SBDs from rough sequence data[26] and the corresponding development of detection algorithms. Apart from this required content in specific amino acids, each SBD has its own sequence and the remaining amino acids, which do not concur to defining the motif, may determine its glycolipid-binding pattern.[8,27] For instance, two vicinal histidine residues in the SBD of Aβ (His-13/His-14) have been shown to be critical for GM1 recognition.[8] A selective list of SBDs used by pathogenic proteins to bind to brain cells is shown in Fig. 13.5. As one can see, this list includes amyloid, viral, and bacterial proteins. Thus the SBD is a common structural feature that mediates the convergence of brain pathogens to the same type of membrane lipids.

At this stage, the pathogenic protein is attached to the glycolipid but does not "know" yet that cholesterol is just below (Fig. 13.2). The interaction of the protein with cholesterol is a two-time process. The first step is the recognition of the OH group of cholesterol by the protein. It requires that this OH group is not already engaged in an interaction with vicinal

Protein	Sphingolipid-binding domain (SBD)	Sphingolipid
HIV-1 gp120 V3 loop (LAI isolate)	CTRPNNNTRKSIRIQRGPGRAFVTIGKIGNMRQAHC	GalCer sulfatide Gb_3 GM3 SM
HIV-1 gp120 V3 loop (89.6 isolate)	CTRPNNNTRRRLSIGPGRAFYARRNIIGDIRQAHC	GM3
HIV-1 gp120 V3 loop (NDK isolate)	CTRPYKYTRQRTSIGLRQSLYTITGKKKKTGYIGQAHC	GalCer sulfatide Gb_3 GM3 SM
HIV-2 gp105 V3 loop (ROD isolate)	CKRPGNKTVKQIMLMSGHVFHSHYQPINKRPRQAWC	GalCer sulfatide Gb_3 GM3
Rabies virus glycoprotein	KSVRTWNEILPSK	GT1b
Neisseria meningitidis adhesin	SISSTRAFLKEKHKAAK	LacCer Asialo-GM1
PrP	KQHTVTTTTKGENFTETDVKMMER	GalCer SM
β-Amyloid Peptide	RHDSGYEVHHQK	GM1 GalCer SM
α-Synuclein	KEGVLYVGSKTK	GM3 Gb_3 GalCer

FIGURE 13.5 **Sphingolipid-binding domains (SBDs) of pathogenic proteins that interact with the plasma membrane of brain cells.** The motif involved in sphingolipid binding is highlighted in yellow. Basic residues appear in blue, aromatic amino acids in red. SM, sphingomyelin. For HIV isolates, the whole sequence of the V3 loop is shown.[29] Note that the loop structure of the V3 domain is locked by a disulfide bridge involving the first and last cysteine residues. The SBD of the PrP protein was first described by Mahfoud et al.,[24] and the minimal SBDs of Alzheimer's β-amyloid peptide and α-synuclein by Fantini and Yahi [27].

lipids. Thus, cholesterol can enter the scene only after the dissociation of the glycolipid dimer. By nature, the binding of the protein to only one OH group of a ligand is a transient process: this type of protein–lipid complex is of low affinity and thus labile. In most cases, an amino acid belonging to the tip of the SBD will fulfill this function.[28] Thereafter, the establishment of a more stable complex between cholesterol and the protein requires that both the polar and

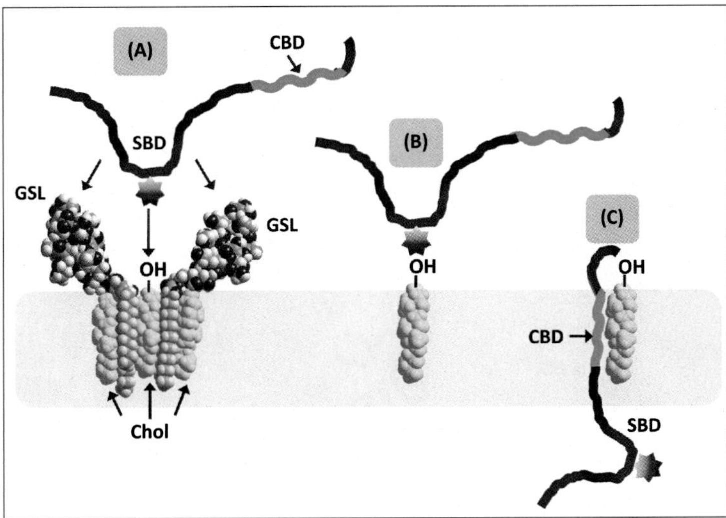

FIGURE 13.6 **Dual use of glycosphingolipids and cholesterol by pathogenic proteins.** (A) In this model, a chalice-shaped dimer of glycosphingolipid (GSL) interacts with three cholesterol (chol) molecules (in yellow). The pathogenic protein inserts its sphingolipid-binding domain (SBD) in the chalice. This primary interaction involves the polar amino acids of the SBD and the sugar groups of the GSL. Residues located at the tip of the SBD loop may also interact with the OH group of cholesterol, as shown for Tyr-39 of α-synuclein.[27] This residue is symbolized by a yellow/red star. At this stage the cholesterol-binding domain (CBD) is still extracellular and requires a protection against water molecules (see Fig. 13.7 for an example). (B) The tip of the SBD is now entering the membrane, pushing the GSLs away and forcing the passage through the cholesterol-enriched domain. The driving force of this insertion mechanism is the clash between the apolar nature of the CBD and the polar extracellular environment. (C) The penetration of the SBD allows the CBD to interact with the whole cholesterol molecule. At the end of the process the SBD has totally passed the membrane and is now facing the cytosol.

apolar domains of cholesterol interact with the protein, which is possible if the protein contains a cholesterol-binding domain. If this is the case, cholesterol-binding occurs outside the membrane, in the extracellular milieu, because at this stage of the mechanism only the SBD has started to penetrate the membrane, and the SBD is a polar domain that cannot accommodate the apolar part of the sterol (Fig. 13.6).

Like most basic peptides, the SBD is prone to rapidly crossing the membrane to find an environment compatible with its polar nature.[10,28] Conversely, the cholesterol-binding domain should not remain extracellular; and it is precisely the apolar nature of the cholesterol-binding domain that provides the main driving force of the membrane penetration of amyloid proteins. An important question then is: If the cholesterol-binding domain is so prone to penetrate the membrane, why would the protein need a primary interaction with glycosphingolipids? The answer is simply because before the glycolipid-binding step, the cholesterol-binding domain is not exposed to the solvent. Thus the binding to cell surface glycolipids is required to induce a conformational change allowing the demasking of the cholesterol-binding domain. Thereafter, thermodynamics will take it over and finish the job: the apolar cholesterol-binding domain, suddenly confronted to an unfriendly water environment will seek a protective apolar phase. At this stage, the closest shelter is provided by the plasma

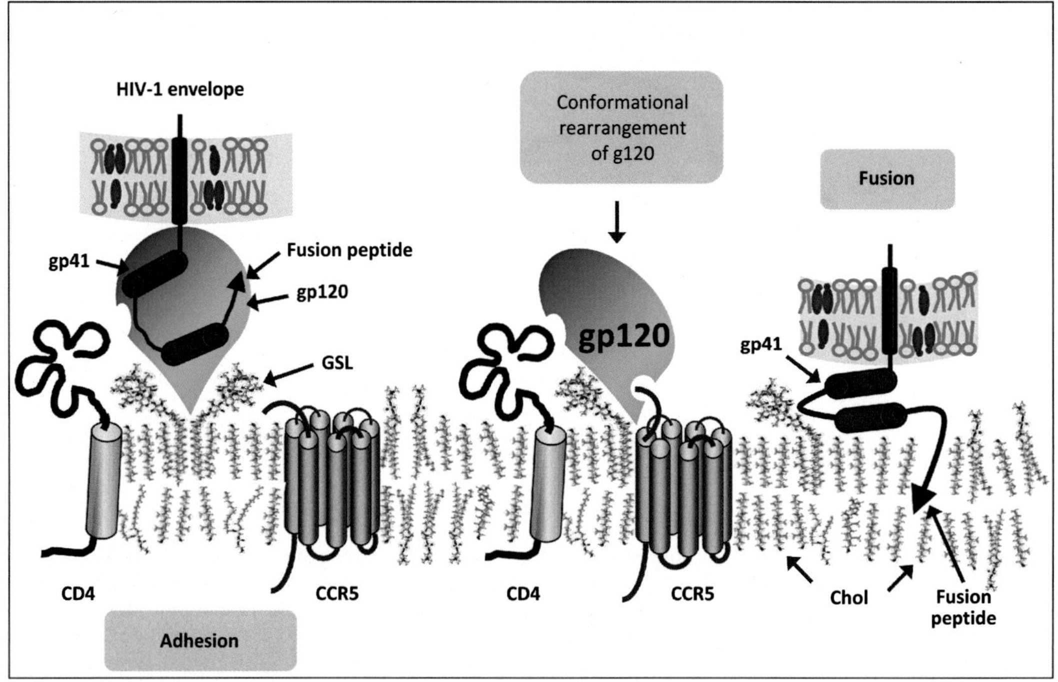

FIGURE 13.7 **HIV-1 fusion requires both glycosphingolipids and cholesterol.** The adhesion of HIV-1 to the surface of $CD4^+$ cells is a lipid raft-dependent process that requires CD4, a coreceptor (CCR5), and a specific glycosphingolipid (GSL) that laterally transports the virus to the coreceptor and favors the conformational changes of gp120 required for the fusion process. The HIV-1 envelope displays two glycoproteins, the surface gp120 that covers the transmembrane gp41 and provides a protective shelter for the apolar fusion peptide located at the N-terminal of gp41. When gp120 binds to the coreceptor, a conformational change releases the fusion peptide, whereas gp120 remains bound to CD4 and CCR5. The fusion peptide inserts into the host cell membrane, which is, at this stage, the closest apolar environment able to relax its apolar structure.

membrane. This striking sequence of events has been fully deciphered for the fusion process of HIV-1 (Fig. 13.7).

The spikes of the HIV-1 virions contain the surface envelope glycoprotein gp120, which totally covers, through noncovalent interactions, the transmembrane glycoprotein gp41. The N-terminal domain of gp41 contains the fusion peptide, an apolar motif that is initially protected from the aqueous environment by gp120.[20] When gp120 binds to the CD4–coreceptor complex, gp120 dissociates from gp41 so that the fusion peptide is suddenly exposed to the water environment. Thus, this hydrophobic domain has no other issue than immediately inserting into the plasma membrane of the host cell that is, at this stage of the mechanism, the closest apolar milieu. Because the fusion occurs in lipid raft domains, the virus must take the classic glycolipid–cholesterol pathway.[20] This implies that the virus interacts first with cell surface glycosphingolipids. This interaction is mediated by the binding of the V3 loop of gp120 (a typical SBD) and a glycosphingolipid cofactor (GM3 in the brain).[29–31] Synthetic peptides mimicking the V3 loop[32–35] and oligosaccharide dendrimers[36] can competitively inhibit the V3–glycosphingolipid interaction. Correspondingly, these compounds can efficiently prevent the infection of various strains of HIV-1 and HIV-2 in $CD4^+$ lymphocytes and macrophages, but also in $CD4^-$ cells.[33] For instance, a polymeric V3 peptide (SPC3) that binds to all glycosphingolipids recognized by HIV-1, including GalCer, can block HIV-1 infection of $CD4^-$/$GalCer^+$ mucosal and neural cells.[35]

After binding to CD4 (e.g., at the surface of a brain macrophage), the virus will seek a coreceptor (CCR5 in the brain),[37] after which the lipid raft on which it sailed disaggregates, leaving the fusion domain of gp41 free to interact with membrane lipids. The virus is attached to a lipid raft, and therefore it is clear that the first lipid encountered by the fusion peptide inside the membrane is cholesterol, whose presence is required for the fusion process.[38] This absolute dependence for cholesterol has opened the chase for identifying a sterol-binding domain in gp41. After a while, a short motif displaying the amino acid sequence LWYIK has been characterized.[39] This sequence fulfills the criteria of the CRAC algorithm[40] and mutational studies have shown that the integrity of the domain is required for the fusion mechanism.[41,42] However, this CRAC motif is located far from the N-terminal fusion peptide (the CRAC domain corresponds to amino acid residues 168–172), and although there is no doubt that it can physically interact with cholesterol,[39] its existence is in no way incompatible with the presence of another cholesterol-binding domain within the fusion peptide. Molecular modeling simulations have shown that the fusion peptide of HIV-1 could indeed interact with cholesterol.[43] A model of this fusion peptide complexed with cholesterol is shown in Fig. 13.8.

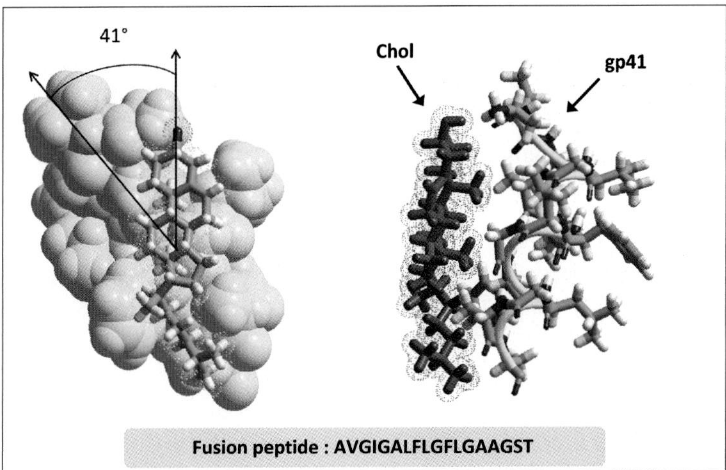

FIGURE 13.8 **Docking of cholesterol onto the fusion peptide of HIV-1 gp41.** The N-terminal sequence of HIV-1 gp41, corresponding to the fusion peptide of HIV-1 is merged with a cholesterol molecule. Molecular dynamics simulations were then conducted until an energy-minimized complex was obtained. In this complex, the apolar part of cholesterol interacts with alanine, glycine, and phenylalanine residues through van der Waals interactions. The OH group of cholesterol forms a hydrogen bond with the α–NH_3^+ group of the peptide (N-terminus group). The tilted orientation of the peptide with respect to cholesterol (41°) is clearly visible in the left panel, where gp41 is in yellow and cholesterol in tube rendition. In the right panel, cholesterol is in purple. The sequence of the fusion peptide (amino acids 1–18 of gp41) is indicated.

The binding of the fusion peptide of gp41 to cholesterol is driven by a series of van der Waals interactions between the apolar part of cholesterol and the aliphatic and aromatic amino acids of the fusion peptide. The only charged chemical group of the fusion peptide is the free N-terminal NH_3^+ of the first residue of gp41 (i.e., alanine). This group can form a hydrogen bond with the OH group of cholesterol, but this interaction is only transient. In fact, the leading cationic amino group at the front line of the peptide–membrane interface is the driving force of the insertion process. A similar mechanism has been described for α-synuclein, which uses a tyrosine residue to find its way through the apolar phase of the plasma membrane.[28] However, in the case of α-synuclein, the insertion stops as soon as the phenolic group of this residue (Tyr-39) has crossed the membrane. The remaining parts of the protein begin an oligomerization process that eventually leads to building a Ca^{2+}-permeable pore.[23] Because all fusion peptides have an N-terminal NH_3^+ group, one can consider that the mechanism described for HIV-1 may apply for other fusion peptides.

We do not really know how gp41 interacts with cholesterol and whether it uses its N-terminal tilted peptide or the CRAC domain, or both, during the fusion process. It remains that in the case of HIV-1, glycolipid recognition and cholesterol binding involve two distinct viral proteins. The surface envelope glycoprotein displays the SBD and binds to specific glycolipids, the transmembrane protein gp41 interacts with cholesterol (Fig. 13.9). In the case of amyloid proteins, the SBD and the cholesterol-binding domains are located in vicinal regions of the same protein[10] (Fig. 13.9).

If we decompose the mechanisms controlling the binding of glycolipids to both viral and amyloid proteins, we can see that these mechanisms are quite similar. In all cases, they can be recapitulated by the interaction of a synthetic SBD peptide and a monolayer of glycolipids.[27,44–47] At the molecular level, the same types of sugar–protein interactions operate for both viral and amyloid proteins. The driving force of these interactions consists in a combination of hydrogen bonds and sugar–aromatic stacking interactions.[48] These interactions also determine the binding of free sugars to lectins.[49] The main difference between a soluble oligosaccharide and the same sugar belonging to a glycolipid is the restriction of the conformation imposed by membrane cholesterol.

13.4 GEOMETRIC ASPECTS OF GLYCOLIPID–PROTEIN AND CHOLESTEROL INTERACTIONS

An important aspect of lipid–protein binding is the geometry of the ligands. This specific issue was discussed in Chapter 6, but we would like to underscore two main points. First, the loop shape of the SBD is consistent with the insertion of the domain in the chalice-like receptacle of ganglioside dimers. If we envision a different geometry of the glycolipid–protein complex, such as the one induced upon protein binding to a condensed complex of gangliosides (Fig. 13.4A), it would not be mandatory for the SBD to adopt a loop shape. If this was the case, the algorithm used to detect SBD motifs in glycolipid-binding proteins would be inoperative, which is, however, heavily contradicted by the experiments. Indeed, synthetic peptides derived from SBD sequences have a high affinity for glycolipids, as predicted by the algorithm. Most importantly, mutating amino acid residues that physically contact the glycolipid in modeling studies systematically result in a loss of interaction in physicochemical

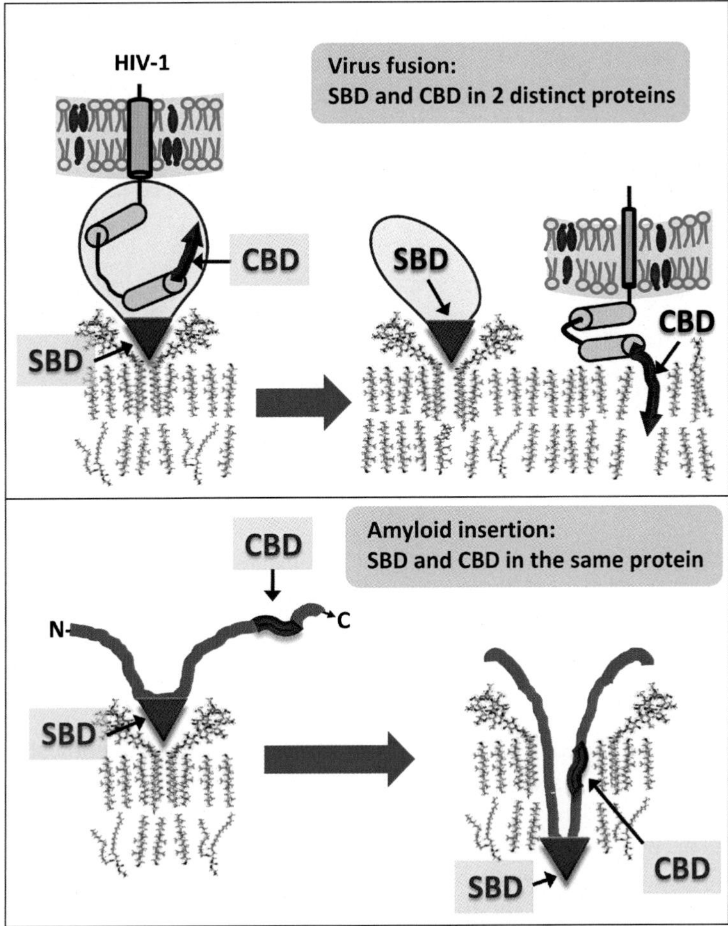

FIGURE 13.9 **Common mechanisms of membrane insertion for viral and amyloid proteins.** Both viral and amyloid proteins display a sphingolipid-binding domain (SBD, symbolized by a triangle) and a cholesterol-binding domain (CBD, red ribbon). The SBD mediates the binding to lipid rafts, whereas the CBD facilitates membrane insertion. In the case of HIV-1 (upper panel), the SBD and the CBD belong to two distinct proteins, respectively the surface envelope glycoprotein gp120 (blue structure) and the transmembrane gp41 (gray/black/red structure inside gp120). Following a conformational rearrangement of gp120, gp41 dissociates from gp120, allowing the CBD (in fact the fusion peptide of gp41) to penetrate the plasma membrane of the target cell. For amyloid proteins (e.g., α-synuclein), the SBD and CBD are two distinct domains of the same protein (lower panel). In this case, the insertion stops when cholesterol interacts with the CBD, allowing the oligomerization of inserted monomers of amyloid proteins into a Ca^{2+}-permeable pore.

experiments.[8,27] On the contrary, mutating the peptides at positions that do not participate in the interaction had no effect on the binding reaction.[27,28] Thus, the combination of *in silico*, mutational, and physicochemical approaches gave a solid background to the loop-shaped SBD concept. Overall, this structurally conserved domain is present on the surface of various microbial and amyloid proteins that use glycolipids to interact with host target cells.

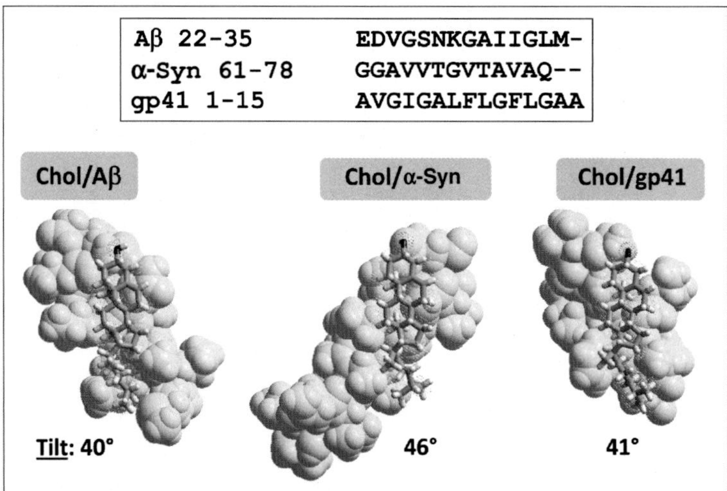

FIGURE 13.10 **The cholesterol-binding domains of amyloid and viral proteins have no sequence homology but are all tilted when bound to membrane cholesterol.** The amino acid sequences of the cholesterol-binding domains of Alzheimer's β-amyloid peptide, α-synuclein, and HIV-1 gp41 fusion peptide are shown in the upper panel. Note the lack of sequence homology. Molecular dynamics simulations indicated that all these domains may adopt a common tilted topology when bound to membrane cholesterol (lower panels). In these models, the peptide is in yellow (sphere rendering) and cholesterol in atom colors (carbon green, oxygen red, and hydrogen white).

Similarly, the interaction of pathogenic proteins with membrane cholesterol follows common geometric rules. It is a fact that several algorithms, chiefly represented by CRAC[40] and CARC[11] consensus motifs, may correctly predict the presence of a linear cholesterol-binding domain. However, we and others have also shown that protein segments that do not fulfill the criteria of these algorithms may also specifically interact with cholesterol.[43,50,51] This is the case for Aβ,[9] α-synuclein,[10] and the fusion peptide of HIV-1 gp41.[43] The sequence alignments of these cholesterol-binding domains (Fig. 13.10) did not reveal obvious common features. Nevertheless, all these non-CRAC/non-CARC cholesterol-binding domains shared the same tilted geometry when bound to cholesterol (Fig. 13.10). If we assumed that the main axis of cholesterol is perpendicular to the plane of the membrane, we can precisely measure the tilted orientation of the cholesterol-binding domains of Aβ (40°), α-synuclein (46°), and gp41 (41°) in the cholesterol-bound state.[9,10,43]

Initially, tilted peptides were defined as short helical protein fragments able to disturb the organization of the membrane into which they insert.[52] These peptides are characterized by an asymmetric distribution of their hydrophobic residues, so that a tilted orientation (~45°) toward the membrane plane is generally spontaneously acquired during the insertion process.[52] In other words, they do not need cholesterol or any other lipid to adopt a tilted orientation in a membrane environment, because the tilt is an intrinsic property of these peptides. However, one should note that the majority of tilted peptides identified so far belong to virus glycoproteins.[53] This association is not surprising since tilted peptides may induce an important distortion of the membrane structure, a property required for virus fusion. Moreover, viral fusion is in most cases a cholesterol-dependent process.[38] Apart from viral glycoproteins, tilted peptides have also been identified in lipid transport proteins[52] and in various amyloid proteins, including α-synuclein,[54] Aβ,[55] and PrP.[56] Our data show that both the cholesterol-binding domains and the tilted peptides of α-synuclein and Aβ share the same location. On this basis we suggest that most if not all tilted peptides may be able to interact with membrane cholesterol and that both the geometry and the cholesterol-binding capability are intrinsic properties of tilted peptides. The fact that amyloid proteins have a tilted peptide in their sequence adds an important item to the long list of common features shared by microbial and amyloid proteins.

13.5 WHY TWO LIPID RECEPTORS ARE BETTER THAN ONE?

Aβ and α-synuclein interact with gangliosides receptors via their N-terminal half.[27] The C-terminal part of these proteins contains the tilted sequence, interacts with cholesterol, and is thus logically assigned to membrane-inserting functions.[23,44] Thus, the binding and insertion of amyloid proteins involve two distinct protein domains. This mechanism is similar to the one used by botulinum and tetanus toxins that penetrate neural cells through a complex and coordinated process involving both glycolipid and protein receptors.[57] Tetanus toxin causes spastic paralysis by blocking presynaptic neurotransmitter release in the spinal cord following its retrograde axonal transport from the presynaptic membrane of the neuromuscular junction.[58] Botulinum toxin is responsible for a flaccid paralysis caused by a blocking of acetylcholine release in peripheral nerves.[59] Both toxins belong to the family of clostridial neurotoxins, which comprises a unique type of tetanus neurotoxin and seven serotypes of botulinum neurotoxins.[60] Tetanus neurotoxin, produced by *Clostridium tetani* shares ~65% sequence homology and ~35% identity with the botulinum neurotoxins produced by *Clostridium botulinum*.[60] In 1986, Montecucco developed the "dual receptor" model to explain how tetanus and botulinum toxins bind to neural cells.[61] According to this model, the toxin first binds to a cell surface ganglioside.[62,63] Thereafter, the membrane-bound ganglioside–toxin complex moves laterally in the bidimensional plane of the membrane to reach and bind to a specific protein receptor. It is clear that the concomitant use of both ganglioside and protein-binding sites significantly increases the avidity of the toxin for the membrane. In this model, the role of the ganglioside is (1) to deliver the neurotoxin to its protein receptor, and (2) to concentrate the neurotoxin and its protein receptor in a small volume. These functions are typical of those fulfilled by lipid raft microdomains. With remarkable intuition, this model was proposed 10 years before the lipid raft concept was developed.[14] As expected, lipid rafts would later be incriminated for providing a landing platform for botulinum toxin.[64] Interestingly, a second lipid raft-binding domain of this neurotoxin was recently localized in a highly conserved region of the N-terminal part of the toxin, but separated from its classical ganglioside-binding domain.[64] This novel membrane-binding domain had been overlooked for a long time because the attention was focused on the ganglioside-binding domain. Interestingly, this new binding site showed a marked specificity for phosphatidylinositol phosphate (PIP). This interaction with PIP could facilitate the membrane insertion of the neurotoxin, just like cholesterol does for viral and amyloid proteins.

These studies revealed that membrane binding and membrane insertion of neural toxins are distinct and successive processes that involve distinct protein domains and at least two types of membrane lipids. In the case of Aβ and α-synuclein, the N-terminal part is responsible for membrane binding. This function is assumed by the SBD, which recognizes specific glycosphingolipids on the surface of neural cells. Then the C-terminal half is responsible for membrane insertion via an interaction between cholesterol and a tilted cholesterol-binding domain. In the case of HIV-1, the SBD and the cholesterol-binding domain are located on two distinct proteins: gp120 has the SBD, and gp41 has the tilted cholesterol-binding domain. Nevertheless, the mechanism is always the same: (1) binding to glycosphingolipids; (2) recognition of a second lipid (usually cholesterol); and (3) insertion via a tilted peptide (Fig. 13.9).

13.6 WHEN CHOLESTEROL PLAYS TWO ROLES

As stated already, in some rare cases the pathogenic protein may take a shortcut and interact directly with cholesterol, which is the case for cholesterol-dependent cytolysins.[65] These bacterial toxins form large oligomeric pores in the plasma membrane of recipient cells. In the first step of this process, toxin monomers bind to the cell membrane. Because cholesterol serves as the initial binding target for these toxins, it can be hypothesized that they preferentially interact with cholesterol-rich domains of host cell membranes.[66,67] However, this assertion has been challenged by several experimental data. First, both binding and pore formation occurred in liposomal membranes containing only phosphatidylcholine and cholesterol, providing that the amount of cholesterol is greater than 30 mol%.[68–72] Given that phosphatidylcholine and cholesterol do not form rafts by themselves[20] and that both lipids coexist in the liquid-disordered phase of biological membranes, these findings suggested that cytolysins do require cholesterol, but not necessarily in a lipid raft domain. Second, the membrane association of perfringolysin was inhibited by the addition of sphingomyelin, a typical component of lipid rafts.[73] This finding is not really surprising because sphingomyelin masks cholesterol via a typical "umbrella" effect.[74] Thus, when it is associated with sphingomyelin (or with a glycosphingolipid), cholesterol is not accessible to most external ligands. In contrast, phosphatidylcholine does not limit the accessibility of cholesterol. Therefore, it is likely that cholesterol-dependent cytolysins have an easier access to cholesterol in the liquid-disordered phase of the membrane than in lipid raft domains. Moreover, the mobility of cholesterol in the bulk membrane is consistent with the recruitment of several molecules of this lipid and ensures the binding and insertion of the numerous toxin monomers required to form a large oligomeric pore.[75] Incidentally, these data may also suggest that the use of perfringolysin as a lipid raft probe[76] should require a careful reassessment.

In any case, it appeared that the exposure of cholesterol at the membrane surface is necessary and sufficient for perfringolysin binding and pore formation.[73] As described in Chapter 12, cholesterol may thus mediate both the binding step and the insertion process of toxin monomers. The binding of perfringolysin to cholesterol involves a conserved pair of branched residues (Leu/Thr) located at the tip of a sensor loop that recognizes the OH group of cholesterol. Once again, it is important to note the myriad of surface-exposed OH groups in lipid rafts domains due to the presence of sphingolipids (sphingomyelin and glycosphingolipids). In contrast, phosphatidylcholine has no OH group so that in the liquid-disordered phase, the OH group of cholesterol is instantaneously detected by external ligands. Following this binding step, the monomer is attracted inside the membrane, presumably through a specific interaction involving this time the whole cholesterol molecule and a functional cholesterol-binding domain (the structure of this domain has been detailed in Chapter 12). In the liquid-disordered phase, cholesterol may provide a rigid annular wall ensuring the correct orientation of the monomers and guiding their oligomerization into a highly structured pore. In conclusion, we can see that in the specific case of cholesterol-dependent cytolysins, cholesterol cumulates the functions of binding receptor and insertion cofactor, both functions that are usually fulfilled by two distinct types of membrane lipids.

13.7 STRUCTURAL DISORDER AS A COMMON TRAIT OF PATHOGENICITY

One of the most exciting concepts to emerge in recent years in biology is the recognition that a significant part of the proteome consists of proteins that do not have a stable structure in solution, but instead fluctuate between a wide range of different conformations.[3] Although they are in flagrant contradiction with the traditional "sequence–structure–function paradigm," these intrinsically disordered proteins (IDPs) play some crucial biological functions including signaling, recognition, and regulation.[77,78] Consistent with these vital functions, IDPs are highly abundant in all species, especially eukaryotes that need sophisticated signaling and regulation mechanisms.[79] However, as suggested by Xue et al.,[80] the level of disorder may represent a strong adaptive trait that reflects pathogenic lifestyle. Indeed, significant percentages of totally or partially disordered proteins have been detected in viruses,[80] prokaryotes,[79] and parasites.[81]

In Chapters 8, 10, and 11, we have extensively discussed the importance of intrinsic disorder in amyloid proteins. Disorder, cardinal feature of viral proteins, applies to a wide variety of viruses, including those able to infect the central nervous system (e.g., influenza, HIV-1, rabies, and measles viruses). Both structural and nonstructural viral proteins are affected by the phenomenon.[80] In fact, viral proteins have a disorder burden comparable to that in eukaryotes. Of particular interest is the study of the surface glycoproteins (hemagglutinins) from influenza virus. During membrane fusion, these hemagglutinins bind to cell surface gangliosides on the host cell surface.[82] The virus is then endocyted and confronted to an acidic environment, which induces a large and irreversible conformational rearrangement of hemagglutinin structure.[83,84] This structural change positions the N-terminal hydrophobic fusion peptide of hemagglutinin in close proximity to the target membrane. Goh et al.[85] have recently compared the hemagglutinin sequence from highly virulent and less virulent strains of influenza virus. They noticed interesting differences in a region of hemagglutinin involved in receptor binding (residues 68–79 of hemagglutinin HA_2). Intrinsic disorder could be predicted in this region among the more virulent strains but not among the less virulent ones. They concluded that the intrinsically disordered regions of hemagglutinin could serve as a predictor of influenza A virulence.[85] The analogy with amyloid proteins is remarkable because these proteins also belong to the unfoldome,[2] bind to cell surface gangliosides,[23] and, following this binding, undergo a conformational rearrangement that precedes their functional membrane insertion. Interestingly, the V3 loop of HIV-1 gp120 is considered as a major virulence factor that determines the tropism of the virus[86] and its ability to induce neurological diseases.[87] This domain is naturally enriched in disorder-promoting residues,[88,89] in particular at the tip of the loop that often displays a typical GPGR motif.[90] Moreover, the sequence of the V3 loop can accumulate several mutations that may potentiate virus infectivity and extend its tropism.[91] In particular, an increase in positively charged (basic) amino acids in the V3 loop has been associated with a more aggressive HIV-1 phenotype.[92] For instance, the highly cytopathic strain HIV-1(NDK),[93] which infects both $CD4^+$ and $CD4^-$ cells,[94,95] has an uncommon surface envelope glycoprotein with nine basic amino acids in its V3 loop.[96] In particular, a stretch of four contiguous lysine residues confers a high cationic charge to the C-terminal part of the loop (Fig. 13.5). Incidentally, intrinsically disordered regions of proteins are also substantially enriched in lysine and/or arginine.[88,89] Therefore, these data suggest that the natural evolution of HIV-1 quasi-species within an infected individual leads to the emergence of more aggressive strains characterized by an increased disorder of the V3 loop domain of HIV-1 gp120. Because the V3 loop is a disulfide-linked domain, these mutations preserve the loop nature, but extend the glycosphingolipid repertory and thus the tropism of the virus. A similar pattern of mutations with basic amino acids, associated with an increased disorder, has been evidenced upon the analysis of highly virulent strains of influenza virus.[85] Overall, these data support the notion that local intrinsic disorder in regions of surface glycoproteins involved in glycosphingolipid binding may reflect the virulence of enveloped viruses. In the case of amyloid proteins, disorder confers conformational plasticity and various possibilities of membrane-induced structuration, enabling proteins without a defined structure in solution to form highly structured pore-like assemblies in the plasma membrane of neural cells.

13.8 KEY EXPERIMENT: PROBES TO STUDY CHOLESTEROL AND/OR GLYCOLIPID-DEPENDENT MECHANISMS

Two main strategies can be used to schematically determine the role of cholesterol and/or glycosphingolipids in a biological function. The first approach is to use live cells and to selectively modulate their lipid content. The second approach is to work with artificial membrane models in which the lipid content is precisely controlled and check the consequence of the absence or presence of the lipids of interest. Modulating the cholesterol content of a cell can be done with metabolic inhibitors of cholesterol biosysnthesis, the most widely used being inhibitors of hydroxylmethylglutaryl-CoA (HMG-CoA) reductase, commonly referred to as statins.[97] The other option is to affect the cholesterol pool of the plasma membrane by using cholesterol oxidase, which transforms the OH group of cholesterol into a ketone,[98] methyl-β-cyclodextrin, which extracts cholesterol from the membrane,[99] or cholesterol-sequestering agents such as filipin.[100] In all cases, recovery experiments that consist of replenishing membrane cholesterol pools by incubation of the cells with cyclodextrin–cholesterol complexes are required. Ideally, the loss of biological function induced by treatments affecting membrane cholesterol should no longer exist upon reintroduction of cholesterol into the plasma membrane. These strategies were successfully used to demonstrate the regulatory function of cholesterol on the cholecystokinin receptor.[101]

The modulation of glycosphingolipid content can be obtained by metabolic inhibitors such as L-cycloserine,[102,103] fumonisin B_1,[104,105] PDMP,[106] PPMP,[107] or, more specifically by siRNAs silencing only one enzyme of the metabolic pathway. For instance, in a comprehensive study aimed at determining the role of ganglioside GM3 in neuronal cell death Sohn et al.[108] have used the latter approach to selectively decrease the amounts of GM3 through RNA interference-mediated silencing of GM3 synthase. Conversely, the overexpression of GM3 synthase was obtained by mRNA microinjection.[108]

It is widely admitted that depleting the cholesterol and/or sphingolipids from the cell membrane is a good approach to identify lipid raft dependent mechanisms.[20] However, this may often be an overinterpretation of the experimental data. For instance, methyl-β-cyclodextrin may have poor access to the cholesterol pool of lipid rafts because cholesterol is masked by the large sphingolipid head groups of sphingomyelin and glycosphingolipid in lipid rafts. This protective umbrella effect has been demonstrated in artificial bilayer and monolayer membranes, using methyl-β-cyclodextrin as a detector of surface accessible cholesterol.[10,109] Thus, a loss of function induced by methyl-β-cyclodextrin treatment as proof that this function is lipid raft-dependent might be in some instances an overstatement. In fact, rare are the drugs or chemicals able to selectively act on the homeostasis of lipid rafts (e.g., by blocking the clustering of lipids and proteins in these membrane microdomains). Here we should cite the striking results obtained with a synthetic glycosphingolipid displaying nonnatural stereochemistry, β-D-lactosyl-*N*-octanoyl-L-*threo*-sphingosine (L-*threo*-LacCer)[110] (Fig. 13.11).

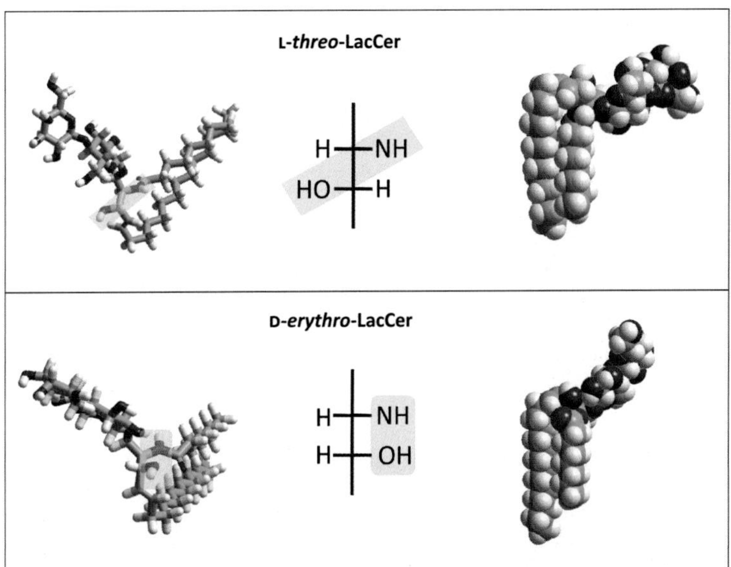

FIGURE 13.11 **Stereochemistry of natural and synthetic LacCer isomers.** The synthetic nonnatural stereoisomer of LacCer (L-*threo*-LacCer) has an inverted configuration of the C3 of sphingosine. In the Fisher representation (middle panels), the OH group is on the right, in a typical D configuration (similar position of the OH group of the D isomer of the reference molecule glyceraldehyde). The NH group carried by the C2 is also on the right, so that in natural sphingosine and all derived sphingolipids both the C3–OH and the C2–NH groups are on the right, which defines an *erythro* configuration (similar to that of the sugar erythrose). In the non-natural synthetic analogue, the C3–OH is on the left, which defines a L configuration. The C2–NH group is still on the right so that this time the C3–OH and the C2–NH are not on the same side of the Fisher representation, but on each side of the carbon chain. This defines the *threo* configuration (similar to that the sugar threose). This explains why natural LacCer is a D-*erythro* compound, and the nonnatural stereoisomer is L-*threo*. Note the distorted structure of the analogue compared with natural LacCer. It is likely that the inhibitory effect of L-*threo*-LacCer on lipid raft dependent mechanisms is due to this structural distortion, which may greatly perturb cholesterol–glycosphingolipid interactions in membrane microdomains.

This compound inhibited the infection of monkey cells by SV40, a virus that uses ganglioside GM1 as cell surface receptor.[111] The interesting point is that L-*threo*-LacCer did not act directly on GM1 but prevented the clustering of microdomains induced by the virus and required for its infection. Thus, the function perturbed by L-*threo*-LacCer is typically an integrated process that depends on the lateral organization of lipid raft components in the bidimensional plane of the membrane. Moreover, the transmembrane signaling events triggered by lipid rafts were inhibited by L-*threo*-LacCer, showing that this function requires a proper organization of raft components. Targeting membrane microdomains[112] may thus represent a common approach for the treatment of neurological disorders caused by bacteria, viruses, or amyloid proteins.

References

1. Uversky VN. Intrinsic disorder in proteins associated with neurodegenerative diseases. *Front Biosci.* 2009;14:5188–5238.
2. Uversky VN. The mysterious unfoldome: structureless, underappreciated, yet vital part of any given proteome. *J Biomed Biotechnol.* 2010;2010:568068.
3. Uversky VN. Introduction to intrinsically disordered proteins (IDPs). *Chem Rev.* 2014;114(13):6557–6560.
4. Uversky VN, Dave V, Iakoucheva LM, et al. Pathological unfoldomics of uncontrolled chaos: intrinsically disordered proteins and human diseases. *Chem Rev.* 2014;114(13):6844–6879.
5. Woods AS, Jackson SN. Brain tissue lipidomics: direct probing using matrix-assisted laser desorption/ionization mass spectrometry. *AAPS J.* 2006;8(2):E391–395.
6. Stehle T, Harrison SC. High-resolution structure of a polyomavirus VP1-oligosaccharide complex: implications for assembly and receptor binding. *EMBO J.* 1997;16(16):5139–5148.
7. Neu U, Maginnis MS, Palma AS, et al. Structure-function analysis of the human JC polyomavirus establishes the LSTc pentasaccharide as a functional receptor motif. *Cell Host Microbe.* 2010;8(4):309–319.
8. Yahi N, Fantini J. Deciphering the glycolipid code of Alzheimer's and Parkinson's amyloid proteins allowed the creation of a universal ganglioside-binding Peptide. *PloS One.* 2014;9(8):e104751.
9. Di Scala C, Yahi N, Lelievre C, Garmy N, Chahinian H, Fantini J. Biochemical identification of a linear cholesterol-binding domain within Alzheimer's beta amyloid peptide. *ACS Chem Neurosci.* 2013;4(3):509–517.
10. Fantini J, Carlus D, Yahi N. The fusogenic tilted peptide (67–78) of alpha-synuclein is a cholesterol binding domain. *Biochim Biophys Acta.* 2011;1808(10):2343–2351.
11. Baier CJ, Fantini J, Barrantes FJ. Disclosure of cholesterol recognition motifs in transmembrane domains of the human nicotinic acetylcholine receptor. *Sci Rep.* 2011;1:69.
12. Harouse JM, Bhat S, Spitalnik SL, et al. Inhibition of entry of HIV-1 in neural cell lines by antibodies against galactosyl ceramide. *Science.* 1991;253(5017):320–323.
13. Radhakrishnan A, Anderson TG, McConnell HM. Condensed complexes, rafts, and the chemical activity of cholesterol in membranes. *Proc Natl Acad Sci USA.* 2000;97(23):12422–12427.
14. Simons K, Ikonen E. Functional rafts in cell membranes. *Nature.* 1997;387(6633):569–572.
15. Levental I, Grzybek M, Simons K. Greasing their way: lipid modifications determine protein association with membrane rafts. *Biochemistry.* 2010;49(30):6305–6316.
16. Veit M, Thaa B. Association of influenza virus proteins with membrane rafts. *Adv Virol.* 2011;2011:370606.
17. Vetrivel KS, Thinakaran G. Membrane rafts in Alzheimer's disease beta-amyloid production. *Biochim Biophys Acta.* 2010;1801(8):860–867.
18. Head BP, Patel HH, Insel PA. Interaction of membrane/lipid rafts with the cytoskeleton: impact on signaling and function: membrane/lipid rafts, mediators of cytoskeletal arrangement and cell signaling. *Biochim Biophys Acta.* 2014;1838(2):532–545.
19. Frisz JF, Lou K, Klitzing HA, et al. Direct chemical evidence for sphingolipid domains in the plasma membranes of fibroblasts. *Proc Natl Acad Sci USA.* 2013;110(8):E613–622.
20. Fantini J, Garmy N, Mahfoud R, Yahi N. Lipid rafts: structure, function and role in HIV, Alzheimer's and prion diseases. *Expert Rev Mol Med.* 2002;4(27):1–22.

21. Fantini J, Yahi N, Garmy N. Cholesterol accelerates the binding of Alzheimer's beta-amyloid peptide to ganglioside GM1 through a universal hydrogen-bond-dependent sterol tuning of glycolipid conformation. *Front Physiol*. 2013;4:120.
22. Yahi N, Aulas A, Fantini J. How cholesterol constrains glycolipid conformation for optimal recognition of Alzheimer's beta amyloid peptide (Abeta1-40). *PloS One*. 2010;5(2):e9079.
23. Fantini J, Yahi N. Molecular insights into amyloid regulation by membrane cholesterol and sphingolipids: common mechanisms in neurodegenerative diseases. *Expert Rev Mol Med*. 2010;12:e27.
24. Mahfoud R, Garmy N, Maresca M, Yahi N, Puigserver A, Fantini J. Identification of a common sphingolipid-binding domain in Alzheimer, prion, and HIV-1 proteins. *J Biol Chem*. 2002;277(13):11292–11296.
25. Fantini J. How sphingolipids bind and shape proteins: molecular basis of lipid-protein interactions in lipid shells, rafts and related biomembrane domains. *Cell Mol Life Sci*. 2003;60(6):1027–1032.
26. Fantini J, Garmy N, Yahi N. Prediction of glycolipid-binding domains from the amino acid sequence of lipid raft-associated proteins: application to HpaA, a protein involved in the adhesion of *Helicobacter pylori* to gastrointestinal cells. *Biochemistry*. 2006;45(36):10957–10962.
27. Fantini J, Yahi N. Molecular basis for the glycosphingolipid-binding specificity of alpha-synuclein: key role of tyrosine 39 in membrane insertion. *J Mol Biol*. 2011;408(4):654–669.
28. Fantini J, Yahi N. The driving force of alpha-synuclein insertion and amyloid channel formation in the plasma membrane of neural cells: key role of ganglioside- and cholesterol-binding domains. *Adv Exp Med Biol*. 2013;991:15–26.
29. Hammache D, Pieroni G, Yahi N, et al. Specific interaction of HIV-1 and HIV-2 surface envelope glycoproteins with monolayers of galactosylceramide and ganglioside GM3. *J Biol Chem*. 1998;273(14):7967–7971.
30. Hammache D, Yahi N, Maresca M, Pieroni G, Fantini J. Human erythrocyte glycosphingolipids as alternative cofactors for human immunodeficiency virus type 1 (HIV-1) entry: evidence for CD4-induced interactions between HIV-1 gp120 and reconstituted membrane microdomains of glycosphingolipids (Gb3 and GM3). *J Virol*. 1999;73(6):5244–5248.
31. Hammache D, Yahi N, Pieroni G, Ariasi F, Tamalet C, Fantini J. Sequential interaction of CD4 and HIV-1 gp120 with a reconstituted membrane patch of ganglioside GM3: implications for the role of glycolipids as potential HIV-1 fusion cofactors. *Biochem Biophys Res Commun*. 1998;246(1):117–122.
32. Delezay O, Hammache D, Fantini J, Yahi N. SPC3, a V3 loop-derived synthetic peptide inhibitor of HIV-1 infection, binds to cell surface glycosphingolipids. *Biochemistry*. 1996;35(49):15663–15671.
33. Yahi N, Fantini J, Baghdiguian S, et al. SPC3, a synthetic peptide derived from the V3 domain of human immunodeficiency virus type 1 (HIV-1) gp120, inhibits HIV-1 entry into CD4+ and CD4- cells by two distinct mechanisms. *Proc Natl Acad Sci USA*. 1995;92(11):4867–4871.
34. Yahi N, Fantini J, Mabrouk K, et al. Multibranched V3 peptides inhibit human immunodeficiency virus infection in human lymphocytes and macrophages. *J Virol*. 1994;68(9):5714–5720.
35. Yahi N, Sabatier JM, Baghdiguian S, Gonzalez-Scarano F, Fantini J. Synthetic multimeric peptides derived from the principal neutralization domain (V3 loop) of human immunodeficiency virus type 1 (HIV-1) gp120 bind to galactosylceramide and block HIV-1 infection in a human CD4-negative mucosal epithelial cell line. *J Virol*. 1995;69(1):320–325.
36. Rosa Borges A, Wieczorek L, Johnson B, et al. Multivalent dendrimeric compounds containing carbohydrates expressed on immune cells inhibit infection by primary isolates of HIV-1. *Virology*. 2010;408(1):80–88.
37. Berger EA, Murphy PM, Farber JM. Chemokine receptors as HIV-1 coreceptors: roles in viral entry, tropism, and disease. *Ann Rev Immunol*. 1999;17:657–700.
38. Schroeder C. Cholesterol-binding viral proteins in virus entry and morphogenesis. *Subcell Biochem*. 2010;51:77–108.
39. Vincent N, Genin C, Malvoisin E. Identification of a conserved domain of the HIV-1 transmembrane protein gp41 which interacts with cholesteryl groups. *Biochim Biophys Acta*. 2002;1567(1–2):157–164.
40. Jamin N, Neumann JM, Ostuni MA, et al. Characterization of the cholesterol recognition amino acid consensus sequence of the peripheral-type benzodiazepine receptor. *Mol Endocrinol*. 2005;19(3):588–594.
41. Epand RM. Proteins and cholesterol-rich domains. *Biochim Biophys Acta*. 2008;1778(7–8):1576–1582.
42. Epand RM, Thomas A, Brasseur R, Epand RF. Cholesterol interaction with proteins that partition into membrane domains: an overview. *Subcell Biochem*. 2010;51:253–278.
43. Fantini J, Barrantes FJ. How cholesterol interacts with membrane proteins: an exploration of cholesterol-binding sites including CRAC, CARC, and tilted domains. *Front Physiol*. 2013;4:31.

44. Di Scala C, Chahinian H, Yahi N, Garmy N, Fantini J. Interaction of Alzheimer's beta-amyloid peptides with cholesterol: mechanistic insights into amyloid pore formation. *Biochemistry*. 2014;53(28):4489–4502.
45. Leblanc RM. Molecular recognition at Langmuir monolayers. *Curr Opin Chem Biol*. 2006;10(6):529–536.
46. Thakur G, Pao C, Micic M, Johnson S, Leblanc RM. Surface chemistry of lipid raft and amyloid Abeta (1–40) Langmuir monolayer. *Colloids Surf B Biointerfaces*. 2011;87(2):369–377.
47. Hammache D, Pieroni G, Maresca M, Ivaldi S, Yahi N, Fantini J. Reconstitution of sphingolipid-cholesterol plasma membrane microdomains for studies of virus-glycolipid interactions. *Methods Enzymol*. 2000;312: 495–506.
48. Nishio M, Umezawa Y, Fantini J, Weiss MS, Chakrabarti P. CH–π hydrogen bonds in biological macromolecules. *Phys Chem Chem Phys*. 2014;16(25):12648–12683.
49. Weis WI, Drickamer K. Structural basis of lectin-carbohydrate recognition. *Ann Rev Biochem*. 1996;65:441–473.
50. Levitan I, Singh DK, Rosenhouse-Dantsker A. Cholesterol binding to ion channels. *Front Physiol*. 2014:5,65.
51. Rosenhouse-Dantsker A, Noskov S, Durdagi S, Logothetis DE, Levitan I. Identification of novel cholesterol-binding regions in Kir2 channels. *J Biol Chem*. 2013;288(43):31154–31164.
52. Brasseur R, Pillot T, Lins L, Vandekerckhove J, Rosseneu M. Peptides in membranes: tipping the balance of membrane stability. *Trends Biochem Sci*. 1997;22(5):167–171.
53. Charloteaux B, Lorin A, Brasseur R, Lins L. The "Tilted Peptide Theory" links membrane insertion properties and fusogenicity of viral fusion peptides. *Protein Pept Lett*. 2009;16(7):718–725.
54. Crowet JM, Lins L, Dupiereux I, et al. Tilted properties of the 67-78 fragment of alpha-synuclein are responsible for membrane destabilization and neurotoxicity. *Proteins*. 2007;68(4):936–947.
55. Pillot T, Goethals M, Vanloo B, et al. Fusogenic properties of the C-terminal domain of the Alzheimer beta-amyloid peptide. *J Biol Chem*. 1996;271(46):28757–28765.
56. Pillot T, Lins L, Goethals M, et al. The 118-135 peptide of the human prion protein forms amyloid fibrils and induces liposome fusion. *J Mol Biol*. 1997;274(3):381–393.
57. Montecucco C, Papini E, Schiavo G. Bacterial protein toxins penetrate cells via a four-step mechanism. *FEBS Lett*. 1994;346(1):92–98.
58. Schwab ME, Suda K, Thoenen H. Selective retrograde transsynaptic transfer of a protein, tetanus toxin, subsequent to its retrograde axonal transport. *J Cell Biol*. 1979;82(3):798–810.
59. Simpson LL. The origin, structure, and pharmacological activity of botulinum toxin. *Pharmacol Rev*. 1981;33(3):155–188.
60. Lacy DB, Stevens RC. Sequence homology and structural analysis of the clostridial neurotoxins. *J Mol Biol*. 1999;291(5):1091–1104.
61. Montecucco C. How do tetanus and botulinum toxins bind to neuronal membranes? *Trends in Biochem Sci*. 1986;:314–317.
62. Van Heyningen WE, Miller PA. The fixation of tetanus toxin by ganglioside. *J Gen Microbiol*. 1961;24:107–119.
63. Kitamura M, Iwamori M, Nagai Y. Interaction between Clostridium botulinum neurotoxin and gangliosides. *Biochim Biophys Acta*. 1980;628(3):328–335.
64. Muraro L, Tosatto S, Motterlini L, Rossetto O, Montecucco C. The N-terminal half of the receptor domain of botulinum neurotoxin A binds to microdomains of the plasma membrane. *Biochem Biophys Res Commun*. 2009;380(1):76–80.
65. Gilbert RJ. Cholesterol-dependent cytolysins. *Adv Exp Med Biol*. 2010;677:56–66.
66. Skocaj M, Bakrac B, Krizaj I, Macek P, Anderluh G, Sepcic K. The sensing of membrane microdomains based on pore-forming toxins. *Curr Med Chem*. 2013;20(4):491–501.
67. Taylor SD, Sanders ME, Tullos NA, et al. The cholesterol-dependent cytolysin pneumolysin from *Streptococcus pneumoniae* binds to lipid raft microdomains in human corneal epithelial cells. *PloS One*. 2013;8(4):e61300.
68. Alving CR, Habig WH, Urban KA, Hardegree MC. Cholesterol-dependent tetanolysin damage to liposomes. *Biochim Biophys Acta*. 1979;551(1):224–228.
69. Rosenqvist E, Michaelsen TE, Vistnes AI. Effect of streptolysin O and digitonin on egg lecithin/cholesterol vesicles. *Biochim Biophys Acta*. 1980;600(1):91–102.
70. Heuck AP, Hotze EM, Tweten RK, Johnson AE. Mechanism of membrane insertion of a multimeric beta-barrel protein: perfringolysin O creates a pore using ordered and coupled conformational changes. *Mol Cell*. 2000;6(5):1233–1242.
71. Ohno-Iwashita Y, Iwamoto M, Ando S, Iwashita S. Effect of lipidic factors on membrane cholesterol topology--mode of binding of theta-toxin to cholesterol in liposomes. *Biochim Biophys Acta*. 1992;1109(1):81–90.

72. Nelson LD, Johnson AE, London E. How interaction of perfringolysin O with membranes is controlled by sterol structure, lipid structure, and physiological low pH: insights into the origin of perfringolysin O-lipid raft interaction. *J Biol Chem*. 2008;283(8):4632–4642.
73. Flanagan JJ, Tweten RK, Johnson AE, Heuck AP. Cholesterol exposure at the membrane surface is necessary and sufficient to trigger perfringolysin O binding. *Biochemistry*. 2009;48(18):3977–3987.
74. Huang J, Feigenson GW. A microscopic interaction model of maximum solubility of cholesterol in lipid bilayers. *Biophys J*. 1999;76(4):2142–2157.
75. Koster S, van Pee K, Hudel M, et al. Crystal structure of listeriolysin O reveals molecular details of oligomerization and pore formation. *Nat Commun*. 2014;5:3690.
76. Ohno-Iwashita Y, Shimada Y, Waheed AA, et al. Perfringolysin O, a cholesterol-binding cytolysin, as a probe for lipid rafts. *Anaerobe*. 2004;10(2):125–134.
77. Dunker AK, Brown CJ, Lawson JD, Iakoucheva LM, Obradovic Z. Intrinsic disorder and protein function. *Biochemistry*. 2002;41(21):6573–6582.
78. Wright PE, Dyson HJ. Intrinsically unstructured proteins: re-assessing the protein structure-function paradigm. *J Mol Biol*. 1999;293(2):321–331.
79. Ward JJ, Sodhi JS, McGuffin LJ, Buxton BF, Jones DT. Prediction and functional analysis of native disorder in proteins from the three kingdoms of life. *J Mol Biol*. 2004;337(3):635–645.
80. Xue B, Blocquel D, Habchi J, et al. Structural disorder in viral proteins. *Chem Rev*. 2014;114(13):6880–6911.
81. Feng ZP, Zhang X, Han P, Arora N, Anders RF, Norton RS. Abundance of intrinsically unstructured proteins in P. falciparum and other apicomplexan parasite proteomes. *Mol Biochem Parasitol*. 2006;150(2):256–267.
82. Matrosovich M, Herrler G, Klenk HD. Sialic acid receptors of viruses. *Top Curr Chem*. 2013.
83. Lin X, Eddy NR, Noel JK, et al. Order and disorder control the functional rearrangement of influenza hemagglutinin. *Proc Natl Acad Sci USA*. 2014;111(33):12049–12054.
84. Xu R, Wilson IA. Structural characterization of an early fusion intermediate of influenza virus hemagglutinin. *J Virol*. 2011;85(10):5172–5182.
85. Goh GK, Dunker AK, Uversky VN. Protein intrinsic disorder and influenza virulence: the 1918 H1N1 and H5N1 viruses. *Virol J*. 2009;6:69.
86. Hwang SS, Boyle TJ, Lyerly HK, Cullen BR. Identification of the envelope V3 loop as the primary determinant of cell tropism in HIV-1. *Science*. 1991;253(5015):71–74.
87. Di Stefano M, Wilt S, Gray F, Dubois-Dalcq M, Chiodi F. HIV type 1 V3 sequences and the development of dementia during AIDS. *AIDS Res Hum Retroviruses*. 1996;12(6):471–476.
88. Campen A, Williams RM, Brown CJ, Meng J, Uversky VN, Dunker AK. TOP-IDP-scale: a new amino acid scale measuring propensity for intrinsic disorder. *Protein Pept Lett*. 2008;15(9):956–963.
89. Uversky VN. Unusual biophysics of intrinsically disordered proteins. *Biochim Biophys Acta*. 2013;1834(5): 932–951.
90. Javaherian K, Langlois AJ, LaRosa GJ, et al. Broadly neutralizing antibodies elicited by the hypervariable neutralizing determinant of HIV-1. *Science*. 1990;250(4987):1590–1593.
91. Kuiken CL, de Jong JJ, Baan E, Keulen W, Tersmette M, Goudsmit J. Evolution of the V3 envelope domain in proviral sequences and isolates of human immunodeficiency virus type 1 during transition of the viral biological phenotype. *J Virol*. 1992;66(7):4622–4627.
92. Milich L, Margolin B, Swanstrom R. V3 loop of the human immunodeficiency virus type 1 Env protein: interpreting sequence variability. *J Virol*. 1993;67(9):5623–5634.
93. Spire B, Sire J, Zachar V, et al. Nucleotide sequence of HIV1-NDK: a highly cytopathic strain of the human immunodeficiency virus. *Gene*. 1989;81(2):275–284.
94. Fantini J, Baghdiguian S, Yahi N, Chermann JC. Selected human immunodeficiency virus replicates preferentially through the basolateral surface of differentiated human colon epithelial cells. *Virology*. 1991;185(2):904–907.
95. Fantini J, Yahi N, Chermann JC. Human immunodeficiency virus can infect the apical and basolateral surfaces of human colonic epithelial cells. *Proc Natl Acad Sci USA*. 1991;88(20):9297–9301.
96. Yahi N, Fantini J, Hirsch I, Chermann JC. Structural variability of env and gag gene products from a highly cytopathic strain of HIV-1. *Arch Virol*. 1992;125(1–4):287–298.
97. Wood WG, Li L, Muller WE, Eckert GP. Cholesterol as a causative factor in Alzheimer's disease: a debatable hypothesis. *J Neurochem*. 2013;129(4):559–572.
98. Mattjus P, Slotte JP. Does cholesterol discriminate between sphingomyelin and phosphatidylcholine in mixed monolayers containing both phospholipids? *Chem Phys Lipids*. 1996;81(1):69–80.

99. Mahammad S, Parmryd I. Cholesterol depletion using methyl-beta-cyclodextrin. *Methods Mol Biol*. 2015;1232: 91–102.
100. Gimpl G, Gehrig-Burger K. Cholesterol reporter molecules. *Biosci Rep*. 2007;27(6):335–358.
101. Gimpl G, Burger K, Fahrenholz F. Cholesterol as modulator of receptor function. *Biochemistry*. 1997;36(36): 10959–10974.
102. Garmy N, Taieb N, Yahi N, Fantini J. Apical uptake and transepithelial transport of sphingosine monomers through intact human intestinal epithelial cells: physicochemical and molecular modeling studies. *Arch Biochem Biophys*. 2005;440(1):91–100.
103. Gerold P, Schwarz RT. Biosynthesis of glycosphingolipids de-novo by the human malaria parasite *Plasmodium falciparum*. *Mol Biochem Parasitol*. 2001;112(1):29–37.
104. Merrill Jr AH, Sullards MC, Wang E, Voss KA, Riley RT. Sphingolipid metabolism: roles in signal transduction and disruption by fumonisins. *Environ Health Perspect*. 2001;109(suppl 2):283–289.
105. Paila YD, Ganguly S, Chattopadhyay A. Metabolic depletion of sphingolipids impairs ligand binding and signaling of human serotonin1A receptors. *Biochemistry*. 2010;49(11):2389–2397.
106. Naslavsky N, Shmeeda H, Friedlander G, et al. Sphingolipid depletion increases formation of the scrapie prion protein in neuroblastoma cells infected with prions. *J Biol Chem*. 1999;274(30):20763–20771.
107. Puri A, Hug P, Munoz-Barroso I, Blumenthal R. Human erythrocyte glycolipids promote HIV-1 envelope glycoprotein-mediated fusion of CD4+ cells. *Biochem Biophys Res Commun*. 1998;242(1):219–225.
108. Sohn H, Kim YS, Kim HT, et al. Ganglioside GM3 is involved in neuronal cell death. *FASEB J*. 2006;20(8): 1248–1250.
109. Ramstedt B, Slotte JP. Interaction of cholesterol with sphingomyelins and acyl-chain-matched phosphatidylcholines: a comparative study of the effect of the chain length. *Biophys J*. 1999;76(2):908–915.
110. Singh RD, Holicky EL, Cheng ZJ, et al. Inhibition of caveolar uptake, SV40 infection, and beta1-integrin signaling by a nonnatural glycosphingolipid stereoisomer. *J Cell Biol*. 2007;176(7):895–901.
111. Tsai B, Gilbert JM, Stehle T, Lencer W, Benjamin TL, Rapoport TA. Gangliosides are receptors for murine polyoma virus and SV40. *EMBO J*. 2003;22(17):4346–4355.
112. Malchiodi-Albedi F, Contrusciere V, Raggi C, et al. Lipid raft disruption protects mature neurons against amyloid oligomer toxicity. *Biochim Biophys Acta*. 2010;1802(4):406–415.

CHAPTER

14

Therapeutic Strategies for Neurodegenerative Diseases

Jacques Fantini, Nouara Yahi

OUTLINE

14.1 PROTEINS INVOLVED IN BRAIN DISEASES CONSIDERED AS INFECTIOUS PROTEINS

Numerous pathogenic proteins have the capability to perturb brain functions and to trigger, maintain, or amplify brain diseases. Schematically, we can class these proteins in two main categories: endogenous brain proteins with amyloid/aggregative properties (the so-called amyloid proteins involved in Alzheimer's and Parkinson's diseases)[1] and infectious microbial proteins such as the surface envelope glycoprotein gp120 that can either play a role in the infection process of brain cells[2–7] or act as a neurotoxin.[8–12] In any case, these proteins

Brain Lipids in Synaptic Function and Neurological Disease. http://dx.doi.org/10.1016/B978-0-12-800111-0.00014-X

have a particular affinity for cell surface glycosphingolipids and/or membrane cholesterol. Glycosphingolipids attract the protein on lipid raft domains and generally induces α-helix structuration.[1] Cholesterol then interacts with a linear domain of the newly folded α-helix[13,14] and promotes the insertion of the protein.[15] In the case of amyloid proteins, cholesterol also facilitates the oligomerization process that eventually leads to the formation of amyloid pores.[1,16] Annular oligomers preassembled in the extracellular milieu can also penetrate the plasma membrane of brain cells and perturb ion fluxes.[17,18] In any case, this process is strikingly analogous to the mechanism of pore formation by bacterial toxins, leading some authors to use the term *infectious proteins* for pore-forming brain proteins involved in neurodegenerative diseases.[19] We believe that this analogy is an interesting angle of approach for developing new therapeutic strategies that could be applied to most (if not all) brain disorders.

Popular wisdom would most likely say it is safe to consider that the best outcome of an infectious protein is to remain outside the body. We can (at least theoretically) limit the contacts with microbial pathogens (e.g., by living in sterile atmosphere), yet obviously we cannot escape the proteins that are synthesized by our own brain. In general, the hazard comes when the infectious protein meets brain membranes. If physiological mechanisms cannot neutralize the proteins outside the cells, membrane insertion and pore formation will occur, and pretty soon we will have disease. Ideally, therapeutic agents should act before this stage. It is thus crucial to identify new drugs able to prevent the interaction of infectious proteins with brain membranes. If the protein is already attached to the membrane, we should develop a drug able to prevent its insertion. If the protein is already inserted, we should urgently find a molecule able to block the oligomerization process. If the amyloid pore is already functional, things are becoming alarming because our last option would be to destroy the pore or extract it from the membrane and clear it from the brain. We will briefly discuss these different possibilities and see which solutions are feasible today.

14.2 HOW TO PREVENT THE INTERACTION OF PATHOGENIC PROTEINS WITH BRAIN MEMBRANES

The two partners are the membrane and the infectious protein (e.g., Alzheimer's β-amyloid peptide Aβ, Parkinson's associated α-synuclein, and HIV-1 gp120) (Fig. 14.1A–C). Each of these proteins faces a complex membrane formed by a collection of lipids with wide biochemical diversity. Hopefully, among the myriad of brain membrane lipids, the protein generally selects a preferred target, most often located in a lipid raft microdomain.[1] For instance, Aβ peptides interact preferentially with ganglioside GM1,[20,21] and α-synuclein with GM3.[22]

This schematic interpretation of lipid–protein interactions can be viewed as useful because it reduces the binding reaction to only two perfectly identified partners: the protein and the glycolipid. However, it is totally misleading. Indeed, the protein we are considering can adopt several distinct conformations. Most often, the protein is intrinsically disordered[23] and so has no defined conformation before its interaction with cell surface glycolipids.[1] Thus, mimicking a protein lacking a precise 3D structure to use it as a molecular decoy might be problematic. However, we will see that this not out of reach. As for the glycolipid, even if we consider a particular one (e.g., GM1 or GM3), the situation is not as simple as it seems. The conformation

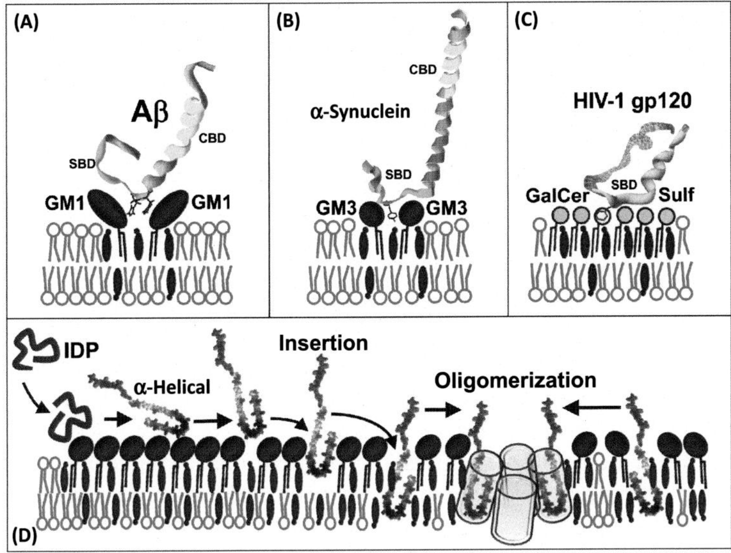

FIGURE 14.1 **Infectious proteins interact with lipid rafts.** (A) Interaction of Alzheimer's β-amyloid peptide with ganglioside GM1. Note the histidine residues of the SBD (sphingolipid-binding domain) that "equilibrate" the peptide in the chalice-shaped dimer of GM1. (B) In the case of α-synuclein, the binding to ganglioside GM3 involves the central aromatic residue Tyr-39. (C) The HIV-1 surface envelope glycoprotein gp120 binds to GalCer and sulfatide via its V3 loop domain, which was the first SBD to be characterized. The V3/GalCer interaction is driven by a sugar–aromatic (CH–π stacking) mechanism involving the central aromatic residue (Phe-20) of the V3 loop and the galactose ring. (D) Sequence of events leading to the insertion of infectious proteins into cholesterol-enriched domains of the plasma membrane. When the protein is initially disordered (IDP), binding to cell surface glycolipids induces α-helical structuration. The insertion process allows the cholesterol-binding domain (CBD) to reach cholesterol (in red). This can eventually lead to the formation of oligomeric Ca^{2+} pores.

of glycolipids is constrained in two ways: laterally and vertically. Lateral interactions between vicinal glycolipids in a lipid raft microdomain are responsible for conformational adjustments that have a decisive effect on the binding capability of the glycolipid for various proteins, including endogenous membrane receptors[24] and infectious proteins.[25] Vertical interactions with cholesterol molecules dipped in the membrane also affect glycolipid conformation, as shown for ganglioside GM1[26] as well as several other glycosphingolipids.[27,28] In particular, the cholesterol-constrained conformation of gangliosides is consistent with the formation of a chalice-shaped dimer that is recognized by various amyloid proteins.[22,25,29] Therefore, if we want to inhibit the interaction of infectious proteins with synthetic glycolipid analogs,[30–32] we should take into account the active conformation of the glycolipid partner generated in lipid raft domains both vertically by cholesterol and laterally by vicinal glycolipids. In conclusion, it might be possible to prevent the interaction of pathogenic proteins with brain membranes by synthetic molecules derived from the partners involved in this interaction. These molecules include synthetic peptides with glycolipid-binding properties[25] and conformationally constrained analogs of glycolipids able to mimic glycolipid clusters[30] as they occur in lipid raft microdomains of brain membranes.

14.3 HOW TO PREVENT THE INSERTION OF PATHOGENIC PROTEINS INTO BRAIN MEMBRANES

The insertion of infectious proteins into brain membranes (Fig. 14.1D) is a cholesterol-dependent process.[13,16] With the exception of preformed annular oligomers of β-structured amyloid proteins that may brutally perforate the membrane and form ion-permeable pores,[17] most if not all individual amyloid proteins (i.e., monomers) acquire an α-helix structure upon binding to lipid rafts.[33–36] This α-helix structuration creates a linear cholesterol-binding domain that has been delineated to amino acid residues 20–35 in Aβ peptides[14] and 67–79 in α-synuclein.[13] Correspondingly, the next step is the progressive insertion of the protein under the control of underlying cholesterol.[16] In this respect, the initial interaction of the infectious protein with glycolipid clusters in lipid raft domains[37] ensures that cholesterol is indeed present underneath these glycolipids.[13] The overall process of protein insertion as shown in Fig. 14.1D is remarkable: (1) the disordered protein starts to interact with selected cell surface glycolipids in a lipid raft; (2) upon binding, the protein acquires a 3D structure dominated by α-helix; (3) a highly specific cholesterol-binding domain is created in a contiguous segment of the α-helix; and (4) the protein then rushes into the membrane until a stable interaction with cholesterol is reached. At this stage, the protein is locked in the membrane in a specific orientation that is compatible with its oligomerization into a functional annular pore crossed with calcium fluxes.[15,16] Due to the prominent role of cholesterol in amyloid pore formation, one of the most rational approaches to block the whole process is to act at the level of protein binding to cholesterol. This could be done by developing short peptide inhibitors able to stick cholesterol and/or by cholesterol analogs that would ideally occupy the cholesterol-binding domain of infectious proteins.

14.4 HOW TO BLOCK AMYLOID PORE FORMATION

Once the protein is inserted in the membrane and already interacting with cholesterol, the last option would be to block the oligomerization process. A cholesterol analog with a higher affinity for the protein than authentic cholesterol could, in theory, displace cholesterol from the protein and take its place. Moreover, once bound to the protein, the analog would not allow further oligomerization. In other words, we are looking for a cholesterol analog able to outclass cholesterol for protein binding but lacking the cholesterol capability to promote oligomeric pore formation. Fine *in silico* approaches will be required to identify such Graal compounds. It is clear that the earlier in the membrane insertion process we hit, the better it is. Combination therapy approaches with several drugs targeting both the glycolipid-dependent surface binding and cholesterol-assisted membrane insertion of infectious proteins could also be considered.

14.5 A UNIVERSAL GANGLIOSIDE-BINDING PEPTIDE

Despite billions of years of molecular evolution, nature has not yet created a protein able to interact with all cell surface gangliosides. Therefore, a universal ganglioside-binding protein is not necessary for life. In fact, one can even postulate a strong resistance against selection

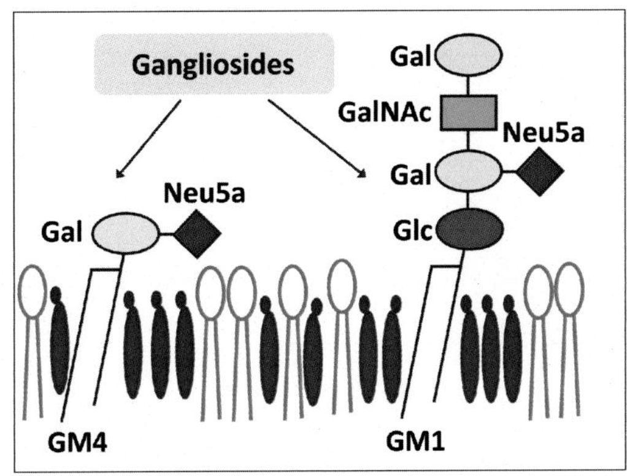

FIGURE 14.2 **Schematic structure of gangliosides.** Cholesterol is colored in red, sphingomyelin in green. The ceramide part of gangliosides interacts with cholesterol, whereas the glycone moiety protrudes from the membrane surface.

of this protein, because, clearly, the goal is not out of reach. Let us briefly recall the schematic structure of gangliosides (Fig. 14.2). Only two modes build gangliosides, according to the first sugar linked to the ceramide moiety. If the sugar is galactose, a unique ganglioside is obtained by addition of sialic acid (i.e., GM4). Indeed, all other gangliosides are derived from GlcCer through a wide diversity of glycone motifs including one, two, or three sialic acid residues.[38] Thus, the schematic structure of all gangliosides can be resumed as a ceramide anchor (mostly dipped in the membrane) linked to a neutral sugar (most often glucose) and bearing at least one negatively charged sialic acid.[39] In this case, both the ceramide-linked neutral sugar and the sialic acid(s) are accessible to proteic ligands, which can be either the extracellular domain(s) of a membrane protein or an infectious protein.

Instead of a universal ganglioside-binding protein, molecular evolution has tinkered with the primary structure of proteins to render them able to specifically interact with a small subset of membrane lipids. Indeed, most proteins that interact with sphingolipids (including gangliosides) carry a surface-exposed loop that is functionally dedicated to sphingolipid recognition.[40] This domain, referred to as the sphingolipid-binding domain (SBD) was discovered in 2002 in viral and amyloidogenic proteins,[41] and has since been found in numerous cellular[42-46] and pathogen proteins.[22,30,47,48] In general, the SBD consists of a hairpin motif with basic residues (lysine, arginine) at both ends and a central aromatic residue (tyrosine, phenylalanine, tryptophan). Glycine and serine residues, which confer a high flexibility to the loop, are often present in the SBD. Otherwise, SBDs exhibit a wide sequence variability, so that the domain is structurally conserved but lacks a consensus amino acid sequence.[25]

Both α-synuclein and Alzheimer's β-amyloid peptide (Aβ) interact with membrane gangliosides and therefore possess the SBD.[22] Studies with synthetic peptides derived from these proteins indicated that the ganglioside-binding properties of α-synuclein (a protein with 140 amino acid residues) totally rely on a short motif lying between amino acids 34 and 45.[22] The ganglioside-binding domain of α-synuclein shares structural homology with a fragment of Aβ (Aβ5–16) also involved in ganglioside recognition.[22] Synthetic peptides derived from the sequence of Aβ5–16 and α-syn34–45 display a high affinity for gangliosides. However,

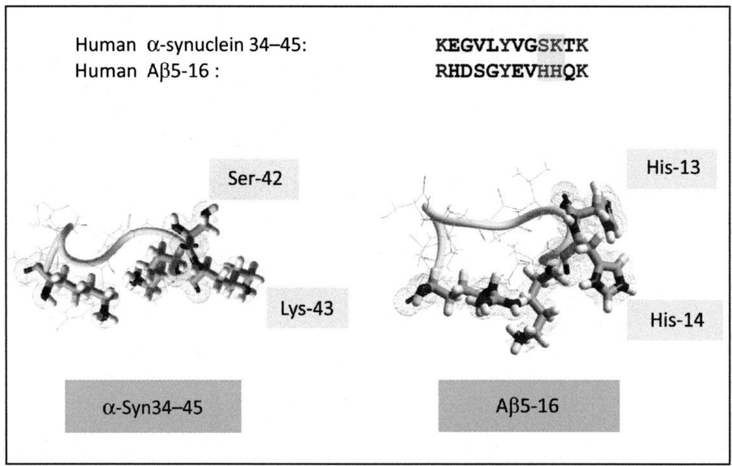

FIGURE 14.3 **Structure of Aβ5–16 and α-syn34–45.** In the upper panel, the basic residues at each end of the peptides are in red and the central tyrosine in blue. The position of His-13/His-14 (Aβ) and Ser-42/Lys-43 (α-synuclein) is framed in yellow. The minimized structures of both peptides are shown in the lower panel. *(Adapted from Yahi and Fantini,[25] with permission.)*

Aβ5–16 preferentially interacts with GM1, and α-syn34–45 with GM3. These interactions are in full agreement with the respective ganglioside-binding specificities of the parental proteins (i.e., GM1 for Aβ and GM3 for α-synuclein).[22] The amino acid sequence of Aβ5–16 and α-syn34–45 is shown in Fig. 14.3. Some interesting information can be deduced by comparing both sequences.

In each case, the motif has a basic residue at each end (Lys-34 and Lys-45 for α-syn, Arg-5 and Lys-16 for Aβ). Another common feature is the presence of a central tyrosine residue (Tyr-39 and Tyr-10, for α-synuclein and Aβ, respectively). Otherwise, all nine remaining residues are specific to each SBD. Studies with mutant peptides indicated that these secondary amino acids are responsible for the ganglioside-binding specificity of Aβ and α-syn.[22] In particular, a pair of contiguous histidine residues in Aβ (His-13 and His-14) plays a critical role in GM1 recognition, whereas at the same positions, the amino acids of α-syn (Ser-42 and Lys-43) are not involved in GM3 binding.[22] The effects of alanine substitutions on the ganglioside-binding properties of α-syn and Aβ are summarized in Fig. 14.4.

It is particularly meaningful (and striking at first glance) that several distinct single mutations of Aβ5–16, including His-13Ala and His-14Ala, resulted in a total loss of interaction with GM1.[25] This loss remained puzzling until molecular modeling simulations could shed some light on this result (Fig. 14.5). In fact, we observed that the formation of a stable complex between Aβ5–16 and GM1 required two GM1 molecules forming a chalice-like receptacle for the peptide.[25,26] In presence of the peptide, this interaction led to the formation of a trimolecular [GM1/Aβ5–16/GM1] complex. The imidazole rings of His-13 and His-14 interacted with each GM1 in a way that could be metaphorically compared to the wings of a butterfly gathering the "chalice" of the ganglioside dimer.[25] Interestingly, this dimeric topology of GM1 gangliosides is under the control of cholesterol, which induces a tilt in the orientation of the sugar part with respect to the main axis of GM1.[27,49] Because cholesterol is enriched in lipid raft

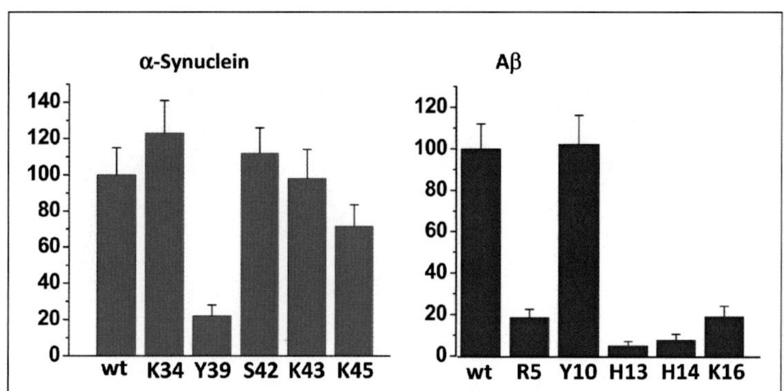

FIGURE 14.4 **Mutations affecting the interaction of α-syn34–45 (left panel) and Aβ5–16 (right panel) with gangliosides.** The data show the response of each wild-type (wt) or mutant peptide with a monolayer of GM1 (Aβ) or GM3 (α-synuclein). The results are expressed as the percentage of the surface pressure increase induced by a given peptide on the ganglioside monolayer, with the relevant wt peptide taken as 100%. Results are expressed as mean ± SD ($n = 4$). The indicated positions in the mutant are replaced by alanine.

domains, such condensed GM1–cholesterol complexes, which have been detected *in vitro*,[50] are likely to occur in the plasma membrane of neural cells.

The fact that the SBD of α-synuclein, which is devoid of histidine, has a marked preference for GM3 versus GM1,[22] has reinforced our opinion that the variations in the amino acid sequence of the SBDs follow a biochemical code controlling glycolipid recognition. Cracking this code has been a long road that took almost 10 years of intense research efforts.

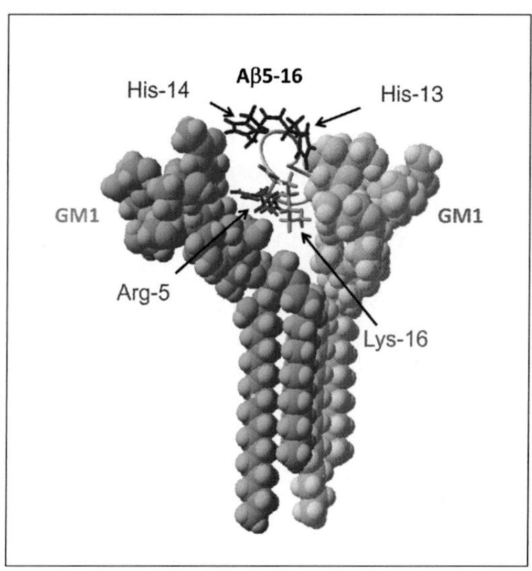

FIGURE 14.5 **Interaction of Aβ5–16 with a chalice-shaped dimer of ganglioside GM1.** *(Reproduced from Yahi and Fantini,[25] with permission.)*

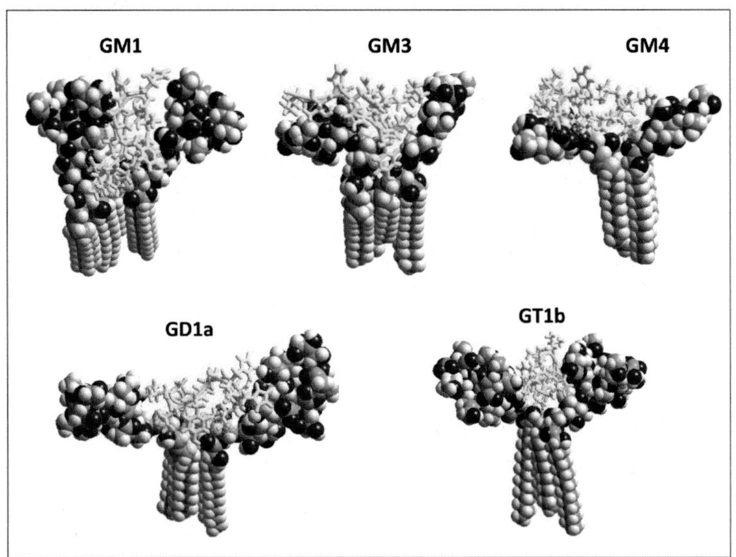

FIGURE 14.6 **Interaction of the chimeric α-syn34–45/HH peptide with major brain gangliosides.** The chimeric peptide interacts with chalice-shaped ganglioside dimers. The models were obtained with the HyperChem software.

Nevertheless, it was all worth it because in the end we could create a chimeric peptide combining the ganglioside-binding properties of both α-synuclein and Aβ.[25] Our strategy was to introduce the pair of histidine residues that allow Aβ to interact with GM1 in the frame of α-syn34–45, thus replacing Ser-42 and Lys-43 by two contiguous His residues (i.e., transforming the α-syn34–45 sequence **KEGVLYVGSKTK** into a chimeric syn–Aβ creation **KEGVLYVGHHTK**). We demonstrated that this chimeric peptide retained the property of α-synuclein to recognize GM3 with high affinity and acquired the Aβ-inherited capacity to recognize GM1.[25] Moreover, it turned out that the chimeric peptide could in fact interact with all gangliosides tested so far, including the main gangliosides expressed in human brain (GM1, GM3, GM4, GD1a, GD1b, GT1b). Examples of molecular complexes between the chimeric peptide and various ganglioside dimers are shown in Fig. 14.6.

In contrast, the chimeric peptide had a low affinity for neutral glycosphingolipids, including GlcCer, LacCer, and asialo-GM1, in agreement with the absolute requirement of sialic acids in the formation of a histidine-driven complex with chalice-shaped gangliosides.[25] Estimations of the energies of interaction of the chimeric peptide with gangliosides confirmed the prominent role of the histidine pair, ranging from 25% (GT1b) to 41% (GM1) of the total energy of interaction as estimated by molecular modeling simulations. The only exception was GM3 (<9%), consistent with the fact that the amino acid residues at positions 42 and 43 are not particularly involved in the recognition of this ganglioside by α-synuclein.[22,29] Nevertheless, because it combines the ganglioside-binding properties of both α-synuclein and Aβ, the chimeric peptide (referred to as α-syn34–45/HH) is indeed a universal ganglioside-binding peptide.

We propose to use the chimeric α-syn34–45/HH peptide as a competitive inhibitor of the binding of amyloid proteins to lipid raft microdomains of brain cells. As schematized

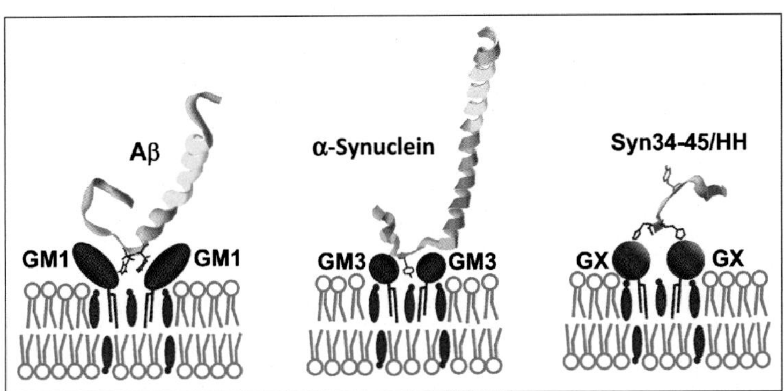

FIGURE 14.7 **How the chimeric peptide blocks the binding of infectious proteins to gangliosides.** Whatever the ganglioside targeted by each type infectious protein (e.g., GM1 for Aβ, GM3 for α-synuclein), and the molecular mechanism that controls each of these interactions (His13/His14 for Aβ, Tyr-39 for α-synuclein), the chimeric peptide is able to bind to any ganglioside dimer (referred to as GX on the far right model). Thus, this universal ganglioside-binding peptide can potentially inhibit the cellular binding of all infectious proteins that target cell surface gangliosides. The colors used are the same as in Fig. 14.1.

in Fig. 14.7, α-syn34–45/HH has the unique capability to interact with ganglioside dimers expressed on the surface of brain cells. Accordingly, in presence of this peptide, neither α-synuclein nor Aβ peptides could reach their ganglioside targets (i.e., respectively, GM3 and GM1). Because GM3 is chiefly expressed in astrocytes[51] and GM1 in neuronal postsynaptic membranes,[52] the chimeric peptide could be able to prevent the infection of neurodegenerative disease-associated proteins in various brain cell types. Because gangliosides are also physically associated with neurotransmitter receptors,[24] one could argue that the chimeric α-syn34–45/HH could affect the physiological functions of these receptors. Although further experiments are required to check this point, such undesirable side effects are not likely because these neurotransmitter receptors interact with ganglioside monomers.[24] Moreover, a cell surface ganglioside bound to a membrane receptor may not be accessible to external ligands.[28,49] Thus, it is likely that the chimeric peptide could only bind to ganglioside dimers that are not masked by membrane proteins, and not to gangliosides already engaged in a functional interaction with a neurotransmitter receptor.

Moreover, the N-terminal domains of α-synuclein and Aβ peptides, from which the chimeric peptide has been designed, are not toxic from neural cells.[25] Nevertheless, it was important to verify that this peptide does not display unexpected neurotoxic properties. As a matter of fact, the chimeric α-syn34–45/HH peptide did not induce significant toxicity in cultured neural cells at concentrations up to 100 μM. Most importantly, the chimeric peptide induced a dose-dependent inhibition of the binding of full-length Aβ1–42 peptide on the surface of cultured neuroblastoma SH-SY5Y cells.[25] Given the specific affinity of the peptide for cell surface gangliosides and the fact that these gangliosides are physically involved in the attachment of Aβ to neural cells, it can be reasonably concluded that the chimeric peptide prevented the binding of Aβ by occupying the gangliosides recognized by Aβ. From these data, we concluded that α-syn34–45/HH has promising anti-Alzheimer and anti-Parkinson properties that warrants further evaluation. In this respect, the chimeric α-syn34–45/HH peptide opens a new route that could lead to a common treatment of various neurodegenerative diseases. Obviously we are not yet at this stage, but these *in vitro* findings demonstrate that the goal is not out of reach. In any case, it is now essential to take into account the role of brain lipids (especially gangliosides) in neural diseases for developing new therapeutic strategies. We should underscore here that the chimeric α-syn34–45/HH peptide is the first compound ever designed with this lipid-oriented approach.

14.6 A UNIVERSAL SQUATTER OF CHOLESTEROL-BINDING SITES

Another angle of approach is to prevent the infectious proteins from interacting with cholesterol. In this respect, one should wonder why the long-time known association of neurodegenerative diseases with cholesterol[53–55] has not led yet to an arsenal of efficient medications. Obviously, one could cite statins (metabolic inhibitors of cholesterol biosynthesis) as popular anti-Alzheimer drugs, but the benefit of such treatments for Alzheimer patients has been questioned.[55] Moreover, even if we consider that statins slow down the progression of the disease, the mechanisms by which these effects could be mediated remain puzzling. Indeed, most statins do not cross the blood–brain barrier and thus have virtually no effect on the brain's cholesterol contents.[55] In fact, statins could simply improve brain oxygenation through their classical antiatherosclerosis activity (prevention of cholesterol deposition in blood vessels), as proposed by Martins et al.[56]

Studies of the physical interaction between cholesterol and Aβ have allowed the characterization of a functional cholesterol-binding domain consisting of a linear fragment (Aβ20–35) located in the apolar C-terminal domain of Aβ.[14] Similarly, a cholesterol-binding domain has been detected in α-synuclein (α-syn67–79).[13] There is no sequence homology between both motifs. Moreover, classical algorithms used for identifying cholesterol-binding motifs (CRAC and CARC)[57] failed to find anything in the amino acid sequence of those proteins, except a low-affinity binding site for cholesterol within the SBD of α-synuclein.[13] Instead, binding to cholesterol was predicted by *in silico* approaches and confirmed by physicochemical studies with wild-type and mutant peptides.[16] Nevertheless, as for the SBD, the cholesterol-binding domains of Aβ and α-synuclein appeared to share a common structural feature. Indeed, these fragments were shown to adopt a tilted topology in lipid membranes.[13,15] As explained in Chapter 8, the tilted topology is both a hallmark and a structural requirement of the oligomerization process of infectious proteins that form annular pores in brain membranes.[16] The cholesterol-constrained tilted topology of α-synuclein and Aβ is illustrated in Fig. 14.8.

Would it be possible to prevent the interaction between pore-forming infectious proteins and cholesterol? Once again, *in silico* approaches can considerably facilitate the design of drugs able to occupy the cholesterol-binding domain of these proteins. We recently identified bexarotene (Targretin), an anticancer drug, as one potential candidate.[58] Our interest for this drug stems from a study published in 2012 by Cramer et al.[59] These authors have shown that bexarotene could rapidly restore some cognitive functions in animal models of Alzheimer's disease. In parallel, the drug seemed to stimulate the clearance of amyloid plaques in brain tissues. Later on, other authors have found that bexarotene did not affect the burden of amyloid plaques but instead reduced the levels of soluble Aβ peptides[60,61] and, in some cases, of

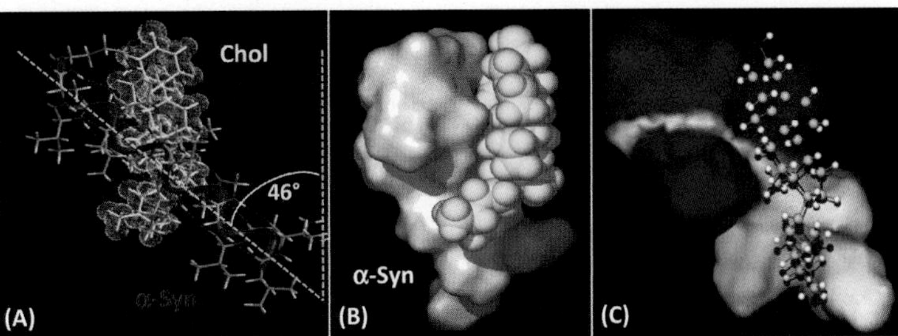

FIGURE 14.8 **The cholesterol-binding domains of α-synuclein (A, B) and Aβ (C) are tilted.** Cholesterol is in yellow spheres (A, B) or in balls and sticks rendering (C). The surface rendering of the protein indicates that the cholesterol-binding domain of α-synuclein is globally apolar (B), whereas in the case of Aβ (C) it contains both apolar and polar residues (white apolar, blue positive, and red negative). *(The models in subparts A and B are reproduced from Fantini et al.*[13] *and the model in subpart C from Di Scala et al.,*[15] *with permission.)*

Aβ oligomers.[60] Conflicting results have also been reported as to the effects of bexarotene on the recovery of cognitive functions. For instance, some authors noted an improvement of social recognition memory in bexarotene-treated APP/PS1 mice,[62] but other authors failed to confirm the effect.[63] These conflicting data did not discourage us from studying the mechanism of action of bexarotene, chiefly because we observed that this drug shares some structural analogy with cholesterol.[58] As shown in Fig. 14.9, bexarotene is an amphipathic compound with a large apolar part and a small polar head group (carboxylate). Like cholesterol, the apolar part of bexarotene contains several hydrocarbon rings. The dimensions of cholesterol and

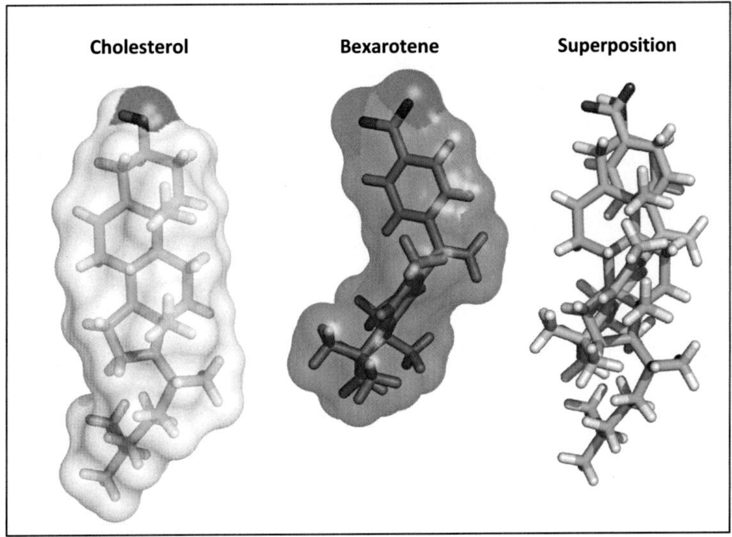

FIGURE 14.9 **Structural comparison of cholesterol and bexarotene.** Note that bexarotene is more tilted than cholesterol. *(Reproduced from Fantini et al.,*[58] *with permission.)*

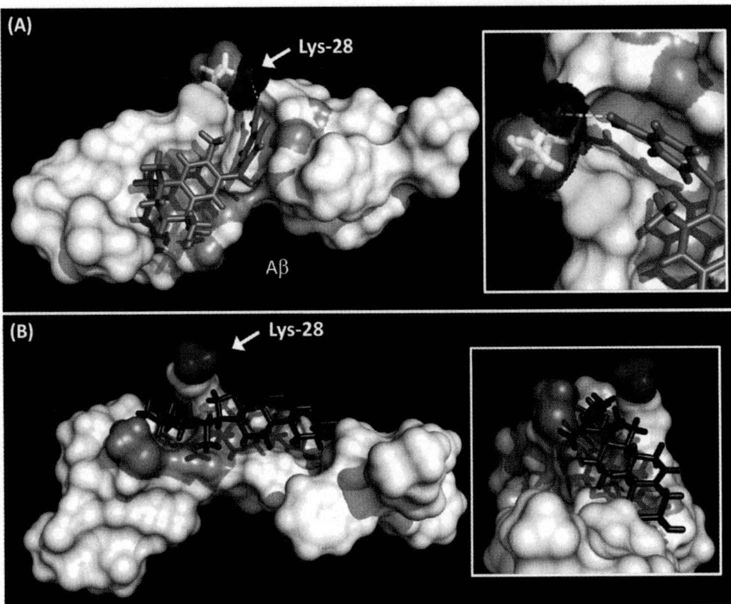

FIGURE 14.10 **Bexarotene (A) and cholesterol (B) bind to the same domain of Aβ1–42.** The initial position of the cholesterol-binding domain is shaded in pink. One can see that, following the formation of a hydrogen bond between bexarotene and Lys-28 (inset of panel A), the Aβ1–42 peptide undergoes a major conformational rearrangement that destroys the cholesterol-binding domain. In the case of cholesterol, it is the apolar chain of Lys-28 that wraps around the cholesterol rings (inset of panel B). In this case, it is impossible to form a hydrogen bond. Lys-28 is in blue, Ile-31 in yellow, and Met-35 in orange. *(Reproduced from Fantini et al.,[58] with permission.)*

bexarotene are also similar: both molecules have the same width (6.5 Å) and a similar length (17.0 Å for cholesterol and 13.5 Å for bexarotene). The structural similarity between cholesterol and bexarotene is well visible when both molecules are superposed (Fig. 14.9). Finally it is important to note that the hinge linking the rings of bexarotene induces an important tilt in the molecule.

Molecular modeling simulations suggested that bexarotene and cholesterol can interact with the same domain of Aβ1–42, located in the C-terminal part of the amyloid peptide.[58] In particular, an electrostatic bridge between the amino group (ε–NH_3^+) of Lys-28 and the carboxylate group of bexarotene stabilizes the complex (Fig. 14.10). Van der Waals contacts with aliphatic residues efficiently contribute to the stabilization of the complexes. The energy of interaction was estimated to -76.8 kJ mol^{-1} and -83.4 kJ mol^{-1} for cholesterol/Aβ1–42 and bexarotene/Aβ1–42, respectively. These energies are of the same magnitude, so that it can be concluded that cholesterol and bexarotene have roughly the same affinity for Aβ1–42. If this is true, how could bexarotene win the competition with cholesterol for binding to Aβ1–42? In fact, upon binding to bexarotene, Aβ1–42 undergoes an important conformational rearrangement that considerably distorts the cholesterol-binding domain (Fig. 14.10). The driving force of this conformational change is the establishment of a hydrogen bond between the carboxylate group of bexarotene and the ε–NH_3^+ of Lys-28. This hydrogen bond requires

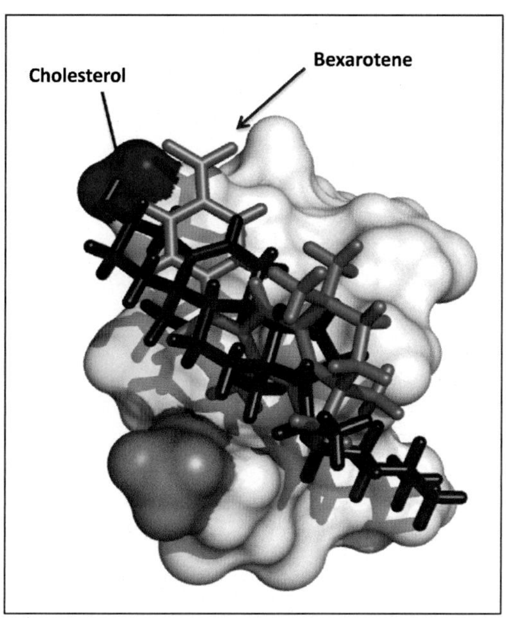

FIGURE 14.11 **Cholesterol and bexarotene share the same binding site on Aβ25–35.** Cholesterol is in red, bexarotene in green. In both cases, Lys-28 can spontaneously form a hydrogen bond with the polar groups of these ligands, so that bexarotene binding does not induce any conformational change in Aβ25–35. Lys-28 is in blue, Ile-31 in yellow, and Met-35 in orange. *(Reproduced from Fantini et al.,[58] with permission.)*

a reorientation of the whole side chain of Lys-28, which in turn modifies the shape of the peptide. Through this molecular mechanism, bexarotene constrains Aβ1–42 to adopt a 3D structure that can no longer interact with cholesterol. Under these conditions, if both cholesterol and bexarotene are simultaneously present in the vicinity of Aβ1–42, bexarotene will occupy and destroy the cholesterol-binding domain, thereby rendering Aβ1–42 totally unable to bind cholesterol.

Our docking studies also showed that both cholesterol and bexarotene could interact with the minimal neurotoxic Aβ25–35 peptide. In contrast with Aβ1–42, bexarotene did not affect the conformation of this short peptide (Fig. 14.11). Nevertheless, the affinity of bexarotene for Aβ25–35 peptide was high in comparison with cholesterol, as indicated by the values of the energy of interaction (–62.4 kJ mol^{-1} for the bexarotene/Aβ25–35 complex and only –37.1 kJ mol^{-1} for the cholesterol/Aβ25–35 complex). Once again, bexarotene would easily beat cholesterol for binding Aβ25–35, this time because of a higher affinity for the peptide.

Physicochemical experiments fully confirmed these predictions based on *in silico* approaches. In particular, we demonstrated that a synthetic peptide encompassing the C-terminal domain of Aβ (Aβ17–40) interacted with both cholesterol and bexarotene.[58] In contrast, Aβ1–16 (N-terminal domain) did not recognize these ligands. Further studies with shorter peptides identified Aβ25–35 as the minimal fragment of Aβ able to interact with both cholesterol and bexarotene. Competition experiments also showed that low concentrations of bexarotene in the aqueous phase could prevent the insertion of Aβ25–35 into a model cholesterol-containing membrane.[58] Due to the high affinity of bexarotene for Aβ25–35, a molecular bexarotene-to-peptide ratio of 1:3 was sufficient to totally prevent the interaction between cholesterol and Aβ25–35. If we extrapolate these physicochemical data, we can expect that in the brain, bexarotene could restrict the insertion of Aβ peptides into cholesterol-rich domains of the plasma membrane of neural cells. Moreover, the amphipathic properties of bexarotene also confer membrane-insertion properties. Thus, bexarotene could also bind to Aβ peptides that have already penetrated the plasma membrane of brain cells. In this case, bexarotene would prevent the association of Aβ with membrane cholesterol and, due to its higher affinity for these peptides, displace preformed Aβ–cholesterol complexes.[16] From this discussion we can see that bexarotene has a large spectrum of Aβ-neutralizing properties that can be summarized as follows: (1) in the extracellular milieu, bexarotene binds to Aβ monomers. This could prevent the oligomerization of Aβ peptides into β-rich annular oligomers before they can perforate the cells.[17] This could also affect the propensity of the peptides to spread onto lipid raft microdomains of brain cells at the site of membrane insertion, resulting in a total block of the whole insertion–oligomerization process. (2) The insertion of bexarotene in the plasma membrane of neural cells would preclude any functional interaction of Aβ with cholesterol.

At this stage, one could wonder whether the Aβ-bexarotene complex could self-organize into oligomeric annular pores just as Aβ-cholesterol complexes do. In fact, it turned out that if cholesterol has a chaperone effect on the formation of amyloid pores, bexarotene has instead antichaperone properties. The mechanisms of these chaperone–antichaperone effects can be explained at the molecular level by studying the respective geometries of Aβ–cholesterol and Aβ–bexarotene complexes. In this respect, the study of Aβ25–35, which forms annular pores in neural membranes through a cholesterol-dependent mechanism, is particularly informative. Indeed, Aβ25–35 is the shortest neurotoxic Aβ peptide recognized by both cholesterol and bexarotene.[16] First, one should keep in mind that bexarotene has a higher affinity for Aβ25–35 than cholesterol, so bexarotene competitively inhibits the binding of cholesterol to the peptide. This difference in affinity is perfectly illustrated by the values of the energy of interaction estimated through molecular docking experiments: –62.4 kJ mol^{-1} and –37.1 kJ mol^{-1} for the bexarotene/Aβ25–35 and cholesterol/Aβ25–35 complexes, respectively.[58] The higher affinity of bexarotene for Aβ25–35 has three main sources: (1) The electrostatic bridge between the negative charge of the carboxylate group of bexarotene with the ε-NH_3^+ group of Lys-28 (ΔG = –25.4 kJ mol^{-1}) is far more energetic than the hydrogen bond linking this charged amino group and the OH group of cholesterol (ΔG = –6.3 kJ mol^{-1}). (2) Once positioned on Aβ25–35 by these polar–polar interactions, the apolar parts of bexarotene and cholesterol have to accommodate the remaining apolar surface of the peptide. In this process, the specific geometry of bexarotene is fully compatible with the shape of the branched aliphatic residues Ile-31 and Ile-32 (for both residues, ΔG = –28.9 kJ mol^{-1}). For comparison, the cholesterol geometry allows a less energetic interaction (–18.4 kJ mol^{-1}) for the contacts with the same Ile residues. (3) The interaction between bexarotene also involves Asn-27 (–4.2 kJ mol^{-1}), whereas this residue is not concerned by cholesterol binding. From a more global point of view, one can remark that the superiority of bexarotene over cholesterol is also due to its intrinsic tilted geometry (Fig. 14.9). Indeed, due to the flexibility of the isooctyl chain, cholesterol can adopt several distinct shapes among which the specific tilted conformers are able to functionally interact with Aβ peptides.[16] Overall, the Aβ–cholesterol interaction proceeds through an induced-fit mechanism during which each partner adjusts its conformation until the formation of the more stable complex is achieved. In contrast with cholesterol, the tilted geometry of bexarotene is locked in a major conformer that is immediately compatible with a tight interaction with Aβ25–35. In other words, bexarotene beats cholesterol both kinetically and thermodynamically. In addition, the poor flexibility of bexarotene is capable of constraining the conformation of larger Aβ peptides such as Aβ1–42, so that following binding to bexarotene these peptides adopt a distorted geometry that totally destroys the cholesterol-binding domain.[16,58]

Now, can we demonstrate (and if so, explain why) that bexarotene cannot be used by Aβ peptides as a surrogate for cholesterol and that these peptides do not form oligomeric pores with bexarotene just as they do with cholesterol? We have developed an assay to detect and quantify the formation of calcium-permeable amyloid pores in live neural cells.[15,58] This assay has been particularly useful to decipher the molecular and cellular mechanisms controlling the formation of these pores with nanomolar concentrations of Aβ peptides. Using such low and physiologically compatible amounts of Aβ peptides allowed us to isolate the formation of amyloid pores and study the mechanisms underlying their formation, independently of any nonspecific neurotoxic effects induced by larger Aβ concentrations. We used neuroblastoma SH-SY5Y cells as a target for amyloid pore formation. In a typical experiment, the cells are loaded with the fluorescent-sensitive dye Fluo-4 AM and subsequently treated with Aβ peptides. Under these conditions, calcium fluxes can be estimated by measuring the variation of fluorescence intensity after peptide injection into the recording chamber directly above an upright microscope objective equipped with an illuminator system MT20 module.[15] We can observe that Aβ25–35 induces a progressive increase of intracellular Ca^{2+} levels (Fig. 14.12A, left panel). This elevation of Ca^{2+} levels in response to Aβ25–35 did not occur instantaneously, but required several minutes. This is consistent with the setting of a coordinated process including membrane attachment, insertion, and oligomerization steps. To ascertain that the Ca^{2+} fluxes were due to amyloid pores, and not nonspecific membrane damage, we also performed the experiment in presence of Zn^{2+}, a classical amyloid channel inhibitor.[18,64] The dramatic inhibitory effect of Zn^{2+} confirmed that Aβ25–35 induced typical amyloid pores, as previously reported by other authors for various Aβ peptides.[18] Similarly, we could determine that the oligomerization process leading to pore formation requires cholesterol, because the cholesterol depleting agent methyl-β-cyclodextrin efficiently prevented Aβ-induced Ca^{2+} fluxes (Fig. 14.12A, middle panel).

Once these control experiments were performed, we could evaluate the capacity of bexarotene to block the formation of amyloid pores induced by Aβ peptides. In these experiments, SH-SY5Y cells were treated with Aβ25–35 (or Aβ1–42) in presence of bexarotene (molar ratio 1:1). Under these conditions, we did not detect any increase of Ca^{2+} fluxes (Fig. 14.12A, right panel). This effect was observed not only when bexarotene and Aβ25–35 were premixed for 1 h before addition to the cells but also when both compounds were simultaneous added into the cells. Finally, we demonstrated that bexarotene can inhibit the elevation of Ca^{2+} entry induced by Aβ1–42 (Fig. 14.12B–C).[58] The photomicrographs taken during the experiments indicated that the inhibitory effect of bexarotene was uniformly distributed over the cell culture (Fig. 14.12B). Overall, these data show that bexarotene is a potent blocker of amyloid pore formation.

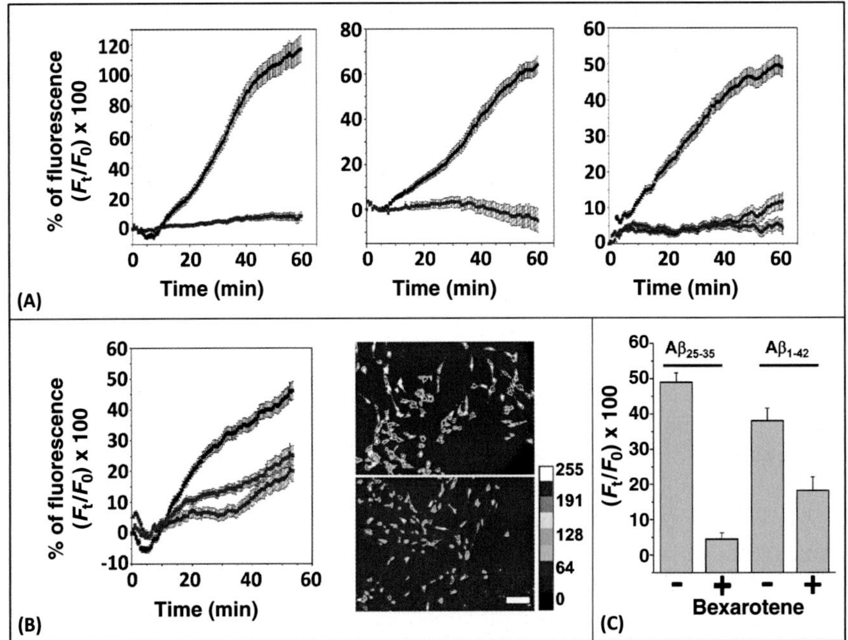

FIGURE 14.12 **Bexarotene blocks the formation of Ca^{2+}-permeable pores in neural cells.** Human neuroblastoma SH-SY5Y cells were loaded with the Ca^{2+} indicator Fluo-4 AM. (A) Studies with Aβ25–35. Left panel: Ca^{2+} fluxes induced by 220 nM of Aβ25–35 (black curve) compared with untreated control cells (red curve). Middle panel: inhibition of the Aβ25–35 effect (control black curve) by the cholesterol-depleting agent methyl-β-cyclodextrin (red curve). Right panel: inhibitory effect of bexarotene premixed with Aβ25–35 and incubated for 1 hr before cell treatment (blue curve), or mixed and immediately injected onto the cells (red curve). In both cases, bexarotene and Aβ25–35 were at equimolar concentrations (220 nM). At this concentration, bexarotene alone did not induce any change of basal Ca^{2+} fluxes. The black curve shows the response of the cells to 220 nM of Aβ25–35. (B) Studies with 220 nM of Aβ1–42 added alone (black curve) or with equimolar concentrations of bexarotene either preincubated for 1 hr (blue curve) or added in competition (red curve) on SH-SY5Y cells. The upper micrographs show the cells incubated with Aβ1–42 alone; the lower graph shows Aβ1–42 in the presence of bexarotene (cf. blue curve). (C) Bexarotene is active on both Aβ25–35 and Aβ1–42. *(Reproduced from Fantini et al.,[58] with permission.)*

It remains to explain why bexarotene and cholesterol, which share the same binding site on cholesterol-binding domain of Aβ peptides, have opposite effects on amyloid formation. One possible explanation is that bexarotene binds to Aβ extracellularly and prevents its insertion in the plasma membrane of SH-SY5Y cells. We also examined the possibility that bexarotene could interact with Aβ peptides already inserted in the membrane. In this case, a true competition would occur for occupying the cholesterol-binding site. As discussed previously, given the high affinity of bexarotene for Aβ, it is likely that the drug would easily win the competition with cholesterol, so that after a while all Aβ peptides would interact with bexarotene. When bound to cholesterol, Aβ25–35 adopts a tilted geometry that favors the oligomerization into an annular Ca^{2+}-permeable pore.[58] In this pore, two vicinal peptides establish a series of interactions that involve residues Gly-25, Asn-27, and Ile-31 on one peptide, and Ser-26, Gly-29, and Gly-33 on its neighbor. When bexarotene occupies the cholesterol-binding site of Aβ25–35, it blocks two of these residues that are critical for the oligomerization process: Asn-27 and Ile-31. Indeed, cholesterol does not interact at all with Asn-27, and although both cholesterol and bexarotene bind to Ile-31, bexarotene has a higher affinity for this residue than cholesterol. Hence, once the peptide is bound to bexarotene in the plasma membrane, two of the key amino acid residues that control the oligomerization of Aβ25–35 are neutralized and unavailable for establishing the specific van der Waals peptide–peptide contacts leading to pore formation.

A last issue to consider is whether bexarotene could also be active on other infectious proteins that form Ca^{2+}-permeable pores, especially α-synuclein. The cholesterol-binding domain of α-synuclein has been identified as a linear fragment encompassing amino acid residues 67–79.[13] Like the cholesterol-binding domain of Aβ peptides, this region adopts a tilted geometry when bound to membrane cholesterol. This is particularly visible in the model shown in Fig. 14.8A. Interestingly, docking studies suggested the cholesterol-binding domain of α-synuclein could also interact with bexarotene. The complex between bexarotene and α-synuclein is shown in Fig. 14.13. Molecular dynamics simulations suggested that bexarotene has a higher affinity for α-synuclein than cholesterol: $\Delta G = -66.4$ kJ mol^{-1}, for bexarotene, compared with -44.3 kJ mol^{-1} for cholesterol.

The tilted geometry induced by cholesterol on the α-synuclein peptide 67–79 (α-syn67–79) is compatible with an oligomeric channel formed by six peptides (Fig. 14.14). In this channel, two vicinal peptides interact through numerous contacts involving chiefly Val-66, Ala-69, and Val-77 on one peptide, and Thr-64, Asn-65, Gly-68, and Val-71 on its neighbor, for a total

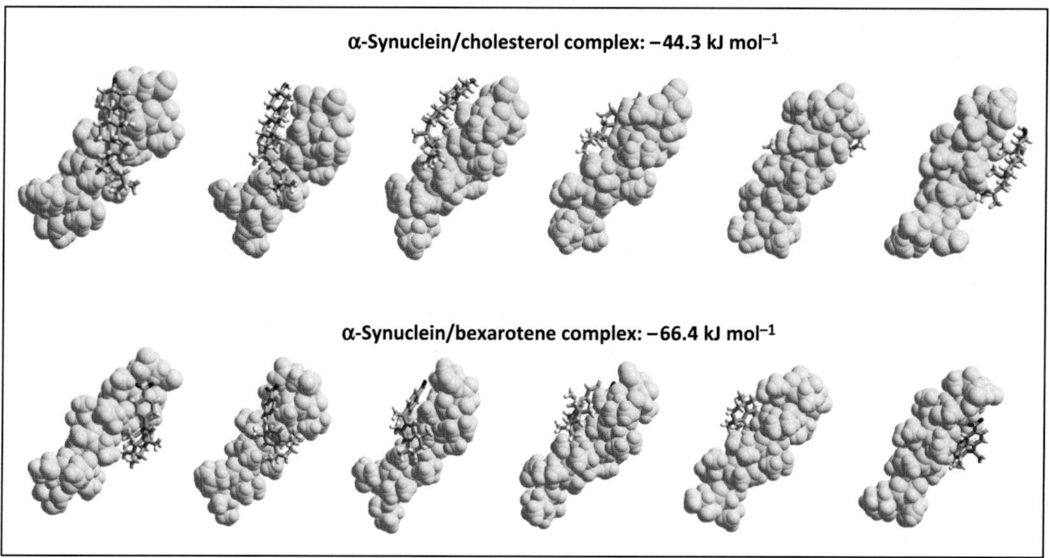

FIGURE 14.13 **Cholesterol and bexarotene share the same binding site on α-synuclein.** Several views of each complex are shown for a better visualization. Note that the intrinsically tilted conformation of bexarotene is particularly well suited for a tight interaction with α-synuclein. Moreover, its aromatic planar surface can adopt the helical form of α-synuclein much better that cholesterol does. This is particularly obvious in the second and third models (from the left).

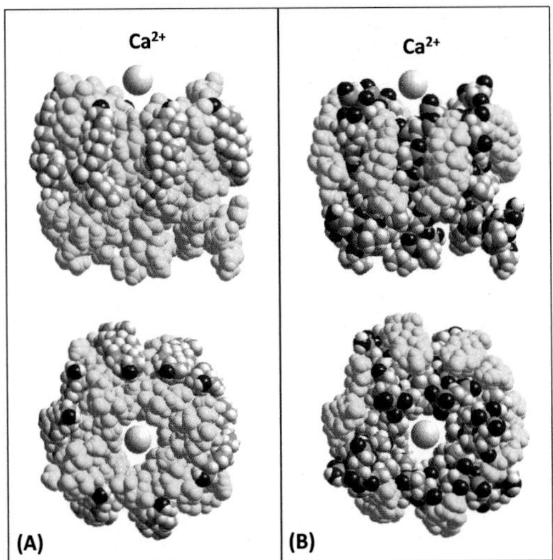

FIGURE 14.14 **Molecular modeling of a Ca^{2+}-permeable pore formed by α-synuclein.** The channel is formed by six monomers of α-syn67–79, each interacting with cholesterol. Two supplemental cholesterol molecules were then added to complete the structure (final stoichiometry: 8 cholesterol/6 α-synuclein peptides). (A) The peptides are in yellow and cholesterol in atom colors. (B) Cholesterol is in yellow and the peptides in atom colors. Two views of the pore are shown (lateral and upside), with a Ca^{2+} ion (gray disk, same scale as the other atoms) positioned at the entry of the pore. The models were obtained by molecular dynamics simulations with the CHARMM force field of the HyperChem program.

energy of interaction of –61.2 kJ mol^{-1}. It is informative to compare these interactions with those involved in the α-syn67–79/cholesterol and the α-syn67–79/bexarotene complexes. In fact, neither cholesterol nor bexarotene seems to interact strongly with any of these specific residues, so that one could conclude that both cholesterol and bexarotene could allow the formation of an oligomeric α-syn67–79 channel. Nevertheless, a careful examination of the α-syn67–79/cholesterol complex in the hexameric channel shows that the tilted geometry of cholesterol bound to the peptide is consistent with the formation of the pore. In contrast, when bexarotene is bound to the peptide, its unique geometry precludes the formation of a channel by preventing the interaction between two vicinal peptides (Fig. 14.15). Indeed, the presence of bexarotene cannot allow the formation peptide–peptide dimers, due to steric hindrance.

These data indicate that bexarotene has broad antichaperone properties on amyloid pore formation. Because it antagonizes the chaperone activity of cholesterol, it is the prototype of a new class of candidate therapeutic agents that could be used for treating various neurodegenerative diseases. Beyond bexarotene, any molecule able to occupy the cholesterol-binding domain of infectious proteins should be evaluated in the Ca^{2+} flux assay to determine whether it displays chaperone or antichaperone effects on amyloid pore formation. In this respect, our strategy based on the sequential use of three methodological approaches, including *in silico* studies, physicochemical experiments, and Ca^{2+} fluxes studies provides a rational strategy to identify new candidate drugs for Alzheimer's, Parkinson's, and other cholesterol-associated impairments of brain functions.

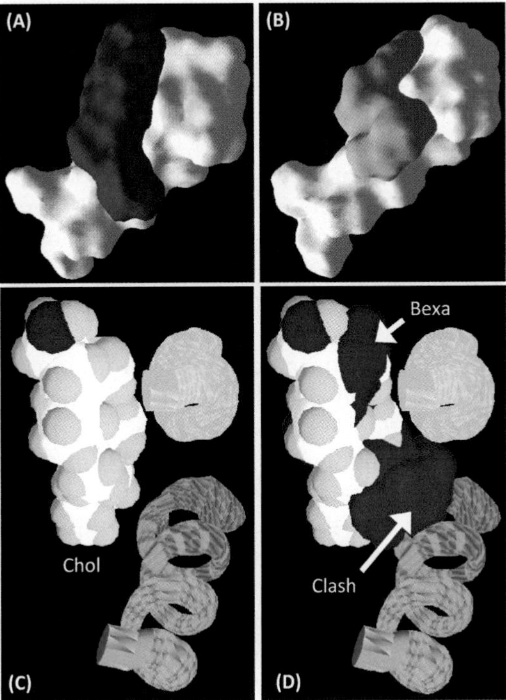

FIGURE 14.15 **A steric clash explains how bexarotene prevents pore formation.** Cholesterol (blue compound in subpart A) and bexarotene (orange drug in subpart B) bind to the same domain of α-synuclein (α-syn67–79). Cholesterol acts as a chaperone for the oligomerization of α-synuclein monomers in the membrane. In subpart C, one can see that the binding of cholesterol to the α-helix of α-syn67–79 (in yellow) orients its neighbor α-syn unit (α-helix in green) in a channel-compatible topology. However, when bexarotene is bound to one α-synuclein peptide, further oligomerization is impossible because bexarotene (in red) evokes a steric clash (in subpart D) which prevents the functional interaction between both α-synuclein peptides. Thus, bexarotene exerts an antichaperone effect on oligomeric pore formation.

14.7 COULD ANTI-HIV DRUGS ALSO BE CONSIDERED FOR THE TREATMENT OF NEURODEGENERATIVE DISEASES?

As discussed in Chapter 13, a striking analogy can be made between the membrane insertion properties of pore-forming amyloid proteins and virus fusion mechanisms.[1,13,39,65] In both cases, a primary interaction with cell surface glycosphingolipids ensures that the infectious protein has immediate access to cholesterol, which promotes the insertion process. Thus, a possible therapeutic strategy to block HIV infection is to prevent both its binding to cell surface glycosphingolipids and its functional interaction with membrane cholesterol. In contrast with amyloid proteins, these two types of interaction with membrane lipids involve not one, but two distinct proteins of the viral envelope: the surface envelope glycoprotein (gp120 for HIV-1), which binds to glycosphingolipids,[2,66-69] and the transmembrane protein (gp41for HIV-1), which is a cholesterol-binding protein.[70]

The glycosphingolipid binding domain of HIV-1 gp120 has been identified as a short variable segment called the V3 loop.[71] For this reason, the situation is more complex than with amyloid proteins that have a unique amino acid sequence. Sequence variability in the V3 loop of gp120 determines distinct cellular tropisms for the virus, according to the nature of its glycosphingolipid partner (e.g., GalCer for intestinal[66] and brain cells,[2] GM3 for monocytes and macrophages, Gb_3 for laboratory-adapted strains that infect T-cell lines).[68] Several attempts have been made to develop universal V3 loop-binding compounds that would have been, in theory, able to neutralize a broad range of HIV-1 strains. One of these compounds is a multibranched peptide based on the consensus sequence of the V3 loop crown (i.e., the hexapeptide GPGRAF) linked to a polylysine dendrimer structure.[72] This multibranched peptide efficiently inhibited the infection of various HIV-1 isolates in both $CD4^-$[73] and $CD4^+$ cells[74] through a unique mechanism pertaining to its ability to interact with cell surface glycosphingolipids. Indeed, this peptide, referred to as SPC3, interacted not only with GalCer and sulfatide expressed by oligodendrocytes but also with neutral glycolipids (LacCer, Gb_3) and gangliosides (GM3, GD3) expressed by lymphocytes and macrophages.[75] Because of its multivalent structure (8 GPGRAF motifs in the multibranched structure), it has been assumed that the SPC3 peptide could cross-link cell-surface glycolipids into "frozen" clusters, thereby preventing the functional interaction of HIV-1 virions with lipid rafts.[75] On the basis of this unique property, it would be interesting to study the effect of SPC3 on amyloid pore formation. We are aware that SPC3 and the universal ganglioside-binding peptide α-syn34–45/HH have distinct patterns of glycosphingolipid recognition (SPC3 recognizes several neutral glycolipids that are totally ignored by α-syn34–45/HH). Nevertheless, both are devoid of toxicity and could perhaps be used in combination to potentiate their antiadhesive properties for infectious proteins that land on lipid raft microdomains. Alternatively, one could try a multibranched formulation of α-syn34–45/HH in a dendrimer and see whether this multivalent presentation of the ganglioside-binding motif is advantageous.

Another V3 loop-targeted agent able to inhibit the interaction of HIV-1 to lipid rafts is suramin,[76] a polyaromatic compound with anticancer properties.[77,78] *In vitro*, suramin binds to the V3 loop of various HIV-1 strains and is a potent inhibitor of HIV-1 infection.[76] Moreover, the drug efficiently prevents the binding of HIV-1 gp120 to GalCer and other cell surface glycosphingolipids.[76] Interestingly, suramin has also been shown to bind to the amyloid core fragment of the human islet amyloid protein (amylin) and to affect its aggregation.[48] Unfortunately, the renal toxicity of suramin[79] does not allow its evaluation as anti-Alzheimer or anti-Parkinson agent, but it would be of interest to study the effects of suramin-like compounds (ideally those devoid of neurotoxicity) on amyloid pore formation. In this respect, compounds like NF036, a nonneurotoxic but fully active suramin analog[80] should be tested in priority.

From the preceding discussion, it can be concluded that some therapeutic strategies initially developed against HIV infection could be considered for the treatment of neurodegenerative diseases. The reverse possibility (i.e., using inhibitors of amyloid pores for preventing HIV-1 infection) could also be considered. Indeed, it would be interesting to determine whether bexarotene, which blocks the cholesterol-dependent pathway of amyloid pore formation,[58] could bind to HIV-1 glycoproteins, especially gp41, and prevent virus infection, a process that also depends on membrane cholesterol.[81–83] Once again, we are facing common molecular mechanisms that could be jammed by similar therapeutic approaches.

14.8 CONCLUSIONS

Until recently, a recurrent option for treating neurodegenerative diseases was to attack amyloid fibers and try to destroy them.[84,85] This was both logical and rational, because a huge number of experiments seemed to convincingly demonstrate that these amyloid aggregates were the main neurotoxic species involved with amyloid proteins.[86–90] The discovery of amyloid oligomers, especially those forming annular Ca^{2+}-permeable pores in the plasma membrane of neural cells, has changed the game.[91–93] Once it has been recognized that breaking amyloid plaques could be harmful because it could in fact generate smaller yet highly toxic oligomers,[94] and that amyloid plaques could arise in elder people with no detectable cognitive dysfunction,[95] a thorough reevaluation of therapeutic strategies was more than necessary.[19] We are now looking for new drugs able to prevent the insertion of amyloid proteins in the plasma membrane of neural cells and to reach amyloid pores inside the membrane.[16,19] Accordingly, it is no longer possible to overlook the important role of membrane lipids, especially gangliosides and cholesterol, in the generation of amyloid pores. The potential therapeutic tracks discussed in this chapter could serve as a first guideline for eventually finding the medicines we are currently lacking for the treatment of Alzheimer's, Parkinson's, and other neurodegenerative diseases.

14.9 A KEY EXPERIMENT: PAMPA-BBB, A LIPID-BASED MODEL FOR THE BLOOD–BRAIN BARRIER[96]

One essential issue for a candidate antineurodegenerative medicine is to check that it is efficiently delivered to the brain. The blood–brain barrier restricts the entry of toxic compounds into the brain, and will also do so for most potentially therapeutic compounds.[97] Therefore, many drugs that are efficient *in vitro* may fail to succeed therapeutically just because they do not cross the blood–brain barrier. Anatomically, the blood–brain barrier is formed by endothelial cells that line cerebral microvessels (Fig. 14.16). A drug has only two possible routes for crossing this cell layer: the transcellular pathway and the paracellular route. The latter one is blocked by intercellular tight junctions that constitute an efficient barrier preventing the free diffusion of molecules. Thus, most therapeutically active drugs use the transcellular pathway. Correspondingly, the drug has to cross the plasma membrane of vascular endothelial cells twice to reach the brain (once for entering the cells and a second time to exit). Predicting drug permeability is a major issue for deciding which drugs, active *in vitro*, warrant further evaluation in animals and humans. Among the models developed to predict the blood–brain barrier penetration potential of drug candidates, those exploiting the lipid composition of endothelial cell membranes are particularly interesting. Mimicking these membranes can be approached in several ways. One popular approach is to use a mixture of polar lipids extracted

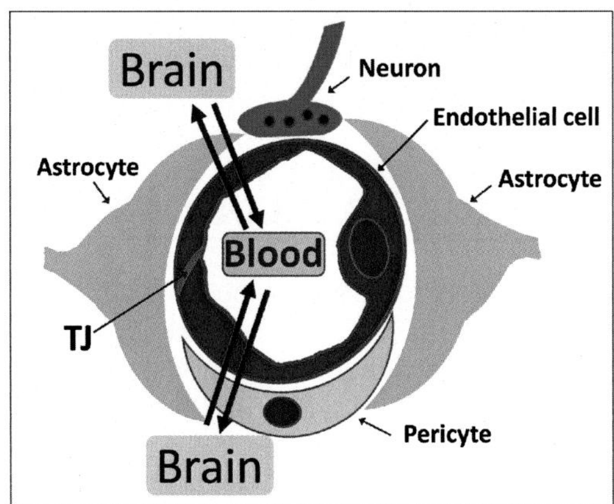

FIGURE 14.16 **Schematic view of the blood–brain barrier.** Endothelial cells form a continuous layer covering the inner surface of blood microvessels. At the point of contacts with brain tissues, these endothelial cells interact with pericytes and astrocytes. Nerve ending with specific neurotransmitters can also be observed. The tight junctions (TJ) that link endothelial cells act as a physical barrier that blocks the paracellular route and forces the molecular traffic to take the transcellular pathway (double arrows).

from whole pig brains.[97,98] The parallel artificial membrane permeability assay (PAMPA) that uses such lipids is a reliable assay for predicting whether a drug will (or will not) cross the blood–brain barrier (BBB).[98]

Nevertheless, one should be aware that the lipid composition of human blood–brain barrier endothelial cell membranes differs from that of the porcine polar brain lipids used in the PAMPA-BBB assay.[97] In particular, the porcine brain lipids used in this assay contain no cholesterol, even though cholesterol represents at least 20% of the lipids found in brain endothelial cell membranes.[99] Moreover, sphingomyelin is more abundant in endothelial brain membranes than in porcine brain lipids.[97] Taken together, these data suggest that the physical properties (fluidity, elasticity, curvature) of artificial membranes made of porcine brain lipids should not reflect the properties of blood–brain barrier membranes. That being said, it remains striking that the PAMPA-BBB system with reconstituted porcine lipid membranes has such good reliability in predicting the capacity of a candidate drug to cross the human blood–brain barrier[98].

From a methodological point of view, the PAMPA-BBB system is quite simple. The lipids, dissolved in dodecane, are soaked with a filter mounted in a two-compartment chamber.[98] The drug is added to the donor compartment (which can be either the upper or lower chamber), and its passage through the artificial membrane is measured in the acceptor compartment (Fig. 14.17). A standard compound with well-characterized permeability properties (e.g., verapamil) is tested in parallel.[100] Compounds that readily cross the blood–brain barrier have an *in vitro* permeability (P_e) > $2.7\ 10^{-6}$ cm s^{-1} in the PAMPA assay.[100] On the opposite, drugs with low blood–brain barrier permeation have a $P_e < 0.7\ 10^{-6}$ cm s^{-1}. Beside the PAMPA-BBB assay, more sophisticated approaches with cultured cells (endothelial cells either tested alone or in coculture with brain cells) have been developed with the aim to better mimic an authentic blood–brain barrier.[101,102]

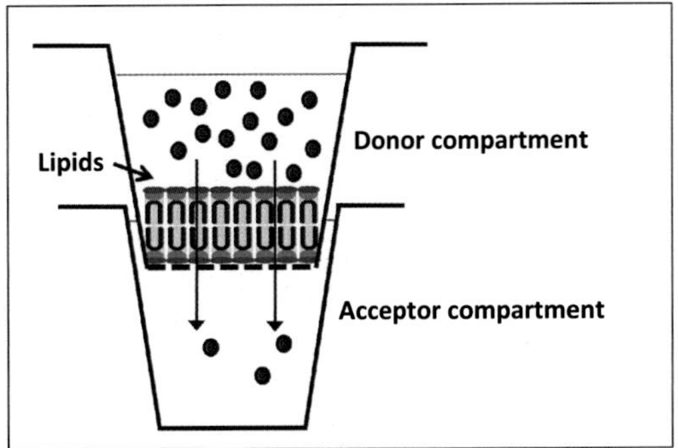

FIGURE 14.17 **Schematic description of the PAMPA-BBB assay.** The compounds of interest (red disks) are introduced in one compartment (referred to as donor) of a two-compartment chamber separated by a reconstituted bilayer of porcine brain lipids spread on a porous membrane. At the end of incubation, the compounds that have crossed the barrier are harvested in the "acceptor" compartment and dosed. Standard compounds with well-characterized permeability properties are tested in parallel.

Finally, it should be underscored that the integrity of the blood–brain barrier can be heavily affected in HIV-infected individuals[103] and in patients with neurodegenerative diseases.[104] In this case, even low-permeating drugs could have access to brain cells.

References

1. Fantini J, Yahi N. Molecular insights into amyloid regulation by membrane cholesterol and sphingolipids: common mechanisms in neurodegenerative diseases. *Expert Rev Mol Med*. 2010;12:e27.
2. Harouse JM, Bhat S, Spitalnik SL, et al. Inhibition of entry of HIV-1 in neural cell lines by antibodies against galactosyl ceramide. *Science*. 1991;253(5017):320–323.
3. Gabuzda D, Wang J. Chemokine receptors and virus entry in the central nervous system. *J Neurovirol*. 1999; 5(6):643–658.
4. Harouse JM, Kunsch C, Hartle HT, et al. CD4-independent infection of human neural cells by human immunodeficiency virus type 1. *J Virol*. 1989;63(6):2527–2533.
5. Harouse JM, Collman RG, Gonzalez-Scarano F. Human immunodeficiency virus type 1 infection of SK-N-MC cells: domains of gp120 involved in entry into a CD4-negative, galactosyl ceramide/3′ sulfo-galactosyl ceramide-positive cell line. *J Virol*. 1995;69(12):7383–7390.
6. Harouse JM, Gonzalez-Scarano F. Infection of SK-N-MC cells, a CD4-negative neuroblastoma cell line, with primary human immunodeficiency virus type 1 isolates. *J Virol*. 1996;70(10):7290–7294.
7. Spudich S, Gonzalez-Scarano F. HIV-1-related central nervous system disease: current issues in pathogenesis, diagnosis, and treatment. *Cold Spring Harb Perspect Med*. 2012;2(6):a007120.
8. Dayanithi G, Yahi N, Baghdiguian S, Fantini J. Intracellular calcium release induced by human immunodeficiency virus type 1 (HIV-1) surface envelope glycoprotein in human intestinal epithelial cells: a putative mechanism for HIV-1 enteropathy. *Cell Calcium*. 1995;18(1):9–18.
9. Lo TM, Fallert CJ, Piser TM, Thayer SA. HIV-1 envelope protein evokes intracellular calcium oscillations in rat hippocampal neurons. *Brain Res*. 1992;594(2):189–196.
10. Ciardo A, Meldolesi J. Effects of the HIV-1 envelope glycoprotein gp120 in cerebellar cultures. $[Ca^{2+}]$i increases in a glial cell subpopulation. *Eur J Neurosci*. 1993;5(12):1711–1718.
11. Dawson VL, Dawson TM, Uhl GR, Snyder SH. Human immunodeficiency virus type 1 coat protein neurotoxicity mediated by nitric oxide in primary cortical cultures. *Proc Natl Acad Sci USA*. 1993;90(8):3256–3259.
12. Nath A, Padua RA, Geiger JD. HIV-1 coat protein gp120-induced increases in levels of intrasynaptosomal calcium. *Brain Res*. 1995;678(1–2):200–206.

13. Fantini J, Carlus D, Yahi N. The fusogenic tilted peptide (67–78) of alpha-synuclein is a cholesterol binding domain. *Biochim Biophys Acta*. 2011;1808(10):2343–2351.
14. Di Scala C, Yahi N, Lelievre C, Garmy N, Chahinian H, Fantini J. Biochemical identification of a linear cholesterol-binding domain within Alzheimer's beta amyloid peptide. *ACS Chem Neurosci*. 2013;4(3):509–517.
15. Di Scala C, Troadec JD, Lelievre C, Garmy N, Fantini J, Chahinian H. Mechanism of cholesterol-assisted oligomeric channel formation by a short Alzheimer beta-amyloid peptide. *J Neurochem*. 2014;128(1):186–195.
16. Di Scala C, Chahinian H, Yahi N, Garmy N, Fantini J. Interaction of Alzheimer's beta-amyloid peptides with cholesterol: mechanistic insights into amyloid pore formation. *Biochemistry*. 2014;53(28):4489–4502.
17. Shafrir Y, Durell S, Arispe N, Guy HR. Models of membrane-bound Alzheimer's Abeta peptide assemblies. *Proteins*. 2010;78(16):3473–3487.
18. Jang H, Arce FT, Ramachandran S, et al. Truncated beta-amyloid peptide channels provide an alternative mechanism for Alzheimer's disease and Down syndrome. *Proc Natl Acad Sci USA*. 2010;107(14):6538–6543.
19. Jang H, Connelly L, Arce FT, et al. Alzheimer's disease: which type of amyloid-preventing drug agents to employ? *Phys Chem Chem Phys*. 2013;15(23):8868–8877.
20. Yanagisawa K. Role of gangliosides in Alzheimer's disease. *Biochim Biophys Acta*. 2007;1768(8):1943–1951.
21. Yanagisawa K. Pathological significance of ganglioside clusters in Alzheimer's disease. *J Neurochem*. 2011; 116(5):806–812.
22. Fantini J, Yahi N. Molecular basis for the glycosphingolipid-binding specificity of alpha-synuclein: key role of tyrosine 39 in membrane insertion. *J Mol Biol*. 2011;408(4):654–669.
23. Uversky VN. Intrinsic disorder in proteins associated with neurodegenerative diseases. *Front Biosci*. 2009; 14:5188–5238.
24. Fantini J, Barrantes FJ. Sphingolipid/cholesterol regulation of neurotransmitter receptor conformation and function. *Biochim Biophys Acta*. 2009;1788(11):2345–2361.
25. Yahi N, Fantini J. Deciphering the glycolipid code of Alzheimer's and Parkinson's amyloid proteins allowed the creation of a universal ganglioside-binding peptide. *PloS One*. 2014;9(8):e104751.
26. Fantini J, Yahi N, Garmy N. Cholesterol accelerates the binding of Alzheimer's beta-amyloid peptide to ganglioside GM1 through a universal hydrogen-bond-dependent sterol tuning of glycolipid conformation. *Front Physiol*. 2013;4:120.
27. Yahi N, Aulas A, Fantini J. How cholesterol constrains glycolipid conformation for optimal recognition of Alzheimer's beta amyloid peptide (Abeta1-40). *PloS One*. 2010;5(2):e9079.
28. Lingwood D, Binnington B, Rog T, et al. Cholesterol modulates glycolipid conformation and receptor activity. *Nat Chem Biol*. 2011;7(5):260–262.
29. Fantini J, Yahi N. The driving force of alpha-synuclein insertion and amyloid channel formation in the plasma membrane of neural cells: key role of ganglioside- and cholesterol-binding domains. *Adv Exp Med Biol*. 2013;991:15–26.
30. Fantini J. Interaction of proteins with lipid rafts through glycolipid-binding domains: biochemical background and potential therapeutic applications. *Curr Med Chem*. 2007;14(27):2911–2917.
31. Lund N, Branch DR, Mylvaganam M, et al. A novel soluble mimic of the glycolipid, globotriaosyl ceramide inhibits HIV infection. *AIDS*. 2006;20(3):333–343.
32. Mahfoud R, Mylvaganam M, Lingwood CA, Fantini J. A novel soluble analog of the HIV-1 fusion cofactor, globotriaosylceramide (Gb(3)), eliminates the cholesterol requirement for high affinity gp120/Gb(3) interaction. *J Lipid Res*. 2002;43(10):1670–1679.
33. Zakharov SD, Hulleman JD, Dutseva EA, Antonenko YN, Rochet JC, Cramer WA. Helical alpha-synuclein forms highly conductive ion channels. *Biochemistry*. 2007;46(50):14369–14379.
34. Ji SR, Wu Y, Sui SF. Cholesterol is an important factor affecting the membrane insertion of beta-amyloid peptide (A beta 1-40), which may potentially inhibit the fibril formation. *J Biol Chem*. 2002;277(8):6273–6279.
35. Martinez Z, Zhu M, Han S, Fink AL. GM1 specifically interacts with alpha-synuclein and inhibits fibrillation. *Biochemistry*. 2007;46(7):1868–1877.
36. Sanghera N, Pinheiro TJ. Binding of prion protein to lipid membranes and implications for prion conversion. *J Mol Biol*. 2002;315(5):1241–1256.
37. Matsubara T, Iijima K, Yamamoto N, Yanagisawa K, Sato T. Density of GM1 in nanoclusters is a critical factor in the formation of a spherical assembly of amyloid beta-protein on synaptic plasma membranes. *Langmuir*. 2013;29(7):2258–2264.

38. Schnaar RL, Gerardy-Schahn R, Hildebrandt H. Sialic acids in the brain: gangliosides and polysialic Acid in nervous system development, stability, disease, and regeneration. *Physiol Rev*. 2014;94(2):461–518.
39. Fantini J, Garmy N, Mahfoud R, Yahi N. Lipid rafts: structure, function and role in HIV, Alzheimer's and prion diseases. *Expert Rev Mol Med*. 2002;4(27):1–22.
40. Fantini J. How sphingolipids bind and shape proteins: molecular basis of lipid-protein interactions in lipid shells, rafts and related biomembrane domains. *Cell Mol Life Sci*. 2003;60(6):1027–1032.
41. Mahfoud R, Garmy N, Maresca M, Yahi N, Puigserver A, Fantini J. Identification of a common sphingolipid-binding domain in Alzheimer, prion, and HIV-1 proteins. *J Biol Chem*. 2002;277(13):11292–11296.
42. Aubert-Jousset E, Garmy N, Sbarra V, Fantini J, Sadoulet MO, Lombardo D. The combinatorial extension method reveals a sphingolipid binding domain on pancreatic bile salt-dependent lipase: role in secretion. *Structure*. 2004;12(8):1437–1447.
43. Chakrabandhu K, Huault S, Garmy N, et al. The extracellular glycosphingolipid-binding motif of Fas defines its internalization route, mode and outcome of signals upon activation by ligand. *Cell Death Differen*. 2008;15(12):1824–1837.
44. Hamel S, Fantini J, Schweisguth F. Notch ligand activity is modulated by glycosphingolipid membrane composition in Drosophila melanogaster. *J Cell Biol*. 2010;188(4):581–594.
45. Chattopadhyay A, Paila YD, Shrivastava S, Tiwari S, Singh P, Fantini J. Sphingolipid-binding domain in the serotonin(1A) receptor. *Adv Exp Med Biol*. 2012;749:279–293.
46. Heuss SF, Tarantino N, Fantini J, et al. A glycosphingolipid binding domain controls trafficking and activity of the mammalian notch ligand delta-like 1. *PloS One*. 2013;8(9):e74392.
47. Fantini J, Garmy N, Yahi N. Prediction of glycolipid-binding domains from the amino acid sequence of lipid raft-associated proteins: application to HpaA, a protein involved in the adhesion of *Helicobacter pylori* to gastrointestinal cells. *Biochemistry*. 2006;45(36):10957–10962.
48. Levy M, Garmy N, Gazit E, Fantini J. The minimal amyloid-forming fragment of the islet amyloid polypeptide is a glycolipid-binding domain. *FEBS J*. 2006;273(24):5724–5735.
49. Krengel U, Bousquet PA. Molecular recognition of gangliosides and their potential for cancer immunotherapies. *Front Immunol*. 2014;5:325.
50. Radhakrishnan A, Anderson TG, McConnell HM. Condensed complexes, rafts, and the chemical activity of cholesterol in membranes. *Proc Natl Acad Sci USA*. 2000;97(23):12422–12427.
51. Asou H, Hirano S, Uyemura K. Ganglioside composition of astrocytes. *Cell Struct Funct*. 1989;14(5):561–568.
52. Hansson HA, Holmgren J, Svennerholm L. Ultrastructural localization of cell membrane GM1 ganglioside by cholera toxin. *Proc Nat Acad Sci USA*. 1977;74(9):3782–3786.
53. Yanagisawa K. Cholesterol and amyloid beta fibrillogenesis. *Subcell Biochem*. 2005;38:179–202.
54. Harris JR, Milton NG. Cholesterol in Alzheimer's disease and other amyloidogenic disorders. *Subcell Biochem*. 2010;51:47–75.
55. Wood WG, Li L, Muller WE, Eckert GP. Cholesterol as a causative factor in Alzheimer's disease: a debatable hypothesis. *J Neurochem*. 2013.
56. Martins IJ, Berger T, Sharman MJ, Verdile G, Fuller SJ, Martins RN. Cholesterol metabolism and transport in the pathogenesis of Alzheimer's disease. *J Neurochem*. 2009;111(6):1275–1308.
57. Fantini J, Barrantes FJ. How cholesterol interacts with membrane proteins: an exploration of cholesterol-binding sites including CRAC, CARC, and tilted domains. *Front Physiol*. 2013;4:31.
58. Fantini J, Di Scala C, Yahi N, et al. Bexarotene blocks calcium-permeable ion channels formed by neurotoxic Alzheimer's β-amyloid peptides. *ACS Chem Neurosci*. 2014.
59. Cramer PE, Cirrito JR, Wesson DW, et al. ApoE-directed therapeutics rapidly clear beta-amyloid and reverse deficits in AD mouse models. *Science*. 2012;335(6075):1503–1506.
60. Fitz NF, Cronican AA, Lefterov I, Koldamova R. Comment on "ApoE-directed therapeutics rapidly clear beta-amyloid and reverse deficits in AD mouse models". *Science*. 2013;340(6135):924-c.
61. Veeraraghavalu K, Zhang C, Miller S, et al. Comment on "ApoE-directed therapeutics rapidly clear beta-amyloid and reverse deficits in AD mouse models". *Science*. 2013;340(6135):924-f.
62. Tesseur I, Lo AC, Roberfroid A, et al. Comment on "ApoE-directed therapeutics rapidly clear beta-amyloid and reverse deficits in AD mouse models". *Science*. 2013;340(6135):924-e.
63. LaClair KD, Manaye KF, Lee DL, et al. Treatment with bexarotene, a compound that increases apolipoprotein-E, provides no cognitive benefit in mutant APP/PS1 mice. *Mol Neurodegen*. 2013;8:18.

64. Arispe N, Pollard HB, Rojas E. Zn^{2+} interaction with Alzheimer amyloid beta protein calcium channels. *Proc Natl Acad Sci USA*. 1996;93(4):1710–1715.
65. Posse de Chaves E, Sipione S. Sphingolipids and gangliosides of the nervous system in membrane function and dysfunction. *FEBS Lett*. 2010;584(9):1748–1759.
66. Yahi N, Baghdiguian S, Moreau H, Fantini J. Galactosyl ceramide (or a closely related molecule) is the receptor for human immunodeficiency virus type 1 on human colon epithelial HT29 cells. *J Virol*. 1992;66(8):4848–4854.
67. Hammache D, Pieroni G, Yahi N, et al. Specific interaction of HIV-1 and HIV-2 surface envelope glycoproteins with monolayers of galactosylceramide and ganglioside GM3. *J Biol Chem*. 1998;273(14):7967–7971.
68. Hammache D, Yahi N, Maresca M, Pieroni G, Fantini J. Human erythrocyte glycosphingolipids as alternative cofactors for human immunodeficiency virus type 1 (HIV-1) entry: evidence for CD4-induced interactions between HIV-1 gp120 and reconstituted membrane microdomains of glycosphingolipids (Gb3 and GM3). *J Virol*. 1999;73(6):5244–5248.
69. Fantini J, Hammache D, Pieroni G, Yahi N. Role of glycosphingolipid microdomains in CD4-dependent HIV-1 fusion. *Glycoconj J*. 2000;17(3–4):199–204.
70. Vincent N, Genin C, Malvoisin E. Identification of a conserved domain of the HIV-1 transmembrane protein gp41 which interacts with cholesteryl groups. *Biochim Biophys Acta*. 2002;1567(1–2):157–164.
71. Cook DG, Fantini J, Spitalnik SL, Gonzalez-Scarano F. Binding of human immunodeficiency virus type I (HIV-1) gp120 to galactosylceramide (GalCer): relationship to the V3 loop. *Virology*. 1994;201(2):206–214.
72. Yahi N, Fantini J, Baghdiguian S, et al. SPC3, a synthetic peptide derived from the V3 domain of human immunodeficiency virus type 1 (HIV-1) gp120, inhibits HIV-1 entry into CD^{4+} and CD^{4-} cells by two distinct mechanisms. *Proc Natl Acad Sci USA*. 1995;92(11):4867–4871.
73. Yahi N, Sabatier JM, Baghdiguian S, Gonzalez-Scarano F, Fantini J. Synthetic multimeric peptides derived from the principal neutralization domain (V3 loop) of human immunodeficiency virus type 1 (HIV-1) gp120 bind to galactosylceramide and block HIV-1 infection in a human CD^{4}-negative mucosal epithelial cell line. *J Virol*. 1995;69(1):320–325.
74. Yahi N, Fantini J, Mabrouk K, et al. Multibranched V3 peptides inhibit human immunodeficiency virus infection in human lymphocytes and macrophages. *J Virol*. 1994;68(9):5714–5720.
75. Delezay O, Hammache D, Fantini J, Yahi N. SPC3, a V3 loop-derived synthetic peptide inhibitor of HIV-1 infection, binds to cell surface glycosphingolipids. *Biochemistry*. 1996;35(49):15663–15671.
76. Yahi N, Sabatier JM, Nickel P, Mabrouk K, Gonzalez-Scarano F, Fantini J. Suramin inhibits binding of the V3 region of HIV-1 envelope glycoprotein gp120 to galactosylceramide, the receptor for HIV-1 gp120 on human colon epithelial cells. *J Biol Chem*. 1994;269(39):24349–24353.
77. Fantini J, Rognoni JB, Roccabianca M, Pommier G, Marvaldi J. Suramin inhibits cell growth and glycolytic activity and triggers differentiation of human colic adenocarcinoma cell clone HT29-D4. *J Biol Chem*. 1989; 264(17):10282–10286.
78. McGeary RP, Bennett AJ, Tran QB, Cosgrove KL, Ross BP. Suramin: clinical uses and structure-activity relationships. *Mini Rev Med Chem*. 2008;8(13):1384–1394.
79. Figg WD, Cooper MR, Thibault A, et al. Acute renal toxicity associated with suramin in the treatment of prostate cancer. *Cancer*. 1994;74(5):1612–1614.
80. Baghdiguian S, Nickel P, Fantini J. Double screening of suramin derivatives on human colon cancer cells and on neural cells provides new therapeutic agents with reduced toxicity. *Cancer Lett*. 1991;60(3):213–219.
81. Ablan S, Rawat SS, Viard M, Wang JM, Puri A, Blumenthal R. The role of cholesterol and sphingolipids in chemokine receptor function and HIV-1 envelope glycoprotein-mediated fusion. *Virol J*. 2006;3:104.
82. Rawat SS, Viard M, Gallo SA, Blumenthal R, Puri A. Sphingolipids, cholesterol, and HIV-1: a paradigm in viral fusion. *Glycoconj J*. 2006;23(3–4):189–197.
83. Viard M, Parolini I, Sargiacomo M, et al. Role of cholesterol in human immunodeficiency virus type 1 envelope protein-mediated fusion with host cells. *J Virol*. 2002;76(22):11584–11595.
84. Golde TE. Open questions for Alzheimer's disease immunotherapy. *Alzheimers Res Ther*. 2014;6(1):3.
85. Iqbal K, Liu F, Gong CX. Alzheimer disease therapeutics: focus on the disease and not just plaques and tangles. *Biochem Pharmacol*. 2014;88(4):631–639.
86. Duda JE, Lee VM, Trojanowski JQ. Neuropathology of synuclein aggregates. *J Neurosci Res*. 2000;61(2):121–127.
87. Parihar MS, Hemnani T. Alzheimer's disease pathogenesis and therapeutic interventions. *J Clin Neurosci*. 2004;11(5):456–467.
88. Ghahghaei A. Review: structure of amyloid fibril in diseases. *J Biomed Sci Eng*. 2009;2(5):345–358.

89. Lundvig D, Lindersson E, Jensen PH. Pathogenic effects of alpha-synuclein aggregation. *Brain Res*. 2005;134(1): 3–17.
90. Iversen LL, Mortishire-Smith RJ, Pollack SJ, Shearman MS. The toxicity *in vitro* of beta-amyloid protein. *Biochem J*. 1995;311(Pt 1):1–16.
91. Kawahara M. Disruption of calcium homeostasis in the pathogenesis of Alzheimer's disease and other conformational diseases. *Curr Alzheimer Res*. 2004;1(2):87–95.
92. Lin H, Bhatia R, Lal R. Amyloid beta protein forms ion channels: implications for Alzheimer's disease pathophysiology. *FASEB J*. 2001;15(13):2433–2444.
93. Arispe N, Pollard HB, Rojas E. beta-amyloid Ca($^{2+}$)-channel hypothesis for neuronal death in Alzheimer disease. *Mol Cell Biochem*. 1994;140(2):119–125.
94. Rosenblum WI. Why Alzheimer trials fail: removing soluble oligomeric beta amyloid is essential, inconsistent, and difficult. *Neurobiol Aging*. 2014;35(5):969–974.
95. Esparza TJ, Zhao H, Cirrito JR, et al. Amyloid-beta oligomerization in Alzheimer dementia versus high-pathology controls. *Ann Neurol*. 2013;73(1):104–119.
96. Bicker J, Alves G, Fortuna A, Falcao A. Blood–brain barrier models and their relevance for a successful development of CNS drug delivery systems: a review. *EurJ Pharm Biopharm*. 2014;87(3):409–432.
97. Campbell SD, Regina KJ, Kharasch ED. Significance of lipid composition in a blood–brain barrier-mimetic PAMPA assay. *J Biomol Screen*. 2014;19(3):437–444.
98. Di L, Kerns EH, Fan K, McConnell OJ, Carter GT. High throughput artificial membrane permeability assay for blood–brain barrier. *Eur J Med Chem*. 2003;38(3):223–232.
99. Siakotos AN, Rouser G. Isolation of highly purified human and bovine brain endothelial cells and nuclei and their phospholipid composition. *Lipids*. 1969;4(3):234–239.
100. Marco-Contelles J, Leon R, de los Rios C, et al. Tacripyrines, the first tacrine-dihydropyridine hybrids, as multitarget-directed ligands for the treatment of Alzheimer's disease. *J Med Chem*. 2009;52(9):2724–2732.
101. Wilhelm I, Krizbai IA. *In vitro* models of the blood–brain barrier for the study of drug delivery to the brain. *Mol Pharm*; 2014;11(7):1949–1963.
102. Takeda S, Sato N, Morishita R. Systemic inflammation, blood–brain barrier vulnerability and cognitive/non-cognitive symptoms in Alzheimer disease: relevance to pathogenesis and therapy. *Front Aging Neurosci*. 2014;6:171.
103. Strazza M, Pirrone V, Wigdahl B, Nonnemacher MR. Breaking down the barrier: the effects of HIV-1 on the blood–brain barrier. *Brain Res*. 2011;1399:96–115.
104. Kook SY, Seok Hong H, Moon M, Mook-Jung I. Disruption of blood–brain barrier in Alzheimer disease pathogenesis. *Tissue Barriers*. 2013;1(2):e23993.

Glossary

Acyl chain When a fatty acid is condensed with an alcohol (e.g., glycerol or ceramide), its carbon chain is no longer an acid but an acyl group.

Amphipathic The property of a molecule displaying both a polar and an apolar group. Lipids are generally amphipathic molecules.

Amphiphile An amphipathic molecule.

Amyloid plaque A protein aggregate. Initially found in the brain of patients with Alzheimer's disease and considered as a pathogenesis hallmark of this disease. However, amyloid plaques have also been observed in the brain of healthy individuals, so that they are no longer considered as pathological manifestation of neurodegenerative diseases.

Amyloid pore An oligomer of amyloid proteins that forms a channel, generally permeable to calcium ions, in the plasma membrane of brain cells in patients with neurodegenerative diseases. Pore oligomers are considered as the more toxic forms of amyloid proteins.

Amyloid protein A protein that self-aggregates into organized fibrillar structures; one of the main components of amyloid plaques.

Anionic lipid A lipid displaying at least one net negative charge at physiological pH. Phosphatidic acid, phosphatidylserine, sulfatide, and gangliosides are anionic lipids.

Apolipoprotein A lipid-binding protein that mediates the transport of lipid molecules through aqueous media. The association of apolipoproteins with lipids form a lipoprotein particle.

CARC Linear cholesterol binding domain. Corresponds to an inverted form of the CRAC algorithm. The consensus sequence of CARC is: (K/R)–X_{1-5}–(Y/F/W)–X_{1-5}–(L/V). Found for the first time in the nicotinic acetylcholine receptor.

CRAC (cholesterol recognition/interaction amino acid consensus) Linear cholesterol binding domain. The consensus sequence of CRAC is: (L/V)–X_{1-5}–(Y)–X_{1-5}–(K/R). Found for the first time in peripheral-type benzodiazepine receptor (now referred to as translocator protein TSPO).

Diacylglycerol (DAG) Also named *diglyceride*, a lipid formed by the condensation of two fatty acids on glycerol. Known as a lipid second messenger.

Endocannabinoid Endocannabinoids are endogenous ligands of cannabinoid receptors (CB1 and CB2). An endocannabinoid is a derivative of arachidonic acid linked to either ethanolamine (anandamide) or glycerol (2-arachidonoylglycerol, 2-AG). These neurotransmitters are most often produced by the postsynaptic neuron, and transported through the synaptic cleft in the retrograde direction to reach CB1 receptors in the plasma membrane of the presynaptic neuron.

Galactocerebroside Old name for galactosylceramide (GalCer).

Galactosylceramide (GalCer) A glycosphingolipid with galactose as unique sugar. Abundantly expressed in myelin, together with its sulfated companion galactosylsulfatide (usually called *sulfatide*).

Ganglioside A glycosphingolipid (GSL) whose glycone part contains at least one sialic acid residue. Gangliosides are classified according to the number of sialic acid residues they contain: 1 (mono), 2 (di), 3 (tri), 4 (tetra), and so on. Gangliosides are also designated by a number that refers to chromatographic mobility. For instance, ganglioside GM1 is the monosialyl (M) ganglioside (G) that migrates in first position (1) on thin layer chromatography plates. When two gangliosides with the same number of sialic acids migrate in the same area, they are designated by *a* and *b* letters (e.g., GD1a and GD1b).

Glucocerebroside Old name for glucosylceramide (GlcCer).

Glycerolipid A lipid derived from glycerol.

Glycerophospholipid A glycerolipid displaying a phosphate group.

Brain Lipids in Synaptic Function and Neurological Disease. http://dx.doi.org/10.1016/B978-0-12-800111-0.00015-1

Glycosphingolipid (GSL) A lipid with a carbohydrate head group linked to a ceramide. The sugar part can be neutral, acidic, sulfated, or cationic. Acidic GSLs containing at least one sialic acid are referred to as gangliosides.

Hydroxy fatty acid (HFA) A fatty acid with a OH group linked to carbon C2 (also referred to as α). Ceramide with a hydroxyacyl chain is referred to as ceramide-HFA. If the acyl chain is nonhydroxylated, the ceramide is referred to as ceramide-NFA.

Intrinsically disordered protein (IDP) A protein without a stable three-dimensional structure. IDPs oscillate between a wide range of extended conformations. In fact, these proteins do not have a sufficient proportion of apolar amino acid residues to adopt a globular structure in water. IDPs have a high conformational plasticity that allows them to recognize and adopt their shape to various ligands. Amyloid proteins are generally IDPs.

Langmuir monolayer The structural organization of amphiphiles at the air–water interface. This system is used for studies of protein–lipid interactions through surface pressure measurements.

Lecithins Common name for phosphatidylcholine lipids.

Lipid chaperone A lipid that modulates the conformation, and thus the function, of a membrane protein. Cholesterol and sphingolipids are typical chaperone lipids.

Lipid raft A membrane area enriched in cholesterol and sphingolipids, also referred to as membrane microdomains or sphingolipid microdomains. Lipid rafts are involved in signal transduction functions. Most brain pathogens use lipid raft as portal of entry in brain cells.

Liquid-disordered (Ld) phase The physical state of the bulk membrane enriched in phosphatidylcholine and with low cholesterol levels.

Liquid-ordered (Lo) phase The physical state of cholesterol-/sphingolipid-rich plasma membrane microdomains (lipid rafts).

Nonhydroxy fatty acid (NFA) A fatty acid without OH group linked to carbon C2; opposite of hydroxyl fatty acid (HFA).

Packing parameter A quantitative approach to lipid shape developed by J. Israealchvili. Lipids with a cylinder shape (e.g., phosphatidylcholine) have a packing parameter P close to 1. If $P < 1$, the lipid has an inverted cone shape and a large polar head group (e.g., gangliosides). If $P > 1$, the lipid adopts a cone shape, with a small head group and a large apolar domain (e.g., cholesterol).

Phosphatidic acid A lipid formed by phosphorylation of diacylglycerol; belongs to the category of anionic lipids.

Phosphatidylethanolamine (PE) One of the most abundant glycerophospholipids, more abundant in the intracellular than in the extracellular leaflet of the plasma membrane. Formed by the addition of ethanolamine on a phosphatidic acid backbone.

Phosphatidylcholine (PC) One of the most abundant membrane lipids, especially abundant in the extracellular leaflet of the plasma membrane. Formed by the addition of choline on a phosphatidic acid backbone.

Phosphatidylinositol bisphosphate (PIP2) Belongs to the category of anionic lipids and is involved in signal transduction as a classical lipid second messenger.

Phosphatidylserine (PS) One of the most abundant membrane lipids, whose expression is normally restricted to the intracellular leaflet of the plasma membrane, except for apoptotic cells. Formed by the addition of serine acid on a phosphatidic acid backbone; belongs to the category of anionic lipids.

Polyunsaturated fatty acid A fatty acid with several double bonds. When incorporated in membrane lipids, polyunsaturated acyl chains increase membrane fluidity.

Sphingolipid All lipids derived from sphingosine, the "enigmatic" lipid (in reference to the Sphinx enigma) discovered by Thudichum. Includes sphingosine, ceramides, sphingomyelins, and glycosphingolipids.

Sphingolipid-binding domain (SBD) A structural hairpin motif found in sequence-unrelated proteins that interact with sphingolipids. Discovered for the first time in HIV-1 surface envelope glycoprotein gp120.

Sphingomyelin (SM) A lipid with a phosphocholine group linked to a ceramide; shares the same head group as phosphatidylcholine (PC) but displays different physicochemical properties. SM and PC do not normally mix in biological membranes, because SM is associated with cholesterol in plasma membrane microdomains such as lipid rafts.

Sphingomyelin signature motif A consensus motif conferring sphingomyelin recognition by transmembrane domains. The consensus sequence of the motif is (VxxTLxxIY). Discovered for the first time in the COPI machinery protein p24.

Sulfatide The sulfated form of GalCer (thus more correctly named *galactosylsulfatide*), with a sulfonated group linked to carbon 3 of the galactosyl ring.

Surface pressure When present at the air–water interface, amphiphiles exert a pressure on the water surface that is proportional to their effect on surface tension. A simple relationship links the surface tension and the surface pressure: $\pi = \gamma_0 - \gamma$ where π is the surface pressure induced by the amphiphiles, γ_0 the surface tension of pure water (78.2 mN m^{-1}), and γ the surface tension measured in the presence of amphiphiles. The surface pressure of a biological membrane has been estimated to 30 mN m^{-1}.

Surface tension The force that constrains the water surface to contract to the minimal area. Surface tension is responsible for the shape of liquid droplets. The surface tension of pure water (γ_0) is 78.2 mN m^{-1}.

Zwitterion A charged molecule displaying one positive and one negative charge (the net electric charge is null). Phosphatidylcholine is a zwitterionic lipid.

Subject Index

B

F

G

H

M

T

U

V

W

Z